Integrated Series in Information Systems

Volume 30

For further volumes:
http://www.springer.com/series/6157

Hsinchun Chen

Dark Web

Exploring and Data Mining the Dark Side of the Web

Hsinchun Chen
Department of Management Information Systems
University of Arizona
Tuscon, AZ, USA
hchen@eller.arizona.edu

ISSN 1571-0270
ISBN 978-1-4614-1556-5 e-ISBN 978-1-4614-1557-2
DOI 10.1007/978-1-4614-1557-2
Springer New York Dordrecht Heidelberg London

Library of Congress Control Number: 2011941611

Printed on acid-free paper

Springer is part of Springer Science+Business Media (www.springer.com)

Preface

Aims

The University of Arizona Artificial Intelligence Lab (AI Lab) Dark Web project is a long-term scientific research program that aims to study and understand the international terrorism (jihadist) phenomena via a computational, data-centric approach. We aim to collect "ALL" web content generated by international terrorist groups, including web sites, forums, chat rooms, blogs, social networking sites, videos, virtual world, etc. We have developed various multilingual data mining, text mining, and web mining techniques to perform link analysis, content analysis,web metrics (technical sophistication) analysis, sentiment analysis, authorship analysis, and video analysis in our research. The approaches and methods developed in this project contribute to advancing the field of Intelligence and Security Informatics (ISI). Such advances will help related stakeholders perform terrorism research and facilitate international security and peace.

Dark Web research has been featured in many national, international and local press and media, including: National Science Foundation press, Associated Press, BBC, Fox News, National Public Radio, Science News, Discover Magazine, Information Outlook, Wired Magazine, The Bulletin (Australian), Australian Broadcasting Corporation, Arizona Daily Star, East Valley Tribune, Phoenix ABC Channel 15, and Tucson Channels 4, 6, and 9. As an NSF-funded research project, our research team has generated significant findings and publications in major computer science and information systems journals and conferences. We hope our research will help educate the next generation of cyber/Internet-savvy analysts and agents in the intelligence, justice, and defense communities.

This monograph aims to provide an overview of the Dark Web landscape, suggest a systematic, computational approach to understanding the problems, and illustrate research progress with selected techniques, methods, and case studies developed by the University of Arizona AI Lab Dark Web team members.

Audience

This book aims to provide an interdisciplinary and understandable monograph about Dark Web research. We hope to bring useful knowledge to scientists, security professionals, counter-terrorism experts, and policy makers. The proposed work could also serve as a reference material or textbook in graduate level courses related to information security, information policy, information assurance, information systems, terrorism, and public policy.

The primary audience for the proposed monograph will include the following:

- IT Academic Audience: College professors, research scientists, graduate students, and select undergraduate juniors and seniors in computer science, information systems, information science, and other related IT disciplines who are interested in intelligence analysis and data mining and their security applications.
- Security Academic Audience: College professors, research scientists, graduate students, and select undergraduate juniors and seniors in political sciences, terrorism study, and criminology who are interested in exploring the impact of the Dark Web on society.
- Security Industry Audience: Executives, managers, analysts, and researchers in security and defense industry, think tanks, and research centers that are actively conducting IT-related security research and development, especially using open source web contents.
- Government Audience: Policy makers, managers, and analysts in federal, state, and local governments who are interested in understanding and assessing the impact of the Dark Web and their security concerns.

Scope and Organization

The book consists of three parts. In Part I, we provide an overview of the research framework and related resources relevant to intelligence and security informatics (ISI) and terrorism informatics. Part II presents ten chapters on computational approaches and techniques developed and validated in the Dark Web research. Part III presents nine chapters of case studies based on the Dark Web research approach. We provide a brief summary of each chapter below.

Part I. Research Framework: Overview and Introduction

- **Chapter 1. Dark Web Research Overview**
  The AI Lab Dark Web project is a long-term scientific research program that aims to study and understand the international terrorism (jihadist) phenomena via a computational, data-centric approach. We aim to collect "ALL" web content generated by international terrorist groups, including web sites, forums, chat rooms, blogs, social networking sites, videos, virtual world, etc. We have developed various multilingual data mining, text mining, and web mining techniques to perform link analysis, content analysis,web metrics (technical sophistication)

analysis, sentiment analysis, authorship analysis, and video analysis in our research.

- **Chapter 2. Intelligence and Security Informatics (ISI): Research Framework**
  In this chapter we review the computational research framework that is adopted by the Dark Web research. We first present the security research context, followed by description of a data mining framework for intelligence and security informatics research. To address the data and technical challenges facing ISI, we present a research framework with a primary focus on KDD (Knowledge Discovery from Databases) technologies. The framework is discussed in the context of crime types and security implications.

- **Chapter 3. Terrorism Informatics**
  In this chapter we provide an overview of selected resources of relevance to "Terrorism Informatics," a new discipline that aims to study the terrorism phenomena with a data-driven, quantitative, and computational approach. We first summarize several critical books that lay the foundation for studying terrorism in the new Internet era. We then review important terrorism research centers and resources that are of relevance to our Dark Web research.

Part II. Dark Web Research: Computational Approach and Techniques

- **Chapter 4. Forum Spidering**
  In this study we propose a novel crawling system designed to collect Dark Web forum content. The system uses a human-assisted accessibility approach to gain access to Dark Web forums. Several URL ordering features and techniques enable efficient extraction of forum postings. The system also includes an incremental crawler coupled with a recall improvement mechanism intended to facilitate enhanced retrieval and updating of collected content.

- **Chapter 5. Link and Content Analysis**
  To improve understanding of terrorist activities, we have developed a novel methodology for collecting and analyzing Dark Web information. The methodology incorporates information collection, analysis, and visualization techniques, and exploits various web information sources. We applied it to collecting and analyzing information of selected jihad web sites and developed visualization of their site contents, relationships, and activity levels.

- **Chapter 6. Dark Network Analysis**
  Dark networks such as terrorist networks and narcotics-trafficking networks are hidden from our view yet could have a devastating impact on our society and economy. Based on analysis of four real-world "dark" networks, we found that these covert networks share many common topological properties with other types of networks. Their efficiency in communication and flow of information, commands, and goods can be tied to their small-world structures characterized by small average path length and high clustering coefficient. In addition, we found that because of the small-world properties dark networks are more vulnerable to attacks on the bridges that connect different communities than to attacks on the hubs.

- **Chapter 7. Interactional Coherence Analysis**
  Despite the rapid growth of text-based computer-mediated communication (CMC), its limitations have rendered the media highly incoherent. Interactional coherence analysis (ICA) attempts to accurately identify and construct interaction networks of CMC messages. In this study, we propose the Hybrid Interactional Coherence (HIC) algorithm for identification of web forum interaction. HIC utilizes both system features, such as header information and quotations, and linguistic features, such as direct address and lexical relation. Furthermore, several similarity-based methods, including a Lexical Match Algorithm (LMA) and a sliding window method, are utilized to account for interactional idiosyncrasies.

- **Chapter 8. Dark Web Attribute System**
  In this study we propose a Dark Web Attribute System (DWAS) to enable quantitative Dark Web content analysis from three perspectives: technical sophistication, content richness, and web interactivity. Using the proposed methodology, we identified and examined the Internet usage of major Middle Eastern terrorist/extremist groups. In our comparison of terrorist/extremist web sites to U.S. government web sites, we found that terrorists/extremist groups exhibited levels of web knowledge similar to that of U.S. government agencies. Moreover, terrorists/extremists had a strong emphasis on multimedia usage and their web sites employed significantly more sophisticated multimedia technologies than government web sites.

- **Chapter 9. Authorship Analysis**
  In this study we addressed the online anonymity problem by successfully applying authorship analysis to English and Arabic extremist group web forum messages. The performance impact of different feature categories and techniques was evaluated across both languages. In order to facilitate enhanced writing style identification, a comprehensive list of online authorship features was incorporated. Additionally, an Arabic language model was created by adopting specific features and techniques to deal with the challenging linguistic characteristics of Arabic, including an elongation filter and a root clustering algorithm.

- **Chapter 10. Sentiment Analysis**
  In this study the use of sentiment analysis methodologies is proposed for classification of web forum opinions in multiple languages. The utility of stylistic and syntactic features is evaluated for sentiment classification of English and Arabic content. Specific feature extraction components are integrated to account for the linguistic characteristics of Arabic. The Entropy Weighted Genetic Algorithm (EWGA) is also developed, which is a hybridized genetic algorithm that incorporates the information gain heuristic for feature selection. The proposed features and techniques are evaluated on U.S. and Middle Eastern extremist web forum postings.

- **Chapter 11. Affect Analysis**
  Analysis of affective intensities in computer-mediated communication is important in order to allow a better understanding of online users' emotions and preferences. In this study we compared several feature representations for affect analysis,

including learned n-grams and various automatically- and manually-crafted affect lexicons. We also proposed the support vector regression correlation ensemble (SVRCE) method for enhanced classification of affect intensities. Experiments were conducted on U.S. domestic and Middle Eastern extremist web forums.

- **Chapter 12. CyberGate Visualization**
  Computer-mediated communication (CMC) analysis systems are important for improving participant accountability and researcher analysis capabilities. However, existing CMC systems focus on structural features, with little support for analysis of text content in web discourse. In this study we propose a framework for CMC text analysis grounded in Systemic Functional Linguistic Theory. Our framework addresses several ambiguous CMC text mining issues, including the relevant tasks, features, information types, feature selection methods, and visualization techniques. Based on it, we have developed a system called CyberGate, which includes the Writeprint and Ink Blot techniques. These techniques incorporate complementary feature selection and visualization methods in order to allow a breadth of analysis and categorization capabilities.

- **Chapter 13. Dark Web Forum Portal**
  The Dark Web Forum Portal provides web-enabled access to critical international jihadist web forums. The focus of this chapter is on the significant extensions to previous work including: increasing the scope of our data collection; adding an incremental spidering component for regular data updates; enhancing the searching and browsing functions; enhancing multilingual machine translation for Arabic, French, German and Russian; and advanced Social Network Analysis. A case study on identifying active jihadi participants in web forums is shown at the end.

Part III. Dark Web Research: Case Studies

- **Chapter 14. Jihadi Video Analysis**
  This chapter presents an exploratory study of jihadi extremist groups' videos using content analysis and a multimedia coding tool to explore the types of video, groups' modus operandi, and production features that lend support to extremist groups. The videos convey messages powerful enough to mobilize members, sympathizers, and even new recruits to launch attacks that are captured (on video) and disseminated globally through the Internet. The videos are important for jihadi extremist groups' learning, training, and recruitment. In addition, the content collection and analysis of extremist groups' videos can help policy makers, intelligence analysts, and researchers better understand the extremist groups' terror campaigns and modus operandi, and help suggest counter-intelligence strategies and tactics for troop training.

- **Chapter 15. Extremist YouTube Videos**
  In this study, we propose a text-based framework for video content classification of online video-sharing web sites. Different types of user-generated data (e.g., titles, descriptions, and comments) were used as proxies for online videos, and

three types of text features (lexical, syntactic, and content-specific features) were extracted. Three feature-based classification techniques (C4.5, Naïve Bayes, and SVM) were used to classify videos. To evaluate the proposed framework, we developed a testbed based on jihadi videos collected from the most popular video-sharing site, YouTube.

- **Chapter 16. Improvised Explosive Devices (IED) on Dark Web**
  This chapter presents a cyber-archaeology approach to social movement research. Cultural cyber-artifacts of significance to the social movement are collected and classified using automated techniques, enabling analysis across multiple related virtual communities. Approaches to the analysis of cyber-artifacts are guided by perspectives of social movement theory. A Dark Web case study on a broad group of related IED virtual communities is presented to demonstrate the efficacy of the framework and provide a detailed instantiation of the proposed approach for evaluation.

- **Chapter 17. Weapons of Mass Destruction (WMD) on Dark Web**
  In this chapter we propose a research framework that aims to investigate the capability, accessibility, and intent of critical high-risk countries, institutions, researchers, and extremist or terrorist groups. We propose to develop a knowledge base of the Nuclear Web that will collect, analyze, and pinpoint significant actors in the high-risk international nuclear physics and weapons communities. We also identify potential extremist or terrorist groups from our Dark Web testbed who might pose WMD threats to the U.S. and the international community. Selected knowledge mapping and focused web crawling techniques and findings from a preliminary study are presented.

- **Chapter 18. Bioterrorism Knowledge Mapping**
  In this research we propose a framework to identify the researchers who have expertise in the bioterrorism agents/diseases research domain, the major institutions and countries where these researchers reside, and the emerging topics and trends in bioterrorism agents/diseases research. By utilizing knowledge mapping techniques, we analyzed the productivity status, collaboration status, and emerging topics in the bioterrorism domain. The analysis results provide insights into the research status of bioterrorism agents/diseases and thus allow a more comprehensive view of bioterrorism researchers and ongoing work.

- **Chapter 19. Women's Forums on the Dark Web**
  In this study, we develop a feature-based text classification framework to examine the online gender differences between female and male posters on web forums by analyzing writing styles and topics of interests. We examine the performance of different feature sets in an experiment involving political opinions. The results of our experimental study on this Islamic women's political forum show that the feature sets containing both content-free and content-specific features perform significantly better than those consisting of only content-free features.

- **Chapter 20. US Domestic Extremist Groups**
  U.S. domestic extremist groups have increased in number and are intensively utilizing the Internet as an effective tool to share resources and members with limited regard for geographic, legal, or other obstacles. In this study, we develop automated and semi-automated methodologies for capturing, classifying, and organizing domestic extremist web site data. We found that by analyzing the hyperlink structures and content of domestic extremist web sites and constructing social network maps, their inter-organizational structure and cluster affinities could be identified.

- **Chapter 21. International Falun Gong Movement on the Web**
  In this study, we developed a cyber-archaeology approach and used the international Falun Gong (FLG) movement as a case study. The FLG is known as a peaceful international social movement, unlike the more violent jihadi movement. We employed Social Network Analysis and Writeprint to analyze FLG's cyber-artifacts from the perspectives of links, web content, and forum content. In the link analysis, FLG's web sites linked closely to Chinese democracy and human rights social movement organizations (SMOs), reflecting FLG's historical conflicts with the Chinese government after the official ban in 1999.

- **Chapter 22. Botnets and Cyber Criminals**
  In the last several years, the nature of computer hacking has completely changed. Cybercrime has risen to unprecedented sophistication with the evolution of botnet technology, and an underground community of cyber criminals has arisen, capable of inflicting serious socioeconomic and infrastructural damage in the information age. This chapter serves as an introduction to the world of modern cybercrime and discusses information systems to investigate it. We investigated the command and control (C&C) signatures of major botnet herders using data collected from the ShadowServer Foundation, a nonprofit research group for botnet research. We also performed exploratory population modeling of the bots and cluster analysis of selected cyber criminals.

Tuscon, Arizona, USA Hsinchun Chen

About the Author

Dr. Hsinchun Chen is the McClelland Professor of Management Information Systems at the University of Arizona. He received a B.S. degree from the National Chiao-Tung University in Taiwan, an MBA degree from SUNY Buffalo, and his Ph.D. degree in Information Systems from New York University. Dr. Chen has served as a Scientific Counselor/Advisor of the National Library of Medicine (USA), Academia Sinica (Taiwan), and National Library of China (China). Dr. Chen is a Fellow of IEEE and AAAS. He received the IEEE Computer Society 2006 Technical Achievement Award and the INFORMS Design Science Award in 2008. He has an h-index score of 50. He is author/editor of 20 books, 25 book chapters, 210 SCI journal articles, and 140 refereed conference articles covering web computing, search engines, digital library, intelligence analysis, biomedical informatics, data/text/web mining, and knowledge management. His recent books include: *Infectious Disease Informatics* (2010); *Mapping Nanotechnology Knowledge and Innovation (2008)*, *Digital Government: E-Government Research, Case Studies, and Implementation (2007); Intelligence and Security Informatics for International Security: Information Sharing and Data Mining (2006)*; and *Medical Informatics: Knowledge Management and Data Mining in Biomedicine (2005),* all published by Springer. Dr. Chen was ranked #8 in publication productivity in Information Systems (CAIS 2005) and #1 in Digital Library research (IP&M 2005) in two bibliometric studies. He is Editor in Chief (EIC) of the new *ACM Transactions on Management Information Systems* (*ACM TMIS*) and Springer *Security Informatics (SI)* Journal, and the Associate EIC of *IEEE Intelligent Systems*. He serves on ten editorial boards including: *ACM Transactions on Information Systems, IEEE Transactions on Systems, Man, and Cybernetics, Journal of the American Society for Information Science and Technology, Decision Support Systems,*

and *International Journal on Digital Library*. He has been an advisor for major NSF, DOJ, NLM, DOD, DHS, and other international research programs in digital library, digital government, medical informatics, and national security research. Dr. Chen is the founding director of the Artificial Intelligence Lab and Hoffman E-Commerce Lab. The UA Artificial Intelligence Lab, which houses 20+ researchers, has received more than $30M in research funding from NSF, NIH, NLM, DOD, DOJ, CIA, DHS, and other agencies. Dr. Chen has also produced 25 Ph.D. students who are placed in major academic institutions around the world. The Hoffman E-Commerce Lab, which has been funded mostly by major IT industry partners, features one of the most advanced e-commerce hardware and software environments in the College of Management. Dr. Chen was conference co-chair of ACM/IEEE Joint Conference on Digital Libraries (JCDL) 2004 and has served as the conference/program co-chair for the past eight International Conferences of Asian Digital Libraries (ICADL), the premiere digital library meeting in Asia that he helped develop. Dr. Chen is also (founding) conference co-chair of the IEEE International Conference on Intelligence and Security Informatics (ISI) 2003-present. The ISI conference, which has been sponsored by NSF, CIA, DHS, and NIJ, has become the premiere meeting for international and homeland security IT research. Dr. Chen's COPLINK system, which has been quoted as a national model for public safety information sharing and analysis, has been adopted in more than 3,500 law enforcement and intelligence agencies. The COPLINK research had been featured in the *New York Times, Newsweek, Los Angeles Times, Washington Post, Boston Globe,* and *ABC News*, among others. The COPLINK project was selected as a finalist by the prestigious International Association of Chiefs of Police (IACP)/Motorola 2003 Weaver Seavey Award for Quality in Law Enforcement in 2003. COPLINK research has recently been expanded to border protection (BorderSafe), disease and bioagent surveillance (BioPortal), and terrorism informatics research (Dark Web), funded by NSF, DOD, CIA, and DHS. In collaboration with selected international terrorism research centers and intelligence agencies, the Dark Web project has generated one of the largest databases in the world about extremist/terrorist-generated Internet contents (web sites, forums, blogs, and multimedia documents). Dark Web research supports link analysis, content analysis, web metrics analysis, multimedia analysis, sentiment analysis, and authorship analysis of international terrorism contents. The project has received significant international press coverage, including: *Associated Press, USA Today, The Economist, NSF Press, Washington Post, Fox News, BBC, PBS, Business Week, Discover magazine, WIRED magazine, Government Computing Week, Second German TV (ZDF), Toronto Star,* and *Arizona Daily Star*, among others. Dr. Chen is also a successful entrepreneur. He is the founder of Knowledge Computing Corporation (KCC), a university spin-off IT company and a market leader in law enforcement and intelligence information sharing and data mining. KCC was acquired by a major private equity firm for $40M in the summer of 2009 and merged with I2, the industry leader in crime analytics. The combined I2/KCC company was acquired by IBM for $420M in 2011. Dr. Chen has also received numerous awards in information technology and knowledge management education and research including: AT&T Foundation Award,

SAP Award, the Andersen Consulting Professor of the Year Award, the University of Arizona Technology Innovation Award, and the National Chiao-Tung University Distinguished Alumnus Award. He was also named Distinguished Alumnus by SUNY Buffalo. Dr. Chen has served as a keynote or invited speaker in major international security informatics, medical informatics, information systems, knowledge management, and digital library conferences and major international government meetings (NATO, UN, EU, FBI, CIA, DOD, DHS). He is a Distinguished/Honorary Professor of several major universities in Taiwan and China and was named the Distinguished University Chair Professor of the National Taiwan University. Dr. Chen recently served as the Program Chair of the International Conference on Information Systems (ICIS) 2009, held in Phoenix, Arizona.

Contents

Part I
Research Framework: Overview and Introduction

Chapter 1
Dark Web Research Overview

1 Introduction

Gabriel Weimann of Haifa University in Israel estimated that there are about 5,000 terrorist web sites as of 2006 (Weimann 2006). Based on our actual spidering experience over the past 8 years, we believe there are about 100,000 sites of extremist and terrorist content as of 2010, including: web sites, forums, blogs, social networking sites, video sites, and virtual world sites (e.g., Second Life). The largest increase since 2006–2007 is in various new Web 2.0 sites (forums, videos, blogs, virtual world, etc.) in different languages (i.e., for homegrown groups, particularly in Europe). We have found significant terrorism content in more than 15 languages.

We collect (using computer programs) various web contents every 2–3 months; we started spidering in 2002. Currently, we only collect the complete contents of about 1,000 sites, in Arabic, Spanish, and English languages. We also have partial contents of about another 10,000 sites. In total, our collection is about 15 TBs in size, with close to 2,000,000,000 pages/files/postings from more than 10,000 sites. *We believe our Dark Web collection is the largest open-source extremist and terrorist collection in the academic world.* Researchers can have graded access to our collection by contacting our research center. We present below a summary of important Dark Web contents.

1.1 Web Sites

Our web site collection consists of the complete contents of about 1,000 sites, in various static (html, pdf, Word) and dynamic (PHP, JSP, CGI) formats. We collect every single page, link, and attachment within these sites. We also collect partial information from about 10,000 related (linked) sites. Some large well-known sites contain more than 10,000 pages/files in 10+ languages (in selected pages).

H. Chen, *Dark Web: Exploring and Data Mining the Dark Side of the Web*,
Integrated Series in Information Systems 30, DOI 10.1007/978-1-4614-1557-2_1,

1.2 Forums

We collect the complete contents (authors, headings, postings, threads, time tags, etc.) of about 300 terrorist forums. We also perform periodic updates. Some large radical sites include more than 30,000 members with close to 1,000,000 messages posted. We have also developed the Dark Web Forum Portal, which provides beta search access to several international jihadist "Dark Web" forums collected by the Artificial Intelligence Lab at the University of Arizona. Users may search, view, translate, and download messages (by forum member name, thread title, topic, keyword, etc.).

Preliminary social network analysis visualization is also available.

1.3 Blogs, Social Networking Sites, and Virtual Worlds

We have identified and extracted many smaller, transient (meaning the sites appear and disappear very quickly) blogs and social networking sites, mostly hosted by terrorist sympathizers and "wannabes." We have also identified more than 30 (self-proclaimed) terrorist or extremist groups in virtual world sites. (However, we are still unsure whether they are "real" terrorist/extremists or just playing the roles in virtual games).

1.4 Videos and Multimedia Content

Terrorist sites are extremely rich in content, with heavy usage of multimedia formats. We have identified and extracted about 1,000,000 images and 100,000 videos from many terrorist sites and specialty multimedia file-hosting third-party servers. More than 50% of our videos are IED (Improvised Explosive Devices) related.

2 Computational Techniques (Data Mining, Text Mining, and Web Mining)

Our computational tools are grouped into two categories: (1) collection and (2) analysis and visualization. Significant deep web spidering, computational linguistic analysis, sentiment analysis, social network analysis, and social media analysis and visualization research has been conducted by members of the AI Lab over the past 8 years. We summarize selected approaches below. More details about specific techniques and case studies will be presented in subsequent chapters.

2.1 Dark Web Collection

2.1.1 Web Site Spidering

We have developed various focused spiders/crawlers for collecting deep web content on our previous digital library research. Our spiders can access password-protected sites and perform randomized (human-like) fetching. Our spiders are trained to fetch all html, pdf, and Word files; links; PHP, CGI, and ASP files; images; audios; and videos in a web site. To ensure freshness, we spider selected web sites every 2–3 months.

2.1.2 Forum Spidering

Our forum spidering tool recognizes 15+ forum-hosting software and their formats. We collect the complete forum, including authors, headings, postings, threads, time tags, etc., which allows us to reconstruct participant interactions. We perform periodic forum spidering and incremental updates based on research needs. We have collected and processed forum contents in Arabic, English, Spanish, French, and Chinese using selected computational linguistics techniques.

2.1.3 Multimedia (Image, Audio, and Video) Spidering

We have developed specialized techniques for spidering and collecting multimedia files and attachments from web sites and forums. We plan to perform stenography research to identify encrypted images in our collection and multimedia analysis (video segmentation, image recognition, voice/speech recognition) to identify unique terrorist-generated video contents and styles.

2.2 Dark Web Analysis and Visualization

2.2.1 Social Network Analysis (SNA)

We have developed various SNA techniques to examine web site and forum posting relationships. We have used various topological metrics (betweenness, degree, etc.) and properties (preferential attachment, growth, etc.) to model terrorist and terrorist site interactions. We have developed several clustering (e.g., blockmodeling) and projection (e.g., multidimensional scaling, spring embedder) techniques to visualize their relationships. Our focus is on understanding "Dark Networks" (unlike traditional "bright" scholarship, e-mail, or computer networks) and their unique properties (e.g., hiding, justice intervention, rival competition, etc.).

2.2.2 Content Analysis

We have developed several detailed (terrorism-specific) coding schemes to analyze the contents of terrorist and extremist web sites. Content categories include: recruiting, training, sharing ideology, communication, propaganda, etc. We have also developed computer programs to help automatically identify selected content categories (e.g., web master information, forum availability, etc.).

2.2.3 Web Metrics Analysis

Web metrics analysis examines the technical sophistication, media richness, and web interactivity of extremist and terrorist web sites. We examine technical features and capabilities (e.g., their ability to use forms, tables, CGI programs, multimedia files, etc.) of such sites to determine their level of "web-savviness." Web metrics provides a measure for terrorists/extremists' capability and resources. All terrorist site web metrics are extracted and computed using computer programs.

2.2.4 Sentiment and Affect Analysis

Not all sites are equally radical or violent. Sentiment (polarity: positive/negative) and affect (emotion: violence, racism, anger, etc.) analysis allows us to identify radical and violent sites that warrant further study. We also examine how radical ideas become "infectious" based on their contents and senders and their interactions. We rely heavily on recent advances in opinion mining – analyzing opinions in short web-based texts. We have also developed selected visualization techniques to examine sentiment/affect changes in time and among people. Our research includes several probabilistic multilingual affect lexicons and selected dimension reduction and projection (e.g., principal component analysis) techniques.

2.2.5 Authorship Analysis and Writeprint

Grounded in authorship analysis research, we have developed the (cyber) Writeprint technique to uniquely identify anonymous senders based on the signatures associated with their forum messages. We expand the lexical and syntactic features of traditional authorship analysis to include system (e.g., font size, color, web links) and semantic (e.g., violence, racism) features of relevance to online texts of extremists and terrorists. We have also developed advanced Ink blot and Writeprint visualizations to help visually identify web signatures. Our Writeprint technique has been developed for Arabic, English, and Chinese languages. The Arabic Writeprint consists of more than 400 features, all automatically extracted from online messages using computer programs. Writeprint can achieve an accuracy level of 95%.

2.2.6 Video Analysis

Based on previous terrorism ontology research, we have developed a unique coding scheme to analyze terrorist-generated videos based on the contents, production characteristics, and metadata associated with the videos. We have also developed a semiautomated tool to allow human analysts to quickly and accurately analyze and code these videos.

2.2.7 IEDs in Dark Web Analysis

We have conducted several systematic studies to identify IED-related content generated by terrorist and insurgency groups in the Dark Web. A smaller number of sites are responsible for distributing a large percentage of IED-related web pages, forum postings, training materials, explosive videos, etc. We have developed unique signatures for those IED sites based on their contents, linkages, and multimedia file characteristics. Much of the content needs to be analyzed by military analysts. Training materials also need to be developed for troops before their deployment ("seeing the battlefield from your enemies' eyes").

2.2.8 Dark Web Forum Portal

For several years, we have monitored and collected many international jihadist forums. These online discussion sites are dedicated to topics relating primarily to Islamic ideology and theology. The Lab now provides search access to these forums through its Dark Web Forum Portal, and in its beta form, the portal provides access to 29 forums, which together comprise nearly 13,000,000 messages from 340,000 participants in four different languages (English, Arabic, German, and Russian). The Portal also provides statistical analysis, download, translation, and social network visualization functions for each selected forum. It is accessible at http://128.196.40.222:8080/CRI_Indexed_new/index.jsp.

3 Dark Web Project Structure and Resources

The Dark Web project is a multiyear, multiinstitutional, and multidisciplinary research program that spans information systems, computer science, terrorism study, intelligence analysis, international relations, etc. Many AI Lab researchers and students have helped contribute to the success of the project.

3.1 Team Members (Selected)

Dr. Hsinchun Chen is project PI and director of the Artificial Intelligence Lab. He is in charge of all aspects of the Dark Web project. Ms. Cathy Larson is AI Lab associate director and a Dark Web project lead, in charge of project coordination, partnership, and user studies. Selected Dark Web research team members and their expertise include (in alphabetical order):

- Dr. Ahmed Abbasi, affect analysis and visualization
- Enrique Arevelo, system development
- Alfonso A. Bonillas, system development and Spanish content collection
- Yida Chen, social network analysis
- Dr. Wingyan Chung, web portal development and evaluation
- Oscar de Ita, system development and Spanish content collection
- Carrie Fang, database development
- Tianjun Fu, forum spidering and coherence analysis
- Dr. Sidd Kaza, network analysis
- Dr. Danning Hu, dynamic network analysis
- Dr. Guanpi Lai (Greg), web portal development
- Dr. Jiexun Jason Li, English and Chinese authorship analysis
- Dr. Dan McDonald, computational linguistic analysis
- Dr. Jialun Qin, web attribute system research
- Dr. Edna Reid, terrorism knowledge mapping
- Arab Salim, video analysis and Arabic language processing
- Lu Tseng, system lead
- Shing Ka Wu, system interface design
- Wei Xi, system development
- Dr. Jennifer Jie Xu, Dark Network analysis and visualization
- Lijun Yan, system development
- Dr. Rong Zheng, English and Chinese authorship analysis
- Dr. Yilu Zhou, computational linguistic analysis

3.2 Press Coverage and Interest

Dark Web research has been featured in many national, international, and local press and media, including: Associated Press, USA Today, The Economist, NSF Press, Washington Post, Fox News, BBC, PBS, Business Week, National Public Radio, Science News, Discover Magazine, WIRED Magazine, Government Computing Week, Second German TV (ZDF), Toronto Star, Bulletin (Australian), Australian Broadcasting Corporation, Arizona Daily Star, East Valley Tribune, Phoenix ABC Channel 15, Tucson TV Channels 4, 6, and 9, among others. It has been considered a model of advanced computational research of significant societal and international relevance.

As an NSF-funded research project, our research team has generated significant findings and publications in major computer science and information systems journals and conferences. However, we have taken great care not to reveal sensitive group information or technical implementation details.

We also wish to make some comments regarding civil liberties and human rights, based on concerned comments made by readers after hearing about the Dark Web project in the press. A few readers and reporters have cautioned about potential misuse of Dark Web contents by government agencies and authorities.

The Dark Web project is unlike Total Information Awareness (TIA). This is not a secretive government project conducted by spooks. We perform scientific, longitudinal, hypothesis-guided terrorism research like other terrorism researchers. However, we are clearly more computationally oriented, unlike other traditional terrorism research that relies on sociology, communications, and policy-based methodologies. Our contents are open-source in nature (similar to Google's contents), and our major research targets are international, jihadist groups, not regular US citizens. Our researchers are primarily computer and information scientists from all over the world. We develop computer algorithms, tools, and systems. Our research goal is to study and understand the international extremism and terrorism phenomena and the associated web-enabled "social movements" in some regions and communities. Some people may refer to this as understanding the "root cause of terrorism." It can also be considered a "soft power" approach to understanding the social and geopolitical landscape of the Leaderless Jihad.

3.3 The IEEE Intelligence and Security Informatics Conference

The Dark Web project has been frequently reported in the major IEEE ISI conference, held annually in the USA and internationally. The ISI conference was initiated by Dr. Hsinchun Chen in 2003 with initial funding support from various government agencies, including NSF, DHS, CIA, and DOJ. Since then, the meeting has been held in Tucson, San Diego, Atlanta, New Brunswick, Taipei (Taiwan), Dallas, and Vancouver (Canada) (see http://ai.eller.arizona.edu/news/events.asp). The IEEE ISI Conference has also been expanded to Asia (Pacific Asia ISI, held in Singapore; Chengdu, China; Taipei, Taiwan; Bangkok, Thailand; and Hyderabad, India) and Europe (EuroISI held in Denmark). The conference typically draws 100–200 participants in academia (social and computer sciences), industry, and governments who are interested in adopting advanced IT solutions to security problems. More recently, a new Springer journal, *Security Informatics* (SI), has been created to publish high-quality and high-impact research in the field (see: http://ai.eller.arizona.edu/ISI/index.asp). Dr. Hsinchun Chen is serving as editor in chief of Springer SI.

3.4 Dark Web Publications

The Dark Web team has been extremely productive in generating and sharing significant scientific findings and academic publications in books, proceedings, and journal and conference papers over the past 8 years. We include significant publications below such that readers can find more information about our project. Selected research findings are also summarized in subsequent chapters, with more details on approaches, findings, and references.

Books (Monograph, Edited Volumes, and Proceedings): Most Dark Web research has been published in the IEEE ISI conference proceedings and selected edited volumes.

- C. Yang, M. Chau, J. Wang, and H. Chen (Eds.), "Security Informatics," *Annals of Information Systems*, Springer, 2010.
- H. Chen, M. Dacier, et al. (Eds.), Proceedings of the ACM SIGKDD Workshop on CyberSecurity and Intelligence Informatics, Paris, France, June 2009.
- D. Zeng, L. Khan, L. Zhou, M. Day, C. Yang, B. Thuraisingham, and H. Chen (Eds.), Proceedings of the 2009 IEEE International Conference on Intelligence and Security Informatics, ISI 2009, Dallas, Texas, June 2009.
- H. Chen, C. Yang, M. Chau, and S. Li (Eds.), Intelligence and Security Informatics, Proceedings of the Pacific-Asia Workshop, PAISI 2009, Bangkok, Thailand, Lecture Notes in Computer Science (LNCS 5477), Springer-Verlag, 2009.
- C. Yang, H. Chen, et al., "Intelligence and Security Informatics," IEEE ISI 2008 International Workshops: PAISI, PACCF, and SOCO, Taipei, Taiwan, June 2008, Proceedings, Lecture Notes in Computer Science (LNCS 5075), Springer-Verlag, 2008.
- D. Zeng, H. Chen, H. Rolka, and B. Lober (Eds.), Biosurveillance and BioSecurity, International Workshop, BioSecure 2008, Springer-Verlag, December 2008.
- H. Chen and C. Yang (Eds.), Intelligence and Security Informatics: Techniques and Applications, Springer, 2008.
- H. Chen, E. Reid, J. Sinai, A. Silke, and B. Ganor (Eds.), Terrorism Informatics: Knowledge Management and Data Mining for Homeland Security, Springer, 2008.
- H. Chen, T. S. Raghu, R. Ramesh, A. Vinze, and D. Zeng (Eds.), Handbooks in Information Systems – National Security, Elsevier Scientific, 2007.
- C. Yang, D. Zeng, M. Chau, K. Chang, Q. Yang, X. Cheng, J. Wang, F. Wang, and H. Chen (Eds.), Intelligence and Security Informatics, Proceedings of the Pacific-Asia Workshop, PAISI 2007, Lecture Notes in Computer Science (LNCS 4430), Springer-Verlag, 2007.
- S. Mehrotra, D. Zeng, H. Chen, B. Thursaisingham, and F. Wang (Eds.), Intelligence and Security Informatics, Proceedings of the IEEE International Conference on Intelligence and Security Informatics, ISI 2006, Lecture Notes in Computer Science (LNCS 3975), Springer-Verlag, 2006.
- H. Chen, F. Wang, C. Yang, D. Zeng, M. Chau, and K. Chang (Eds.), Intelligence and Security Informatics, Proceedings of the Workshop on Intelligence and Security Informatics, WISI 2006, Lecture Notes in Computer Science (LNCS 3917), Springer-Verlag, 2006.

- H. Chen, Intelligence and Security Informatics for International Security: Information Sharing and Data Mining, Springer, 2006.
- P. Kantor, G. Muresan, F. Roberts, D. Zeng, F. Wang, H. Chen, and R. Merkle (Eds.), Intelligence and Security Informatics, Proceedings of the IEEE International Conference on Intelligence and Security Informatics, ISI 2005, Lecture Notes in Computer Science (LNCS 3495), Springer-Verlag, 2005.
- H. Chen, R. Moore, D. Zeng, and J. Leavitt (Eds.), Intelligence and Security Informatics, Proceedings of the Second Symposium on Intelligence and Security Informatics, ISI 2004, Lecture Notes in Computer Science (LNCS 3073), Springer-Verlag, 2004.
- H. Chen, R. Miranda, D. Zeng, T. Madhusudan, C. Demchak, and J. Schroeder (Eds.), Intelligence and Security Informatics, Proceedings of the First NSF/NIJ Symposium on Intelligence and Security Informatics, ISI 2003, Lecture Notes in Computer Science (LNCS 2665), Springer-Verlag, 2003.

Journal Articles (Published and Forthcoming): Selected Dark Web research has been published in major, SCI-indexed IT journals.

2010

- D. Zimbra, A. Abbasi, and H. Chen, "A Cyber-archeology Approach to Social Movement Research: Framework and Case Study," *Journal of Computer-Mediated Communication*, forthcoming, 2010.
- J. Xu, D. Hu, and H. Chen, "Dynamics of Terrorist Networks," *Journal of Homeland Security and Emergency Management*, forthcoming, 2010.
- A. Abbasi, H. Chen, and Z. Zhang, "Selecting Attributes for Sentiment Classification using Feature Relation Networks," *IEEE Transactions on Knowledge and Data Engineering*, forthcoming, 2010.
- T. J. Fu, A. Abbasi, and H. Chen, "A Focused Crawler for Dark Web Forums," *Journal of the American Society for Information Science and Technology*, Volume 61, Number 6, Pages 1213–1231, 2010.

2009

- Y. Chen, A. Abbasi, and H. Chen, "Framing Social Movement Identity with Cyber-Artifacts: A Case Study of the International Falun Gong Movement," *Annals of Information Systems,* Volume 9, Pages 1–24, 2009.
- Y. Dang, Y. Zhang, H. Chen, P. Hu, S. Brown, and C. Larson, "Arizona Literature Mapper: An Integrated Approach to Monitor and Analyze Global Bioterrorism Research Literature," *Journal of the American Society for Information Science and Technology*, Volume 60, Number 7, Pages 1466–1485, July 2009.
- D. Hu, S. Kaza, and H. Chen, "Identifying Significant Facilitators of Dark Network Evolution," *Journal of the American Society for Information Science and Technology*, Volume 60, Number 4, Pages 655–665, April 2009.

2008

- A. Abbasi and H. Chen, "CyberGate: A System and Design Frame-work for Text Analysis of Computer Mediated Communication," *MIS Quarterly (MISQ)*, Special Issue on Design Science Research, Vol. 32, No. 4, Pages 811–837, December 2008.
- A. Abbasi, H. Chen, S. Thoms, T. Fu, "Affect Analysis of Web Forums and Blogs Using Correlation Ensembles," *IEEE Transactions on Knowledge and Data Engineering*, Volume 20, Number 9, Pages 1168–1180, September 2008.
- H. Chen, W. Chung, J. Qin, E. Reid, M. Sageman, and G. Weinmann, "Uncovering the Dark Web: A Case Study of Jihad on the Web," *Journal of the American Society for Information Science and Technology*, Volume 59, Number 8, Pages 1347–1359, 2008.
- A. Abbasi, H. Chen, H. A. Salem, "Sentiment Analysis in Multiple Languages: Feature Selection for Opinion Classification in Web Forums," *ACM Transactions on Information Systems*, Vol. 26, No. 3, Article 12, June 2008.
- A. Abbasi and H. Chen, "Writeprints: A Stylometric Approach to Identity-Level Identification and Similarity Detection in Cyberspace," *ACM Transactions on Information Systems*, Vol. 26, No. 2, Article 7, March 2008.

2007

- E. Reid and H. Chen, "Mapping the Contemporary Terrorism Research Domain," *International Journal of Human-Computer Studies*, 65, Pages 42–56, 2007.
- J. Qin, Y. Zhou, E. Reid, G. Lai, and H. Chen, "Analyzing Terror Campaigns on the Internet: Technical Sophistication, Content Richness, and Web Interactivity," *International Journal of Human-Computer Studies*, 65, Pages 71–84, 2007.
- E. Reid and H. Chen, "Internet-Savvy U.S. and Middle Eastern Extremist Groups," *Mobilization: An International Quarterly*, 12(2), Pages 177–192, 2007.
- R. Schumaker and H. Chen, "Leveraging Question Answer Technology to Address Terrorism Inquiry," *Decision Support Systems*, Volume 43, Number 4, Pages 1419–1430, 2007.
- E. Reid and H. Chen, "Internet-savvy U.S. and Middle Eastern Extremist Groups," *Mobilization: An International Quarterly Review*, Volume 12, Number 2, Pages 177–192, 2007.
- T. S. Raghu and H. Chen, "Cyberinfrastructure for Homeland Security: Advances in Information Sharing, Data Mining, and Collaboration Systems," *Decision Support Systems*, Volume 43, Number 4, Pages 1321–1323, 2007.

2006

- E. Reid and H. Chen, "Extremist Social Movements Groups and Their Online Digital Libraries," *Information Outlook*, Volume 10, Number 6, Pages 57–65, June 2006

- H. Chen and J.Xu, "Intelligence and Security Informatics for National Security: A Knowledge Discovery Perspective," *Annual Review of Information Science and Technology (ARIST)*, Volume 40, Pages 229–289, 2006.
- J. Li, R. Zheng, and H. Chen, "From Fingerprint to Writeprint," *Communications of the ACM*, Volume 49, Number 4, Pages 76–82, April 2006.
- R. Zheng, J. Li, H. Chen, and Z. Huang, "A Framework for Authorship Identification of Online Messages: Writing-Style Features and Classification Techniques," *Journal of the American Society for Information Science and Technology,* Volume 57, Number 3, Pages 378–393, 2006.

2005

- H. Chen and F. Wang, "Artificial Intelligence for Homeland Security," *IEEE Intelligent Systems,* Special Issue on Artificial Intelligence for National and Homeland Security, Pages 12–16, September/October 2005.
- H. Chen, "Applying Authorship Analysis to Extremist-Group Web Forum Messages," *IEEE Intelligent Systems*, Special Issue on Artificial Intelligence for National and Homeland Security, Pages 67–75, September/October 2005.
- Y. Zhou, E. Reid, J. Qin, G. Lai, and H. Chen, "U.S. Domestic Extremist Groups on the Web: Link and Content Analysis," *IEEE Intelligent Systems*, Special Issue on Artificial Intelligence for National and Homeland Security, Pages 44–51, September/October 2005.

Conference Papers (Published): Dark Web research can also be found in past IEEE ISI and PAISI proceedings.

2010

- Y. Zhang, S. Zeng, C. Huang, L. Fan, X. Yu, Y. Dang, C. Larson, D. Denning, N. Roberts, and H. Chen, "Developing a Dark Web Collection and Infrastructure for Computational and Social Sciences," Proceedings of the 2010 IEEE International Conference on Intelligence and Security Informatics, ISI 2010, Vancouver, Canada, May 2010.
- D. Zimbra and H. Chen, "Comparing the Virtual Linkage Intensity and Real World Proximity of Social Movements," Proceedings of the 2010 IEEE International Conference on Intelligence and Security Informatics, ISI 2010, Vancouver, Canada, May 2010.
- Y. Ku, C. Chiu, Y. Zhang, L. Fan, and H. Chen, "Global Disease Surveillance using Social Media: HIV/AIDS Content Intervention in Web Forums," Proceedings of the 2010 IEEE International Conference on Intelligence and Security Informatics, ISI 2010, Vancouver, Canada, May 2010.

2009

- Y. Zhang, Y. Dang, and H. Chen, "Gender Difference Analysis of Political Web Forums: An Experiment on an International Islamic Women's Forum," in Proceedings of the IEEE International Intelligence and Security Informatics Conference, Dallas, TX, June 2009.
- Y. Zhang, S. Zeng, L. Fan, Y. Dang, C. Larson, and H. Chen, "Dark Web Forums Portal: Searching and Analyzing Jihadist Forums," in Proceedings of the IEEE International Intelligence and Security Informatics Conference, Dallas, TX, June 2009.
- H. Chen, "IEDs in the Dark Web: Lexicon Expansion and Genre Classification," in Proceedings of the IEEE International Intelligence and Security Informatics Conference, Dallas, TX, June 2009.
- T. Fu, C. Huang, and H. Chen, "Identification of Extremist Videos in Online Video Sharing Sites," in Proceedings of the IEEE International Intelligence and Security Informatics Conference, Dallas, TX, June 2009.

2008

- H. Chen and the Dark Web Team, "IEDs in the Dark Web: Genre Classification of Improvised Explosive Device Web Pages," in Proceedings of the IEEE International Intelligence and Security Informatics Conference, Taipei, Taiwan, June 2008. Springer Lecture Notes in Computer Science.
- H. Chen and the Dark Web Team, "Discovery of Improvised Explosive Device Content in the Dark Web," in Proceedings of the IEEE International Intelligence and Security Informatics Conference Taipei, Taiwan, June 2008. Springer Lecture Notes in Computer Science.
- H. Chen and the Dark Web Team, "Sentiment and Affect Analysis of Dark Web Forums: Measuring Radicalization on the Internet," in Proceedings of the IEEE International Intelligence and Security Informatics Conference, Taipei, Taiwan, June 2008. Springer Lecture Notes in Computer Science.
- H. Chen, S. Thoms, T. Fu. "Cyber Extremism in Web 2.0: An Exploratory Study of International Jihadist Groups," in Proceedings of the 2008 IEEE Intelligence and Security Informatics Conference, Taiwan, June 2008.

2007

- H. Chen, "Interaction Coherence Analysis for Dark Web Forums," in Proceedings of the 2007 IEEE Intelligence and Security Informatics Conference, New Brunswick, NJ, May 2007, pp. 342–349.

- H. Chen, "Categorization and Analysis of Text in Computer Mediated Communication Archives Using Visualization," in Proceedings of the 2007 Joint Conference on Digital Libraries (JCDL), Vancouver, BC, Canada, June 2007, pp. 11–18.

2006

- H. Chen, "Visualizing Authorship for Identification," in Proceedings of the Intelligence and Security Informatics: IEEE International Conference on Intelligence and Security Informatics (ISI 2006), San Diego, CA, USA, May 2006.
- H. Chen, "A Framework for Exploring Gray Web Forums: Analysis of Forum-Based Communities in Taiwan," in Proceedings of the Intelligence and Security Informatics: IEEE International Conference on Intelligence and Security Informatics (ISI 2006), San Diego, CA, USA, May 2006.
- Y. Zhou, J. Qin, G. Lai, E. Reid, and H. Chen, "Exploring the Dark Side of the Web: Collection and Analysis of U.S. Extremist Online Forums," in Proceedings of the Intelligence and Security Informatics: IEEE International Conference on Intelligence and Security Informatics (ISI 2006), San Diego, CA, USA, May 2006.
- E. Reid and H. Chen, "Content Analysis of Jihadi Extremist Groups' Videos," in Proceedings of the Intelligence and Security Informatics: IEEE International Conference on Intelligence and Security Informatics (ISI 2006), San Diego, CA, USA, May 2006.
- J. Xu, H. Chen, Y. Zhou, and J. Qin, "On the Topology of the Dark Web of Terrorist Groups," in Proceedings of the Intelligence and Security Informatics: IEEE International Conference on Intelligence and Security Informatics (ISI 2006), San Diego, CA, USA, May 2006.

2005

- Y. Zhou, J. Qin, G. Lai, E. Reid and H. Chen, "Building Knowledge Management System for Researching Terrorist Groups on the Web," Proceedings of the AIS Americas Conference on Information Systems (AMCIS 2005), Omaha, NE, USA, August 2005.
- E. Reid and H. Chen. "Mapping the Contemporary Terrorism Research Domain: Researchers, Publications, and Institutions Analysis," ISI Conference 2005, Atlanta, GA, May 2005.
- H. Chen, "Applying Authorship Analysis to Arabic Web Content," ISI Conference 2005, Atlanta, GA, May 2005.

- E. Reid, J. Qin, Y. Zhou, G. Lai, M. Sageman, G. Weimann, and H. Chen, "Collecting and Analyzing the Presence of Terrorists on the Web: A Case Study of Jihad Websites," IEEE International Conference on Intelligence and Security (ISI 2005), Atlanta, Georgia, 2005.
- D. McDonald, H. Chen, and R. Schumaker, "Transforming Open-Source Documents to Terror Networks: The Arizona TerrorNet," American Association for Artificial Intelligence Conference Spring Symposia (AAAI-2005), Stanford, CA, March 2005.

2004

- H. Chen, J. Qin, E. Reid, W. Chung, Y. Zhou, W. Xi, G. Lai, A. Bonillas, and M. Sageman, "The Dark Web Portal: Collecting and Analyzing the Presence of Domestic and International Terrorist Groups on the Web," Proceedings of the 7th International Conference on Intelligent Transportation Systems (ITSC), Washington D.C., October 2004.
- E. Reid, J. Qin, W. Chung, J. Xu, Y. Zhou, R. Schumaker, M. Sageman, and H. Chen, "Terrorism Knowledge Discovery Project: A Knowledge Discovery Approach to Addressing the Threats of Terrorism," Proceedings of the Second Symposium on Intelligence and Security Informatics, Tucson, AZ, June 2004, pp. 125–145.

2003

- H. Chen, "The Terrorism Knowledge Portal: Advanced Methodologies for Collecting and Analyzing Information from the Dark Web and Terrorism Research Resources," Presented at the Sandia National Laboratories, August 14, 2003.

3.5 Dark Web Project Funding and Acknowledgments

We thank the following agencies for providing research funding support. However, the opinions and results generated from the Dark Web research are entirely those of the University of Arizona Artificial Intelligence Lab and do not represent the position of the funding agencies.

There has been significant interest from various intelligence, justice, and defense agencies in our computational methodologies, tools, and systems. We are glad that we have been able to offer assistance in these technical areas of interest. However, we do not perform (security) clearance-level work nor do we conduct targeted cybercrime or intelligence investigations. Our research staff members are primarily computer and information scientists from all over the world and have expertise in many languages. We perform academic research, write papers, and develop computer programs. We do hope that our work can contribute to international security and peace. Selected Dark Web research projects are listed in Table 1.1.

Table 1.1 Funding of Dark Web research projects (selected)

Agency and project title	Funding period
National Science Foundation	Sept 2003–Aug 2011
– (CRI: CRD) Developing a Dark Web Collection and Infrastructure for Computational and Social Sciences (NSF # CNS-0709338)	
– (EXP-LA) Explosives and IEDs in the Dark Web: Discovery, Categorization, and Analysis (NSF # CBET-0730908)	
– (SGER) Multilingual Online Stylometric Authorship Identification: An Exploratory Study (NSF # IIS-0646942)	
– (ITR, Digital Government) COPLINK Center for Intelligence and Security Informatics Research (partial support) (NSF # EIA-0326348)	
Air Force Research Lab	Aug 2008–May 2009
– Dark Web WMD-Terrorism Study (Subcontract No FA8650–02)	
Defense Threat Reduction Agency	Jul 2009–Jul 2012
– WMD Intent Identification and Interaction Analysis Using the Dark Web (HDTRA1-09-1-0058)	
Dept. of Homeland Security/CNRI	Oct 2003–Sept 2005
– Border Safe Initiative (partial support)	
Library of Congress	Jul 2005–Jun 2008
– Capture of Multimedia, Multilingual Open Source Web-based At-Risk Content	

4 Partnership Acknowledgments

We thank the following academic partners and colleagues for their support, help, and comments. Many of our terrorism research colleagues have taught us much about the significance and intricacy of this important domain. They also help guide us in the development of our scientific, computational approach.

- Officers and domain experts of Tucson Police Department, Arizona Department of Customs and Border Protection, and San Diego Automated Regional Justice Information System (ARJIS) Program
- Dr. Shlomo Argamon, Illinois Institute of Technology
- Dr. Michael Chau, University of Hong Kong
- Rick Eaton, Simon Wiesenthal Center
- Chip Ellis, Memorial Institute for the Prevention of Terrorism (MIPT)
- Dr. Paul J. Hu, University of Utah
- Rex Hudson, Library of Congress
- Drs. Henrik Larsen and Nasrullah Memon, Aalborg University, Denmark
- Dr. Mark Last, Ben-Gurion University, Israel
- Dr. Ee-Peng Lim, Singapore Management University, Singapore
- Dr. Edna Reid, Federal Bureau of Investigation
- Dr. Johnny Ryan, The Institute of International and European Affairs (IIEA)
- Dr. Marc Sageman, University of Pennsylvania
- Dr. Joshua Sinai, The Analysis Corporation
- Dr. Katrina von Knop, George Marshall Center, Germany

- Dr. Feiyue Wang, The University of Arizona and the Chinese Academy of Sciences, China
- Dr. Jau-Hwang Wang and Robert Chang, Central Police University, Taiwan
- Dr. Gabriel Weimann, University of Haifa, Israel
- Dr. Chris Yang, Drexel University
- Dr. Daniel Zeng, University of Arizona

Reference

G. Weimann, "Terror on the Internet: The New Arena, the New Challenges," US Institute of Peace Press, Washington, DC, 2006.

Chapter 2
Intelligence and Security Informatics (ISI): Research Framework

1 Information Technology and National Security

The tragic events of September 11 and the following anthrax contamination of letters caused drastic effects on many aspects of society. Terrorism became the most significant threat to national security because of its potential to bring massive damage to our infrastructure, economy, and people.

In response to this challenge, federal authorities are actively implementing comprehensive strategies and measures in order to achieve the three objectives identified in the "National Strategy for Homeland Security" report (Office of Homeland Security 2002): (1) preventing future terrorist attacks, (2) reducing the nation's vulnerability, and (3) minimizing the damage and recovering from attacks that occur. State and local law enforcement agencies, likewise, are becoming more vigilant about the criminal activities that harm public safety and threaten national security.

Academics in the fields of natural sciences, computational science, information science, social sciences, engineering, medicine, and many others have been called upon to help enhance the government's ability to fight terrorism and other crimes. Science and technology have been identified in the "National Strategy for Homeland Security" report as the keys to win the new counterterrorism war (Office of Homeland Security 2002). It is widely believed that information technology will play an indispensable role in making our nation safer (National Research Council 2002) by supporting intelligence and knowledge discovery through collecting, processing, analyzing, and utilizing terrorism- and crime-related data (Chen 2006). Based on the crime and intelligence knowledge discovered, the federal, state, and local authorities can make timely decisions to select effective strategies and tactics as well as allocate the appropriate amount of resources to detect, prevent, and respond to future attacks.

Six critical mission areas have been identified where information technology can contribute to the accomplishment of the three strategic national security objectives

H. Chen, *Dark Web: Exploring and Data Mining the Dark Side of the Web*,
Integrated Series in Information Systems 30, DOI 10.1007/978-1-4614-1557-2_2,

identified in the "National Strategy for Homeland Security" report (Office of Homeland Security 2002):

- *Intelligence and warning*. Although terrorism depends on surprise to damage its targets (Office of Homeland Security 2002), terrorist activities are not random and impossible to track. Terrorists must plan and prepare before the execution of an attack by selecting a target, recruiting and training executors, acquiring financial support, and traveling to the country where the target is located (Sageman 2004). To avoid being preempted by authorities, they may hide their true identities and disguise attack-related activities. Similarly, criminals may use falsified identities during police contacts (Wang et al. 2004a). Although it is difficult, detecting potential terrorist attacks or crimes is possible and feasible with the help of information technology. By analyzing the communication and activity patterns among terrorists and their contacts (i.e., terrorist networks), detecting deceptive identities, or employing other surveillance and monitoring techniques, intelligence and warning systems may issue timely, critical alerts and warnings to prevent attacks or crimes from occurring.
- *Border and transportation security*. Terrorists enter a targeted country through an air, land, or sea port of entry. Criminals in narcotics rings travel across borders to purchase, carry, distribute, and sell drugs. Information, such as travelers' identities, images, fingerprints, vehicles used, and other characteristics, is collected from customs, border, and immigration authorities on a daily basis. Counterterrorism and crime-fighting capabilities can be greatly improved by the creation of a "smart border," where information from multiple sources is shared and analyzed to help locate wanted terrorists or criminals. Technologies such as information sharing and integration, collaboration and communication, biometrics, and image and speech recognition will be greatly needed in such smart borders.
- *Domestic counterterrorism*. As terrorists, both international and domestic, may be involved in local crimes, state and local law enforcement agencies are also contributing to the missions by investigating and prosecuting crimes. Terrorism, like gangs and narcotics trafficking, is regarded as a type of organized crime in which multiple offenders cooperate to carry out offenses. Information technologies that help find cooperative relationships between criminals and their interactive patterns would also be helpful for analyzing terrorism. Monitoring activities of domestic terrorist and extremist groups using advanced information technologies will also be helpful to public safety personnel and policy makers.
- *Protecting critical infrastructure and key assets*. Roads, bridges, water supplies, and many other physical service systems are critical infrastructure and key assets of a nation. They may become the target of terrorist attacks because of their vulnerabilities (Office of Homeland Security 2002). Moreover, virtual (cyber) infrastructure such as the Internet may also be vulnerable to intrusions and inside threats. Criminals and terrorists are increasingly using cyberspace to conduct illegal activities, share ideology, solicit funding, and recruit. In addition to physical devices such as sensors and detectors, advanced information technologies are needed to model the normal behaviors of the usage of these systems and then use

the models to distinguish abnormal behaviors from normal behaviors. Protective or reactive measures can be selected based on the results to secure these assets from attacks.

- *Defending against catastrophic terrorism.* Terrorist attacks can cause devastating damage to a society through the use of chemical, biological, or nuclear weapons. Biological attacks, for example, may cause contamination, infectious disease outbreaks, and significant loss of life. Information systems that can efficiently and effectively collect, access, analyze, and report data about catastrophe-leading events can help prevent, detect, respond to, and manage these attacks.
- *Emergency preparedness and responses.* In case of a national emergency, prompt and effective responses are critical to reduce the damage resulting from an attack. In addition to the systems that are prepared to defend against catastrophes, information technologies that help design and experiment with optimized response plans, identify experts, train response professionals, and manage consequences are beneficial in the long run. Moreover, information systems that facilitate social and psychological support to the victims of terrorist attacks can also help society recover from disasters.

Although it is important for the critical missions of national security, the development of information technology for counterterrorism and crime-fighting applications faces many problems and challenges.

1.1 Problems and Challenges

Currently, intelligence and security agencies are gathering large amounts of data from various sources. Processing and analyzing such data, however, has become increasingly difficult. By treating terrorism as a form of organized crime, these challenges can be categorized into three types:

- *Characteristics of criminals and crimes.* Some crimes may be geographically diffused and temporally dispersed. In organized crimes such as transnational narcotics trafficking, criminals often live in different countries, states, and cities. Drug distribution and sales occur in different places at different times. Similar situations exist in other organized crimes (e.g., terrorism, armed robbery, and gang-related crime). As a result, an investigation must cover multiple offenders who commit criminal activities in different places at different times. This can be fairly difficult given the limited resources that intelligence and security agencies have. Moreover, as computer and Internet technologies advance, criminals are utilizing cyberspace to commit various types of cybercrimes under the guise of ordinary online transactions and communications.
- *Characteristics of security- and intelligence-related data.* A significant source of challenge is information stovepipe and overload resulting from diverse data sources, multiple data formats, and large data volumes. Unlike other domains

such as marketing, finance, and medicine in which data can be collected from particular sources (e.g., sales records from companies, patient medical history from hospitals), the intelligence and security domain does not have a well-defined data source. Both authoritative information (e.g., crime incident reports, telephone records, financial statements, immigration and customs records) and open source information (e.g., news stories, journal articles, books, web pages) need to be gathered for investigative purposes. Data collected from these different sources often are in different formats ranging from structured database records to unstructured text, image, audio, and video files. Important information such as criminal associations may be available but contained in unstructured, multilingual texts and remains difficult to access and retrieve. Moreover, as data volumes continue to grow, extracting valuable and credible intelligence and knowledge becomes a difficult problem.

- *Characteristics of security and intelligence analysis techniques*. Current research on the technologies for counterterrorism and crime-fighting applications lacks a consistent framework addressing the major challenges. Some information technologies, including data integration, data analysis, text mining, image and video processing, and evidence combination, have been identified as being particularly helpful (National Research Council 2002). However, the question of how to employ them in the intelligence and security domain and use them to effectively address the critical mission areas of national security remains unanswered.

Facing the critical missions of national security and various data and technical challenges, we believe there is a pressing need to develop the science of "Intelligence and Security Informatics" (ISI) (Chen 2006), with its main objective being the "*development of advanced information technologies, systems, algorithms, and databases for national security-related applications, through an integrated technological, organizational, and policy-based approach.*"

1.2 Intelligence and Security Informatics Versus Biomedical Informatics: Emergence of a Discipline

Comparing ISI with biomedical informatics, an established academic discipline addressing information management issues in biological and medical applications (Shortliffe and Blois 2000; Chen et al. 2005), we found tremendous analogies between these two disciplines. Table 2.1 summarizes the similarities and differences between ISI and biomedical informatics.

In terms of data characteristics, they both face the information stovepipe and information overload problem. In terms of technology development, they both are searching for new approaches, methods, and innovative use of existing techniques. In terms of scientific contributions, they both may add new insights and knowledge to various academic disciplines.

Table 2.1 Analogies between ISI and biomedical informatics

		Biomedical informatics	ISI
Challenges	Domain-specific	• Complexity and uncertainty associated with organisms and diseases	• Geographically diffused and temporally dispersed organized crimes
		• Critical decisions regarding patient well-being and biomedical discoveries	• Cybercrimes on the Internet
			• Critical decisions related to public safety and homeland security
	Data	• Information stovepipe and overload	• Information stovepipe and overload
		• HL7 XML standard	• Justice XML standard
		• PHIN MS messaging	• Criminal incident records
		• Patient records, disease data, medical images	• Multilingual intelligence open sources
	Technology	• Ontologies and linguistic parsing	• Information integration
		• Information integration	• Criminal network analysis
		• Data and text mining	• Data, text, and web mining
		• Medical decision support systems and techniques	• Identity management and deception detection
Methodology		KDD	KDD
Contributions	Scientific	• Computer and information science, sociology, policy, legal	• Computer and information science, sociology, policy, legal
		• Clinical medicine and biology	• Criminology, terrorism research
	Practical	• Public health	• Crime investigation and counterterrorism
		• Patient well-being	• National and homeland security
		• Biomedical treatment and discovery	

Most importantly, as a consistent research framework based on knowledge management and data mining has begun to emerge in biomedical informatics (Chen et al. 2005), ISI also needs a framework to guide its research. Facing the unique challenges (and associated opportunities) of information overload and the pressing need for advanced criminal and intelligence analyses and investigations, we believe that the Knowledge Discovery from Databases (KDD) methodology (Fayyad and Uthurusamy 2002), which has achieved significant success in other information-intensive, knowledge-critical domains including business, engineering, biology, and medicine, could be critical in addressing the challenges and problems facing ISI.

1.3 Research Opportunities

The emergence of a new discipline such as ISI would require careful cultivation and development by many top-notch researchers and practitioners from many different disciplines, including (but not limited to): computer science, information science, information systems, electrical engineering, social science, law, public policy, criminal justice, terrorism research, psychology, behavioral and economic sciences, management science, bioinformatics, public health, etc. There is an abundance of opportunities for developing new and innovative funded ISI-related projects.

Regardless of which funding programs you may be considering for your research, there are some common characteristics among successfully funded and (eventually, after execution) high-impact projects:

- *Unique and critical scientific or engineering innovations*: You need to clearly distinguish your research from others.
- *Important problems and significant partners*: You need to address important national security problems and demonstrate your commitment to address these problems with the help and support of your local, state, and federal agency partners.
- *From small to large*: Most funded projects began humbly with proof-of-concept level funding.
- *A multidisciplinary team*: After initial success, a multidisciplinary team of computer scientists, system developers, social scientists, policy and legal experts, domain (intelligence and security) experts, and such will be needed to implement a full-scale, multiphased, complex national security-related project.
- *Aim high*: The following tangible (but somewhat lofty) project goals are always good for your project team to aim at: (1) publishing your project findings in *Science, Nature*, or *Proceedings of the Academy of Science* (for its scientific contributions) and (2) being featured in a *New York Times* or *USA Today* front-page article (for its societal impact).

2 ISI Research Framework

Crime is an act or the commission of an act that is forbidden, or the omission of a duty that is commanded by a public law and that makes the offender liable to punishment by that law. The more threat a crime type poses on public safety, the more likely it is to be of national security concern. Some crimes such as traffic violations, theft, and homicide are mainly in the jurisdiction of local law enforcement agencies. Some other crimes need to be dealt with by both local law enforcement and national security authorities. Identity theft and fraud, for instance, are relevant at both the local and national level – criminals may escape arrest by using false identities; drug smugglers may enter the United States by holding counterfeited passports or visas.

Table 2.2 Crime types and security concerns

(Arrow: Increasing public influence ↓)

Crime type	Local law enforcement level	National security level
Traffic violations	Driving under influence (DUI), fatal/personal injury/property damage, traffic accident, road rage	–
Sex crime	Sexual offenses, sexual assaults, child molesting	Organized prostitution, people smuggling
Theft	Robbery, burglary, larceny, motor vehicle theft, stolen property	Theft of national secrets or weapon information
Fraud	Forgery and counterfeiting, fraud, embezzlement, identity deception	Transnational money laundering, identity fraud, transnational financial fraud
Arson	Arson on buildings, apartments	–
Organized crime	Narcotic drug offenses (sales or possession), gang-related offenses	Transnational drug trafficking, terrorism (bioterrorism, bombing, hijacking, etc.)
Violent crime	Criminal homicide, armed robbery, aggravated assault, other assaults	Terrorism
Cybercrime	Internet fraud (e.g., credit card fraud, advance fee fraud, fraudulent web sites), illegal trading, network intrusion/hacking, virus spreading, hate crimes, cyber-piracy, cyber-pornography, cyber-terrorism, theft of confidential information	

Organized crimes, such as terrorism and narcotics trafficking, are often diffuse geographically, resulting in common security concerns across cities, states, and countries. Cybercrimes can pose threats to public safety across multiple jurisdictional areas due to the widespread nature of computer networks.

Table 2.2 summarizes the different types of crimes sorted by the degree of their respective public influence. International and domestic terrorism, in particular, often involves multiple crime types (e.g., identity theft, money laundering, arson and bombing, organized and violent activities, and cyber-terrorism) and causes great damage.

We believe that KDD techniques can play a central role in improving counterterrorism and crime-fighting capabilities of intelligence, security, and law enforcement agencies by reducing the cognitive and information overload. Knowledge discovery refers to nontrivial extraction of implicit, previously unknown, and potentially useful knowledge from data. Knowledge discovery techniques promise easy, convenient, and practical exploration of very large collections of data for organizations and users and have been applied in marketing, finance, manufacturing, biology, and many other domains (e.g., predicting consumer behaviors, detecting credit card

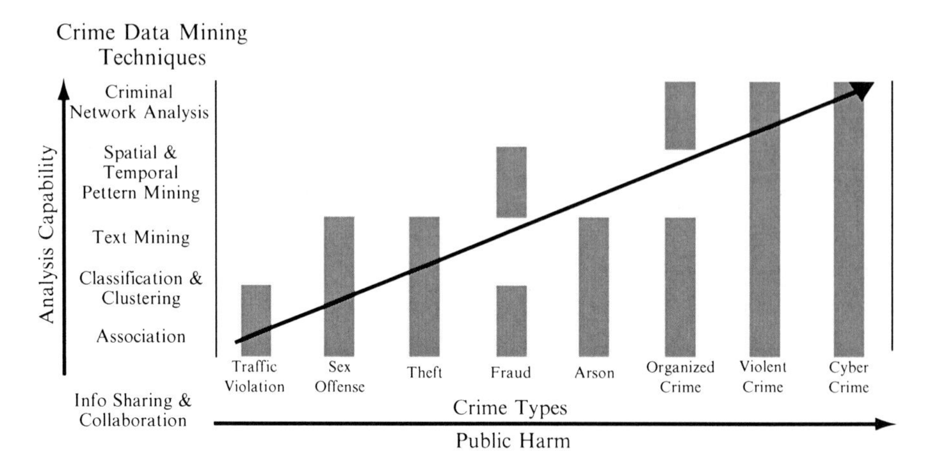

Fig. 2.1 A knowledge discovery research framework for ISI

frauds, or clustering genes that have similar biological functions) (Fayyad and Uthurusamy 2002). Traditional knowledge discovery techniques include association rules mining, classification and prediction, cluster analysis, and outlier analysis. As natural language processing (NLP) research advances, (multilingual) text mining approaches that automatically extract, summarize, categorize, and translate text documents have also been widely used (Chen 2006).

Many of these KDD technologies could be applied in ISI studies. Keeping in mind the special characteristics of crimes, criminals, and security-related data, we categorize existing ISI technologies into six classes: *information sharing and collaboration, crime association mining, crime classification and clustering, intelligence text mining, spatial and temporal crime mining,* and *criminal network mining*.

These six classes are grounded on traditional knowledge discovery technologies with a few new approaches added, including spatial and temporal crime pattern mining and criminal network analysis, which are more relevant to counterterrorism and crime investigation. Although information sharing and collaboration are not data mining per se, they help prepare, normalize, warehouse, and integrate data for knowledge discovery and thus are included in the framework.

In Fig. 2.1, we present our proposed research framework, with the horizontal axis being the crime types and the vertical axis being the six classes of techniques (Chen 2006). The shaded regions on the chart show promising research areas, i.e., that a certain class of techniques is relevant to solving a certain type of crime. Note that more serious crimes may require a more complete set of knowledge discovery techniques. For example, the investigation of organized crimes such as terrorism may depend on criminal network analysis technology, which requires the use of other knowledge discovery techniques such as association mining and clustering. An important observation about this framework is that the high-frequency occurrences and strong association patterns of severe and organized crimes such as terrorism and narcotics present a unique opportunity and potentially high rewards for adopting such a knowledge discovery framework.

Several unique classes of data mining techniques are of great relevance to ISI research. *Text mining* is critical for extracting key entities (people, places, narcotics, weapons, time, etc.) and their relationships presented in voluminous police incident reports, intelligence reports, open source news clips, etc. Some of these techniques need to be multilingual in nature, including the abilities for machine translation and crosslingual information retrieval (CLIR). *Spatial and temporal mining and visualization* are often needed for geographic information systems (GIS) and temporal analysis of criminal and terrorist events. Most crime analysts are well trained in GIS-based crime mapping tools; however, automated spatial and temporal pattern mining techniques (e.g., hotspot analysis) have not been adopted widely in intelligence and security applications.

Organized criminals (e.g., gangs and narcotics) and terrorists often form interconnected covert networks for their illegal activities. Often referred to as "dark networks," these organizations exhibit unique structures, communication channels, and resilience to attack and disruption. New computational techniques, including social network analysis, network learning, and network topological analysis (e.g., random network, small-world network, and scale-free network), are needed for the systematic study of those complex and covert networks. We broadly consider these techniques under *criminal network analysis* in Fig. 2.1.

2.1 Caveats for Data Mining

Before we review in detail relevant ISI-related data mining techniques, applications, and literature in the next chapter, we wish to briefly discuss the legal and ethical caveats regarding crime and intelligence research.

The potential negative effects of intelligence gathering and analysis on the privacy and civil liberties of the public have been well publicized (Cook and Cook 2003). There exist many laws, regulations, and agreements governing data collection, confidentiality, and reporting, which could directly impact the development and application of ISI technologies. We strongly recommend that intelligence and security agencies and ISI researchers be aware of these laws and regulations in research. Moreover, we also suggest that a hypothesis-guided, evidence-based approach be used in crime and intelligence analysis research. That is, there should be probable and reasonable causes and evidence for targeting particular individuals or datasets for analysis. Proper investigative and legal procedures need to be strictly followed. It is neither ethical nor legal to "fish" for potential criminals from diverse and mixed crime-, intelligence-, and citizen-related data sources. The well-publicized Defense Advanced Research Program Agency (DARPA), Total Information Awareness (TIA) Program, and the Multistate Anti-Terrorism Information Exchange (MATRIX) system, for example, have recently been shut down due to their potential misuse of citizen data and impairment of civil liberties (American Civil Liberties Union 2004; O'Harrow 2005).

2.2 Domestic Security, Civil Liberties, and Knowledge Discovery

In an important recent review article by Strickland, Baldwin, and Justsen (Strickland et al. 2005), the authors provide an excellent historical account of government surveillance in the United States. The article presents new surveillance initiatives in the age of terrorism (including the passage of the USA Patriot Act), discusses in great depth the impact of technology on surveillance and citizen's rights, and proposes balancing between needed secrecy and oversight. We believe this is one of the most comprehensive articles addressing civil liberties issues in the context of national security research. We summarize some of the key points made in the article in the context of our proposed ISI research.

Framed in the context of domestic security surveillance, the paper considers surveillance as an important intelligence tool that has the potential to contribute significantly to national security but also to infringe civil liberties. As faculty of the University of Maryland Information Science Department, the authors believe that information science and technology has drastically expanded the mechanisms by which data can be collected, and knowledge extracted and disseminated through some automated means.

An immediate result of the tragic events of September 11, 2001, was the extraordinarily rapid passage of the USA Patriot Act in late 2001. The legislation was passed by the Senate on October 11, 2001, by the House on October 24, 2001, and signed by the President on October 26, 2001. The continuing legacy of the then-existing consensus and the lack of detailed debate and considerations created a bitter ongoing national argument as to the proper balance between national security and civil liberties. The Patriot Act contains ten titles in 131 pages. It amends numerous laws, including, for example, expansion of electronic surveillance of communications in law enforcement cases, authorizing sharing of law enforcement data with intelligence, expansion of the acquisition of electronic communications as well as commercial records for intelligence use, and creation of new terrorism-related crimes.

However, as new data mining and/or knowledge discovery techniques become mature and potentially useful for national security applications, there are great concerns of violating civil liberties. Both the DARPA's TIA Program and the Transportation Security Administration's (TSA) Computer Assisted Passenger Prescreening Systems (CAPPS II) were cited as failed systems that faced significant media scrutiny and public opposition. Both systems were based on extensive data mining of commercial and government databases collected for one purpose and to be shared and used for another purpose, and both systems were sidetracked by a widely perceived threat to personal privacy. Based on much of the debate generated by these programs, the authors suggest that data mining using public or private sector databases for national security purposes must proceed in two stages – first, the search for general information must ensure anonymity; second, the acquisition of specific identity, if required, must be by court order under appropriate standards (e.g., in terms of "special needs" or "probable causes").

In their concluding remarks, the authors cautioned that secrecy in any organization could pose a real risk of abuse and must be constrained through effective checks and balances. Moreover, information science and technology professionals are ideally situated to provide the tools and techniques by which the necessary intelligence is collected, analyzed, and disseminated, while civil liberties are protected through established laws and policies.

In addition to the review article by Strickland et al., readers are also referred to an excellent book entitled "No Place to Hide," written by Washington Post reporter Robert O'Harrow (2005). He reveals how the government is creating a national intelligence infrastructure with the help of private information, security, and technology companies. The book examines in detail the potential impact of this new national security system on our traditional notions of civil liberties, autonomy, and privacy.

2.3 *Research Opportunities*

National security research poses unique challenges and opportunities. Much of the established data mining and knowledge discovery literature, findings, and techniques need to be reexamined in light of the unique data and problem characteristics in the law enforcement and intelligence community. New text mining, spatial and temporal pattern mining, and criminal network analysis of relevance to national security are among some of the most pressing research areas. However, researchers cannot conduct research in a vacuum. Partnerships with local, state, and federal agencies need to be formed to obtain relevant test data and necessary domain expertise for ISI research. Only after rigorous testing with scrubbed or anonymous data can select techniques be field examined and verified by domain experts (i.e., law enforcement personnel, intelligence analysts, and policy makers). These techniques should be used in actual investigations only after experts have confirmed their potential value. At this stage, the researcher-designed algorithms or systems are often much improved and refined, and are often operated and controlled by the domain experts with their own heuristics, know-how, and judgment.

References

American Civil Liberties Union. (2004). *MATRIX: Myths and Reality*. Retrieved July 27, 2004, from the World Wide Web: http://www.aclu.org/Privacy/Privacy.cfm?ID=14894&c=130.

Chen, H. (2006). *Intelligence and Security Informatics for International Security: Information Sharing and Data Mining*, Springer.

Chen, H., Fuller, S. S., Friedman, C., and Hersh, W. (Eds.) (2005). *Medical Informatics: Knowledge Management and Data Mining in Biomedicine*. Berlin: Springer.

Cook, J. S. and Cook, L. L. (2003). Social, ethical and legal issues of data mining. In J. Wang (Ed.), *Data Mining: Opportunities and Challenges* (pp. 395–420). Hershey, PA: Idea Group Publishing.

Fayyad, U. M. and Uthurusamy, R. (2002). Evolving data mining into solutions for insights. *Communications of the ACM, 45*(8), 28–31.

National Research Council. (2002). *Making the Nation Safer: The Role of Science and Technology in Countering Terrorism*. Washington, DC: National Academy Press.

O'Harrow, R. (2005). *No Place to Hide*. New York: Free Press.

Office of Homeland Security. (2002). *National Strategy for Homeland Security*. Washington D.C.: Office of Homeland Security.

Sageman, M. (2004). *Understanding Terror Networks*. Philadelphia: University of Pennsylvania Press.

Shortliffe, E. H. and Blois, M. S. (2000). The computer meets medicine and biology: Emergence of a discipline. In K. J. Hannah and M. J. Ball (Eds.), *Health Informatics* (pp. 1–40). New York: Springer-Verlag.

Strickland, L. S., Baldwin, D. A., and Justsen, M. (2005). Domestic security surveillance and civil liberties. In B. Cronin (Ed.), *Annual Review of Information Science and Technology (ARIST)*, Volume 39. Medford, New Jersey: Information Today, Inc.

Wang, G., Chen, H., and Atabakhsh, H. (2004a). Automatically detecting deceptive criminal identities. *Communications of the ACM, 47*(3), 71–76.

Chapter 3
Terrorism Informatics

1 Introduction

Terrorism informatics is defined as the "application of advanced methodologies and information fusion and analysis techniques to acquire, integrate, process, analyze, and manage the diversity of terrorism-related information for national/international and homeland security-related applications" (Chen et al. 2008). These techniques are derived from disciplines such as computer science, informatics, statistics, mathematics, linguistics, social sciences, and public policy. Because the study of terrorism involves copious amounts of information from multiple sources, data types, and languages, information fusion and analysis techniques such as data mining, text mining, web mining, data integration, language translation technologies, and image and video processing are playing key roles in the future prevention, detection, and remediation of terrorism. Although there has been substantial investment and research in the application of computer technology to terrorism research, much of the literature in this emerging area is fragmented and often narrowly focused within specific domains. There is a critical need to develop a multidisciplinary approach to answering important terrorism-related research questions.

2 Terrorism and the Internet

Terrorism is the systematic use of terror especially as a means of coercion. At present, the international community has been unable to formulate a universally agreed, legally binding, criminal law definition of terrorism. Common definitions of terrorism refer only to those violent acts which are intended to create fear (terror), are perpetrated for an ideological goal, and deliberately target or disregard the safety of noncombatants (civilians) (http://en.wikipedia.org/wiki/Terrorism). Rooted in political science, terrorism study has attracted researchers from many social science disciplines, from international relations to communications, and from defense analysis

H. Chen, *Dark Web: Exploring and Data Mining the Dark Side of the Web*,
Integrated Series in Information Systems 30, DOI 10.1007/978-1-4614-1557-2_3,

to intelligence study. Bruce Hoffman of the Georgetown University School of Foreign Service and Brian Jenkins of the RAND Corporation are some of the prominent scholars in terrorism and counterinsurgency study.

More recently, several terrorism scholars have begun to look into modern terrorism with a more data and network centric perspective. Marc Sageman's two critically acclaimed books, *Understanding Terror Networks* (2004) and *Leaderless Jihad* (2008), are good examples. As stated on the book cover of *Understanding Terror Networks*:

> For decades, a new type of terrorism has been quietly gathering ranks in the world. American ability to remain oblivious to the new movements ended on September 11, 2001. The Islamic fanatics in the global Salafi jihad (the violent, revivalist social movement of which al Qaeda is a part) target the West, but their operations mercilessly slaughter thousands of people of all races and religions throughout the world. Marc Sageman challenges conventional wisdom about terrorism, observing that the key to mounting an effective defense against future attacks is a thorough understanding of the networks that allow the new terrorist to proliferate.
>
> (Sageman 2004).

Based on intensive data collection and analysis of documents from international press and court hearings on 172 important jihadists, Sageman was able to look into the social bonds, predating ideological commitment. Many important network-based observations about these groups were identified with striking examples, including small-world network and clique formation, network robustness, wide geographical distribution, fuzzy boundaries, and the strength of weak bonds. The Internet has also been found to affect the global jihad by making possible a new type of relationship between an individual and a virtual community (Sageman 2004). In his 2007 book "Leaderless Jihad," Sageman continued his rigorous and systematic analysis of his detailed, evidence-based terrorist (500+ members) database. He described that "The process of radicalization that generates small, local, self-organized groups in a hostile habitat but linked through the Internet leads to a disconnected global network, the leaderless jihad" (Sageman 2008). He urged terrorism study to go from anecdote to data and from journalism to social sciences. Among his findings, he found that before 2004, face-to-face interactions were more common among the average 26-year-old jihadi members; while after 2004, most interactions were on the Internet, and the average member was 20 years of age. In a matter of 3–4 years and consistent with the modern information communication and technology (ICT) evolution, the Internet and the social media are helping the radical Islamists create a global, virtual social movement.

In addition to Sageman's influential work, several scholars have also examined the impact of the Internet on the proliferation and radicalization of the global jihadi movement. In his seminal book in 2006, "Terror on the Internet," Gabriel Weimann (2006) of the Haifa University Department of Communications in Israel reported his 8-year study of Internet use by terrorist organizations and their supporters. Sophisticated web sites have been found to help these organizations raise funds, recruit members, plan and launch attacks, and publicize their chilling results. Weimann describes the Internet as the new media for promoting new terrorism to a new generation of audience and for winning the war over minds. In Johnny Ryan's 2007 book "Countering Militant Islamist Radicalization on the Internet," he presented the EU's perspective on such developments, especially for the Europe-based homegrown radical groups. Due to the ubiquity and scale of the Internet, he suggested pooling technical and linguistic resources to monitor extremism on the Internet for the EU member states. He also suggested disseminating moderate opinions of credible Muslim scholars and web opinion leaders and encouraged user-driven content in social media to counter radicalization and violence.

As summarized in Chap. 1, the Dark Web research program of the University of Arizona Artificial Intelligence Lab is a complementary effort in the emerging discipline of terrorism informatics (Chen 2006). Unlike the social sciences approach adopted by the abovementioned scholars, the Dark Web project adopts data, system, and computational approaches to studying terror and terrorism on the Internet. By collecting and analyzing a large-scale, longitudinal, and fluid collection of terrorist-generated content using computer programs, we offer our complementary perspective and approach in understanding the overwhelmingly complex international terrorism landscape. Hsinchun Chen's 2006 book "Intelligence and Security Informatics" reports selected examples from the Dark Web project at its early stage. This book serves to report significant findings and observations from the recent Dark Web developments.

The edited volume of "Terrorism Informatics" (2008) by Chen, Reid, Sinai, Silke, and Ganor became the first manuscript dedicated to the terrorism informatics topic. The book is highly interdisciplinary, with editors and contributors from both social sciences and computational science. The goal of the book is to present terrorism informatics along two highly intertwined dimensions: methodological issues in terrorism research, including information diffusion techniques to support terrorism prevention, detection, and response; and legal, social, privacy, and data confidentiality challenges and approaches.

3 Terrorism Research Centers and Resources

Terrorism informatics relies heavily on terrorism domain knowledge and databases. We summarize some critical terrorism research centers and resources below based on our Dark Web experience. They are grouped into three major categories: (1) think tanks and intelligence resources, (2) terrorism databases and online resources, and (3) higher education research institutes. Clearly, the summary is not intended to be exhaustive. We only hope to help (terrorism) community outsiders (like ourselves or other computer scientists) get a glimpse of possible terrorism-related resources to familiarize themselves with this complex area. For each research unit, we also provide a web link for readers to find additional detailed information.

3.1 Think Tanks and Intelligence Resources

The RAND Corporation is a US nonprofit research and development outfit. It has evolved from a think tank during World War II to an independent corporation.

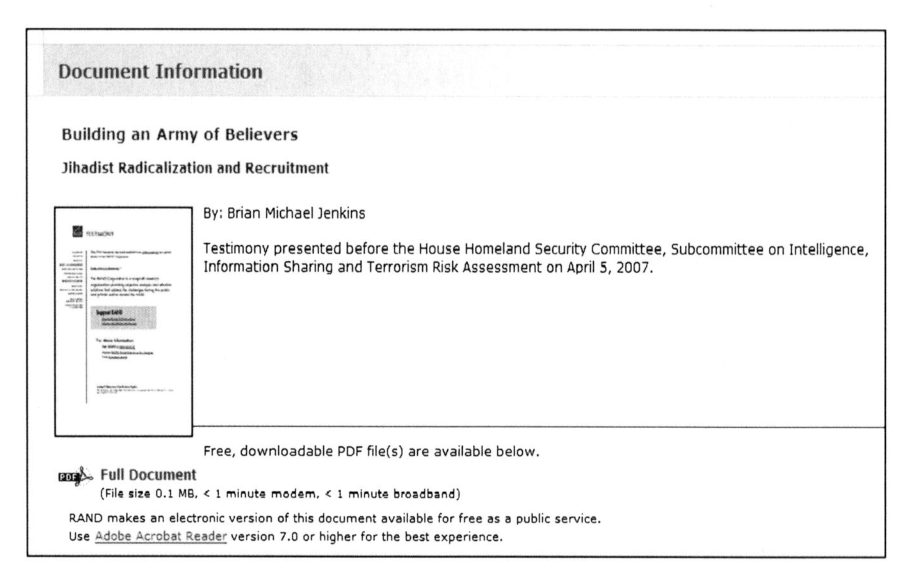
Document Information

Building an Army of Believers

Jihadist Radicalization and Recruitment

By: Brian Michael Jenkins

Testimony presented before the House Homeland Security Committee, Subcommittee on Intelligence, Information Sharing and Terrorism Risk Assessment on April 5, 2007.

Free, downloadable PDF file(s) are available below.

Full Document
(File size 0.1 MB, < 1 minute modem, < 1 minute broadband)

RAND makes an electronic version of this document available for free as a public service.
Use Adobe Acrobat Reader version 7.0 or higher for the best experience.

Fig. 3.1 Sample RAND report

Two prominent terrorism scholars, Brian Jenkins and Bruce Hoffman, are affiliated with the RAND Corporation. The company has been influential with its many excellent, timely, and thorough reports of international relations, political violence, and terrorism studies. For examples, see: http://www.rand.org/media/experts/policy_areas/homeland_security_and_terrorism/index.html (Fig. 3.1).

As part of the West Point military academy, the Combating Terrorism Center (CTC) has provided counterterrorism strategic analyses independent from academy curricula, Pentagon tactics, or US government politics since 2003. Among its notable topical reports are the Harmony project (making sense of DoD al-Qaeda document database) and the Islamic imagery project. For more detail, see: http://ctc.usma.edu/harmony/harmony_docs.asp (Fig. 3.2).

The International Institute for Counter-Terrorism (ICT), located in Herzliya, Israel, provides coverage of Middle Eastern events from an Israeli perspective. The institute regularly produces reports, commentaries, and multimedia contents. Its highly successful annual conference typically draws 400–800 participants from all over the world. For more information, see: http://www.ict.org.il/ (Fig. 3.3).

3.2 *Terrorism Databases and Online Resources*

Internet Haganah (Hebrew word for “defense”) is heavily involved in monitoring and disabling terror web sites on the web. It alerts counterterrorism vigilantes and indirectly fosters information “warfare.” Its Open Source Intelligence (OSINT) gathers and catalogs available intelligence documents of relevance to terrorism. For more information, see: http://internet-haganah.com/haganah/internet.html (Fig. 3.4).

Fig. 3.2 The CTC harmony reports

Fig. 3.3 Sample ICT multimedia content

Middle East Media Research Institute (MEMRI) is a nonprofit Washington, D.C.–based organization, with branches in Europe, Japan, and Israel. It regularly translates Arabic, Persian, and Turkish media and annotates videos, news articles, and web sites in the region. It also provides Islamic reformers a platform by translating their ideas and thoughts. For more information, see: http://memri.org/index.html (Fig. 3.5).

The Memorial Institute for the Prevention of Terrorism (MIPT) is a nonprofit organization funded by the US Department of Homeland Security (DHS). Based in Oklahoma City, Oklahoma, the 1995 federal building bombing spurred interest in a terrorism information repository, which resulted in creation of the MIPT Terrorism Knowledge Base (TKB). TKB contains two separate terrorist incident databases, the RAND Terrorism Chronology 1968–1997 and the RAND-MIPT Terrorism Incident database (1998–Present). The TKB ceased operations on March 31, 2008. For more information, see: http://www.mipt.org/ (Fig. 3.6).

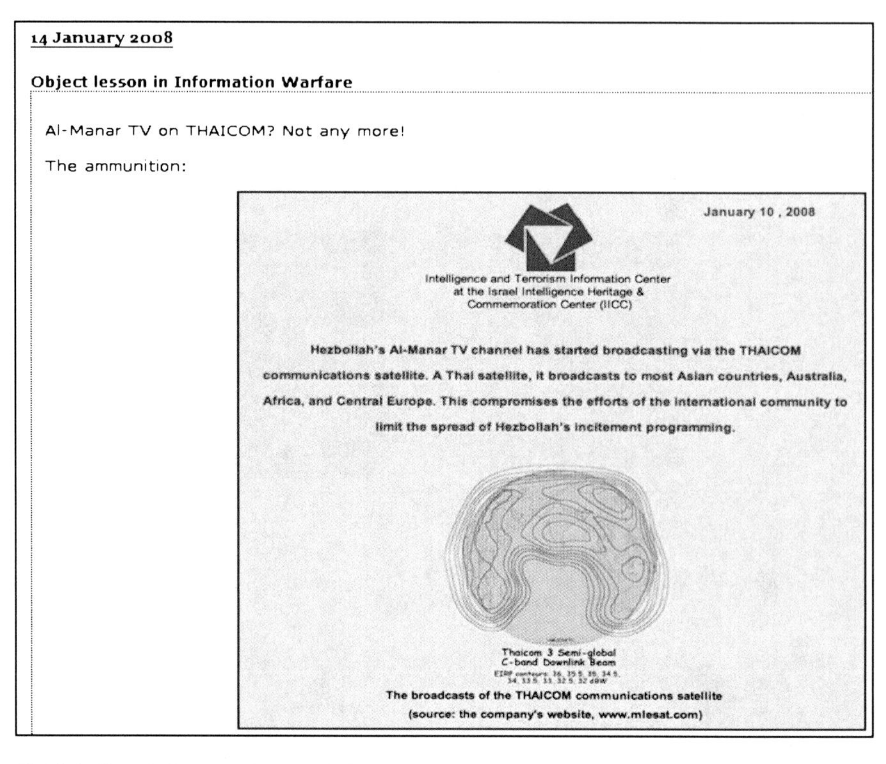

Fig. 3.4 Sample Internet Haganah information on the web

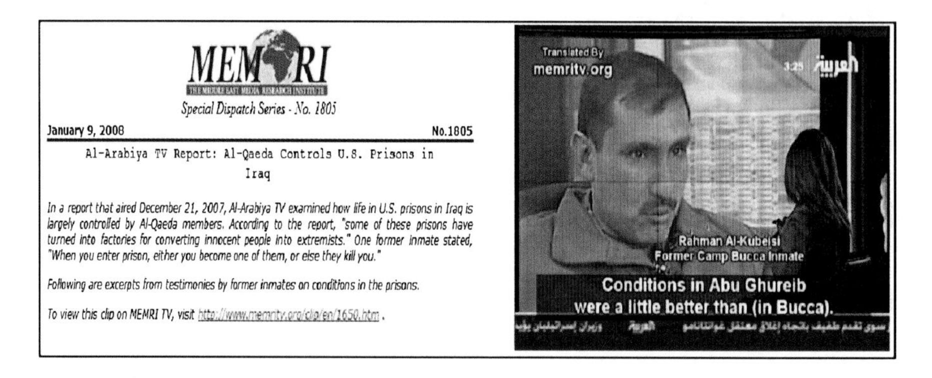

Fig. 3.5 Sample MEMRI web content

Also funded by the DHS, the University of Maryland's National Consortium for the Study of Terrorism and Responses to Terrorism (START) has extensive research about terrorist group formation and recruitment, persistence and dynamics, and societal responses to terrorist threats and attacks. It provides a searchable open source Global Terrorism Database (GTD), presenting information on terrorist events

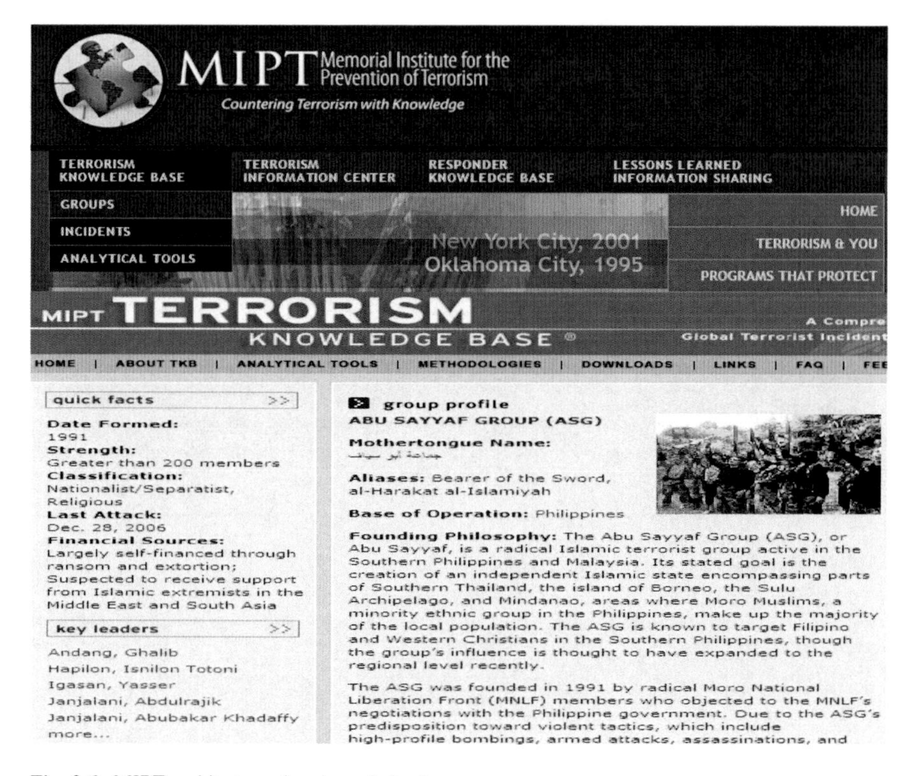

Fig. 3.6 MIPT and its terrorism knowledge base

A CENTER OF EXCELLENCE OF THE U.S. DEPARTMENT OF HOMELAND SECURITY BASED AT THE UNIVERSITY OF MARYLAND

Fig. 3.7 START and its global terrorism database

around the world since 1970 (currently updated through 2007), including data on where, when, and how each of over 80,000 terrorist events occurred. For more information, see: http://www.start.umd.edu/start/ (Fig. 3.7).

Originally founded to document the crimes of the Holocaust and find criminals responsible for the genocide, the Simon Wiesenthal Center has a targeted tolerance education mission which it carries out through the Snider Social Action Institute

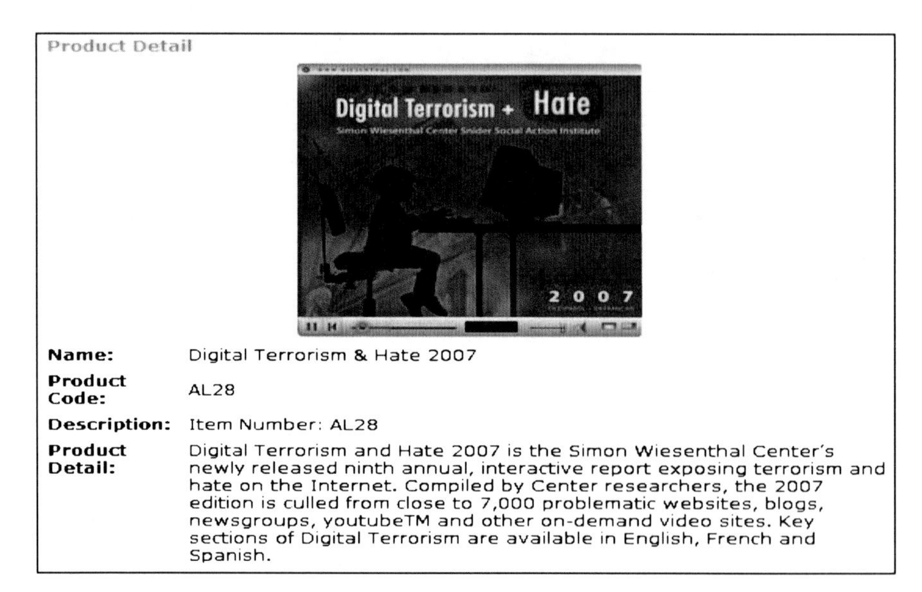

Fig. 3.8 Simon Wiesenthal Center and its digital terrorism DVDs

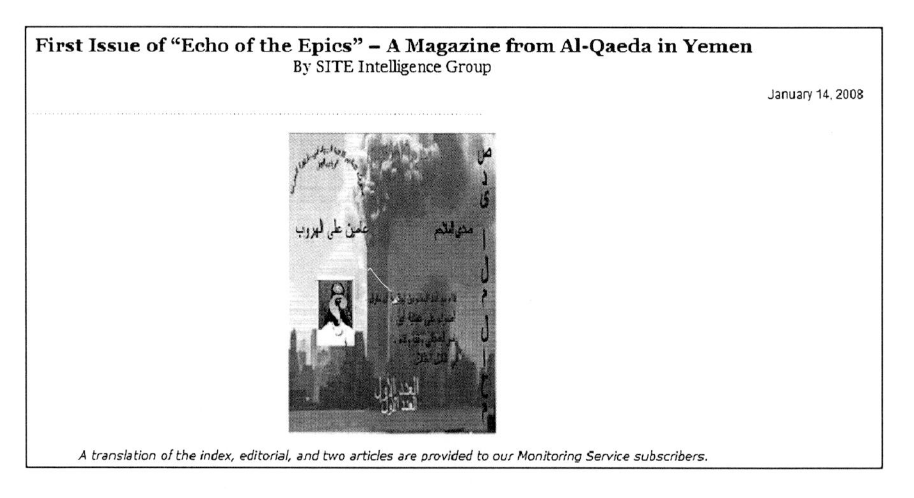

Fig. 3.9 Sample SITE report

and its Museum of Tolerance. It also produces the Digital Terrorism DVDs with extremist web site snapshots as part of its larger educational mission. For more information, see: http://www.wiesenthal.com (Fig. 3.8).

SITE (Search for International Terrorist Entities) was founded in 2002 by undercover activist Rita Katz. The Site Intelligence Group, a for-profit organization, is now monitoring terrorist activities. It keeps translations of terrorist media and documents and makes them available via subscription to media, governments, and corporations. For more information, see: http://www.siteintelgroup.org/ (Fig. 3.9).

Latest additions to CSTPV e-library (from Open Sources)

Title	Author
Tackling terror Svengalis on web a priority	Johnson, Philip
Al Qaidas "MySpace": Terrorist Recruitment on the Internet	Kohlmann , Evan F.
Al Qaidas Extensive Use of the Internet	Weimann, Gabriel
The Changing Face of SalafiJihadi Movements in the United Kingdom	Brandon, James
Evolution of Jihadism in Spain Following the 3/11 Madrid Terrorists Attacks	Jordan, Javier
The Modern Terrorist Threat to Aviation Security	Forest, James J.F.
The radical dawa in transition. The rise of Islamic neoradicalism in the Netherlands	
Turkey's Other War on Terrorism	Jenkins, Gareth
South Waziri Tribesmen Organize Counterinsurgency Lashkar	McGregor, Andrew
Insurrection in Iranian Balochistan	Zambelis, Chris

Fig. 3.10 Sample CSTPV reports

3.3 *Higher Education Research Institutes*

The University of St. Andrews Centre for the Study of Terrorism and Political Violence (CSTPV) provides a political science perspective to terrorism. The program offers subscription-based access to political analyses and awards distance learning terrorism study certificates. Many influential political science–grounded terrorism scholars were trained at St. Andrews. For more information, see: http://www.st-andrews.ac.uk/~wwwir/research/cstpv/ (Fig. 3.10).

The International Centre for Political Violence and Terrorism Research (ICPVTR) is a research and education center within the S. Rajaratnam School of International Studies (RSIS) at Nanyang Technological University, Singapore. ICPVTR conducts research, training, and outreach programs aimed at reducing the threat of politically motivated violence and at mitigating its effects on the international system. Its Global Pathfinder System is a one-stop repository for information on the current and emerging terrorist threat. The database focuses on terrorism and political violence in the Asia-Pacific region – comprising Southeast Asia, North Asia, South Asia, Central Asia, and Oceania. For more information, see: http://www.pvtr.org/ (Fig. 3.11).

Sponsored by many US government agencies, the Dartmouth Institute for Security Technology Studies (ISTS) focuses on cyber security, trust, and cyberterrorism. Its project topics include hardening IT infrastructure against attack, image and video forensics, trusted digital certificate provision, information infrastructure risk assessment, etc. For more information, see: http://www.ists.dartmouth.edu/library.php (Fig. 3.12).

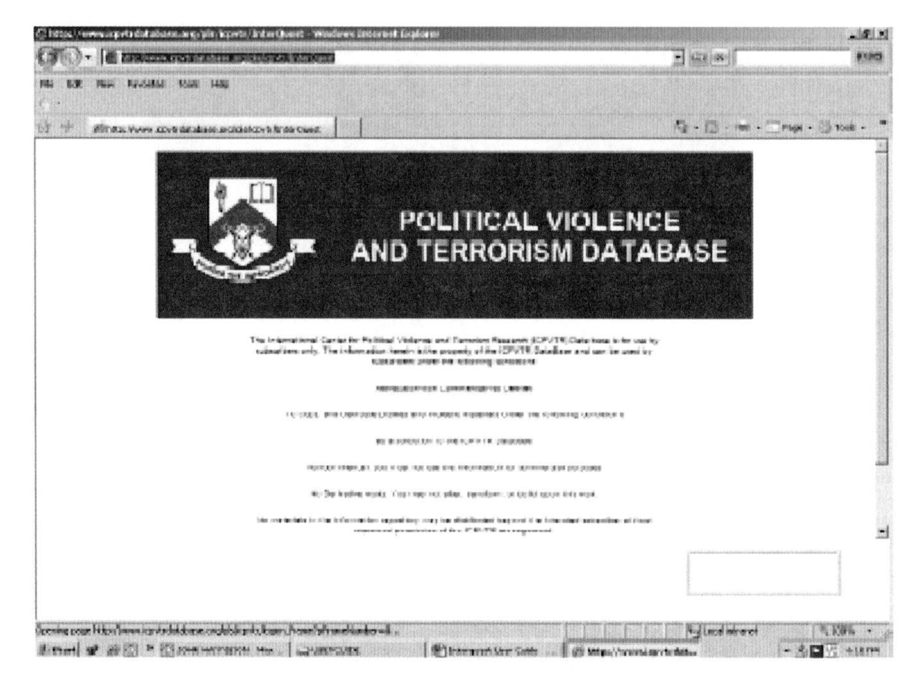

Fig. 3.11 ICPVTR terrorism database

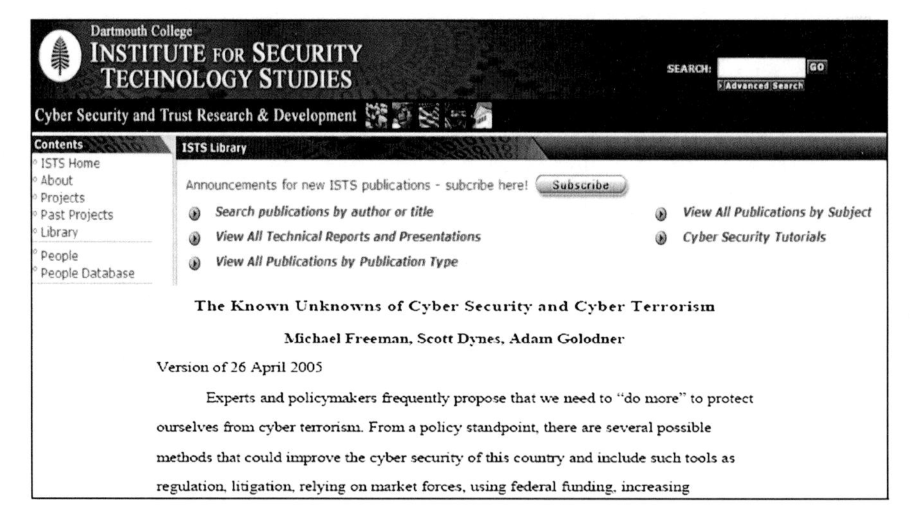

Fig. 3.12 The Dartmouth ISTS web site

4 Conclusions

As the field of terrorism informatics continues to grow and evolve, we anticipate broader collaboration between social scientists and computational researchers who are interested in counterterrorism research. We also expect new methodologies, terrorism databases, and computational approaches to emerge and mature based on the rich content and complex interactions produced by the terrorists and extremists on the Internet.

References

Chen, H. "Intelligence and Security Informatics for International Security: Information Sharing and Data Mining," Springer, 2006.

Chen, H., Reid, E., Sinai, J., Silke, A., and Ganor, B. (Eds.), "Terrorism Informatics: Knowledge Management and Data Mining for Homeland Security," Springer, 2008.

Ryan, J. "Countering Militant Islamist Radicalization on the Internet: A User Driven Strategy to Recover the Web," Institute of European Affairs, 2007.

Sageman, M. "Understanding Terror Networks," University of Pennsylvania Press, Philadelphia, Pennsylvania, 2004.

Sageman, M. "Leaderless Jihad," University of Pennsylvania Press, Philadelphia, Pennsylvania, 2008.

Weimann, G. "Terror on the Internet: The New Arena, the New Challenges," US Institute of Peace Press, Washington, DC, 2006.

Part II
Dark Web Research: Computational Approach and Techniques

Chapter 4
Forum Spidering

1 Introduction

The Internet acts as an ideal method for information and propaganda dissemination (Whine 1997; Gustavson and Sherkat 2004). Computer-mediated communication offers a quick, inexpensive, and anonymous means of communication for extremist groups (Crilley 2001). Extremist groups frequently use the web to promote hatred and violence (Glaser et al. 2002). This problematic facet of the Internet is often referred to as the Dark Web (Chen 2006). An important component of the Dark Web is extremist forums hidden deep within the Internet. Many have stated the need for collection and analysis of Dark Web forums (Burris et al. 2000; Schafer 2002). Dark Web materials have important implications for intelligence and security informatics–related applications (Chen 2006). The collection of such content is also important for studying and understanding the diverse social and political views present in these online communities.

The unprecedented growth of the Internet has resulted in considerable focus on web crawling/spidering techniques in recent years. Crawlers are defined as "software programs that traverse the World Wide Web information space by following hypertext links and retrieving web documents by standard HTTP protocol" (Cheong 1996). They are programs that can create a local collection or index of large volumes of web pages (Cho and Garcia-Molina 2000). Crawlers can be used for general-purpose search engines or for domain-specific collection building. The latter are referred to as focused or topic-driven crawlers (Chakrabarti et al. 1999; Pant et al. 2002).

There is a need for a focused crawler that can collect Dark Web forums. Many previous focused crawlers have focused on collecting static English web pages from the "surface web." A Dark Web forum–focused crawler faces several design challenges. One major concern is *accessibility*. Web forums are dynamic and often require memberships. They are part of the "hidden web" (Florescu et al. 1998; Raghavan and Garcia-Molina 2001) which is not easily accessible through normal web navigation or standard crawling. There are also *multilingual* web

H. Chen, *Dark Web: Exploring and Data Mining the Dark Side of the Web*,
Integrated Series in Information Systems 30, DOI 10.1007/978-1-4614-1557-2_4,

mining considerations. More than 30% of the web is in non-English languages (Chen and Chau 2003). Consequently, the Dark Web also encompasses numerous languages. Another important concern is *content richness*. Dark Web forums contain rich content used for routine communication and propaganda dissemination (Abbasi and Chen 2005; Zhou et al. 2005; Qin et al. 2005). These forums contain static and dynamic text files, archive files, and various forms of multimedia (e.g., images, audio, and video files). Collection of such diverse content types introduces many unique challenges not encountered with standard spidering of indexable (text-based) files.

In this chapter, we propose the development of a focused crawler that can collect Dark Web forums. Our spidering system uses breadth- and depth-first (BFS and DFS) traversal based on URL tokens, anchor text, and link levels for crawl space URL ordering. We also utilize incremental crawling for collection updating using wrappers to identify updated content. The system also includes design elements intended to overcome the previously mentioned accessibility, multilingual, and content richness challenges. For accessibility, we use a human-assisted approach (Raghavan and Garcia-Molina 2001) for attaining Dark Web forum membership. Our system also includes tailored spidering parameters and proxies for each forum in order to improve accessibility. The crawler uses language-independent features for crawl space URL ordering in order to negate any complications attributable to the presence of numerous languages. We also incorporate iterative collection of incomplete downloads and relevance feedback for improved multimedia collection.

The remainder of this chapter is organized as follows: Section 4.2 presents a review of related work on focused and hidden web crawling. Section 4.3 describes research gaps and our related research questions. Section 4.4 describes a research design geared toward addressing those questions. Section 4.5 presents a detailed description of our Dark Web forum spidering system. Section 4.6 describes experimental results evaluating the efficacy of our human-assisted approach for gaining access to Dark Web forums as well as the incremental update procedure that uses recall improvement. This section also highlights the Dark Web forum collection statistics for data gathered using the proposed system. Section 4.7 contains concluding remarks.

2 Related Work: Focused and Hidden Web Crawlers

Focused crawlers "seek, acquire, index, and maintain pages on a specific set of topics that represent a narrow segment of the web" (Chakrabarti et al. 1999). The need to collect high-quality, domain-specific content results in several important characteristics for such crawlers that are also relevant to collection of Dark Web forums. Some of these characteristics are specific to focused and/or hidden web crawling while others are relevant to all types of spiders. We review previous research pertaining to these important considerations, which include accessibility, collection type and content richness, URL ordering features and techniques, and collection update procedures.

2.1 *Accessibility*

Most search engines cover what is referred to as the "publicly indexable web" (Lawrence and Giles 1998; Raghavan and Garcia-Molina 2001). This is the part of the web easily accessible with traditional web crawlers (Sizov et al. 2003). As noted by Lawrence and Giles (1998), a large portion of the Internet is dynamically generated. Such content typically requires users to have prior authorization, fill out forms, or register (Raghavan and Garcia-Molina 2001). This covert side of the Internet is commonly referred to as the hidden/deep/invisible web. Hidden web content is often stored in specialized databases (Lin and Chen 2002). For example, the IMDB movie review database contains a plethora of useful information regarding movies, yet standard crawlers cannot access this information (Sizov et al. 2003). A study conducted in 2000 found that the invisible web contained 400–550 times the information present in the traditional surface web (Bergman 2000; Lin and Chen 2002).

Two general strategies have been introduced to access the hidden web via automated web crawlers. The first approach entails use of automated form-filling techniques. Several different automated query generation approaches for querying such "hidden web" databases and fetching the dynamically generated content have been proposed (e.g., Barbosa and Freire 2004; Ntoulas et al. 2005). Other techniques keep an index of hidden web search engines and redirect user queries to them (Lin and Chen 2002) without actually indexing the hidden databases. However, many automated approaches ignore/exclude collection or querying of pages requiring log-in (e.g., Lage et al. 2002). Thus, automated form-filling techniques seem problematic for Dark Web forums where log-in is often required.

A second alternative for accessing the hidden web is a task-specific human-assisted approach (Raghvan and Garcia-Molina 2001). This approach provides a semiautomated framework that allows human experts to assist the crawler in gaining access to hidden content. The amount of human involvement is dependent on the complexity of the accessibility issues faced. For example, many simple forms asking for name, e-mail address, etc., can be automated with standardized responses. Other more complex questions require greater expert involvement. Such an approach seems more suitable for the Dark Web, where the complexity of the access process can vary significantly.

2.2 *Collection Type*

Previous focused crawling research has been geared toward collecting web sites, blogs, and web forums. There has been considerable research on collection of standard web sites and pages relating to a particular topic, often for portal building. Srinivasan et al. (2002) and Chau and Chen (2003) fetched biomedical content from the web. Sizov et al. (2003) collected web pages pertaining to handicrafts and movies. Pant et al. (2002) evaluated their topic crawler on various keyword queries (e.g., "recreation").

There has also been work on collecting weblogs. BlogPulse (Glance et al. 2004) is a blog analysis portal. The site contains analysis of key discussion topics/trends for roughly 100,000 spidered weblogs. Such blogs can also be useful for marketing intelligence (Glance et al. 2005). Blogs containing product reviews analyzed using sentiment analysis techniques can provide insight into how people feel about various products.

Web forum crawling presents a unique set of difficulties. Discovering web forums is challenging due to the lack of a centralized index (Glance et al. 2005). Furthermore, web forums require information extraction wrappers for derivation of metadata (e.g., authors, messages, time stamps, etc.). Wrappers are important for data analysis and incremental crawling (respidering only those threads containing newly posted messages). Incremental crawling is discussed in greater detail in the "Collection Update" section. There has been limited research on web forum spidering. BoardPulse (Glance et al. 2005) is a system for harvesting messages from online forums. It has two components: a crawler and a wrapper. Limanto et al. (2005) developed a web forum information extraction engine that includes a crawler, wrapper generator, and extractor (i.e., application of generated wrapper). Yih et al. (2004) created an online forum mining system composed of a crawler and information extractor for mining deal forums. There has been no prior research on collecting Dark Web forums.

2.3 Content Richness

The web is rich in indexable and multimedia files. Indexable files include static text files (e.g., HTML, Word, and PDF documents) and dynamic text files (e.g., .asp, .jsp, and .php). Multimedia files include images, animation, audio, and video files. Difficulties in indexing make multimedia content difficult to accurately collect (Baeza-Yates 2003). Multimedia file sizes are typically significantly larger than indexable files, resulting in longer download time and frequent time-outs. Heydon and Najork (1999) fetched all MIME file types (including image, video, audio, and .exe files) using their Mercator crawler. They noted that collecting such files increased the overall spidering time and doubled the average file size as compared to just fetching HTML files. Consequently, many previous studies have ignored multimedia content altogether (e.g., Pant et al. 2002).

2.4 URL Ordering Features

Aggarwal et al. (2001) pointed out four categories of features for crawl space URL ordering. These include links, URL and/or anchor text, page text, and page levels. *Link*-based features have been used considerably in previous research. Many studies have used in-/back-links and out-links (Cho et al. 1998; Pant et al. 2002). Sibling links

(Aggarwal et al. 2001) consider sibling pages (ones with shared parent in-link). Context graphs (Diligenti et al. 2000) derive back-links for each seed URL and use these to construct a multilayer context graph. Such graphs can be used to extract paths leading up to relevant nodes (target URLs). Focused/topical crawlers often use bag-of-words (BOW) found in the web *page text* (Aggarwal et al. 2001; Pant et al. 2002). For instance, Srinivasan et al. (2002) used BOW for biomedical text categorization in their focused crawler. While page text features are certainly very effective, they are also language dependent and can be harder to apply in situations where the collection is composed of pages in numerous languages. Other studies have also used *URL/anchor text*. Word tokens found within the URL anchor have been used effectively to help control the crawl space (Cho et al. 1998; Ester et al. 2001). URL tokens have also been incorporated in previous focused crawling research (Aggarwal et al. 2001; Ester et al. 2001). Another important category of features for URL ordering is page *levels*. Diligenti et al. (2000) trained text classifiers to categorize web pages at various levels away from the target. They used this information to build path models that allowed consideration of irrelevant pages as part of the path to attain target pages. A potential path model may consider pages one or two levels away from a target, known as tunneling (Ester et al. 2001). Ester et al. (2001) used the number of slashes "/" or levels from the domain as an indicator of URL importance. They argued that pages closer to the main page are likely to be of greater importance.

2.5 URL Ordering Techniques

Previous research has typically used breadth-, depth-, and best-first search for URL ordering. Depth-first search (DFS) has been used in crawling systems such as Fish Search (De Bra and Post 1994). Breadth-first search (BFS) (Cho et al. 1998; Ester et al. 2001; Najork and Wiener 2001) is one of the simplest strategies. It has worked fairly well in comparison with more sophisticated best-first search strategies (Cho et al. 1998; Najork and Wiener 2001). However, BFS is typically not employed by focused crawlers that are concerned with identifying topic-specific web pages using the aforementioned URL ordering features.

Best-first uses some criterion for ranking URLs in the crawl space, such as *link analysis* or *text analysis*. Numerous link analysis techniques have been used for URL ordering. Cho et al. (1998) evaluated the effectiveness of Page Rank and back-link counts. Pant et al. (2002) also used Page Rank. Aggarwal et al. (2001) used the number of relevant siblings. They considered pages with a higher percentage of relevant siblings more likely to also be relevant. Sizov et al. (2003) used the HITS algorithm to compute authority scores, while Chakrabarti et al. (1999) used a modified HITS. Chau and Chen (2003) used a Hopfield Net crawler that collected pages related to the medical domain based on link weights.

Text analysis methods include similarity scoring approaches and machine learning algorithms. Aggarwal et al. (2001) used similarity equations with page content and URL tokens. Others have used the vector space model and cosine similarity measure (Pant et al. 2002; Srinivasan et al. 2002). Sizov et al. (2003) used support vector machines (SVM) with BOW for document classification. Srinivasan et al. (2002) used BOW and link structures with a neural net for ordering URLs based on the prevalence of biomedical content. Chen et al. (1998a; 1998b) used a genetic algorithm to order the URL crawl space for the collection of topic-specific web pages based on bag-of-word representations of pages.

2.6 Collection Update Procedure

Two approaches for collection updating are periodic and incremental crawling (Cho and Garcia-Molina 2001). *Periodic crawling* entails building of a brand-new collection for updating. This is commonly done since it is often easier than figuring out which pages to refresh. Periodic crawling is inefficient from a spidering perspective (more time consuming). However, multiple versions of a collection may improve overall recall. *Incremental crawling* gathers new and updated content. In the case of web sites, this often requires some form of change frequency estimation (Cho and Garcia-Molina 2003) in order to determine which pages need to be updated. For web forums, this entails fetching only those threads that have been updated (Yih et al. 2003; Glance et al. 2005) since we only want to fetch newly posted messages. This requires the use of a wrapper that can parse out the "last updated" dates for threads and compare them against the previous collection to determine which pages need to be collected.

2.7 Summary of Previous Research

Table 4.1 provides a summary of selected previous research on focused crawling.

The majority of studies have focused on collection of indexable files from the surface web. There have only been a few studies that performed focused crawling on the hidden web. Similarly, only a few studies have collected content from web forums. Most previous research on focused crawling has used bag-of-word (BOW), link, or URL token features coupled with a best-first search strategy for crawl space URL ordering. Furthermore, most prior research ignored the multilingual dimension, only collecting content in a single language (usually English). Collection of Dark Web forums entails retrieving rich content (including indexable and multimedia files) from the hidden web in multiple languages. Dark Web forum crawling is therefore at the cross section of several important areas of crawling research, many of which have received limited attention in prior research. The following section summarizes these important research gaps and provides a set of related research questions which are addressed in the remainder of this chapter.

Table 4.1 Selected previous research on focused crawling

System name and study	Access	Collection type	Content richness	URL ordering features	URL ordering techniques
GA spider (Chen et al. 1998a; 1998b)	Surface web	Topic-specific web pages	Indexable files only	BOW	Best-first: genetic algorithm
Focused crawler (Chakrabarti et al. 1999)	Surface web	Topic-specific web pages	Indexable files only	BOW and links	Hypertext classifier and modified HITS algorithm
Context-focused (Diligenti et al. 2000)	Surface web	Topic-specific web pages	Indexable files only	BOW and context graphs	Best-first: vector space, Naïve Bayes, and path models
Intelligent crawler (Aggarwal et al. 2001)	Surface web	Topic-specific web pages	Indexable files only	BOW, URL tokens, anchor text, links	Best-first: similarity scores and link analysis
Ariadne (Ester et al. 2001)	Surface web	Topic-specific web pages	Indexable files only	BOW, URL tokens, anchor text, links, user feedback, levels	Relevance scoring and text classifier
Hidden web exposer (Raghavan and Garcia-Molina, 2001)	Hidden web	Dynamic search forums	Indexable files only	URL tokens	Rule-based: crawler stayed within target sites
InfoSpiders (Srinivasan et al. 2002)	Surface web	Biomedical pages and documents	Indexable files only	BOW and links	Best-first: vector space model and neural net
NetScan (Smith, 2002)	Surface web	USENET web forums	Indexable files only	n/a	n/a
Topic crawler (Pant et al. 2002)	Surface web	Topic-specific web pages	Indexable files only	BOW	Best-N-first: vector space model
Hopfield net crawler (Chau and Chen 2003)	Surface web	Medical-domain web pages	Indexable files only	Links	Best-first: Hopfield net
BINGO! (Sizov et al. 2003)	Surface and hidden web	Handicraft and movie web pages	Indexable files only	BOW and links	Best-first: SVM and HITS
BlogPulse (Glance et al. 2004)	Surface web	Weblogs for various topics	Indexable files only	Weblog text	Differencing algorithm
Hot deal crawler (Yih et al. 2004)	Surface web	Online deal forums	Indexable files only	URL tokens, thread date	Date comparison
BoardPulse (Glance et al. 2005)	Surface web	Product web forums	Indexable files only	URL tokens, thread date	Wrapper learning of site structure
Web forum spider (Limanto et al. 2005)	Surface web	Web forums	Indexable files only	Web page text and URL tokens	Machine learning classifier
Board forum crawler (Guo et al. 2006)	Surface web	Board web forums	Indexable files only	Web page text and URL tokens	Rule-based: uses URL tokens and text
Recipe crawler (Li et al. 2006)	Surface web	Recipe sites, blogs, and web forums	Indexable files only	Web page text	Best-first: tree edit distance similarity scores

3 Research Gaps and Questions

Based on our review of previous literature, we have identified several important research gaps.

3.1 Focused Crawling of the Hidden Web

There has been limited focused crawling work on the hidden web. Most focused crawler studies developed crawlers for the surface web (Raghavan and Garcia-Molina 2001). Prior hidden web research mostly focused on automated form filling or query redirection to hidden databases, that is, accessibility issues. There has been little emphasis on building topic-specific web page collections from these hidden sources. We are not aware of any attempts to automatically collect Dark Web content pertaining to hate and extremist groups.

3.2 Content Richness

Most previous research has focused on indexable (text-based) files. Large multimedia files (e.g., videos) can be hundreds of megabytes. This can cause connection time-outs or excessive server loads, resulting in partial/incomplete downloads. Furthermore, the challenges in indexing multimedia files pose problems. It is difficult to assess the quality of collected multimedia items. As Baeza-Yates (2003) noted, automated multimedia indexing is more of an image retrieval challenge than an information retrieval problem. Nevertheless, given the content richness of the Internet in general and the Dark Web in specific (Chen 2006), there is a need to capture multimedia files.

3.3 Web Forum Collection Update Strategies

There has been considerable research on evaluating various collection update strategies for web sites (e.g., Cho and Garcia-Molina 2000). However, there has been little work done on comparing the effectiveness of periodic versus incremental crawling for web forums. Most web forum research has assumed an incremental approach. Given the accessibility concerns associated with Dark Web forums, periodic and incremental approaches both provide varying benefits. Periodic crawlers can improve collection recall by allowing multiple attempts at capturing previously uncollected pages. This may be less of a concern for surface web forums but is important for the Dark Web. In contrast, incremental crawlers can improve collection efficiency and reduce redundancy. There is a need to evaluate the effectiveness of periodic and incremental crawling applied to Dark Web forums.

3.4 Research Questions

Based on the gaps described, we propose the following research questions:

1. How effectively can Dark Web forums be identified and accessed for collection purposes?
2. How effectively can Dark Web content (indexable and multimedia) be collected?
3. Which collection update procedure (periodic or incremental) is more suitable for Dark Web forums? How can recall improvement further enhance the update process?
4. How can analysis of extracted information from Dark Web forums improve our understanding of these online communities?

4 Research Design

4.1 Proposed Dark Web Forum Crawling System

In this chapter, we propose a Dark Web forum spidering system. Our proposed system consists of an accessibility component that uses a human-assisted registration approach to gain access to Dark Web forums. Our system also utilizes multiple dynamic proxies and forum-specific spidering parameter settings to maintain forum access.

Our URL ordering component uses language-independent URL ordering features to allow spidering of Dark Web forums across languages. We plan to focus on groups from three regions: US domestic, Middle East, and Latin America/Spain. Additionally, a rule-based URL ordering technique coupled with BFS and DFS crawl space traversal is utilized. Such a technique is employed in order to minimize the amount of irrelevant web pages collected.

We also propose the use of an incremental crawler that uses forum wrappers to determine the subset of threads that need to be collected. Our system will include a recall improvement procedure that parses the spidering log and reinserts incomplete downloads into the crawl space. Finally, the system features a collection analyzer that checks multimedia files for duplicate downloads and generates collection statistics at the forum, region, and overall collection levels.

4.2 Accessibility

As noted by Raghavan and Garcia-Molina (2001), the most important evaluation criterion for hidden web crawling is how effectively the content was accessed. They developed an accessibility metric as follows: *databases accessed/total attempted*.

We intend to evaluate the effectiveness of the task-specific, human-assisted approach in comparison with not using such a mechanism. Specifically, we would also like to evaluate our system's ability to access Dark Web forums. This translates into measuring the percentage of attempted forums accessed.

4.3 Incremental Crawling for Collection Updating

We plan to evaluate the effectiveness of our proposed incremental crawler in comparison with periodic crawling. The incremental crawler will obviously be more efficient in terms of spidering time and data redundancy. However, a periodic crawling approach gets multiple attempts to collect each page, which can improve overall collection recall. Evaluation of both approaches is intended to provide additional insight into which collection update technique is more suitable for Dark Web forum spidering.

5 System Design

Based on our research design, we implemented a focused crawler for Dark Web forums.

Our system consists of four major components (shown in Fig. 4.1):

- *Forum identification*: to identify the list of extremist forums to spider.
- *Forum preprocessing*: includes accessibility and crawl space traversal issues as well as forum wrapper generation.
- *Forum spidering*: consists of an incremental crawler and recall improvement mechanism.
- *Forum storage and analysis*: to store and analyze the forum collection.

5.1 Forum Identification

The forum identification phase has three components:

Step 1: Identify extremist groups
Sources for the US domestic extremist groups include the Anti-Defamation League, FBI, Southern Poverty Law Center, Militia Watchdog, and the Google Web Directory (as a supplement). Sources for the international extremist groups include the United States Committee for a Free Lebanon, Counter-Terrorism Committee of the U.N. Security Council, US State Department report, Official Journal of the European Union, as well as government reports from the United Kingdom, Australia, Japan, and P. R. China. Due to regional and language constraints, we chose to focus on

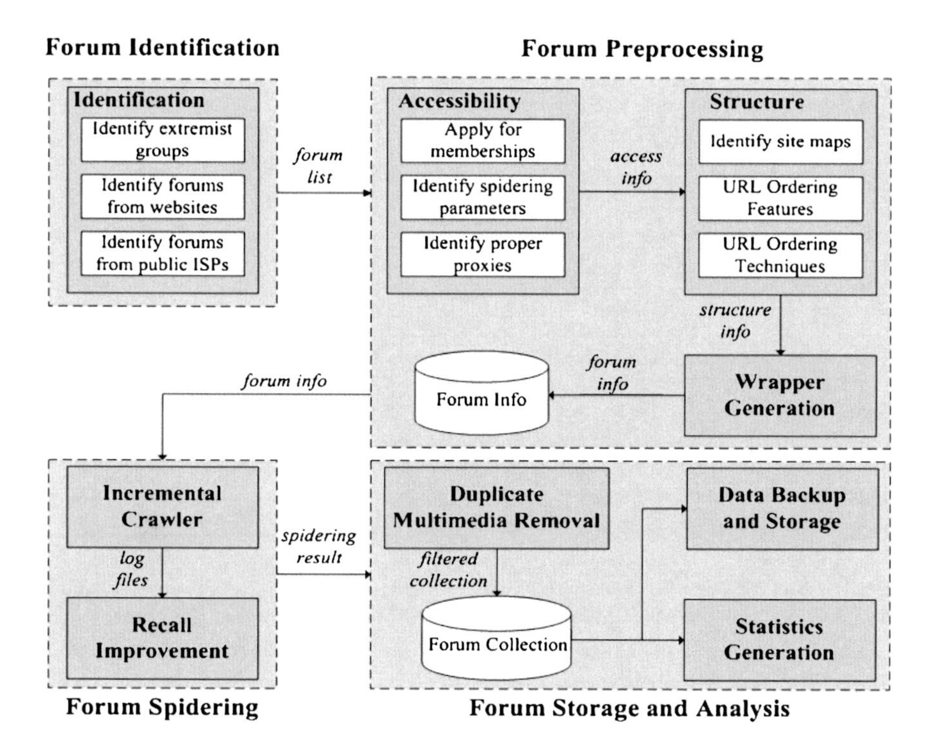

Fig. 4.1 Dark Web forum crawling system design

groups from three areas: North America (English), Latin America (Spanish), and the Middle East. These groups are all significant for their sociopolitical importance. Furthermore, collection and analysis of Dark Web content from these three regions can facilitate a better understanding of the relative social and cultural differences between these groups. In addition to obvious linguistic difference, groups from these regions also display different web design tendencies and usage behaviors (Abbasi and Chen 2005) which provide a unique set of collection and analysis challenges.

Step 2: Identify forums from extremist web sites
We identify an initial set of extremist group URLs and then use link analysis for expansion purposes as shown in Fig. 4.2.

The initial set of URLs is identified from three sources. Firstly, we use search engines coupled with a lexicon containing extremist organization name(s), leader(s)' and key members' names, slogans, and special keywords used by extremists. Secondly, we utilize government reports. Finally, we reference research centers. A link analysis approach is used to expand the initial list of URLs. We incorporate a back-link search using Google, which has been shown to be effective in prior research (Diligenti et al. 2000), and also search out-links. The identified web forums are manually checked.

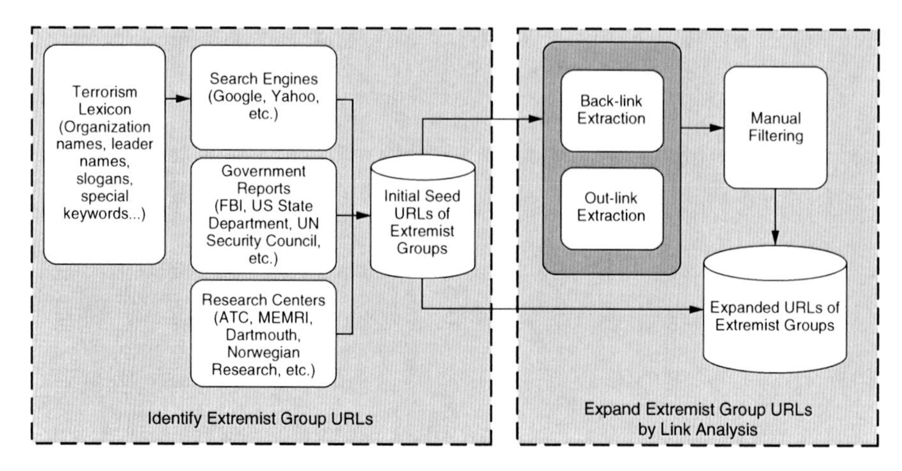

Fig. 4.2 Dark Web forum crawling system design

Step 3: Identify forums hosted on major web sites
We also identify forums hosted by other web sites and public Internet service providers (ISPs) that are likely to be used by Dark Web groups, for example, MSN groups or AOL Groups. Public ISPs are searched with our Dark Web domain lexicon for a list of potential forums.

The above three steps help identify a seed set of Dark Web forums. Once the forums have been identified, several important preprocessing issues must be resolved before spidering in order to develop proper features and techniques for managing the crawl space. These include accessibility concerns and identification of forum structure.

5.2 *Forum Preprocessing*

The forum preprocessing phase has three components: accessibility, structure, and wrapper generation. The accessibility component deals with acquiring and maintaining access to Dark Web forums. The structure component is designed to identify the forum URL mapping and devise the crawl space URL ordering using the relevant features and techniques.

5.2.1 Forum Accessibility

Step 1: Apply for membership
Many Dark Web forums do not allow anonymous access (Zhou et al. 2006). In order to access and collect information from those forums, one must create a user ID and password, send an application request to the web master, and wait to get permission/

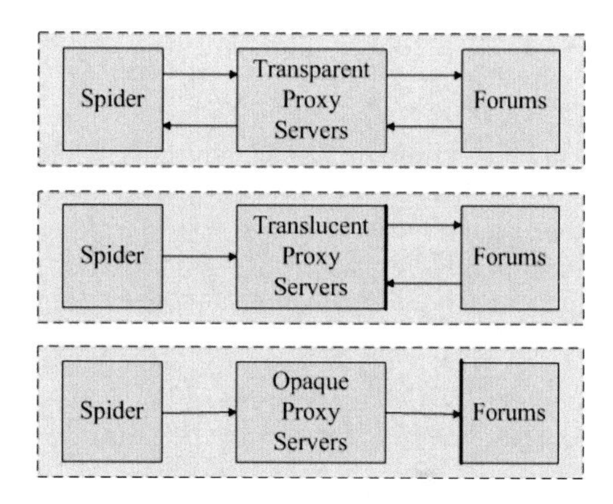

Fig. 4.3 Proxies used for Dark Web forum crawling

registration to access the forum. In certain forums, web masters are very selective. It can take a couple of rounds of e-mails to get access privilege. For such forums, human expertise is invaluable. Nevertheless, in some cases, access cannot be attained.

Step 2: Identify appropriate spidering parameters
Spidering parameters such as number of connections, download intervals, time-out, speed, etc., need to be set appropriately according to server and network limitations and the various forum blocking mechanisms. Dark Web forums are rich in terms of their content. Multimedia files are often fairly large in volume (particularly compared to indexable files). The spidering parameters should be able to handle download of larger files from slow servers. However, we may still be blocked based on our IP address. Therefore, we use proxies to increase not only our recall but also our anonymity.

Step 3: Identify appropriate proxies
We use three types of proxy servers, as shown in Fig. 4.3. *Transparent* proxy servers are those that provide anyone with your real IP address. *Translucent* proxy servers hide your IP address or modify it in some way to prevent the target server from knowing about it. However, they let anyone know that you are surfing through a proxy server. *Opaque* proxy servers (preferred) hide your IP address and do not even let anyone know that you are surfing through a proxy server. There are several criteria for proxy server selection, including the latency (the smaller the better), reliability (the higher the better), and bandwidth (the faster the better). We update our list of proxy servers periodically from various sources.

5.2.2 Forum Structure

Step 1: Identify site maps
Forums typically have hierarchical structures with boards, threads, and messages (Yih et al. 2004; Glance et al. 2005). They also contain considerable additional information such as message posting interfaces, search, and calendar pages. We first identify the site map of the forum based on the forum software packages. Glance et al. (2005) noted that although there are only a handful of commonly used forum software packages, they are highly customizable.

Step 2: URL ordering features
Our spidering system uses two types of language-independent URL ordering features, URL tokens and page levels. With respect to *URL tokens*, for web forums, we are interested in URLs containing words such as "board," "thread," "message," etc. (Glance et al. 2005). Additional relevant URL tokens include domain names of third-party file hosting web sites. These third parties often contain multimedia files. File extension tokens (e.g., ".jpg" and ".wmv") are also important. URLs that contain phrases such as "sort=voteavg" and "goto=next" are also found in relevant pages. However, these are not unique to board, thread, and message pages; hence, such tokens are not considered significant. The set of relevant URL tokens differs based on the forum software being used. Such tokens are language independent yet software specific.

Page *levels* are also important as evidenced by prior focused crawling research (Diligenti et al. 2000; Ester et al. 2001). URL level features are important for Dark Web forums due to the need to collect multimedia content. Multimedia files are often stored on third-party host sites that may be a few levels away from the source URL. In order to capture such content, we need to use a rule-based approach that allows the crawler to go a few additional levels. For example, if the URL or anchor text contains a token that is a multimedia file extension or the domain name for a common third-party file carrier, we want to allow the crawler to "tunnel" a few links.

Step 3: URL ordering techniques
As mentioned in the previous section, we use rules based on URL tokens and levels to control the crawl space. Moreover, to adapt to different forum structures, we need to use different crawl space traversal strategies.

Breadth-first (BFS) is used for board page forums while depth-first (DFS) is used for Internet service provider (ISP) forums. DFS is necessary for ISP forums since these forums often require traversing an ad page in order to get to the message page (typically, the ad pages have a link to the actual message page). Figure 4.4 illustrates how the BFS and DFS are performed for each forum type. Only the colored pages are fetched while the number indicates the order in which the pages are traversed by the crawler.

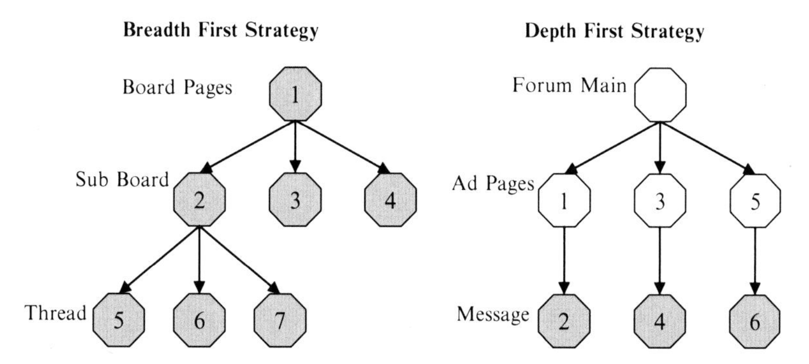

Fig. 4.4 URL traversal strategies

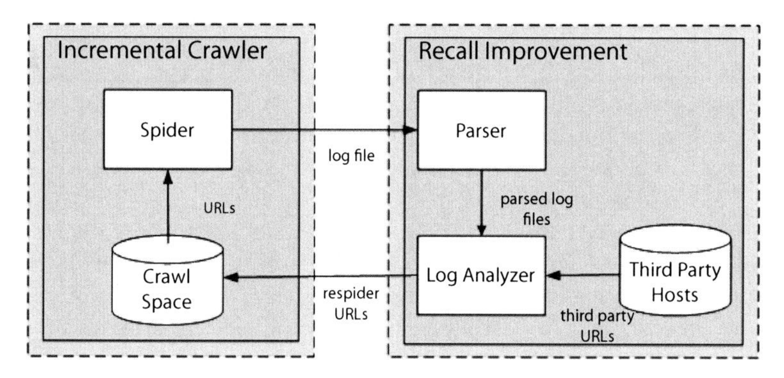

Fig. 4.5 Spidering process

5.2.3 Wrapper Generation

Forums are dynamic archives that keep historical messages. It is beneficial to only spider newly posted content when updating the collection. This is achieved by generating wrappers that can parse web forum board and thread pages (Glance et al. 2005). Board pages tell us when each thread was last updated with new messages. Using this information, one may respider only those thread pages containing new postings.

5.3 Forum Spidering

Figure 4.5 shows the spidering process. The incremental crawler fetches only new and updated threads and messages. A log file is sent to the recall improvement component. The log shows the spidering status of each URL. A parser is used to determine the overall status for each URL (e.g., "download complete" and "connection timed out"). The parsed log is sent to the log analyzer which evaluates all files that were not downloaded. It determines whether the URLs should be respidered.

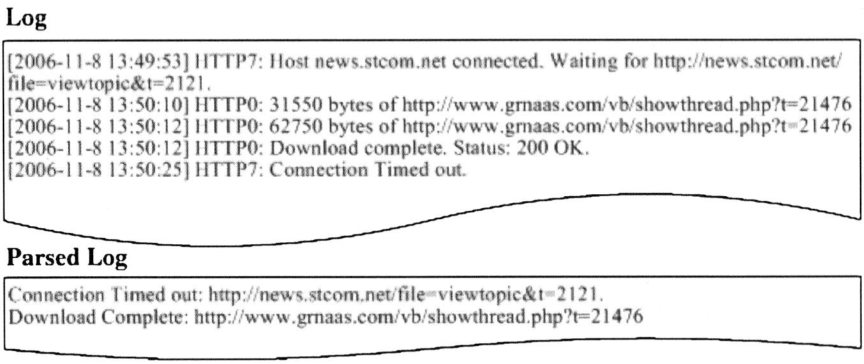

Log

```
[2006-11-8 13:49:53] HTTP7: Host news.stcom.net connected. Waiting for http://news.stcom.net/
file=viewtopic&t=2121.
[2006-11-8 13:50:10] HTTP0: 31550 bytes of http://www.grnaas.com/vb/showthread.php?t=21476
[2006-11-8 13:50:12] HTTP0: 62750 bytes of http://www.grnaas.com/vb/showthread.php?t=21476
[2006-11-8 13:50:12] HTTP0: Download complete. Status: 200 OK.
[2006-11-8 13:50:25] HTTP7: Connection Timed out.
```

Parsed Log

```
Connection Timed out: http://news.stcom.net/file=viewtopic&t=2121.
Download Complete: http://www.grnaas.com/vb/showthread.php?t=21476
```

Fig. 4.6 Example log and parsed log entries

Figure 4.6 shows sample entries from the original and parsed log. The original log file shows the download status for each file (URL). The parsed log shows the overall status as well as the reason for download failure (in the case of undownloaded files). Blue-colored entries relate to downloaded files while red-colored entries relate to undownloaded files. The log analyzer determines the appropriate course of action based on this cause of failure. "File Not Found" URLs are removed (not added to respidering list) while "Connection Timed Out" URLs are respidered. The recall improvement phase also checks the file sizes of collected web pages for partial/incomplete downloads. Multimedia file downloads are occasionally manually downloaded, particularly larger video files that may otherwise time out.

5.4 Forum Storage and Analysis

The forum storage and analysis phase consists of statistics generation and duplicate multimedia removal components.

5.4.1 Statistics Generation

Once files have been collected, they must be stored and analyzed. The statistics consist of four major categories:

- Indexable files: HTML, Word, PDF, Text, Excel, PowerPoint, XML, and Dynamic files (e.g., PHP, ASP, JSP).
- Multimedia files: Image, Audio, and Video files.
- Archive files: RAR and ZIP.
- Nonstandard files: unrecognized file types.

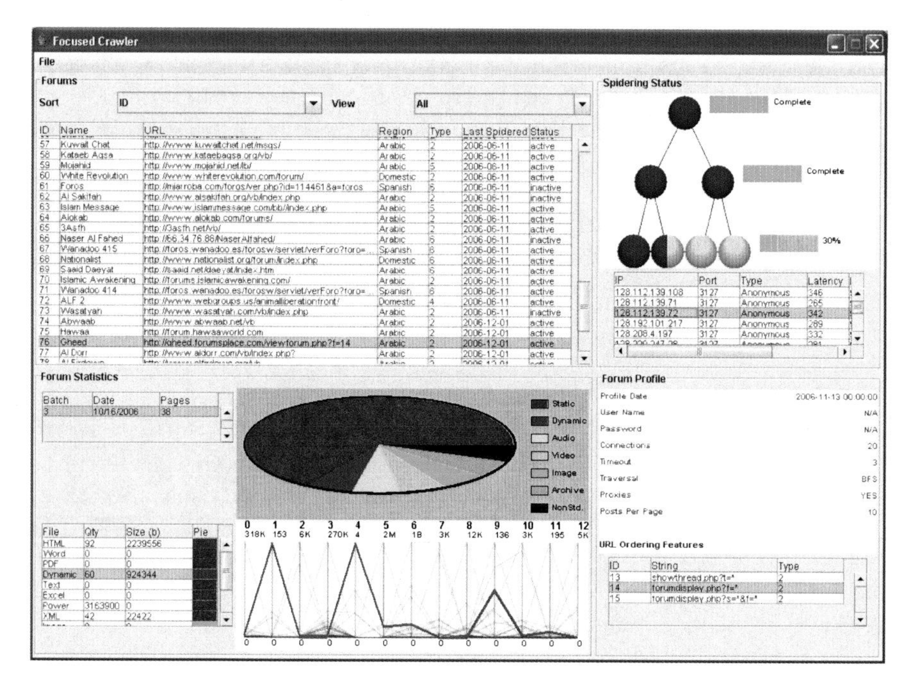

Fig. 4.7 Dark Web forum crawling system interface

5.4.2 Duplicate Multimedia Removal

Dark Web forums often share multimedia files, but the names of those files may be changed. Moreover, some multimedia files' suffixes are changed to other file types' suffixes, and vice versa. For example, an HTML file may be named as a ".jpg." Therefore, simply relying on file names results in inaccurate multimedia file statistics. We use an open-source duplicate multimedia removal software tool that identifies multimedia files by their metadata encoded into the file, instead of their suffixes (file extensions). It compares files based on their MD5 values, which are the same for duplicate video files collected from various Internet sources. MD5 (Message-Digest algorithm 5) is a widely used cryptographic hash function with a 128-bit hash value. Therefore, it can more accurately differentiate multimedia files from other types of files.

5.5 Dark Web Forum Crawling System Interface

Figure 4.7 shows the interface for the proposed Dark Web forum spidering system. The interface has four major components. The "Forums" panel in the top left shows the spidering queue in a table that also provides information such as the forum name, URL, region, when it was last spidered, and whether the forum is still active. The "Spidering Status" panel in the top right corner displays information about the percentage of board, subboard, and thread pages collected for the current forum being spidered. The "Forum Statistics" panel in the bottom left shows the quantity and size of the various file types collected for each forum, using tables, pie charts, and parallel coordinates. The "Forum Profile" panel in the bottom right shows each forum's membership information and forum spidering parameters, including the number of crawlers, URL ordering technique (i.e., BFS or DFS), and URL ordering features (e.g., URL tokens and keywords) used to control the crawl space.

6 Evaluation

We conducted two experiments to evaluate our system. The first experiment involved assessing the effectiveness of our human-assisted accessibility mechanism. Raghavan and Garcia-Molina (2001) noted that accessibility is the most important evaluation criterion for hidden web research. We describe how effectively we were able to access Dark Web forums in our collection efforts using the human-assisted approach in comparison with standard spidering without any accessibility mechanism.

The second experiment entailed evaluating the proposed incremental spidering approach that uses recall improvement as a collection updating procedure. We performed an evaluation of the effectiveness of periodic crawling as compared to standard incremental crawling and our incremental crawler which uses iterative recall improvement for Dark Web forum collection updating.

6.1 Forum Accessibility Experiment

Table 4.2 below presents results on our ability to access Dark Web forums with and without a human-assisted accessibility mechanism. Using the human-assisted accessibility approach, we were able to access over 82% of Dark Web forums hosted by various Internet service providers and virtually all of the attempted stand-alone forums. The overall results (over 91% accessibility) indicate that the use of a human-assisted accessibility mechanism provided good results for Dark Web forums.

Table 4.2 Dark Web forum accessibility statistics

	Human-assisted accessibility			Standard spidering		
	Hosted forums	Stand-alone forums	Total forums	Hosted forums	Stand-alone forums	Total forums
Total attempted	52	67	119	52	67	119
Accessed/collected	43	66	109	25	56	71
Inaccessible	9	1	10	27	11	48
% Collected	*82.69*	*98.51*	*91.60*	*48.08*	*83.58*	*59.66*

Table 4.3 Dark Web forum accessibility statistics

	Human-assisted accessibility versus standard spidering
Hosted forums	<0.001
Stand-alone forums	<0.001
Total forums	<0.001

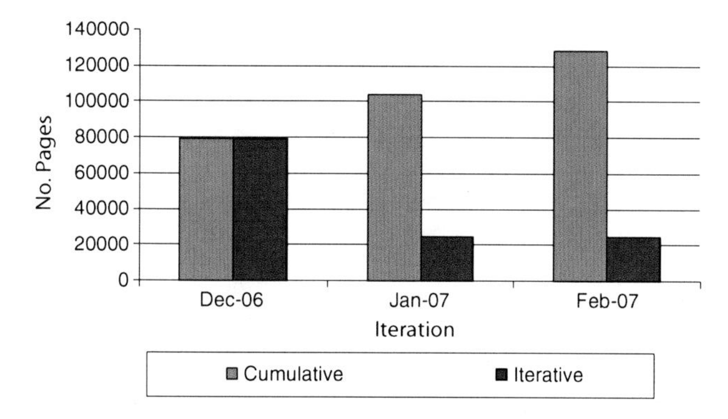

Fig. 4.8 Number of web pages in test bed across 3 months/iterations

In contrast, using standard spidering without any accessibility mechanism resulted in only 59.66% of the forums being accessible to collect. The biggest impact of the accessibility approach occurred on the hosted forums, where lack of usage of human-assisted accessibility resulted in a 34% drop in the number of forums collected (18 forums).

Table 4.3 shows the p values for the pairwise t tests conducted to assess the improved access performance of the human-assisted accessibility mechanism as compared to a standard spidering scheme devoid of any special accessibility method. The improved performance was statistically significant at alpha = 0.01 for total performance as well as both forum types.

6.2 *Forum Collection Update Experiment*

In order to evaluate the effectiveness of the proposed incremental crawling with recall improvement approach (referred to as incremental + RI) for collection updating, we conducted a simulated experiment in which 40 Dark Web forums were spidered three times over a 3-month period between December 2006 and February 2007. Figure 4.8 shows the number of cumulative web pages and the amount of new pages appearing in the 40 test bed forums across the 3-month period. There were approximately 128,000 unique web pages in the test bed, which were used as the gold standard for precision, recall, and F-measure computation. We collected the pages on a monthly basis (a total of three iterations) using a periodic, incremental,

Table 4.4 Macrolevel results for different update procedures

Update procedure	Precision	Recall	F-measure	Time (min)
Periodic	74.32	69.03	71.58	6,101
Incremental	57.80	53.69	55.67	*4,855*
Incremental + RI	*79.59*	*74.74*	*77.09*	5,758

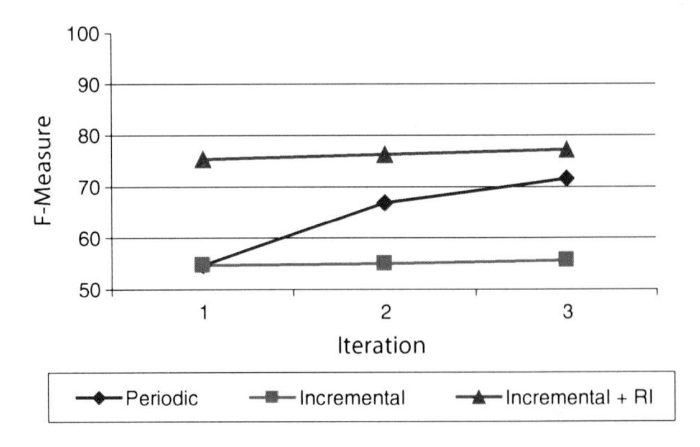

Fig. 4.9 Results by iteration for various collection update procedures

and incremental + RI collection update procedure. The periodic crawler collected all pages in each iteration (the cumulative amounts in Fig. 4.8) while the incremental crawler only collected the new pages for each iteration (the iterative amounts in Fig. 4.8). The advantage of periodic crawling is the ability to ascertain multiple versions of a page, which can improve the likelihood of gathering pages uncollected in the previous round at the expense of collection time and server congestion. The incremental + RI also collected the new pages but used a recall mechanism that allowed improperly retrieved pages to be refetched n number of times. The recall improvement phase, which identifies uncollected pages based on their spidering status and file size, is intended to retrieve uncollected pages in an efficient manner (i.e., without putting excessive burden on the forum servers). Consequently, a value of $n = 2$ was utilized since we have found that excessive attempts (i.e., larger values of n) typically decrease performance due to server congestion.

Performance was evaluated using the precision, recall, and F-measures. Precision was defined as the percentage of pages downloaded that were correctly collected. Correctly collected pages included all relevant pages completely downloaded. Incorrect pages were those that were partial/incomplete or irrelevant. Recall was defined as the percentage of relevant pages collected.

Table 4.4 shows the experimental results for the three collection procedures. The incremental + RI method achieved the highest precision, recall, and F-measure in a more efficient manner than the periodic approach. The incremental update without recall improvement was the most efficient timewise; however, it only had an F-measure of roughly 55%. The results suggest that Dark Web forums require the use of a spidering strategy that entails multiple attempts to fetch uncollected pages.

Figure 4.9 shows the overall F-measure for the three collection updating procedures after each spidering iteration. The diagram exemplifies the impact of making

Table 4.5 Dark Web forum collection statistics

	Hosted forums	Stand-alone forums	Total forums
Middle East	21	50	71
Latin America	6	3	9
US domestic	16	13	29
Total	*43*	*66*	*109*

multiple attempts to collect unfetched pages. We can see that the overall performance of periodic crawling improves dramatically during the second and third iterations since many of the previously uncollected web pages are gathered. Since the incremental + IR method retrieves such pages immediately, it maintains a consistently higher level of performance as compared to the other two methods.

6.3 Forum Collection Statistics

We used our spidering system for collection of Dark Web Forums in three regions (USA, Middle East, Latin America). The spider was initially run incrementally for a 20-month period between April 2005 and December 2006. The spider collected indexable, multimedia, archive (e.g., .zip and .rar), and nonstandard files (e.g., those with unknown/unrecognized file extensions).

Table 4.5 shows the number of forums collected per region. The collection consists of stand-alone and hosted forums. In general, the Middle Eastern groups tend to make greater use of stand-alone forums while the US domestic forums are more evenly distributed between hosted and stand-alone forums.

Table 4.6 shows the detailed collection statistics categorized by file type. Our system was able to collect a rich assortment of indexable and multimedia files.

It is interesting to note the large quantities of dynamic and multimedia files. Static HTML files, which were predominant on the Internet 10 years ago, have a minimal amount of usage in the Dark Web forums. Dynamic files outnumber static HTML files by a ratio of 10:1 while multimedia files (particularly images) are also present more often. This is partially attributable to the use of various forum software packages that generate dynamic thread pages (typically .php files).

Table 4.7 lists sample forums incorporated into our system after additional rounds of spidering. Forum "Al-Firdaws" is a general forum but has subsections containing discussions of radical Islamic ideologies and supporting Salafi-Jihadi organizations. Forum "Alokab" is dedicated to Islamic theology with some radical content. Forum "Hawaaworld" is dedicated to Muslim women. Some of its members have shown their sympathy to certain radical groups. Among all seven forums, "Hawaaworld," "Montada," and "Alsayra" have many more registered members than the other four forums. Some forums are extremely popular and have close to a million messages.

Table 4.6 Dark Web forum collection file statistics

	# of files	Volume (bytes)
Indexable files	*3,001,742*	*140,878,063,124*
HTML files	283,578	2,942,658,681
Word files	2,108	46,649,107
PDF files	16	8,168,345
Dynamic files	2,715,354	137,178,574,841
Text files	657	2,249,471,937
Excel files	1	177,152
PowerPoint files	2	528,834
XML Files	26	466,706
Multimedia files	*433,749*	*25,833,258,770*
Image files	422,155	8,554,125,848
Audio files	5,479	3,664,642,638
Video files	6,115	13,614,490,284
Archive files	*801*	*621,721,139*
Nonstandard files	*443,244*	*17,303,588,746*
Total	*3,878,735*	*185,017,574,960*

Table 4.7 Sample Dark Web forum statistics

Name	Time span	Number of members	Number of threads	Number of messages
Alokab	04/10/2005–03/19/2008	1,232	3,699	30,480
Al-Firdaws	01/02/2005–12/06/2007	2,189	9,359	39,775
Montada	09/28/2000–07/01/2007	31,654	93,548	866,693
Hdrmut	11/26/2000–05/18/2008	1,707	9,030	45,937
Alsayra	04/05/2001–06/03/2008	39,230	42,329	348,933
Hawaaworld	03/27/200–07/01/2008	59,842	20,278	975,695
Islamic Network	06/09/2004–05/07/2008	1,578	12,003	87,769

All forums listed are in Arabic with the exception of Islamic Network, which is in English

7 Conclusions and Future Directions

In this chapter, we developed a focused crawler for collecting Dark Web forums. We used a human-assisted accessibility mechanism to access identified forums with a success rate of over 90%. Our crawler uses language-independent features, including URL tokens, anchor text, and level features, in order to allow effective collection of content in multiple languages. It also uses forum software–specific traversal strategies and wrappers to support incremental crawling. The system uses an incremental crawling approach coupled with a recall improvement mechanism that continually respiders uncollected pages. Such an update approach outperformed the use of a standard incremental update strategy as well as the traditional periodic update method in a head-to-head comparison in terms of precision, recall, and computation time.

The system has been able to maintain up-to-date collections of 109 forums in multiple languages from three regions: US domestic supremacist, Middle Eastern extremist, and Latin groups. We believe that the proposed forum crawling system allows important entry to Dark Web forums which facilitates better accessibility for the analysis of these online communities. The collection of such content has significant academic and scientific value for intelligence and security informatics as well as various other research communities interested in analyzing the social characteristics of Dark Web forums.

We have identified several important directions for future research. We plan to improve the Dark Web forum accessibility mechanism in order to attain higher access rates. We also plan to expand our collection efforts to also include weblogs and chatting log archives. Additionally, we intend to evaluate the effectiveness of multimedia categorization techniques to enhance our ability to collect relevant image and video content.

Acknowledgments This research has been supported in part by the following grants: (1) NSF Digital Government Program, "COPLINK Center: Social Network Analysis and Identity Deception Detection for Law Enforcement and Homeland Security," October 2004–September 2007; (2) NSF/CIA, Knowledge Discovery and Dissemination (KDD) Program, "Detecting Identity Concealment," September 2005–August 2007; and (3) Library of Congress, "Capture of Open Source Web Based Multimedia Multilingual Terrorist Content," February 2007–February 2008.

References

Abbasi, A. and Chen, H. (2005). Identification and Comparison of Extremist-Group Web Forum Messages using Authorship Analysis. IEEE Intelligent Systems, 20(5), 67–75.

Aggarwal, C. C., Al-Garawi, F., and Yu, P. S. (2001). Intelligent Crawling on the World Wide Web with Arbitrary Predicates. In Proceedings of the 10th World Wide Web Conference, Hong Kong.

Baeza-Yates, R. (2003). Information Retrieval in the Web: Beyond Current Search Engines. International Journal of Approximate Reasoning, 34, 97–104.

Barbosa, L. and Freire, J. (2004). Siphoning Hidden-Web Data through Keyword-Based Interfaces. In Proceedings of the SBBD.

Bergman, M. K. (2000). The Deep Web: Surfacing Hidden Value. BrightPlanet.com.

Burris, V., Smith, E., and Strahm, A. (2000). White Supremacist Networks on the Internet. Sociological Focus, 33(2), 215–235.

Chakrabarti, S., Van Den Berg, M., and Dom, B. (1999). Focused Crawling: A New Approach to Topic-Specific Resource Discovery. In Proceedings of the Eighth World Wide Web Conference, Toronto, Canada.

Chau, M. and Chen, H. (2003). Comparison of Three Vertical Search Spiders. IEEE Computer, 36(5), 56–62.

Chen, H. Chung, Y., Ramsey, M., and Yang, C. (1998a). A Smart Itsy Bitsy Spider for the Web. Journal of the American Society for Information Science, 49(7), 604–619.

Chen, H. Chung, Y., Ramsey, M., and Yang, C. (1998b). An Intelligent Personal Spider (Agent) for Dynamic Internet/Intranet Searching. Decision Support Systems, 23(1), 41–58.

Chen, H. and Chau, M. (2003). Web Mining: Machine Learning for Web Applications. Annual Review of Information Science and Technology, (37), 289–329.

Chen, H. (2006). Intelligence and Security Informatics for International Security: Information Sharing and Data Mining. London: Springer Press.

Cheong, F. C. (1996). Internet Agents: Spiders, Wanderers, Brokers, and Bots. Indianapolis, IN: New Riders Publishing.

Cho, J., Garcia-Molina, H., and Page, L. (1998). Efficient Crawling Through URL Ordering. In Proceedings of the 7th World Wide Web Conference, Brisbane, Australia.

Cho, J and Garcia-Molina, H. (2000). The Evolution of the Web and Implications for an Incremental Crawler. In Proceedings of the 26th International Conference on Very Large Databases.

Cho, J. and Garcia-Molina, H. (2003). Estimating Frequency of Change. ACM Transactions on Internet Technology, 3(3), 256–290.

Crilley, K. (2001). Information Warfare: New Battle Fields Terrorists, Propaganda, and the Internet. In Proceedings of the Association for Information Management, 53(7), 250–264.

Diligenti, M., Coetzee, F. M., Lawrence, S., Giles, C. L., and Gori, M. (2000). Focused Crawling Using Context Graphs. In Proceedings of the 26th Conference on Very Large Databases, Cairo, Egypt.

Ester, M., Grob, M., and Kriegel, H. (2001). Focused Web Crawling: A Generic Framework for Specifying the User Interest and for Adaptive Crawling Strategies. In Proceedings of the International Conference on Very Large Databases.

Florescu, D., Levy, A. Y., and Mendelzon, A. O. (1998). Database Techniques for the World-Wide Web: A Survey. SIGMOD Record, 27(3), 59–74.

Glance, N., Hurst, M., and Tomokiyo, T. (2004). BlogPulse: Automated Trend Discovery for Weblogs. In Proceedings of the 13th International World Wide Web Conference, New York, New York.

Glance, N., Hurst, M., Nigam, K. Siegler, M., Stockton, R. and Tomokiyo, T. (2005). Analyzing Online Discussion for Marketing Intelligence, In Proceedings of the 14th International World Wide Web Conference, Chicago, Illinois.

Glaser, J., Dixit, J., and Green, D. P. (2002). Studying Hate Crime with the Internet: What Makes Racists Advocate Racial Violence? Journal of Social Issues, 58(1), 177–193.

Gustavson, A.T. and Sherkat, D.E. (2004). Elucidating the Web of Hate: The Ideological Structuring of Network Ties among White Supremacist Groups on the Internet. Paper presented at Annual Meeting of American Sociological Association.

Heydon, A. and Najork, M. (1999). Mercator: A Scalable, Extensible Web Crawler. In Proceedings of the International Conference on the World Wide Web, 219–229.

Lage, J. P., Da Silva, A. S., Golgher, P. B., and Laender, A. H. F. (2002). Collecting Hidden Web Pages for Data Extraction. In Proceedings of WIDM.

Lawrence, S. and Giles, C. L. (1998). Searching the World Wide Web. Science, 280(5360), 98.

Leuski, A. and Allan, J. (2000). Lighthouse: Showing the Way to Relevant Information. In Proceedings of the IEEE Symposium on Information Visualization, 125–130.

Limanto, H. Y., Giang, N. N., Trung, V. T., Huy, N. Q., and He, J. Z. Q. (2005). An Information Extraction Engine for Web Discussion Forums. In Proceedings of the 14th International Conference on the World Wide Web, Chiba, Japan.

Lin, K. and Chen, H. (2002). Automatic Information Discovery from the "Invisible Web." In Proceedings of the International Conference on Information Technology: Coding and Computing.

Najork, M. and Wiener, J. L. (2001). Breadth-First Search Crawling Yields High-Quality Pages. In Proceedings of the World Wide Web Conference, Hong Kong.

Ntoulas, A., Zerfos, P., and Cho, J. (2005). In Proceedings of the Joint Conference on Digital Libraries, Denver, Colorado.

Pant, G., Srinivasan, P., and Menczer, F. (2002). Exploration versus Exploitation in Topic Driven Crawlers. In Proceedings of the WWW Workshop on Web Dynamics.

Raghavan, S. and Garcia-Molina, H. (2001). Crawling the Hidden Web. In Proceedings of the 27th International Conference on Very Large Databases.

Schafer, J. (2002). Spinning the Web of Hate: Web-Based Hate Propagation by Extremist Organizations. Journal of Criminal Justice and Popular Culture, 9(2), 69–88.

Sizov, S., Graupmann, J., and Theobald, M. (2003). From Focused Crawling to Expert Information: An Application Framework for Web Exploration and Portal Generation. In Proceedings of the 29th International Conference on Very Large Databases, Berlin, Germany.

Srinivasan, P., Mitchell, J., Bodenreider, O., Pant, G., and Menczer, F. (2002). Web Crawling Agents for Retrieving Biomedical Information. In Proceedings of the International Workshop on Agents in Bioinformatics (NETTAB), Bologna, Italy.

Whine, M. (1997). The Governance of Cyberspace: Politics, Technology, and Global Restructuring., London, U.K: Routledge.

Yih, W., Chang, P., and Kim, W. (2004). Mining Online Deal Forums for Hot Deals. In Proceedings of the Web Intelligence Conference.

Zhou, Y., Reid, E., Qin, J., Chen, H., and Lai, G. (2005). U.S. Extremist Groups on the Web: Link and Content Analysis. IEEE Intelligent Systems, 20(5), 44–51.

Chapter 5
Link and Content Analysis

1 Introduction

The Internet has evolved to become a global platform through which anyone can conveniently disseminate, share, and communicate ideas. Despite many advantages, misuse of the Internet has become ever more serious. Terrorist organizations, extremist groups, hate groups, and racial supremacy groups are using the Web to promote their ideology, to facilitate internal communications, to attack their enemies, and to conduct criminal activities. There have been warnings that terrorists may launch attacks on such critical infrastructure as major e-commerce sites and governmental networks (Gellman 2002). Insurgents in Iraq have posted Web messages asking for munitions, financial support, and volunteers (Blakemore 2004). It therefore has become important to obtain intelligence from the Web that permits better understanding and analysis of terrorist and extremist groups. We define this reverse side of the Web as a "Dark Web," the portion of the World Wide Web used to help achieve the sinister objectives of terrorists and extremists.

Currently, intelligence from the Dark Web is scattered in diverse information repositories through which investigators need to browse manually to be aware of their content. Much of the information stored in search engine databases could be properly collected and analyzed for transformation into intelligence and knowledge that would enhance understanding of terrorists' activities. However, search engines often overwhelm users by producing laundry lists of irrelevant results and creating information overload problems. Related but unfocused information makes it difficult to obtain a comprehensive description of a terrorist group or a terrorism topic. Many Web resources contain information *about* terrorism, but a relatively small proportion comes from terrorist groups themselves, and data on the Web often are not persistent and may be misleading. Many terrorist Web sites do not use English, so investigators who do not know its language may be unable to understand a site's content.

H. Chen, *Dark Web: Exploring and Data Mining the Dark Side of the Web*, Integrated Series in Information Systems 30, DOI 10.1007/978-1-4614-1557-2_5,

In this chapter, we have addressed the aforementioned problems by proposing and implementing a semiautomated methodology for collecting and analyzing Dark Web information. Leveraging human preciseness and machine efficiency, the methodology consists of various steps, including collection, filtering, analysis, and visualization of Dark Web information. We used this comprehensive methodology to collect and analyze data from 39 Arabic terrorist Web sites and conducted an evaluation of the results. This research aimed to study to what extent the methodology can assist terrorism analysts in collecting and analyzing Dark Web information. From a broader perspective, this research contributes to the development of the new science of "Intelligence and Security Informatics (ISI)," the study of the use and development of advanced information technologies, systems, algorithms, and databases for national security–related applications through an integrated technological, organizational, and policy-based approach (Chen 2005; Strickland and Hunt 2005). We believe that many existing computer and information systems techniques need to be reexamined and adapted for this unique domain to create new insights and innovations.

The rest of this chapter is structured as follows: Sect. 2 presents a review of terrorists' use of information technologies to facilitate terrorism, information services for studying terrorism, and advanced techniques for collecting and analyzing terrorism information. Section 3 describes a methodology for collecting and analyzing Dark Web information. Section 4 illustrates the use of the methodology in a case study of jihad on the Web (where "jihad" is an Islamic term referring to a holy war waged against enemies) and discusses the evaluation results. Section 5 concludes the study and discusses future directions.

2 Literature Review

2.1 Terrorists' Use of the Web

Recent studies have shown how terrorists use the Web to facilitate their activities. Tsfati and Weimann used the names of terrorist organizations to search six search engines and found 16 relevant sites in 1998 and 29 such sites in 2002 (Tsfati and Weimann 2002). Their analysis of site content revealed heavy use of the Web by terrorist organizations to share ideology, to provide news, and to justify use of violence. Relying on open source information (e.g., court testimony, reports, Web sites), researchers at the Institute for Security Technology Studies identified five categories of terrorist use of the Web (Technical Analysis Group 2004): propaganda (to disseminate radical messages), recruitment and training (to encourage people to join the jihad and get online training), fundraising (to transfer funds, conduct credit card fraud, and other money laundering activities), communications (to provide instruction, resources, and support via e-mail, digital photographs, and chat session), and targeting (to conduct online surveillance and identify vulnerabilities of potential targets such as airports). Among these, using the Web as a propaganda tool has been widely observed.

Identified by the US Government as a terrorist site, Alneda.com called itself the "Center for Islamic Studies and Research," a bogus name, and provided information for al-Qaeda (Thomas 2003). To group members (insiders), terrorists use the Web to share motivational stories and descriptions of operations. To mass media and non-members (outsiders), they provide analysis and commentaries of recent events on their Web sites. For example, Azzam.com urged Muslims to travel to Pakistan and Afghanistan to fight the "Jewish-backed American Crusaders." Qassam.net appealed for donations to purchase AK-47 rifles (Kelley 2002). Al-Qaeda and some humanitarian relief agencies used the same bank accounts via www.explizit-islam.de (Thomas 2003).

Terrorists also share ideologies on the Web that provide religious commentaries to legitimize their actions. Based on a study of 172 members participating in the global Salafi jihad, Sageman concluded that the Internet has created a concrete bond between individuals and a virtual religious community (Sageman 2004). His study reveals that the Web appeals to isolated individuals by easing loneliness through connections to people sharing some commonality. Such virtual community offers a number of advantages to terrorists. It no longer ties to any nation, fostering a priority of fighting against the far enemy (e.g., the USA) rather than the near enemy. Internet chat rooms tend to encourage extreme, abstract, but simplistic solutions, thus attracting most potential jihad recruits who are not Islamic scholars. The anonymity of Internet cafés also protects the identity of terrorists. However, Sageman does not consider the Internet to be a direct contact with jihad because devotion to jihad must be fostered by an intense period of face-to-face interaction. In addition, existing studies about terrorists' use of the Web mostly use a manual approach to analyze voluminous data. Such an approach does not scale up to the rapid growth and frequent change of terrorists' identities on the Web.

2.2 *Information Services for Studying Terrorism*

Despite the public nature of the Web, terrorists often try to prevent authorities from tracing their Web addresses and activities, which has prompted several information services to monitor the Web sites of militant Islamic groups and to provide access to translated versions of information posted there. *The Jihad and Terrorism Project* was developed by the Middle East Media Research Institute to bridge the language gap between the West and the Middle East by providing timely translations of Arabic, Farsi, and Hebrew documents (Middle East Media Research Institute 2004). The *Project for the Research of Islamist Movements* (www.e-prism.org) studies radical Islam and Islamist movements, focusing primarily on Arabic sources. These projects provide access to an array of information such as translated news stories, transcripts, video clips, and training documents produced by terrorists but fall short of supporting analysis and visualization of terrorist data from the Dark Web (Project for the Research of Islamist Movements 2004).

2.3 Advanced Information Technologies for Combating Terrorism

Since the 9/11 attacks, there has been increased interest in using information technologies to counter terrorism. A study conducted by the US Defense Advanced Research Projects Agency shows that their collaboration, modeling, and analysis tools speeded analysis (Popp et al. 2004), but these tools were not tailored to collecting and analyzing Web information. Although new approaches to terrorist network analysis have been called for (Carley et al. 2001), existing efforts have remained mostly small scale; they have used manual analysis of a specific terrorist organization and did not include resources generated by terrorists in their native languages. For instance, Krebs manually collected data from English news releases after the 9/11 attacks and studied the network surrounding the 19 hijackers (Krebs 2001). Although automated social network analysis techniques have been proposed to analyze and portray criminal networks, it is not clear whether the techniques are applicable to the mostly unstructured data in terrorist Web sites that contain textual and multimedia data (Xu and Chen 2005). Their use of structured data in a police department database also does not help understand terrorist Web sites. Other advanced information technologies having potential to help analyze terrorist data on the Web include information visualization and Web mining.

Information visualization technologies have been used in many domains (Zhu and Chen 2005) such as criminal analysis (Chung et al. 2005a) and business stakeholder analysis (Chung 2007). For example, multidimensional scaling (MDS) algorithms consist of a family of techniques that portray a data structure in a spatial fashion, where the coordinates of data points are calculated by a dimensionality reduction procedure (Young 1987). MDS has been used in many different applications. Chung and his colleagues developed a new browsing method based on MDS to depict the competitive landscape of businesses on the Web (Chung et al. 2005b). He and Hui applied MDS to display author cluster maps in their author co-citation analysis (He and Hui 2002). Eom and Farris applied MDS to author co-citation in decision support systems (DSS) literature from 1971 through 1990 in order to find contributing fields to DSS (Eom and Farris 1996). Kealy applied MDS to studying changes in knowledge maps of groups over time to determine the influence of a computer-based collaborative learning environment on conceptual understanding (Kealy 2001). Although much has been done in different domains to visualize relationships of objects using MDS, no attempts to apply it to discovering terrorists' use of the Web have been found.

Web mining is the use of data mining techniques to automatically discover and extract information from Web documents and services (Chen and Chau 2004; Etzioni 1996). Chen et al. (2001) showed that the approach of integrating metasearching with textual clustering tools achieved high precision in searching the Web. Web page classification, a process of automatically assigning Web pages into predefined categories, can be used to assign pages into meaningful classes (Mladenic 1998). Web page clustering, a process of identifying naturally occurring subgroups among a set of Web pages, can be used to discover trends and patterns within a large number of pages (Chen et al. 1996). Although a number of Web mining technologies exist (e.g., Chen and Chau 2004; Last et al. 2006), there has not yet been a comprehensive methodology to address problems of collecting and analyzing terrorist data on the Web. Unfortunately, existing frameworks using data and text mining techniques (e.g., Nasukawa and Nagano 2001; Trybula 1999) do not address issues specific to the Dark Web.

To our knowledge, few studies have used advanced Web and data mining technologies to collect and analyze terrorist information on the Web, though these technologies have been widely applied in such other domains as business and scientific research (e.g., Chung et al. 2004; Marshall et al. 2004). New approaches to collecting and analyzing terrorist information on the Web are needed.

3 A Methodology for Collecting and Analyzing Dark Web Information

3.1 The Methodology

To address threats from the wide range of information sources that terrorists and extremists use to spread their ideas and to conduct destructive activities, we have proposed a semiautomated methodology integrating various information collection and analysis techniques and human domain knowledge. Figure 5.1 shows the methodology aiming to effectively assist human investigators to obtain Dark Web intelligence using information sources, collection methods, filtering, and analysis.

- *Information sources* consist of a wide range of providers of terrorist or terrorism information on the Web. Some of these are readily accessible (e.g., search engines), while some, like terrorism incident databases and Web sites developed and maintained by terrorists and their supporters, can only be reached with the help of domain experts.
- *Collection methods* make possible automatic searching, browsing, and harvesting of information from identified sources. *Domain spidering* starts with a set of relevant seed URLs and relies on an automatic Web page collection program, often called a spider or crawler, to harvest Web pages linked to the seed URLs. *Backlink search*, supported by some search engines such as Google (www.google.com) and AltaVista (www.altavista.com, acquired by Overture that was then acquired by Yahoo! in 2003), allows searching of Web pages that have hyperlinks pointing to a target Web domain or page. It helps investigators trace activities of terrorist supporters and sympathizers, whose Web pages often reference terrorist sites (e.g., glorify martyrs' actions, show a concurrence of terrorist attacks). *Group/personal profile search*, exemplified by major Web portals such as Yahoo! (members.yahoo.com) and MSN (groups.msn.com), reveals the profiles of groups or individuals who share the same interests. Terrorists and their supporters may

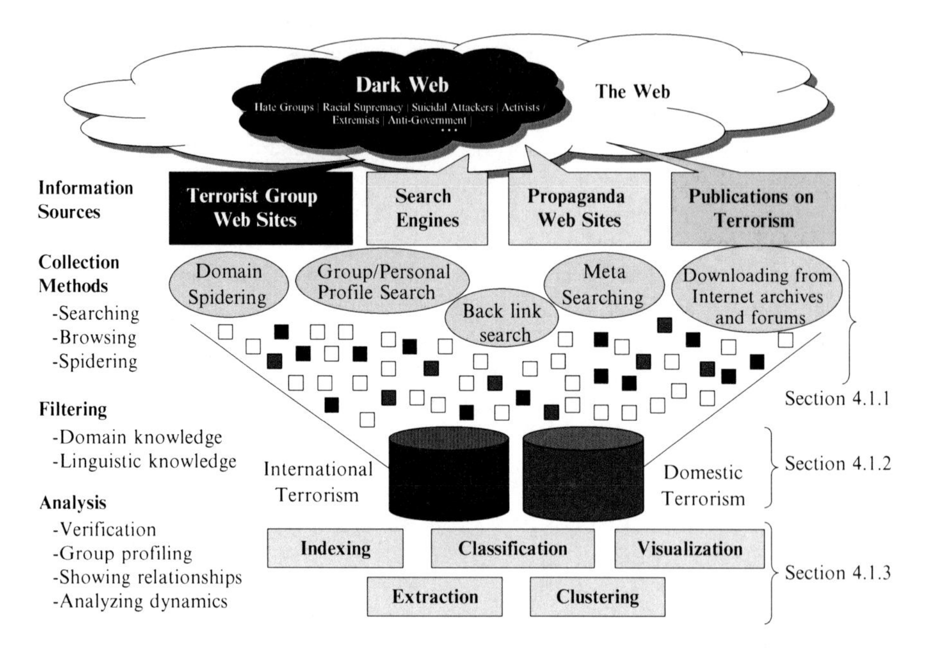

Fig. 5.1 A methodology for collecting and analyzing Dark Web information

perhaps put "hot links" in their profiles, which allow investigators to discover hidden linkages. *Meta-searching* uses related keywords as input to query multiple search engines, from which investigators or automated programs can collate top-ranked results and filter out duplicates to obtain highly pertinent URLs of terrorist Web sites. With careful formulation of search terms and appropriate linguistic knowledge, they can obtain highly relevant results. For example, searching the Arabic name of "Osama bin-Laden" (اسامة بن لادن) in multiple search engines returns mixed results about terrorist news articles and terrorist Web sites, while augmenting "Osama bin-Laden" with the keyword "Sheikh" (the head of tribe or leader in Arabic), which is frequently used by al-Qaeda to refer to bin-Laden, can give more relevant terrorist and supporter Web sites. *Downloading from Internet archives and forums* exploits the temporal dimension of Web information. For instance, the Internet Archive (www.archive.org) offers access to historical snapshots of Web sites. Usenet discussion forums provide a wealth of textual communication that can be mined for hidden patterns over time.

- *Filtering* involves sifting through collected information and removing irrelevant results, but to perform this task requires domain knowledge and linguistic knowledge. Domain knowledge refers to knowledge about terrorist groups, their relationships with other terrorist and supporter groups, their presence on and usage of the Web, as well as their histories, activities, and missions. Linguistic knowledge deals with terms, slogans, and other textual and symbolic clues in the native languages of the terrorist groups. Filtering can be automatic or manual, depending on requirements for efficiency of process and precision of the results.

Typically, manual filtering achieves high precision but is less efficient and relies on domain experts who have had years of experience in the field. Automatic filtering is very efficient as it often uses computers and machine learning to process large amounts of data, but the results are less precise. Investigators can obtain high-quality data for analysis from filtered repositories.

- *Analysis* provides insights into data and helps investigators identify trends and verify conjectures. Several functions support these analytical tasks. *Indexing* relates textual terms to individual Web pages, thereby supporting precise searching of the pages. *Extraction* identifies meaningful entities such as terrorist names, frequently used slogans, and suspicious terms. *Classification* finds common properties among entities and assigns them to predefined categories to help investigators predict trends of terrorist activities. *Clustering* organizes entities into naturally occurring groups and helps to identify similar terrorist groups and their supporters. *Visualization* presents voluminous data in a format perceivable by human eyes, so investigators can picture the relationship within a network organization of terrorist groups and can recognize their underlying structure.

3.2 Discussion of the Methodology

Although the Internet has been publicly available since the 1990s, the Dark Web emerged only in recent years. A lack of useful methodology designed for Dark Web data collection and analysis has limited the capability to fight against terrorism. As discussed above, the proposed methodology has incorporated various data and Web mining technologies while still allowing human domain knowledge to guide their application. Its semiautomated nature combines machine efficiency with the advantages of human precision, a useful complement to computers that usually fail to detect deception and ambiguity on the Dark Web. Its coverage of wide varieties of data sources and techniques ensures a comprehensive Dark Web data collection, a challenge often faced by terrorism and intelligence analysts. Therefore, the methodology and its integration and application of data and Web mining technologies to Dark Web analysis are novel contributions to the ISI research.

4 Jihad on the Web: A Case Study

To demonstrate the value and usability of our methodology, we have applied it to collecting and analyzing the use of the Web for "jihad," an Islamic term referring to a holy war waged against enemies as a religious duty. Believers contend that those who die in jihad become martyrs and are guaranteed a place in paradise. In recent decades, the concept of jihad has been used as an ideological weapon to combat against Western influences and secular governments and to establish an ideal Islamic society (Encyclopedia Britannica Online 2007). Jihad supporters are closely related to terrorist groups while maintaining anonymity using the Web. For example, prior to the 9/11 attacks, al-Qaeda members sent each other thousands of messages in a password-protected section of an extreme Islamic Web site (Anti-Defamation League 2002). Terrorist groups such as Hamas, Hezbollah, and Palestinian Islamic Jihad also use Web sites as propaganda tools. We describe the steps of applying the methodology as follows (see Fig. 5.1). The data described below were collected in 2004.

4.1 Application of the Methodology

4.1.1 Collection

To collect data, we first identified four suspicious URLs through Web searching, referencing to published terrorism reports, and performing personal profile searches on Yahoo! (For example, we searched "hezbollah" in Google where we found its URL among the top-ranked results.) These URLs are Palestinian Islamic Jihad (PIJ) (www.qudsway.com), Hezbollah (www.hezbollah.org), the military wing of Hamas (www.ezzedeen.net), and an Arabic Web site with a pro-jihad forum (www.al-imam.net). A 2003 US Department of State report confirmed PIJ, Hezbollah, and Hamas to be terrorist or terrorist-affiliated groups (Department of State 2003). Though Al-Imam.net is not classified as a terrorist organization, it contains pro-jihad forums in which messages and links to terrorist Web sites are posted. We then used the back-link search function of Google to obtain several hundred URLs that point to the four suspicious URLs. As Dark Web information can be scattered in many different sources and can be changed quickly over time, the several methods used to identify the four initial URLs enabled us to cover a broader scope and a more timely content than relying only on published reports (e.g., US Department of State's annual report). While different initial URLs and different times of data collection could affect the content of the data collected, we believe these four URLs are representative of the Dark Web. It would be an interesting future direction to study the extent to which data collection affects the quality of analysis results.

4.1.2 Filtering

We conducted two rounds of filtering. First, we manually filtered out unrelated sites, such as news or governmental Web sites that report or discuss only terrorist activities, religious Web sites with no reference to jihad or violence, and political Web sites where there is no mention or approval of terrorist activities. We retained Web sites of terrorist organizations, those of terrorist leaders, and those that praise terrorists or their actions. Forty-six sites remained after this round of filtering.

Second, with the help of a native Arabic speaker (who is not a terrorism expert), we manually added 14 terrorist and supporter sites identified by querying Google with the keywords (in Arabic) that we had found in the terrorist and supporter sites. Such keywords included the leaders' and organizations' names in Arabic ("mojahedin iran," "markaz dawa," "الشيخ المجاهد بن لادن," etc.). To limit the scope of analysis, we considered only the top 50 results returned from the search engine in each query search. In addition, we manually removed 21 sites from the set of all sites obtained based on their relevance to the domain. This round of filtering and refining resulted in 39 Arabic Web sites – 24 terrorist sites and 15 supporter sites.

4.1.3 Analysis

We performed clustering, classification, and visualization on the 94,326 Web pages collected by crawling the 39 terrorist and supporter sites using an exhaustive breadth-first search spidering program (with a maximum depth of 10 levels). The first analysis task we performed was clustering, in which we considered as input the 46 Web sites identified from the first round of filtering (see paragraph 1 of Sect. 4.1.2). The clustering involves calculating a similarity between each pair of Web sites in our collection to uncover hidden Web communities. We define similarity to be a real-valued multivariable function of the number of hyperlinks in one Web site ("A") pointing to another Web site ("B"), and the number of hyperlinks in the latter site ("B") pointing to the former site ("A"). In addition, a hyperlink is weighted proportionally to how deep it appears in the Web site hierarchy. For instance, a hyperlink appearing on the homepage of a Web site is given a higher weight than hyperlinks appearing at a deeper level. Specifically, the similarity between Web sites "A" and "B" is calculated as follows:

$$Similarity\quad(A, B) = \sum_{\substack{\text{All links } L \\ \text{b/w } A \text{ and } B}} \frac{1}{1 + lv(L)}$$

where *lv(L)* is the level of link *L* in the Web site hierarchy, with homepage as level 0 and the level increased by 1 with each level down in the hierarchy. Using these heuristics, a computer program automatically extracted hyperlinks on Web pages and calculated their similarities.

In the second analysis task, we classified the sites by their affiliations with terrorist groups, ideologies, and religions, and by their Web site attributes. Our native Arabic speaker manually identified the affiliations of all the Web sites according to their site content. Although we had the help of the Arabic speaker, the components of the methodology are generic enough to be applicable to other domains. The choice of this Arabic speaker, who is not a terrorism expert, also would not affect the results. Table 5.1 shows the details of the Web sites and their affiliations.

In addition to using affiliations, we classified the sites by indicating how terrorists and their supporters use the Web to facilitate their activities. From our literature review, we identified six types of terrorist use of the Web and 27 unique Web site attributes. Table 5.2 presents these attributes categorized under the six types.

Table 5.1 Analysis of jihad terrorist groups and their supporters' sites

No.	Name	URL[a]	Description[b]	Terrorist group[c]	Religion
Terrorist groups' Web sites (total: 24)					
1.	Special Force	www.specialforce.net	Provides computer game replicating the fighting scenes between Lebanese resistance and Israeli occupiers	Hezbollah	Shi'a Muslim
2.	Palestine Info in Urdu	palestine-info-urdu.com	Hamas news Web site in Urdu	Hamas	Sunni Muslim
3.	Al-Manar	web.manartv.org	The Web site of Al-Manar, the TV channel of Lebanese Hezbollah	Hezbollah	Shi'a Muslim
4.	Abrarway	www.abrarway.com	News Web site of Islamic Jihad of Palestine Guerrilla group	Palestinian Islamic Jihad	Sunni Muslim
5.	Islamic Jihad Mail	www.jimail.com	News Web site of Islamic Jihad of Palestine Guerrilla group	Palestinian Islamic Jihad	Sunni Muslim
6.	Ezz-al-dine Al-Qassam	www.ezzedeen.net	A general portal of Izz-Edeen Al-Qasam	Hamas	Sunni Muslim
7.	Hezbollah	www.hizbollah.tv	The official Web site of Hezbollah Organization	Hezbollah	Shi'a Muslim
8.	Info Palestina	www.infopalestina.com	Hamas information and news Web site in Malay	Hamas	Sunni Muslim
9.	Kataeb Al Aqsa	www.kataebalaqsa.com	The official Web site of Al-Aqsa Martyrs Brigade	Al-Aqsa Martyrs Brigade	Secular
10.	Kavkaz	www.kavkaz.org.uk	The news Web site of Chechen guerrilla fighters	Islamic International Brigade, Special Purpose Islamic Regiment	Sunni Muslim
11.	Moqawama	www.moqawama.tv	Web site of the Hezbollah's support group	Hezbollah	Shi'a Muslim
12.	Nasrollah	www.nasrollah.org	Hezbollah leader's site (Sheikh Hassan Nasrollah)	Hezbollah	Shi'a Muslim
13.	Alshohada	www.b-alshohda.com	Web site of Hamas and Islamic Jihad dedicated to martyrs	Hamas, Palestinian Islamic Jihad	Sunni Muslim
14.	Quds Way	www.qudsway.com	Provides general news of Islamic Jihad of Palestine	Palestinian Islamic Jihad	Sunni Muslim

15.	Rantisi	www.rantisi.net	Web site of Abdel Aziz Al Rantisi, a Hamas leader	Hamas	Sunni Muslim
16.	People's Mojahedin of Iran	www.iran.mojahedin.org	Web site posting statements by the People's Mojahedin Organization	Mujahedin-e Khalq Organization	Secular
17.	National Council of Resistance of Iran	www.iranncrfac.org	Official Web site of the Foreign Affairs Committee of the National Council of Resistance of Iran	Mujahedin-e Khalq Organization	Secular
18.	Iranian People's Fadaee Guerrillas	www.siahkal.com	The memorial Web site of the Iranian People's Fadaee Guerrillas	Mujahedin-e Khalq Organization	Secular
19.	The Organization of Iranian People's Fedaian	www.fadai.org	The Organization of Iranian People's Fedaian (Majority) official Web site	Mujahedin-e Khalq Organization	Secular
20.	Organization of Iranian People's Fedayee Guerrillas	www.fadaian.org	Organization of Iranian People's Fedayee Guerrillas memorial Web site	Mujahedin-e Khalq Organization	Secular
21.	The Union of People's Fedaian of Iran	www.etehadefedaian.org	News and information Web site of the Union of People's Fedaian of Iran	Mujahedin-e Khalq Organization	Secular
22.	Revolutionary Peoples Liberation Front	www.dhkc.net	Revolutionary Peoples Liberation Front official Web site. Provides news and statements of the organization	Revolutionary People's Liberation Army/Front	Secular
23.	DHKC International	www.dhkc.info	Web site of DHKC in Turkish	Revolutionary People's Liberation Army/Front	Secular
24.	Crusade Begins	jorgevinhedo.sites.uol.com.br	The Brazil-based Web site links to Lashkar-e-Taiba – a terrorist organization based in Pakistan	Lashkar-e Tayyiba	Sunni Muslim
Supporters' Web sites (total: 15)					
25.	Al Ansar	www.al-ansar.biz	Provides support to al-Qaeda organization, as well as articles about the Salafi Sunni ideology	al-Qaeda	Sunni Muslim
26.	Alokab	www.alokab.com	Provides articles about the Salafi Sunni ideology and the jihadist movement	al-Qaeda	Sunni Muslim

(continued)

Table 5.1 (continued)

No.	Name	URL[a]	Description[b]	Terrorist group[c]	Religion
27.	Alsakifah Forum	www.alsakifah.org	Provides educational services and a forum dedicated to the discussion of the Salafi ideology	al-Qaeda	Sunni Muslim
28.	Cihad	www.cihad.net	A general jihad Web site providing information about all jihad activities around the world	al-Qaeda	Sunni Muslim
29.	Clear Guidance Forum	www.clearguidance.com	Forum of jihad supporters	al-Qaeda	Sunni Muslim
30.	Sheikh Hamid Bin Abdallah Al Ali	www.h-alali.net	Salafi Educational Web site with some jihad ideas	al-Qaeda	Sunni Muslim
31.	Jihadunspun	www.jihadunspun.com	Pro-jihad news Web site	al-Qaeda	Sunni Muslim
32.	Maktab-Al-Jihad	www.maktab-al-jihad.com	Pro-jihad news Web site	al-Qaeda	Sunni Muslim
33.	Qoqaz	www.qoqaz.com	Jihad news from the Caucasus	Islamic International Brigade, Special Purpose Islamic Regiment	Sunni Muslim
34.	Supporters of Shareeah	www.shareeah.org	A general portal dedicated to the jihadist movement	al-Qaeda	Sunni Muslim
35.	Moltaqa	www.almoltaqa.org	Hamas Forum	Hamas	Sunni Muslim
36.	Saraya	www.saraya.com	Pro-jihad Web site	al-Qaeda	Sunni Muslim
37.	Osama bin-Laden	1osamabinladen.5u.com	A Web site dedicated to Osama bin-Laden	al-Qaeda	Sunni Muslim
38.	Tawhed	www.tawhed.ws	Pro-jihad Web site	al-Qaeda	Sunni Muslim
39.	The Right Word	www.rightword.net	Pro-al-Qaeda Web portal	al-Qaeda	Sunni Muslim

[a]Some of the URLs and sites may have changed at the time of reading due to the rapid change of the Dark Web

[b]The descriptions are obtained from the Web sites

[c]Descriptions of these terrorist groups appear in the US Department of State Report *Pattern of Global Terrorism, 2002*

Table 5.2 Categories of terrorist use of the Web and Web site attributes

Category	Attribute	Description
Communications	E-mail	Any listed e-mail address or feedback form
	Telephone (including Web phone)	Telephone numbers of organization officials
	Multimedia tools	Video clips of bombings and other activities. Video, sound recording, and games (e.g., leader's messages and instructions)
	Online feedback form	Allow the user to give feedback or ask questions to the Web site owners and maintainers
	Documentation	Report, book, letter, memo, and other resources provided (e.g., in pdf, Word, Excel, other formats)
Fundraising	External aid mentioned	Other groups or governments supporting the organization
	Fund transfer	Fund transfer methods
	Donation	Donations under the form of direct bank deposits
	Charity	Donations to religious welfare organizations associated with terrorist organization
	Support groups	Suborganizational structures charged with the fundraising program
	Others	Other attributes belonging to this category
Sharing ideology	Mission	The major goals of the organization (e.g., destruction of an enemy state, liberation of occupied territories)
	Doctrine	The beliefs of the group (e.g., religious, communist, extreme right)
	Justification of the use of violence	Ideology condones the use of violence to accomplish goals (e.g., suicide bombing)
	Pinpointing enemies	Classifies others as either enemies or friends (e.g., USA is enemy, Taliban regime is friendly)
Propaganda (insiders)	Slogans	Short phrases with religious or ideological connotations
	Dates	Mentions dates in the history of the terrorist group such as the date of a major attack
	Martyr's description	Lists the names of members who died in terrorism-related operations or descriptions of the circumstances
	Leader's name(s)	Terrorist groups leader(s) name as claimed by the Web site
	Banner and seal	Banner depicting representative figures, graphical symbols, or seals of the organization
	Narratives of operations and events	Provides narratives of the operations and attacks of the group
	Others	Other attributes belonging to this category
Propaganda (outsiders)	Reference to media coverage of events	For example, the Web site criticizes Western media coverage of events with explicit mention of outlets such as CNN and CBS
	News reporting	Group's own interpretation of events
Virtual community	Listserv	Automatic mailing list server that broadcasts to everyone on the list
	Text chat room	Virtual room where a chat session takes place. Text messaging chat session such as ICQ
	Message board	Allows members to post and read messages online
	Web ring	A series of Web sites linked together in a ring that by clicking through all of the sites in the ring the visitor will eventually come back to the originating site

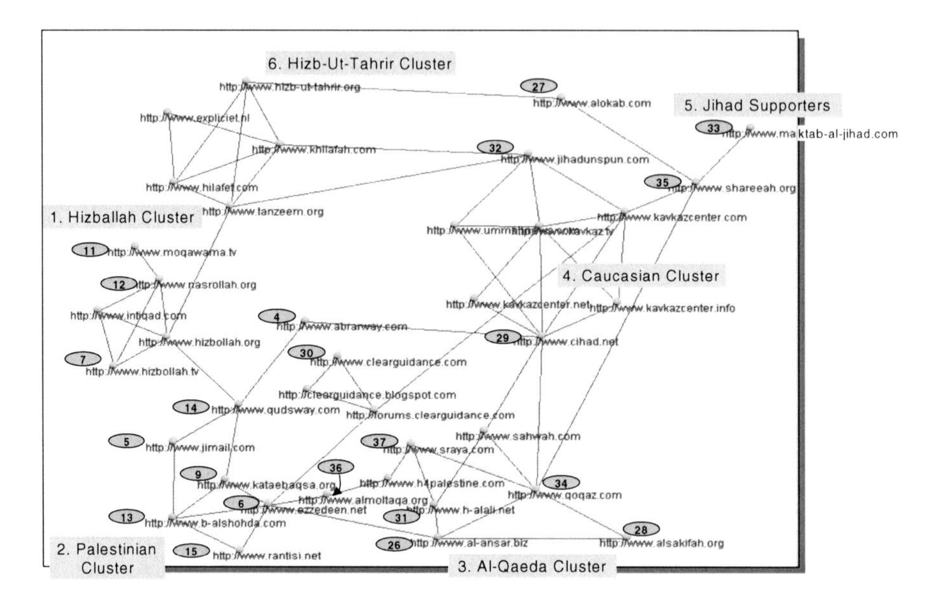

Fig. 5.2 Clustering and visualization of terrorist Web sites (the numbers refer to those appearing in Table 5.1) Some Web sites in Table 5.1 do not appear in this figure because they were added after the first-round filtering

Following this coding scheme, the Arabic speaker manually read through all of the subject Web pages to record terrorist uses of the Web. Similar to that used in studying the openness of government Web sites (La Porte et al. 1999), our coding involved finding whether an attribute existed on the Web sites (i.e., binary scoring). Manual coding of each Web site required 45 min to 1 h.

To reveal patterns of terrorist Web site existence and degree of a site's activities, in the third analysis task, we performed two types of visualization: multidimensional scaling and snowflake visualizations.

Multidimensional scaling visualization provided a high-level picture of all the terrorist groups and their relationships. We used multidimensional scaling (MDS) to transform a high-dimensional similarity matrix to a set of two-dimensional coordinates (Young 1987). While other visualization techniques might have been applicable, we chose MDS because it suits the current data structure and provides a vivid picture summarizing terrorist groups' relationships. Figure 5.2 shows these relationships, in which the sites appear as nodes and the lines connect pairs of sites that have at least one hyperlink pointing from one site to another. Using the similarity matrix as input, the MDS algorithm calculated coordinates of each site and placed the sites on a two-dimensional space where proximity reflects similarity. Upon closer examination of the figure, seven clusters of sites emerge. (The numbers in parentheses refer to the sites in Table 5.1. The URLs were filtered out in the second-round filtering but appeared in the collection after the first-round filtering.):

1. *Hezbollah Cluster* (# 7, 11, 12, hezbollah.org, and intiqad.org) contains the Web site of Hezbollah group (www.hezbollah.org) and its affiliated sites such as Hezbollah E-magazine (www.intiqad.org), Hezbollah Support Association (#11), and the site of Sayyed Hassan Nasrollah (#12), a major leader of Hezbollah.
2. *Palestinian Cluster* (# 4, 5, 6, 9, 13, 14, 15, 36, and h4palestine.com) includes militant groups fighting against Israel (e.g., Al-Aqsa Martyrs Brigade, Hamas). There are links between sites of the same group (e.g., # 4 and 14) and links between sites of different groups (e.g., # 9 and 6).
3. *Al-Qaeda Cluster* (# 26, 28, 31, 35, 37, and sahwah.com) includes Salafi groups' supporters' Web sites that often are linked to each other in their "Other friendly Web sites" section. They use their Web sites heavily to propagate their ideology. For example, Al-ansar.biz posted a video of the beheading of Nicholas Berg, one of the first civilians killed by terrorists (Newman 2004). Alsakifah.org provides an online discussion forum.
4. *Caucasian Cluster* (# 10, 34, kavkazcenter.com, kavkaz.tv, kavkazcenter.net, and kavkazcenter.info) consists of Web sites that link to Chechen rebels and provide news updates from Chechen areas. For example, Qoqaz.com has documented operations against the Russian military.
5. *Jihad Supporters* (# 29, 30, 32, 33, clearguidance.blogspot.com, and ummanews.com) consists of Web sites providing news and general information on the global jihad movement. These sites rarely are linked to each other and often play a propaganda role that targets outsiders.
6. *Hizb-Ut-Tahrir* (# 27, hizb-ut-tahrir.org, expliciet.nl, khilafah.com, and hilafet.com) contains a nonterrorist political group, Hizb-Ut-Tahrir, dedicated to the restoration of Islamic law and Khilafah (global leadership of Muslims). It has a presence in many Arab countries (e.g., Lebanon, Jordan) and some European countries. For instance, Expliciet.nl is a Dutch Web site based in the Netherlands.
7. *Tanzeem-e-Islami Cluster* (tanzeem.org) consists of a single site representing the Pakistani "Tanzeem-e-Islami" party with no clear ties to terrorism.

Snowflake visualization supports analysis of different dimensions (or categories) of activities of a Web site cluster. It originates from a star plot that has been widely used to display multivariate data (Chambers et al. 1983). A snowflake (shown in Fig. 5.3) represents a terrorist site cluster. Figure 5.3 shows five snowflake diagrams, each representing the degree of activity of terrorist/supporter groups in the five terrorist clusters (clusters 1–5) described above (clusters 6 and 7 are not included because they do not contain terrorist sites). The six sides of a snowflake represent the six dimensions of terrorist use of the Web, as shown in Table 5.2 and explained above. Each of these six dimensions represents a normalized scale between 0 and 1

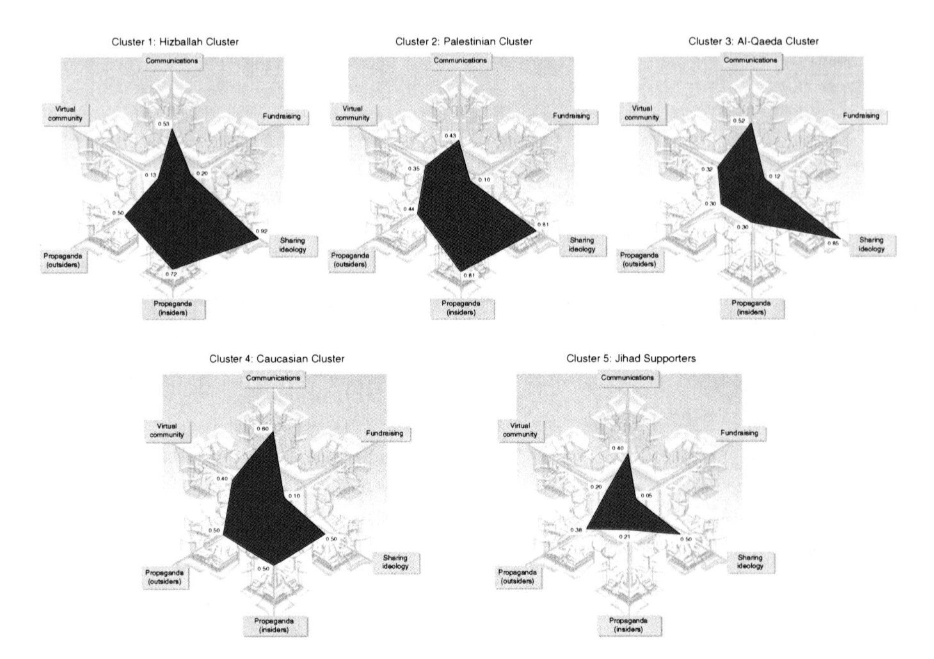

Fig. 5.3 Snowflake visualization of five terrorist site clusters

(activity index), showing the degree of activity on the dimensions. The activity index of cluster c on dimension d was calculated by the following formula:

$$ActivityIndex(c,d) = \frac{\sum_{i}^{n}\sum_{j}^{m} w_{i,j}}{m \times n}$$

where

$$w_{i,j} = \begin{cases} 1 & attribute\, i\, occurs\, in\, website\, j \\ 0 & otherwise \end{cases}$$

n is the total number of attributes in the specified dimension d;
m is the total number of Web sites belonging to the specified cluster c.

The closer the activity index is to 1, the more active on that dimension a cluster is. This index reveals in what areas the terrorist groups are active, and hence provides investigators and analysts with clues about how to devise strategies to combat a group.

5 Results and Discussion

Our preliminary observations show that the methodology yielded promising results. For example, it identified Web sites affiliated with 10 of the 26 groups classified as jihad terrorist organizations in the US State Department report on terrorism. Al-ansar.biz (# 25), the site that posted the beheading video of Nicholas Berg, posted messages from al-Qaeda leaders such as Osama bin-Laden, Ayman Al-Zawahiri, and Al-Zarqawi, praising their attacks on enemies. Another site, Tawhed.com (# 38), posted a poem praising the 9/11 attacks. The rhetoric of the poem commonly appears in many al-Qaeda-affiliated Web sites, referring to the American as crusaders (الصليبيين). Words like Sunna and Jama'h (السنة والجماعة) reflect the branch of Islam to which the Salafi groups belong.

From the snowflake diagrams (Fig. 5.3), we found that terrorists and supporters use the Web heavily to share ideology and to propagate ideas, especially to their members. For example, the Palestinian cluster (cluster 2) actively shares its ideology and heavily uses the Web as a propaganda tool for members. The Web sites in this cluster support liberation of Palestine, pinpoint and criticize their enemies, and describe details of operations and rationales supported by Quaran verses. In contrast, jihad supporters (cluster 5) rarely use the Web for propaganda but share ideology and communicate there. The Hezbollah cluster (cluster 1) resembles the Palestinian cluster in heavy use of the Web for sharing ideology and insider propaganda. For example, the sites in this cluster glorify martyrs and leaders and also were used moderately for outsider propaganda and communications. In the five clusters, we found little evidence of using the Web for fundraising or building a virtual community. Probably such uses have gone underground or do not appear on the Web.

5.1 Expert Evaluation and Results

Based on the above results, we invited a terrorism expert to conduct an evaluation of the methodology. A senior fellow of the US Institute of Peace at Washington, D.C., the expert is a professor of communication in a major research university in Israel. Having expertise in modern terrorism and the Internet, he has published more than 80 refereed journal articles and books and is a frequent speaker at international conferences on counterterrorism. This expert also leads a team of about 16 research assistants who regularly monitor 4,300 sites on the Dark Web for terrorist activities. The approach he and his team use to collect and analyze terrorists' use of the Web is largely manual, relying on laborious human browsing and monitoring of selected Web sites. His experience in manual analysis served to contrast with our methodology that automated part of the Dark Web data collection and analysis. We decided to use expert validation instead of other evaluation methods for two reasons: (1) lab experiment is not suitable because typical experimental subjects do not have much knowledge in the Dark Web, and (2) it is not feasible to invite terrorists to participate in an interview or empirical evaluation. The expert was not involved in writing this paper.

The evaluation was conducted using an unbiased structured questionnaire and a formal procedure. We showed the results to our expert and asked him to provide detailed comments on the categorization of Web sites and attributes, the visualization and clustering of terrorist groups, and the usability of the snowflake visualization. In general, he deemed the results to be very promising and the methodology design to be excellent. He believed that this was the start of very important research that will result in a useful database and a reliable methodology to update and maintain the database.

The expert was greatly impressed by the visualization and clustering capabilities of the methodology and provided valuable comments on our work. However, he said that the 39 Web sites shown in Table 5.1 do not represent the entire population of all terrorist Web sites, the number of which he estimated to be over 4,000. Because we focused only on Middle Eastern terrorist groups (rather than all terrorist groups in the world), we believe that our methodology has yielded representative results and has automated much of the manual work of identifying and analyzing terrorist Web sites. He suggested adding qualitative measures, such as persuasive appeals, rhetoric, and attribution of guilt, to the Web site attributes shown in Table 5.2. We believe that these important attributes are difficult to incorporate into the automated processing of our methodology because of their qualitative nature. He considered the clustering and visualization shown in Fig. 5.2 to be very important because of its usefulness to investigation of terrorist activities on the Web. He called the snowflake visualization very accurate and very useful for investigation of terrorist Web sites but criticized the way we created linkages among Web sites. He suggested considering textual citations and other references in addition to using only hyperlinks.

Overall, the expert agreed that the results were very promising because they offer useful investigation leads and would be very helpful to improve understanding of terrorist activities on the Web. Because of the high qualification and relevant experience of this expert, we believe that the evaluation results accurately reflect the effectiveness of the methodology. These results also contributed to advancing the ISI discipline by showing the applicability of the methodology to Dark Web data collection and analysis.

6 Conclusions and Future Directions

Collecting and analyzing Dark Web information has challenged investigators and researchers because terrorists can easily hide their identities and remove traces of their activities on the Web. The abundance of Web information has made it difficult to obtain a comprehensive picture of terrorists' activities. In this chapter, we proposed a methodology to address these problems. Using advanced Web mining, content analysis, visualization techniques, and human domain knowledge, the methodology exploited various information sources to identify and analyze 39 jihad Web sites. Information visualization was used to help identify terrorist clusters and to understand terrorist use of the Web. Our expert evaluation showed that the methodology yielded promising results that would be very useful to assist investigation of terrorism. The expert considered the visualization results very useful, having potential to guide policy making and intelligence research. Therefore, this research has contributed to developing a useful methodology for collecting and analyzing Dark Web information, applying the methodology to study and analyze 39 jihad Web sites, and providing formal evaluation results of the usability of the methodology.

We are pursuing a number of directions to further our research. As terrorists often change their Web sites to remove traces of their activities, we plan to archive the Dark Web content digitally and to apply our methodology to tracing terrorist activities over time. We will develop scalable techniques to collect such volatile yet valuable content, to visualize large volumes of Dark Web data, and to extract meaningful entities from terrorist Web sites. These efforts will help investigators trace and prevent terrorist attacks.

Acknowledgments This research was partly supported by funding from the US Government Department of Homeland Security and Corporation for National Research Initiatives and by the University of Texas at El Paso. We thank contributing members of the University of Arizona Artificial Intelligence Lab for their support and assistance.

References

Anti-Defamation League. (2002). *Jihad Online: Islamic Terrorists and the Internet*, from http://www.adl.org/internet/jihad_online.pdf.

Blakemore, B. (2004, November 23). Web posting may provide insight into Iraq insurgency. *ABC News*.

Carley, K. M., Lee, J.-S., and Krackhardt, D. (2001). Destabilizing Networks. *Connections*.

Chambers, J., Cleveland, W., Kleiner, B., and Tukey, P. (1983). *Graphical Methods for Data Analysis*: Wadsworth.

Chen, H. (2005). Introduction to the special topic issue: Intelligence and security informatics. *Journal of the American Society for Information Science and Technology, 56*(3), 217–220.

Chen, H. and Chau, M. (2004). Web mining: machine learning for web applications. In M. E. Williams (Ed.), *Annual Review of Information Science and Technology (ARIST),* (Vol. 38, pp. 289–329). Medford, NJ: Information Today, Inc.

Chen, H., Fan, H., Chau, M., and Zeng, D. (2001). MetaSpider: meta-searching and categorization on the web. *Journal of the American Society for Information Science and Technology, 52*(13), 1134–1147.

Chen, H., Schuffels, C., and Orwig, R. (1996). Internet categorization and search: a self-organizing approach. *Journal of Visual Communication and Image Representation, 7*(1), 88–102.

Chung, W. (2007). Visualizing E-Business stakeholders on the web: a methodology and experimental results. *International Journal of Electronic Business, (forthcoming).*

Chung, W., Chen, H., Chaboya, L. G., O'Toole, C., and Atabakhsh, H. (2005). Evaluating event visualization: a usability study of COPLINK Spatio-Temporal Visualizer. *International Journal of Human-Computer Studies, 62*(1), 127–157.

Chung, W., Chen, H., and Nunamaker, J. F. (2005). A visual framework for knowledge discovery on the web: an empirical study on business intelligence exploration. *Journal of Management Information Systems, 21*(4), 57–84.

Chung, W., Zhang, Y., Huang, Z., Wang, G., Ong, T.-H., and Chen, H. (2004). Internet searching and browsing in a multilingual world: an experiment on the Chinese Business Intelligence Portal (CBizPort). *Journal of the American Society for Information Science and Technology, 55*(9), 818–831.

Department of State. (2003). *Patterns of Global Terrorism 2002*: The United States Government.
Encyclopedia Britannica Online. (2007). Jihad. http://www.britannica.com/ebc/article-9368558: Britannica Concise Encyclopedia.
Eom, S. B. and Farris, R. S. (1996). The contributions of organizational science to the development of decision support systems research subspecialties. *Journal of the American Society for Information Science, 47*(12), 941–952.
Etzioni, O. (1996). The World-Wide Web: quagmire or gold mine? *Communications of the ACM, 39*(11), 65–68.
Gellman, B. (2002, June 27). Cyber-attacks by Al Qaeda feared. *Washington Post.*
He, Y., and Hui, S. C. (2002). Mining a web citation database for author co-citation analysis. *Information Processing and Management, 38*(4), 491–508.
Kealy, W. A. (2001). Knowledge maps and their use in computer-based collaborative learning. *Journal of Educational Computing Research, 25*(4), 325–349.
Kelley, J. (2002). Militants Wire Web With Links to Jihad. *USA Today*.
Krebs, V. E. (2001). Mapping network of terrorist cells. *Connections, 24*(3), 43–52.
La Porte, T. M., Jong, M. d., and Demchak, C. C. (1999). *Public Organizations on the World Wide Web: Empirical Correlates of Administrative Openness.* Paper presented at the Proceedings of the 5th National Public Management Research Conference, College Station, TX.
Last, M., Markov, A., and Kandel, A. (2006). *Multi-Lingual Detection of Terrorist Content on the Web.* Paper presented at the Proceedings of the PAKDD'06 International Workshop on Intelligence and Security Informatics, Singapore.
Marshall, B., McDonald, D., Chen, H., and Chung, W. (2004). EBizPort: collecting and analyzing business intelligence information. *Journal of the American Society for Information and Science and Technology, 55*(10), 873–891.
Middle East Media Research Institute. (2004). *Jihad and Terrorism Studies Project.* Retrieved March 2004, from http://www.memri.org/jihad.html.
Mladenic, D. (1998). *Turning Yahoo into an Automatic Web Page Classifier.* Paper presented at the Proceedings of the 13th European Conference on Artificial Intelligence, Brighton, UK.
Nasukawa, T. and Nagano, T. (2001). Text analysis and knowledge mining system. *IBM Systems Journal, 40*(4), 967–984.
Newman, M. (2004, May 11). Video appears to show beheading of American civilian. *The New York Times.*
Popp, R., Armour, T., Senator, T., and Numrych, K. (2004). Countering terrorism through information technology. *Communications of the ACM, 47*(3), 36–43.
Project for the Research of Islamist Movements. (2004). *PRISM*, 2004, from http://www.e-prism.org.
Sageman, M. (2004). *Understanding Terror Networks.* Philadelphia: University of Pennsylvania Press.
Strickland, L. S. and Hunt, L. E. (2005). Technology, security, and individual privacy: new tools, new threats, and new public perceptions. *Journal of the American Society for Information Science and Technology, 56*(3), 221–234.
Technical Analysis Group. (2004). *Examining the cyber capabilities of Islamic terrorist groups.* Hanover, NH: Institute for Security Technology Studies at Dartmouth College.
Thomas, T. L. (2003, Spring). Al Qaeda and the Internet: the danger of cyberplanning. *Parameters,* 112–123.
Trybula, W. J. (1999). Text mining. In M. E. Williams (Ed.), *Annual Review of Information Science and Technology* (Vol. 34, pp. 385–419). Medford, NJ: Information Today, Inc.
Tsfati, Y. and Weimann, G. (2002). www.terrorism.com: terror on the Internet. *Studies in Conflict and Terrorism, 25*, 317–332.
Xu, J. and Chen, H. (2005). Criminal network analysis and visualization. *Communications of the ACM, 48*(6), 100–107.
Young, F. W. (1987). *Multidimensional Scaling: History, Theory, and Applications.* Hillsdale, NJ, USA: Lawrence Erlbaum Associates, Publishers.
Zhu, B. and Chen, H. (2005). Chapter 4: Information Visualization. In B. Cronin (Ed.), *Annual Review of Information Science and Technology* (Vol. 39, pp. 139–177).

Chapter 6
Dark Network Analysis

1 Introduction

In recent years, scientists have revealed the topological properties of a wide variety of complex systems characterized as large-scale networks, such as scientific collaboration networks, the World Wide Web, the Internet, electric power grids, and biological networks, among many others. Despite the enormous variation in their components, functions, and sizes, these networks are surprisingly similar in topology, leading to the conjecture that complex systems are governed by the ubiquitous self-organizing principle.

One missing piece in this picture, however, is the analysis on the topology of "dark" networks that are hidden from view yet could have a devastating impact on our society and economy. Terrorist networks, drug-trafficking rings, arms smuggling networks, gang networks, and many other covert networks are all dark networks. The structure of dark networks is largely unknown due to the difficulty of collecting and accessing reliable data. Do dark networks share the same topological properties with other types of empirical networks? Do they follow the same organizing principle? How do they achieve efficiency under constant surveillance and threats from authorities? How robust are they against attacks? In this chapter, we report the topological properties of several covert criminal- or terrorist-related networks. We hope not only to contribute to the general understanding of structural properties of complex systems in a hostile environment but also to provide authorities with insights regarding disruptive strategies.

2 Topological Analysis of Networks

Topological analysis, which focuses on the statistical characteristics of network structure, is a new methodology for studying large-scale networks (Albert and Barabási 2002; Watts and Strogatz 1998). Large complex networks can be categorized into

H. Chen, *Dark Web: Exploring and Data Mining the Dark Side of the Web*,
Integrated Series in Information Systems 30, DOI 10.1007/978-1-4614-1557-2_6,

Table 6.1 The statistics for studying network topology

Statistics	Description
Average path length, l	The average of the lengths of the shortest paths between all pairs of nodes in a network
Average clustering coefficient, C	The average of all individual clustering coefficients, C_i, which is the number of links that actually exist among node i's neighbors over the possible number of links among these neighbors
Average degree, $<k>$	The average of all individual degrees, k_i, which is the number of links node i has
Degree distribution, $p(k)$	The probability that an arbitrary node has exactly k links
Link density, d	The number of links that actually exist over the possible number of links in a network
Assortativity, r	The Pearson correlation between the degrees of two adjacent nodes
Global efficiency, e	The average of the inverses of the lengths of the shortest paths over all pairs of nodes in a network

three types: random, small-world, and scale-free. A number of statistics have been developed to study the topology of networks. Table 6.1 lists several of these statistics, among which *average path length*, *average clustering coefficient*, and *degree distribution* have been widely used to categorize networks into different types.

In random networks, two arbitrary nodes are connected with a probability p, and as a result, each node has roughly the same number of links. Random networks are characterized by small l, small C, and bell-shaped Poisson distributions (Albert and Barabási 2002). A small l means that an arbitrary node can reach any other node in a few steps. A small C implies that random networks are not likely to contain clusters and groups. Studies have found that most complex systems are not random but present small-world and scale-free properties.

The small-world and scale-free models are different from the random graph model. A small-world network has a significantly larger C than its random network counterpart while maintaining a relatively small l (Watts and Strogatz 1998). Scale-free networks, on the other hand, are characterized by the power-law degree distribution, meaning that while a large percentage of nodes in the network have just a few links, a small percentage of the nodes have a large number of links (Albert and Barabási 2002). It is believed that scale-free networks evolve following the self-organizing principle, where growth and preferential attachment play a key role in the emergence of the power-law distribution. Especially, preferential attachment implies that the more links a node has, the more new links it can attract, manifesting the "rich-get-richer" phenomenon.

The analysis on the topology of complex systems has important implications for our understanding of nature and society. Research has shown that the function of a complex system may be affected to a great extent by its network topology. For instance, the small average path length of the World Wide Web makes cyberspace a very convenient, strongly navigable system, in which any two web pages are on average only 19 clicks away from each other. It has also been shown that the higher tendency for clustering in metabolic networks corresponds to the organization of functional modules in cells, which contributes to the behavior and survival of organisms. In addition, networks with scale-free properties are highly robust against random failures and errors but quite vulnerable under targeted attacks (Holme et al. 2002).

3 Methods and Data

To understand the topology and function of dark networks, we studied four terrorist- and criminal-related networks:

1. The global Salafi jihad (GSJ) terrorist network (Sageman 2004) (see Fig. 6.1), which consists of 366 members including those from Osama bin-Laden's al-Qaeda. These terrorists were connected by kinship, friendship, religious ties, and relations formed after they joined the GSJ network.

   The terrorists belong to one of four groups: al-Qaeda or Central Staff (pink), Core Arabs (yellow), Maghreb Arabs (blue), and Southeast Asians (green). Each circle represents one or more terrorist activities, such as the 9/11 attacks and Bali bombing, which are noted.

   The GSJ data were provided by the author of a recently published book, "Understanding Terror Networks" (Sageman 2004). The network was constructed based entirely on open source data including the documents and transcripts of court proceedings, press and scholarly articles, and web articles of the

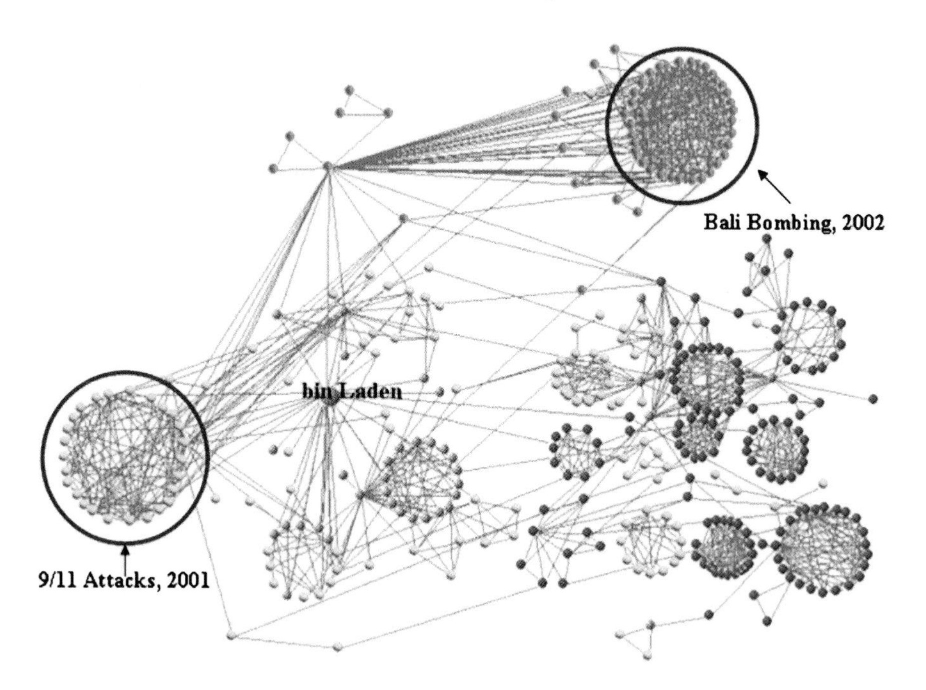

Fig. 6.1 The giant component in the GSJ network (Data courtesy of Marc Sageman)

past few decades. Information about all the nodes (terrorists) and links (relations) was scrutinized and carefully cross-validated. However, as the author pointed out in the book, the data are subject to several limitations. First, the members in the network may not be a representative sample of the global Salafi jihad as a whole. It is biased toward leaders and the members who have been captured or uncovered in executed attacks. Second, because most of the sources were based on retrospective accounts, the data may be subject to self-reported biases. Despite the limitations, the data have revealed stunning insights into the clandestine organizations of terrorists. Interested readers may refer to the book for more details about these terrorist organizations.

2. A narcotics-trafficking criminal network ("Meth World") whose members mainly dealt with methamphetamines (Xu and Chen 2003). The network consists of 1,349 criminals, who have been traced and examined by the Tucson Police Department since 1985. Because no information about the social relationships among these members was directly available, we retrieved from the police databases all the crime incidents in which these criminals were involved from 1985 to 2002. A link was created between two criminals if they committed at least one crime together.

   Although this network had been carefully validated by the crime analysts from the police department (Xu and Chen 2003), the co-occurrence links generated based on crime incidents may not reflect the real relationships between criminals. Two related criminals will appear to be unconnected if they never commit a crime together. On the other hand, a coincidental link may connect two criminals together if they happen to appear in the same incident. These two problems are also common to other types of networks such as the movie actor networks that are constructed based on the co-occurrence of two nodes in the same events or activities.
3. A gang criminal network consisting of 3,917 criminals who were involved in gang-related crimes in Tucson between 1985 and 2002 (Xu and Chen 2003). As in the Meth World, the links in this network were generated using co-occurrence analysis of the crime incidents.
4. A terrorist web site network ("Dark Web") collected based on reliable governmental sources. We identified 104 web sites created by four major international terrorist groups, namely, Al-Gama'a al-Islamiyya, Hezbollah, Al-Jihad, and Palestinian Islamic Jihad and their supporters. All pages from these web sites were fetched, and the hyperlinks were extracted. We created a link between two web sites if at least one hyperlink existed between any two web pages in them.

4 Results and Discussion

4.1 Basic Properties

Table 6.2 presents the basic statistics of the four elicited networks under study. Like many other empirical networks, each network contains many small components and a single giant component. The giant component in a graph is defined as the

Table 6.2 The basic statistics and scale-free properties of the dark networks

	GSJ	Meth World	Gang network	Dark Web
Number of nodes, n	366	1,349	3,917	104
Number of links, m	1,247	4,784	9,051	156
Size of giant component	356 (97.3%)	924 (68.5%)	2,231 (57.0%)	80 (77.9%)
Average degree, $<k>$	6.97	4.62	5.74	3.88
Maximum degree	44 (12.4%)	37 (4.0%)	51 (2.3%)	33 (41.3%)
Link density, d	0.02	0.01	0.003	0.05
Assortativity, r	0.41**	−0.14**	0.17**	−0.24*
Power-law distribution exponent, γ	1.38	1.86	1.95	1.10
Goodness of fit, R^2	0.74	0.89	0.81	0.82

The numbers in parentheses in the third row are the percentages of total nodes included in the giant components. The numbers in parentheses in the fifth row are the percentages of the total nodes that are connected with the highest-degree nodes

**p-value < 0.05

*p-value < 0.01

largest connected subgraph. The separation between the 356 terrorists in the GSJ network and the remaining ten terrorists is because no valid evidence has been found to connect the ten terrorists to the giant component of the network. The giant components in the Meth World and gang networks contain only 68.5% and 57.0% of the nodes, respectively. This may be because the data were collected from a single law enforcement jurisdiction which may not have complete information about all relations between criminals, causing missing links between the giant component and other smaller components. The isolated components in the Dark Web are possibly due to the differences in the terrorist groups' ideologies.

Similar to many other network topology studies (e.g., Barabási et al. 2002), we performed topological analysis only on the giant component in these networks. In Table 6.2, we report the average degrees and maximum degrees of the four networks. It can be seen that some terrorists in the GSJ network and some terrorist web sites in the Dark Web are extremely popular, connecting to more than 10% of the nodes in the networks.

The assortativity indicates the tendency for nodes to connect with others who are similarly popular in terms of degree. The assortativity coefficients of these four networks are all significantly different from 0. The GSJ and the gang networks present positive assortativity, meaning that popular members tend to connect with other popular members. In positively assortative networks, high-degree nodes tend to cluster together as core groups (Newman 2003). This phenomenon is especially evident in the GSJ network in which bin-Laden and his sergeants form the core of the network and issue commands to other parts of the network (Sageman 2004). The Meth World and Dark Web, in contrast, have negative assortativity coefficients – disassortativity. The Meth World consists of drug dealers who sold drugs to many individual buyers; the buyers did not connect with many other buyers or dealers. Further, studies have found that street drug-dealing organizations are led by a few high-level individuals, who connect with a large number of low-level street drug dealers (Levitt and Dubner 2005). Because high-degree nodes connect with low-degree nodes, the

Meth World presents disassortative mixing patterns. The disassortativity in the Dark Web, on the other hand, is due to the fact that the popular Dark Web sites received many inbound hyperlinks from less popular web sites.

4.2 Small-World Properties

To ascertain if the dark networks are small worlds, we calculated their average path lengths, clustering coefficients, and global efficiency (see Table 6.3). For each network, we generated 30 random counterparts that had the same number of nodes and the same number of links as in the corresponding elicited networks. We found that all these networks have significantly high clustering coefficients compared with their random counterparts. In addition, although the differences are statistically significant (greater than three standard deviations), the average path lengths of these networks (except for the gang network) are just slightly greater than their random counterparts.

These small-world properties imply that a terrorist or criminal can connect with any other member in a network through only a few mediators. In addition, these networks are quite sparse with very low link density. These properties have important implications for the communication efficiency of the covert networks. Because the risk of being detected by authorities increases as more people are involved, the small path length and link sparseness can help lower risks and enhance efficiency. As a result, the global efficiency of each network is compatible with their random network counterpart.

On the other hand, a high clustering coefficient contributes to the local efficiency of these dark networks. Previous studies have also shown evidence of groups and teams in these networks (Sageman 2004; Xu and Chen 2003). In these groups and teams, members tend to have denser and stronger relations with one another. The communication between group members becomes more efficient, therefore making a crime or an attack easier to plan, organize, and execute.

We also calculated the path length of other nodes to central nodes. We found that members in the criminal and terrorist networks are extremely close to their leaders. The terrorists in the GSJ network are on average only 2.5 steps away from bin-Laden, meaning that bin-Laden's command can reach an arbitrary member through only two mediators. Similarly, the average path length to the leader in the Meth World (Xu and Chen 2003) is only 3.9. Such a short chain of command also means communication efficiency.

Special attention should be paid to the Dark Web. Despite the small size of its giant component (80), the average path length is 4.70, slightly larger than that (4.20) of the GSJ network, which has almost nine times more nodes. Since hyperlinks help visitors navigate between web pages, and because terrorist web sites are often used for soliciting new members and donations, the relatively big path length may be due to the reluctance of terrorist groups to share potential resources with other terrorist groups.

Table 6.3 The small-world properties of the dark networks

	GSJ		Meth World		Gang network		Dark Web	
	Data	Random	Data	Random	Data	Random	Data	Random
Average path length, l	4.20	3.23 (0.040)	6.49	4.52 (0.056)	9.56	4.59 (0.034)	4.70	3.15 (0.108)
Average clustering coefficient, C	0.55	0.020 (0.0029)	0.60	0.005 (0.0014)	0.68	0.002 (0.0005)	0.47	0.049 (0.0155)
Global efficiency, e	0.28	0.33 (0.004)	0.18	0.23 (0.003)	0.12	0.23 (0.001)	0.30	0.34 (0.019)

For each network, we present the metrics in the elicited network (Data) and those in the random graph counterpart (Random). Numbers in parentheses are standard deviations

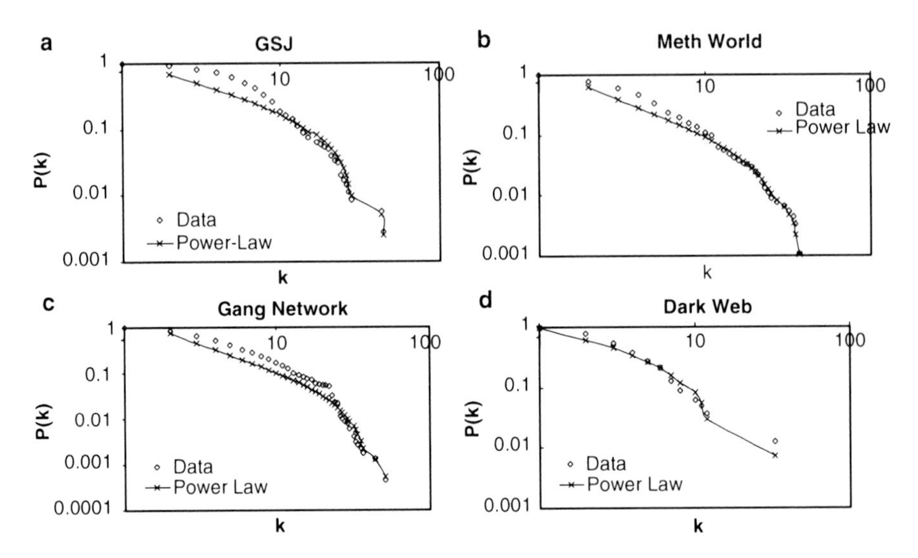

Fig 6.2 The cumulative degree distributions of (**a**) the GSJ network, (**b**) the Meth World, (**c**) the gang network, and (**d**) the Dark Web

4.3 Scale-Free Properties

Moreover, these dark networks present scale-free properties with power-law degree distributions, $p(k) \sim k^{-\gamma}$. Because degree distribution curves often fluctuate a lot, we display the cumulative degree distributions, $P(k)$, in a log-log plot (see Fig. 6.2). $P(k)$ is defined as the probability that an arbitrary node has at least k links. Figure 6.2 also presents the fitted power-law distributions. The last two rows of Table 6.2 report the exponent value, γ, and the goodness of fit, R^2, for each network. It is evident in Fig. 6.2 that all these networks are scale-free networks. The power-law distributions fit especially well at the tails. Note that each of the three human networks displays a two-regime scaling behavior, which has also been observed in some other empirical networks such as scientific collaboration networks.

Two mechanisms have been proposed to account for the emergence of two-regime power-law degree distributions during the evolution of a network (Barabási et al. 2002). First, new links may emerge between existing network members. This implies that criminals or terrorists who were not related before could become connected as time progressed. This is a rather realistic assumption since two previously unacquainted members could become acquaintances through the introduction of a third member who knows both of them. In the GSJ network, 22.6% of the links were postjoining ties which were formed between existing members. Second, an existing link may be rewired, which is a strong possibility in the GSJ and Dark Web. However, this would not affect the Meth World and the gang network, because a co-occurrence link could not be rewired once it was created.

An interesting question follows: what mechanisms have played a role in producing the observed properties of the dark networks, namely, small average path length, high clustering coefficient, and power-law degree distributions with two-regime scaling behavior in the human networks? In other words, can we regenerate the dark networks based on known mechanisms such as growth and preferential attachment? To answer this question, we conducted a series of simulations in which 30 networks were generated for each elicited human network based on three mechanisms:

(a) *Growth*. Starting with a small number of nodes, at each time step, we add a new node to connect with existing nodes in the network.
(b) *Preferential attachment*. The probability that an existing node will receive a link from the new node depends on the number of links the node has already maintained. The more links it has the more likely that it will receive the link.
(c) *New links between existing nodes*. At each time step, a random pair of existing nodes may get connected depending on the number of common neighbors they have. The more common neighbors they share the more likely it is that they will be connected.

Mechanisms (a) and (b) were expected to generate the power-law degree distribution, and mechanism (c) was expected to generate the high clustering coefficient and two-regime scaling behavior. Through the simulations, we found that the power-law degree distributions could be easily regenerated with R^2 ranging from 0.83 (the gang network) to 0.88 (GSJ). The two-regime scaling behavior was also present in the simulated networks for the human networks. However, the highest clustering coefficient from simulation was only 0.24 (GSJ), which was far less than what was obtained from the elicited networks (0.55–0.68). This finding implies that there must have been some other mechanisms that contributed to the substantially high clustering coefficients observed in the dark networks. We suspect that such a mechanism is member recruitment. Because of active recruitment, subgroups of terrorists or criminals could attract new members into their groups. The new members quickly become acquainted with many existing members of the groups, substantially increasing the clustering coefficients.

4.4 Caveats

A point worth paying attention to is that two problems may have affected the three elicited human networks. First, the elicited networks may have missing links that can cause the networks to appear to be less efficient, because there may actually be hidden "shortcuts" connecting distant parts of the networks. Second, the presence of coincidental "fake" links can cause the elicited networks to be more efficient than in actuality since these links are actually not communication channels.

To test how the results would be affected by missing links, we added various percentages of the existing links to the elicited networks based on three effects that had been used in missing link prediction research (Liben-Nowell and Kleinberg 2003):

(a) *Random effect.* A link is added between a randomly selected pair of nodes which are not originally connected.
(b) *Common neighbor effect.* A link is added between a pair of unconnected nodes if they share common neighbors. The more common neighbors they share, the more likely it is that they will be connected.
(c) *Preferential attachment effect.* The probability that a pair of unconnected nodes will be linked together depends on the product of their degrees.

We found that the small-world and scale-free properties of these networks do not change when missing links are added to the networks. For example, when as high as 10% of the links were added, the average path lengths ranged from 3.55 (GSJ, preferential attachment links added) to 9.45 (the gang network, common neighbor links added), the clustering coefficients ranged from 0.45 (GSJ, random links added) to 0.67 (the gang network, common neighbor links added), and the R^2 of power-law degree distributions ranged from 0.61 (GSJ, random links added) to 0.93 (the gang network, preferential attachment links added).

We also randomly removed certain percentages of links to test the impact of "fake" links on the results. We found that the results were still valid even when as high as 10% of the links were removed.

4.5 Network Robustness

Research has found that network topology has a great impact on the network's robustness against failures and attacks and that scale-free networks are quite robust against failures (random removal of nodes) (Holme et al. 2002). Because dark networks have been shown to have scale-free properties, we tested the dark networks' robustness against only targeted attacks.

We simulated two types of attacks represented by node removal: attacks targeting the hubs and attacks targeting the bridges. While hubs are nodes that have many links (high degree), bridges are nodes through which many shortest paths pass (high betweenness (Wasserman and Faust 1994)). When simulating the attacks, we distinguished between two attack strategies: simultaneous removal of a fraction of nodes based on a measure (degree or betweenness) without updating the measure after each removal and progressive removal of nodes with the measure being updated after each removal.

We plotted the changes in S (the fraction of the nodes in the largest component), $<s>$(the average size of remaining components), and average path length after a fraction of nodes are removed. We found that progressive attacks are more devastating than simultaneous attacks. The progressive attacks are similar to "cascading failures" in the Internet where an initial failure can cause a series of failures because unbearably high traffic is redirected to the next bridge node.

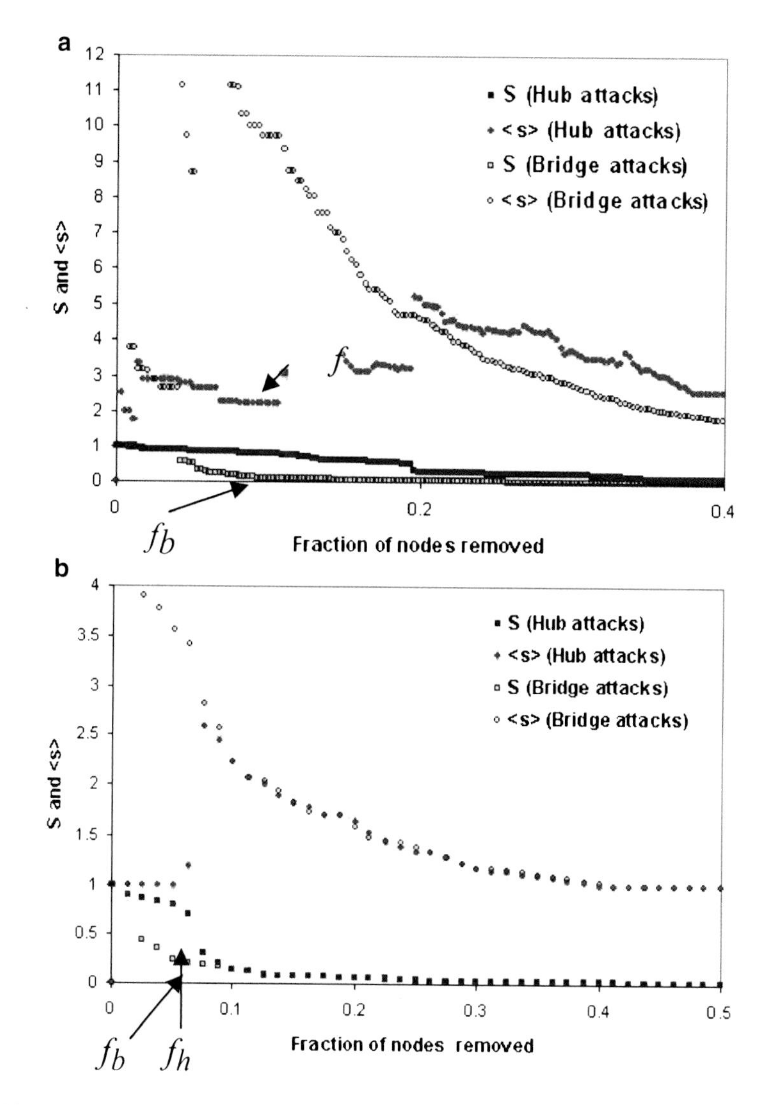

Fig 6.3 Dark networks' robustness against attacks. (**a**) Progressive attacks to the GSJ network. (**b**) Progressive attacks to the Dark Web. Two types of attacks are used: hub attack (*filled markers*) and bridge attack (*empty markers*)

Figure 6.3a–b presents the difference between the network reactions to bridge attacks and hub attacks. The critical points, f, at which the network falls into many small components, are marked on the diagram. The behaviors of the Meth World and the gang network are very similar to that of the GSJ network. It shows that these terrorist and criminal networks are more sensitive to attacks targeting the bridges than those targeting the hubs (fb<fh). In Fig. 6.3b, however, fb and fh are very close; indicating that hub attacks and bridge attacks can be equally effective to disrupt a one-regime scale-free network.

The results are quite consistent with findings from a prior study (Holme et al. 2002) that pure scale-free networks are vulnerable to both hub and bridge attacks, while small-world networks are more vulnerable to bridge attacks. In a small-world network, which consists of communities and groups, there might be many bridges linking different communities together. Intuitively, when these bridges are removed, the network will quickly fall apart. Note that a bridge may not necessarily be a hub since a node that connects two communities can have as few as two links. Small-world networks such as the dark networks thus are more vulnerable to bridge attacks than hub attacks.

In these dark networks, bridges and hubs usually are not the same nodes. The rank order correlations between degree and betweenness in the GSJ, Meth World, and the gang network are 0.63, 0.47, and 0.30, respectively. Note that although bridge attacks are more devastating, strategies targeting the hubs are also fairly effective since these networks also have scale-free properties. Hub attacks and bridge attacks can be equally effective in tearing apart a pure scale-free network (e.g., the Dark Web with a high degree-betweenness rank order correlation, 0.70), in which hubs are also bridges connecting different parts of the network.

5 Conclusions

Dark networks such as terrorist networks and narcotics-trafficking networks are hidden from our view yet could have a devastating impact on our society and economy. Understanding the topology of these dark networks can reveal greater insight into these clandestine organizations and help develop effective disruptive strategies. However, reliable data about these dark networks are extremely difficult to obtain, causing our understanding of covert networks to remain mostly hypothetical. To the best of our knowledge, the datasets used in this chapter, although they are subject to several limitations, are the first sets which allow for statistical analysis of the topologies of dark networks.

We found that these covert networks share many common topological properties with other types of networks. Their efficiency in communication and flow of information, commands, and goods can be tied to their small-world structures, characterized by small average path length and high clustering coefficient. In addition, we found that because of the small-world properties, dark networks are more vulnerable to attacks on the bridges that connect different communities than to attacks on the hubs. This may provide authorities with insight for intelligence and security purposes.

An interesting finding about the three human networks is that their substantially high clustering coefficients, which are not always present in other empirical networks, are difficult to regenerate based only on the known network effects such as preferential attachment and small-world effects. Other mechanisms such as recruitment may have played an important role in the evolution of these networks. Some research has found that alternative mechanisms such as Highly Optimized Tolerance (HOT) may govern the evolution of many complex systems in environments with high risks and uncertainty (Carlson and Doyle 1999). In our future research, we will study the impacts of such alternative mechanisms on the topology of networks. In addition, the findings presented in this chapter are all based on the static views of the networks. That is, we do not consider a large variety of dynamics that might have taken place in the evolution of these networks. Evolution study is definitely in our plan for future research.

We want to point out that people should be careful when interpreting these findings. Because dark networks are covert networks and the underlying actual networks are largely unknown, there may be hidden links which are missing in the elicited networks. These hidden links may play indispensable roles in maintaining the function of the covert organizations. As a result, we must be extremely cautious when any decision is to be made to disrupt these networks.

References

Albert, R. and Barabási, A.-L. Statistical mechanics of complex networks. *Reviews of Modern Physics, 74* (1). 47–97, 2002.

Barabási, A.-L., Jeong, H., Zéda, Z., Ravasz, E., Schubert, A. and Vicsek, T. Evolution of the social network of scientific collaborations. *Physica A, 311*. 590–614, 2002.

Carlson, J.M. and Doyle, J. Highly optimized tolerance: A mechanism for power laws in designed systems. *Physical Review E, 60* (2). 1412–1427, 1999.

Holme, P., Kim, B.J., Yoon, C.N. and Han, S.K. Attack vulnerability of complex networks. *Physical Review E, 65*. 056109, 2002.

Levitt, S.D. and Dubner, S.J. *Freakonomics: A rogue economist explores the hidden side of everything*. William Morrow, New York, NY, 2005.

Liben-Nowell, D. and Kleinberg, J. The link prediction problem for social networks. in *Proceedings of the 12th International Conference on Information and Knowledge Management*, (New Orleans, LA, USA, 2003).

Newman, M.E.J. Mixing patterns in networks. *Physical Review E, 67* (2). 026126, 2003.

Sageman, M. *Understanding Terror Networks*. University of Pennsylvania Press, Philadelphia, PA, 2004.

Wasserman, S. and Faust, K. *Social Network Analysis: Methods and Applications*. Cambridge University Press, Cambridge, 1994.

Watts, D.J. and Strogatz, S.H. Collective dynamics of "small-world" networks. *Nature. 393*. 440–442, 1998.

Xu, J. and Chen, H. Untangling criminal networks: A case study. in *Proceedings of the 1st NSF/NIJ Symposium on Intelligence and Security Informatics (ISI'03)*, (Tucson, AZ, 2003), 232–248.

Chapter 7
Interactional Coherence Analysis

1 Introduction

Computer-mediated communication (CMC) is any form of communication between two or more individuals who interact and/or influence each other via computer-supported media. Text-based modes of CMC include e-mail, listservs, forums, chat rooms, instant messaging, and the World Wide Web (Herring 2002). There is no doubt that the popularity of CMC is continuing to grow. E-mail, Web forums, newsgroups, and chat rooms have already become essential parts of our daily lives, providing a communication medium for various activities (Meho 2006; Radford 2006). Although the ubiquitous nature of CMC provides a convenient mechanism for communication, it is not without its shortcomings. The fragmented, ungrammatical, and interactionally disjointed nature of CMC discourse, attributable to the limitations of the CMC media, has rendered CMC highly incoherent (Hale 1996).

Beaugrande and Dressler (1996) defined coherence in linguistics as a "continuity of senses" and "the mutual access and relevance within a configuration of concepts and relations." For Web discourse, coherence defines the macrolevel semantic structure (Barzilay and Elhadad 1997). Barzilay and Elhadad (1997) further pointed out that "coherence is represented in terms of coherence relations between text segments, such as elaboration, cause and explanation." Coherence of online discourse, correspondingly, is represented in terms of the "reply-to" relations between CMC messages. The "reply-to" relationships can serve several functions, such as elaborating or complementing previous postings, greeting fellow users, answering questions, or oppugning previous messages.

Computer-mediated interaction (CMI) refers to the social interaction between CMC users (Walther et al. 1994). Such social interaction is built through the "reply-to" relationships between messages. Therefore, we also refer to the "reply-to" relationship as the interaction relationship between messages. A social interaction in online discourse happens if a user posts a message that has a "reply-to" relation with other users' messages. Occasionally, a user may interact with other users without specifying the messages he or she responds to. Common greeting messages like

H. Chen, *Dark Web: Exploring and Data Mining the Dark Side of the Web*,
Integrated Series in Information Systems 30, DOI 10.1007/978-1-4614-1557-2_7,

"Hi Jatt" are examples. But we can build fake "reply-to" relationships between such messages with the addressed user's nearest message. This method does not affect the social interaction relationships between the users.

Since the "reply-to" relations between CMC messages can be used to build the social interaction between users, coherence of CMC is also called CMC interactional coherence in previous studies (e.g., Herring 1999). However, current CMC media suffer the "disrupted turn adjacency" problem, and the existing system functionalities do not contain sufficient "reply-to" information. In light of the incoherent and fragmented nature of text-based Web discourse, many researchers have pointed out the importance of automatically identifying CMC interactional coherence. Te'eni (2001) claimed that interactional coherence information is particularly important "when there are several participants" and "when there are several streams of conversation and each stream must be associated with its particular feedback." Users of CMC systems cannot safely assume that they will receive a response to their previous message because of the lack of interactional coherence (Herring 1999). Accurate interaction information is also important to researchers for a plethora of reasons. User interaction in text-based CMC represents one of the fundamental building block metrics for analyzing cyber communities. Interaction-related attributes help identify CMC user roles and user's social and informational value, as well as the social network structure of online communities (Smith and Fiore 2001; Fiore et al. 2002; Barcellini et al. 2005). Moreover, interactional coherence is invaluable for understanding knowledge flow in electronic communities and networks of practice (Osterlund and Carlile 2005; Wasko and Faraj 2005).

Interactional coherence analysis (ICA) attempts to accurately identify the "reply-to" relationships between CMC messages so that we can reconstruct CMC interactional coherence and present the social interaction between CMC users. Previously used ICA features include system-generated attributes such as quotations and message headers, as well as linguistic features such as repetition of keywords across postings (Sack 2001; Spiegel 2001; Yee 2002). Although considerable efforts have been devoted to improving interaction representations using ICA, previous studies suffer from several limitations. Most used a couple of specific features, whereas effective capture of interaction cues entails the use of a larger set of system and linguistic attributes (Nash 2005). Furthermore, the techniques incorporated often ignored noise issues such as typos, misspellings, nicknames, etc., which are prevalent in CMC (Nasukawa and Nagano 2001). In addition, there has been little emphasis on Web forums, a major form of asynchronous online discourse. Previous work has focused on e-mail-based newsgroups and chat rooms. Web forums differ from e-mail and synchronous forms of electronic communication in terms of the types of salient coherence cues, user behavior, and communication dynamics (Hayne et al. 2003).

In this study, we propose the Hybrid Interactional Coherence (HIC) algorithm for Web forum interactional coherence analysis. HIC attempts to address the limitations of previous studies by utilizing a holistic feature set which is composed of both linguistic coherence attributes and CMC system features. The remainder of this chapter is organized as follows: Sect. 2 presents a review of previous ICA research. Section 3 highlights important research gaps and questions. Section 4 presents a system design geared toward addressing the research questions, including the use of the HIC algorithm with an extended set of system and linguistic features. It also provides details of the various components of our HIC algorithm. Experimental results based on evaluations of the HIC algorithm in comparison with previous techniques are described in Sect. 5. Section 6 includes conclusions and future directions.

2 Related Work

CMC interactional coherence is crucial for both researchers and CMC users. Interaction information can be used to identify user roles and messages' values, as well as the social network pertaining to an online discussion. Example applications that can benefit from accurate online discourse interaction information include analyzing the effectiveness of e-mail-based interviewing (Meho 2006) and chat-based virtual reference services (Radford 2006). Interactional coherence analysis provides users and researchers a better understanding of specific online discourse patterns. Unfortunately, deriving interaction information from online discourse can be problematic, as discussed below.

2.1 Obstacles to CMC Interactional Coherence

Two properties of the CMC medium are often cited as obstacles to CMC interactional coherence (Herring 1999): lack of simultaneous feedback and disrupted turn adjacency. Most CMC media are text-based so they lack audio or visual cues prevalent in other communication mediums. Furthermore, text-based messages are sent in their entirety without any overlap. These two characteristics result in a lack of simultaneous feedback. However, advanced CMC media have already provided simple solutions to address this concern. For example, newer versions of instant messaging software include audio and video capabilities in addition to the standard text functionality. These tools also show whether a user is typing a response, thereby providing response cues allowing interaction in a manner more similar to face-to-face communication. Since those solutions perform quite well, lack of simultaneous feedback is no longer a severe problem for CMC interactional coherence.

In contrast, resolving the disrupted turn adjacency problem remains an arduous yet vital endeavor. Disrupted turn adjacency refers to the fact that messages in CMC are often not adjacent to the postings they are responding to. Disrupted adjacency stems from the fact that CMC is "turn-based." As a result, the conversational structure is fragmented, that is, a message may be separated both in time and place from the message it responds to (Herring 1999). Both synchronous (e.g., chat rooms, instant messaging) and asynchronous (e.g., e-mail, forums) forms of CMC suffer from disrupted turn adjacency. Several previous studies have observed and analyzed this phenomenon. Herring and Nix (1997) found that nearly half (47%) of all turns

User	Feature	Message
Ashna	Direct Address	Hi jatt
Dave-G	Direct Address	Kally I was only joking around
Jatt	Direct Address	Ashna: hello?
Kally	Substitution	I don't think so.
Ashna	Direct Address & Co-reference	How are u jatt
LUCKMAN	N/A	SSa all
Dave-G	Co-reference & Conjunction	Therefore we need to talk
Jatt	Lexical relation & Co-reference	Do we know each other? I'm ok how are you

Fig. 7.1 Example of disrupted adjacency

were "off topic" in relation to the previous turn. Recently, Nash (2005) manually analyzed data from an online chat room and found that the gap between a message and its response can be as many as 100 turns.

Figure 7.1 shows an example of disrupted adjacency taken from Paolillo (2006). The disruption is obvious in the example and is attributable to the fact that two discussions are intertwined in a single thread. The lines to the right-hand side indicate the interaction relations among postings: two different widths are used to differentiate the parallel discussions. There is also one message that is not related to any of the other messages, posted by the user "LUCKMAN." The middle column lists the linguistic features used in these messages, which will be introduced in Sect. 2.2.2.

The objective of ICA is to develop techniques to construct the interaction relations such as those shown in the right-hand side of the example. Such message interaction relations can be further used to construct the social network structure of CMC users, leading to a better understanding of CMC and its users and providing necessary information for improving ICA accuracy. A review of previous interactional coherence analysis research is presented in the following section.

2.2 CMC Interactional Coherence Analysis

Common interactional coherence research characteristics include domains, features, noise issues, and techniques. Table 7.1 presents a taxonomy of these vital CMC interactional coherence analysis characteristics. Table 7.2 shows previous CMC interactional coherence studies based on the proposed taxonomy. Header information and quotations (F1 and F2) are system features, whereas features 3–6 (F3–F6) are linguistic features. A dashed line is used to distinguish these feature categories. The taxonomy and related studies are discussed in detail below.

Table 7.1 A taxonomy of CMC interactional coherence research

Category	Description	Label
Domain		
Synchronous CMC	Internet Relay Chat (IRC), MUD, IM, etc.	D1
SMTP-based asynchronous CMC	E-mail, newsgroups	D2
HTTP-based asynchronous CMC	Web forums/BBS, Web blogs	D3
Text document	News, articles, text files, etc.	D4
Feature		
Header information	"Reply-to" information in header or title	F1
Quotation	Copy previous related message in response	F2
Co-reference	Personal, demonstrative, comparative co-reference	F3
Lexical relation	Repetition, synonymy, superordinate	F4
Direct address	Mention username of respondent	F5
Other linguistic features	Substitution, ellipsis, conjunction	F6
Noise		
Typo, misspellings, nicknames, modified quotations		
Technique		
Manual	Manually identify the interaction	T1
Link-based method	Link messages by using CMC system features only	T2
Similarity-based method	Word match, VSM, SVM, lexical chain	T3

Table 7.2 Selected previous CMC interactional coherence studies

		Features							
Previous studies	Domains	F1	F2	F3	F4	F5	F6	Noise	Techniques
Xiong et al. (1998)	SMTP-based	✓						No	Link-based
Bagga and Baldwin (1998)	Text			✓				No	Similarity-based
Choi (2000)	Text				✓			No	Similarity-based
Smith and Fiore (2001)	SMTP-based	✓						No	Link-based
Sack (2001)	SMTP-based	✓	✓					No	Link-based
Spiegel (2001)	Synchronous				✓	✓		No	Similarity-based
Soon et al. (2001)	Text			✓				No	Similarity-based
Newman (2002)	SMTP-based	✓						Yes	Link-based
Yee (2002)	SMTP-based	✓	✓					No	Link-based
Barcellini et al. (2005)	SMTP-based		✓					–	Manual
Nash (2005)	Synchronous			✓	✓	✓	✓	–	Manual

2.2.1 CMC Interactional Coherence Domains

CMC interactional coherence research has been conducted on both synchronous and asynchronous CMC since both of these modes show a high degree of disrupted turn adjacency (Herring 1999). Synchronous CMC, which includes all forms of persistent conversation, suffers from multiple, intertwined topics of conversation (Khan et al. 2002). In comparison, asynchronous CMC has a "thread" function, which is an effective method for categorizing forum postings based on a specific topic. However, the "thread" function is not perfect. First, it does not show message-level

Explicit					Implicit
Header Information	Quotation	Direct Address	Lexical Relation	Co-Reference	Conjunction Substitution Ellipsis

Fig. 7.2 Features' relative explicit/implicit properties

interactions, which are vital for constructing the social network structure of CMC users. Instead, it is just an effort to group related messages together. Second, even in a single thread, subtopics might be generated during the discussion. This phenomenon, which poses severe problems for Web forum information retrieval and content analysis, is called "topic decay/drift" (Herring 1999; Smith and Fiore 2001). Therefore, it is still necessary and important to apply interactional coherence analysis to asynchronous CMC.

Asynchronous CMC modes can be classified into two categories: SMTP-based and HTTP-based. SMTP-based modes (e.g., Usenet) use e-mail to post messages to forums, whereas HTTP-based methods use forms embedded in the Web pages. Previous research often focused on SMTP-based modes because the headers of posted messages contain what is referred to as "reply-to information" that specifically mentions the ID of the message being responded to. Loom (Donath et al. 1999), Conversation Map (Sack 2001), and Netscan (Smith and Fiore 2001) are all well-known tools that have been developed to show interaction networks of Usenet newsgroups (SMTP-based). In contrast, HTTP-based modes such as Web forums and blogs do not contain such useful header information for constructing interaction networks. Consequently, there has been little work on HTTP-based CMC as illustrated by Table 7.2.

We also incorporate text documents into our taxonomy because they experience some problems similar to CMC incoherence, such as co-reference resolution (Bagga and Baldwin 1998; Soon et al. 2001) and text segmentation (Choi 2000). Techniques used for text document co-resolution, such as sliding windows (Hearst 1994), lexical chains (Morris 1988), and entity repetition (Khan et al. 1998), are applicable to all forms of text and can provide utility for CMC interactional coherence research.

2.2.2 CMC Interactional Coherence Research Features

Two categories of features have been used by previous CMC researchers and system developers. The first category is system features, which are functionalities provided by the CMC systems. The second one is linguistic features, which are interpersonal language cues.

Nash (2005) defined explicit features as those that "make fewer assumptions about what information is activated for the recipients." Figure 7.2 shows features' relative explicit/implicit properties. Features on the left side are more explicit than those on the right side. Explicit features are generally easier to use for deriving interaction patterns. In contrast, implicit features such as conjunctions and ellipsis are far more difficult to accurately incorporate for interactional coherence analysis. The various features are described in detail below.

CMC System Features

CMC system features are usually only provided by asynchronous CMC systems. Header information and quotations are two kinds of CMC system features that can be used to construct interaction networks of asynchronous online discourse. Lewis and Knowles (1997) pointed out that SMTP-based asynchronous CMC systems will "automatically insert into a reply message two kinds of header information: unique message IDs of parent messages and a subject line of the parent (copied to the reply message's subject line)." Unique message IDs of the parent message are intuitively useful for interaction identification. In contrast, subject lines of messages are less useful because different conversations in the same thread may have similar subject lines. Unfortunately, for HTTP-based modes, only the second type of header information is available. As shown in Table 7.2, most previous studies for SMTP-based asynchronous CMC systems relied on header information (F1 column) to construct interaction networks (e.g., Sack 2001; Barcellini et al. 2005).

Quotations (F2 column), a context-preserving mechanism used in online discussions (Eklundh 1998), are less frequently used to represent online conversations. Conversation Map (Sack 2001) and Zest (Yee 2002) are among the few previous studies that used automatic quotation identification to address disrupted adjacency. Barcellini et al. (2005) manually analyzed quotations and used them to identify participants' conversation roles.

Although header information and quotations are effective for identifying interaction and should result in high precision intuitively, in reality, they suffer several drawbacks. From the systems' point of view, only asynchronous CMC systems contain such features. Moreover, header information provided by HTTP-based asynchronous CMC systems is of little value in many cases where the subject lines of all subsequent messages are similar or even identical. Furthermore, from the users' point of view, some participants do not use system features, and others may not use system functions correctly (Lewis and Knowles 1997; Eklundh and Rodriguez 2004). For instance, interaction cues may appear in the message body. Finally, some messages can interact with multiple previous messages, and system features may not be able to capture such multiple interactions. As a result, using system features alone fails to consider such idiosyncratic user behavior, resulting in an incomplete representation of CMC interaction.

As is shown in Table 7.2, previous research on SMTP-based asynchronous CMC relied mostly on system features to construct the interaction network. CMC systems incorporating system and linguistic features for identification of interaction patterns, such as the Conversation Map system proposed by Sack (2001), are a rarity. The Conversation Map system also constructs interaction networks primarily using system features, but then uses the message content to construct semantic networks, which display the discussion themes for interacting messages (Sack 2001).

The content of messages, which can be represented by various linguistic features, may be useful to complement system features in constructing CMC interactions and in many cases may be even more important (Nash 2005). Therefore, our approach utilizes both CMC system and linguistic features to construct the interaction network with the intention of creating a more accurate representation of

CMC interactional coherence and its social network structure. Important linguistic features are discussed in the following section.

Linguistic Features

Linguistic features are interpersonal language cues and content-based features. Previous research on synchronous CMC systems had to rely on linguistic features to construct interaction networks since no system features were available. Several linguistic features for online communication have been identified by previous research. Three prevalent features are direct address, lexical relations, and co-reference (Halliday and Hasan 1976; Herring 1999; Spiegel 2001; Nash 2005).

Direct address takes place when a user mentions the username of another user whom he or she is addressing in the message. Coterie (Spiegel 2001), a visualization tool for conversation within Internet Relay Chat, looks for direct addresses of specific people to construct the interaction network. It is important to note that addressing someone is different from referencing someone. Take the following sentence as an example: "John, take care of your brother Tom." The speaker is addressing (and interacting) with "John" only, although "Tom" is also referenced.

Lexical relations occur when a lexical item refers to another lexical item by having common meanings or word stems. Its most common forms are repetition and synonymy (Nash 2005). Lexical relations have also been widely used in previous studies of synchronous CMC systems. For example, Choi (2000) used repetition of keywords to identify relationships between messages. Techniques that compare text similarities are often used for identifying lexical relations, where two messages are considered to have an interaction if their similarity is above some predefined threshold (Bagga and Baldwin 1998).

Co-reference also occurs when a lexical item refers to another one; however, such a relationship can only be identified by the context instead of the word meanings or stems. Personal co-reference is most commonly used in CMC. For example, the word "you" is frequently used to refer to the person a message addresses. Other co-references include demonstrative co-reference, which is made on the basis of proximity, and comparative co-reference, which uses words such as "same," "similar," and "different" (Nash 2005).

Some other linguistic features identified by previous studies include: conjunctions (e.g., but, however, therefore), substitution (e.g., "I think so."), ellipsis (e.g., "Guess that would not be easy."), etc. (Nash 2005). These features have rarely been incorporated in previous studies due to the difficulty in identifying such features and their lack of prevalence in online discourse. Figure 7.1 shows an example that includes most linguistic features mentioned here.

Looking back to Table 7.2, we can see that most previous studies only utilized one or two specific features. Only Nash (2005) manually identified multiple linguistic features for an online chat room and found three of them to be dominant. Lexical relations covered 51% of the interaction pattern, whereas direct address and co-reference covered 28% and 15%, respectively.

2.2.3 CMC Interactional Coherence Analysis Techniques

In light of the fact that several types of features can be used for interactional coherence analysis, many different techniques have previously been used to construct interaction patterns. These can be classified into three major categories: manual analysis, link-based techniques, and similarity-based techniques.

Eklundh and Rodriguez (2004) manually identified lexical relations, direct address, and co-reference for one specific online discussion. Similarly, Nash (2005) identified and extracted six linguistic features for an English chat room. Barcellini et al. (2005) manually analyzed quotations and used them to identify participants' conversation roles. Manual analysis of CMC interactional coherence has the obvious advantage of accuracy. However, its disadvantage is also obvious: it is difficult to apply to large datasets and is labor intensive.

Link-based techniques construct interaction patterns using system features or rules based on message sequences. These techniques are highly prevalent in previous research because of their representational simplicity as compared to techniques that focus on linguistic features. Direct linkage techniques link messages based on header information and quotations. For residual messages unidentified by direct linkage, naïve linkage (Commer and Peterson 1986) has been used. Naïve linkage is a rule-based technique which proposes that a message is related to all prior messages in the same discussion or the first message in the same discussion. The advantage of link-based techniques is that they are easy to implement. However, link-based techniques depend on the assumption that CMC users utilize system features correctly. Moreover, naïve linkage is of low accuracy and often overgeneralizes participation patterns due to its simplistic rule-based properties.

Similarity-based techniques typically use content similarity to construct interaction patterns. These techniques focus on uncovering interaction cues found in the message texts to provide insight into interactional coherence. The simplest method is exact match or direct match, which tries to identify repetition of words, word phrases, or even sentences (Choi 2000; Spiegel 2001). More advanced similarity-based techniques include vector space model, which has been used for the cross-document co-reference solution (Bagga and Baldwin 1998) as well as to identify quoted messages (Lewis and Knowles 1997), and lexical chains, which are often created using WordNet for text summarization and interaction identification (Barzilay and Elhadad 1997; Sack 2001). Similarity-based techniques are effective for identifying certain linguistic features (e.g., lexical relations and direct address). Some have been successfully applied in research related to text documents. However, similarity-based techniques are susceptible to noise and require careful selection of parameters.

3 Research Gaps and Questions

Based on our review of previous literature, we have identified several important research gaps. First, little interactional coherence analysis has been conducted for HTTP-based asynchronous CMC. Previous research focused on Usenet newsgroups and e-mail, the headers of which contain accurate interaction information, rendering the use of system features sufficient for accurately capturing a large proportion of the interaction patterns. However, many Web site and ISP forums (e.g., Yahoo!, MSN) do not use the e-mail protocol. Relying only on system features for such CMC modes can result in a significant amount of neglected interaction information. Second, little previous research has implemented techniques that use both CMC system features and linguistic attributes for interactional coherence analysis. The use of a more holistic feature set comprised of features occurring in message headers and bodies could greatly improve interaction recall. Finally, there has been little emphasis in previous research that takes into account the impact of noise in CMC interaction networks.

Based on the research gaps identified, we propose the following research questions:

1. How effectively can we analyze interactional coherence for HTTP-based Web forums using automated techniques?
2. How can techniques that use both CMC system and linguistic features improve interaction representational accuracy as compared to methods that only utilize a single feature category?
3. What impact do forum dynamics (i.e., user system usage behavior) exert on interaction representational accuracy?

4 System Design: Hybrid Interactional Coherence System

In order to address these research questions, we developed the Hybrid Interactional Coherence (HIC) algorithm to perform more accurate interactional coherence analysis, that is, to identify the "reply-to" relationships between CMC messages. The algorithm has three major components: system feature match, linguistic feature match, and residual match. System feature match and the direct address part of the linguistic feature matching component are used to identify interactions stemming from relatively more explicit features (such as headers, quotations, and direct addresses). The lexical relation match and rule-based module (which derive interaction patterns from relatively implicit cues) are only utilized when more explicit features are not present in a posting.

Several major types of noise have also been addressed.

System features used in our implementation include both headers and quotations. With header information, unique IDs of parent messages are checked first. Message subject lines are also analyzed and used. With quotations, our algorithm can identify not only normal quotations but also two special types of quotation, that is, multiple quotations and nested quotation (Barcellini et al. 2005). The algorithm overcomes quotation noise by using a sliding window method, which compares part of the quotation to previous messages. The sliding window method has been successfully used in text similarity detection and authorship discrimination (Nahnsen

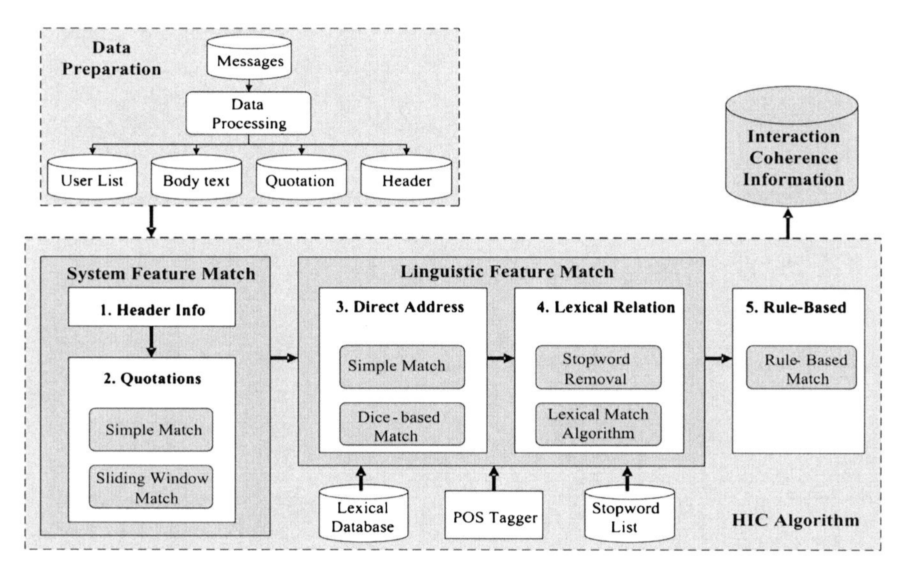

Fig. 7.3 HIC system design

et al. 2005; Abbasi and Chen 2006). Compared with the sentence-level matching approach adopted by Newman (2002), the sliding window is better at dealing with quotation modifications made by systems or users because it is a character-level method (i.e., it compares substrings).

With respect to linguistic features, our algorithm mainly uses direct address and lexical relations. For direct address, besides traditional simple name match, our algorithm uses Dice's equation to overcome noise such as typos, misspellings, and nicknames. Dice's equation uses character-level n-gram matching to identify semantically related pairs of words (Adamson and Boreham 1974). We also differentiate common words and usernames by using a lexical database and automatically generated part-of-speech (POS) tags. For lexical relations, a Lexical Match Algorithm (LMA), developed based on the vector space model frequently used in information retrieval (Salton and McGill 1986), is adopted.

A comprehensive residual matching mechanism is developed for the remaining messages. It improves the naïve linkage method (Commer and Peterson 1986) by matching messages based on their context and co-reference features. Figure 7.3 shows our system design. Details of each component are presented below.

4.1 Data Preparation

The data preparation component is designed to extract messages and their associated metadata from Web forums. All relevant header information is extracted first. Then each message's quotation part and body text are separated using a parser program.

The parser program was also designed to deal with two special types of quotation, nested quotation and multiple quotations. Nested quotation happens when a message which already contains quotations is quoted. The parser program only parses the quotation that is nearest to the message. Sometimes, users respond to different quotations in one message, which is referred to as "multiple quotation." The parser program parses all the related quotations. The final step of data preparation is to extract other relevant information from each message, such as author screen names, date stamps, message subjects, etc.

4.2 *HIC Algorithm: System Feature Match*

4.2.1 Header Information Match

In header information match, unique message IDs of parent messages, if available, are used to identify interaction. Subject lines of messages in the same thread are often consistently similar with each other if they are automatically generated by CMC systems. However, if CMC users intentionally embed interaction cues within them, subject lines can be used to identify interaction patterns as well.

4.2.2 Quotation Match

In quotation match, the quotation part of each message is compared with the body text of previous messages. As previously mentioned, CMC systems may modify the format of quotations (Newman 2002), whereas CMC users may modify quotations to save space (Eklundh 1998). Therefore, in our system, the quotation part of each message is first searched for in the body text of all previous messages, referred to as "simple match." If simple match fails due to the various aforementioned forms of noise, a sliding window method is triggered.

A sliding window method breaks up a text into overlapping windows (substrings) and compares each window against previous body texts (Kjell et al. 1994; Nahnsen et al. 2005; Abbasi and Chen 2006). The system assigns the message (i.e., creates an interaction link) to the quoted message with the highest number of matching windows. The following example shows how a sliding window method with a window size of ten characters and a jump interval of two characters can be used to identify modified quotations (Fig. 7.4).

Original Message	Quoted Content	Message Text Windows	Quoted Text Windows
"What do you prefer?"	"...do you prefer?"	"What do yo"	"...do you "
		"at do you "	".do you pr"
		" do you pr?"	**"o you pref"**
		"o you pref"	**"you prefer"**
		"you prefer"	

Fig. 7.4 Example of sliding window method breaking a text into overlapping windows (substrings)

4.3 *HIC Algorithm: Linguistic Feature Match*

Linguistic features are used to complement system features in constructing CMC interaction patterns. Nash (2005) found that direct address, lexical relations, and co-reference were three dominant linguistic features. Therefore, our Hybrid Interactional Coherence algorithm mainly uses direct address and lexical relations in linguistic feature match, whereas the co-reference feature is indirectly used in residual match.

4.3.1 Direct Address Match

In direct address match, each word of a message is compared to the screen names of previous messages' authors. By only considering authors that have appeared in prior postings within the same thread, we reduce the possibility of incorrectly considering username references to be direct addresses. For the previous example "John, take care of your brother Tom," if user "Tom" has not already appeared in the thread, an interaction between the current message's user and Tom will not be assigned. In situations where a direct address-based interaction is found, the message containing the interaction cue is assumed to have a "reply-to" relation with the addressed user's most recent posting. Initially, a simple match is performed in order to detect messages containing the exact same author screen names. If no simple matches are found, a Dice-based character-level n-gram matching technique is used to compensate for the effect of prevalent direct address noise in CMC such as typos, misspellings, and nicknames. The technique first uses the following Dice equation, which has been successfully used in identifying semantically related pairs of words (Adamson and Boreham 1974; De Roeck and Al-Fares 2000), to estimate the similarity between a word and an author's screen name:

$$Dice\ Score = \frac{2 \times (number\ of\ shared\ unique\ n-grams)}{Total\ unique\ n-grams}$$

A preestablished experiment-based threshold is applied to improve the accuracy of direct address match. However, since many CMC users choose common English words as their screen names, word-sense disambiguation methods need to be applied to differentiate common usages of a word with the use of a word as a screen name. Our HIC algorithm makes use of WordNet (Miller 1990), which has already been widely used in word-sense identification (Voorhees 1993; Resnik 1995), to identify the meaning of words, and a POS tagger (McDonald et al. 2004) to generate the part-of-speech tags. Details of our direct address match are presented below:

1. For each screen name in the author list, query WordNet for meanings.
2. For each word in a message, do the following:

   2.1 Use Dice equation to find the most similar screen name appeared before.

2.2 If the highest Dice score is greater than a predefined threshold, query WordNet for the meanings of the word and do the following:

2.2.1 If neither the word nor the screen name has meanings, assign direct address.
2.2.2 Else, get POS tag for the word. If the word is a noun or noun phrase, assign direct address.
2.2.3 Else, do not assign direct address for the word.

4.3.2 Lexical Relations: The Lexical Match Algorithm

Lexical relation match assumes an interaction between the two messages that are most similar. It calculates the lexical similarities among stopword-removed messages when more explicit interactional coherence features such as quotations and direct address are not found. The key to lexical relation match is to develop an appropriate formula to calculate the similarity score. We propose a Lexical Match Algorithm (LMA) for lexical relation match. The LMA is designed to identify lexical relation-based interactions between postings while taking into consideration the unique characteristics of CMC interaction, such as topic drift/decay and various forms of noise (e.g., misspellings, idiosyncrasies, etc.). The algorithm measures the similarity between messages based on the content as well as turn proximity and levels of inflection and/or idiosyncratic literary variation. LMA integrates the vector space model with Dice's equation and a turn-based proximity scoring mechanism.

Vector space model (VSM) is one of the most popular methods used to identify lexical similarities (Salton and McGill 1986). By using word stems, VSM can also identify morphological word changes. However, in order to identify typos, misspellings, abbreviated references, and other forms of creative user behavior, the Dice equation (Adamson and Boreham 1974; De Roeck and Al-Fares 2000) is adopted in LMA to complement the traditional VSM.

Additionally, a high degree of topic decay/drift has been found in asynchronous CMC (Herring 1999; Smith and Fiore 2001). Nash (2005) also noticed that most CMC interactions happen within three turns. Therefore, CMC interactions represent a "closeness" characteristic, which means two closer messages are more likely to interact than two messages further away. A topic decay factor calculated by the distance (number of turns) between two messages is adopted in our LMA formula to address this "closeness" characteristic.

Here is our LMA formula for lexical similarity:

$$\sum_{i=0}^{LenX}\sum_{j=0}^{LenY}\frac{Tf_{Xi}+Tf_{yj}}{Df_{Xi}+Df_{Yj}}_{if\,(Dice(Xi,Yj))>0.55)}\times(LenX\times LenY)^{-1}\times(Distance(X,Y)+C)^{-1}$$

X and *Y* are the two compared messages. *LenX* and *LenY* are the number of unique non-stopword terms in the two messages, *Xi* refers to the *i*th non-stopword word in message *X*, and *Yj* the *j*th non-stopword term in message *Y*. *Tf* is the term frequency and *Df* is the document frequency. *Distance*(*X*, *Y*) refers to the number of turns or messages between two compared messages. If there are *N* messages between the two compared messages, their distance is $N+1$. *C* is a constant used to control the impact of message proximity on the overall similarity between two messages.

In the formula, *Dice*(*Xi*, *Yj*) is used to compare two non-stopword terms. If their similarity is greater than 0.55, which is a predefined experiment-based threshold, a combined "tf-idf" score is calculated. $(LenX \times LenY)^{-1}$ is the length normalization factor and $(Distance(X, Y)+C)^{-1}$ is the topic decay factor mentioned before. If the highest score calculated by our formula is greater than 0.002, another threshold we use, an interaction is identified. Otherwise, residual match is used. The value of constant *C* and the two thresholds are developed based on a manual analysis of ten other threads in the LNSG forum. These ten threads are not included in our evaluation.

4.4 HIC Algorithm: Residual Match

Residual match is used for messages which do not contain obvious clues for automatic interaction identification. It is utilized to help enhance interaction recall by assigning interactions based on common communication patterns. Prior residual matches have used variants of the naïve linkage method. One such implementation assigns each remaining posting (i.e., one with no identified interaction) to the first message in the thread (Commer and Peterson 1986). Other versions of naïve linkage assign each posting to the preceding message. The intuition behind assigning each remaining post to the prior one is that messages are likely to interact with predecessors in close proximity, given the turn-based nature of CMC (Herring 1999). Since residual matching techniques use very general assignment rules, they tend to have lower precision as compared to other techniques which use system and/or linguistic interaction cues. We propose a new rule-based residual match method which considers the message proximity as well as the conversation structure and context. The details for our residual match are provided below:

X: The residual message of author A

Y: Previous message of author A

Z: Messages of other authors which are posted between Y and X and are replies to messages of author A

1. If Y does not exist, X replies to the first message in the discussion.
2. If Y exists and Z exists, X replies to Z.
3. If Y exists and Z does not exist, X replies to what Y replies to.

The first rule is to apply the improved naïve linkage method when the residual message is the first message the author has posted in the thread. The other two rules are generated based on two human communication characteristics, which can also be found in CMC. If people give feedback or raise questions to our proposed ideas and statements, it is natural for us to comment on the feedback or answer the questions, which is characterized by the second rule. On the other hand, even if no feedback is given, people tend to strengthen or make clear their previous statements, characterized by the third rule.

5 Evaluation

In order to evaluate the effectiveness of our HIC algorithm, an experiment was conducted. The experiment compared the HIC algorithm against the link- and similarity-based methods. The test bed and experimental design are described in detail below.

5.1 *Test Bed*

Our test bed consisted of a large extremist Web forum. The forum was the Libertarian National Socialist Green Party (LNSG) Forum (http://www.nazi.org/community/forum/). Analysis of such social online communities is important in order to improve our understanding of these groups and organizations (Burris et al. 2000; Schafer 2002; Chen 2005). For the forum, several of the longest threads were studied (shown in Table 7.3).

All threads were manually tagged first by a single annotator to identify their interactional coherence. A sample of 100 messages from the annotator was also tagged by a second coder to check the accuracy of the tagging. Both independent annotators were graduate students with strong linguistic backgrounds. The annotators determined a correct interaction by looking for interaction cues in every message. The cues included features found in message headers (e.g., an "RE:" in the subject line), quoted content from another message, linguistic cues inherent in the message body (e.g., direct address and lexical relations), as well as those based on the thread context (i.e., residual rule matching based on previous postings and interaction).

Table 7.3 Details for datasets in test bed

Forum	Thread no.	Thread subject	No. of messages	No. of users
LNSG forum	4	Idea for banner/icon	148	24
	5	Blue eyes, blond hair	62	22
	6	Greetings	85	14
	7	Race mixing	143	39

Table 7.4 Interaction feature breakdowns across threads

Forum	Thread no.	No. of messages	Quotation (%)	Direct address (%)	Lexical relation (%)	Others (%)
LNSG forum	4	148	16.2	16.2	41.9	25.7
	5	62	9.7	9.7	53.2	27.4
	6	85	21.2	24.7	35.3	18.8
	7	143	33.6	8.4	33.6	24.4
	Overall	438	21.9	14.4	39.5	24.2

The annotators utilized the guidelines proposed by Nash (2005) for manually identifying linguistic interaction cues. Figure 7.2 provided examples of how interactional coherence could be derived using linguistic features. The inter-coder reliability across the 100 messages had a kappa statistic of 0.88 which is considered to be reliable. The tagging results were used as our gold standard. The interaction feature breakdowns across threads based on the manual tagging are presented in Table 7.4.

5.2 *Comparison of Techniques*

5.2.1 Experiment Setup

In the first experiment, we compared our HIC algorithm with a link-based method that relies on system features, as well as against a similarity-based method, which relies on linguistic features. These comparison techniques were incorporated since variations of the link-based method and similarity-based method have been adopted in previous studies (Spiegel 2001; Soon et al. 2001; Newman 2002; Yee 2002). The purpose of this experiment was to study the effectiveness of the combined usage of system features and linguistic features, as done in the proposed HIC algorithm, over techniques mostly utilizing a single category of features.

The link-based method uses the quotations in the header information for interactional coherence identification (Yee 2002). If a quotation exactly matches previous messages, the interaction is noted between the two postings. For remaining messages, the naïve linkage method is used, which assumes that the remaining messages are replies to the first message.

The similarity-based method consists of two parts: simple direct address match and vector space model match (Bagga and Baldwin 1998). The first part identifies interactional coherence when a word is an exact match with other authors' screen names. The second part uses the traditional "tf-idf" score to identify lexical similarity. Threshold 0.2, shown as the best threshold by Bagga and Baldwin (1998), is used for this traditional VSM match. Precision, recall, and F-measure at both the forum and thread level were used to evaluate the performance of these methods.

Table 7.5 Experimental results for experiment 1

Forum	Technique	Precision	Recall	F-measure
LNSG forum	HIC algorithm	0.711	0.711	0.711
	Link-based	0.560	0.551	0.555
	Similarity-based	0.584	0.678	0.625

Table 7.6 *p* values for pairwise *t* tests on accuracy for experiment 1

Forum	Techniques	*p* values
LNSG forum	HIC vs. link-based	<0.001*
	HIC vs. similarity-based	<0.001*
	Link-based vs. similarity-based	<0.001*

*p values significant at alpha = 0.01

$$\text{Precision} = \frac{\textit{Number of Correctly Identified Interactions}}{\textit{Total Number of Identified Interactions}}$$

$$\text{Recall} = \frac{\textit{Number of Correctly Identified Interactions}}{\textit{Total Number of Interactions}}$$

$$\text{F - measure} = \frac{2 \times precision \times recall}{precision + recall}$$

5.2.2 Hypotheses

Given the presence of system and linguistic interaction cues in online discourse, we believe that interactional coherence identification techniques incorporating both feature types are likely to provide better performance. Therefore, we propose the following hypotheses.

H1a: The HIC algorithm will outperform the link-based method for Web forum interactional coherence analysis.
H1b: The HIC algorithm will outperform the similarity-based method for Web forum interactional coherence analysis.

5.2.3 Experimental Results

Table 7.5 shows the experimental results for all three methods. Our HIC algorithm had the best performance on both the forums in terms of precision, recall, and F-measure.

5.2.4 Hypotheses Results

Table 7.6 shows the *p* values for the pairwise *t* tests conducted on the interactional coherence identification accuracies to measure the statistical significance of the results. Bolded values indicate statistically significant outcomes in line with our hypotheses. Both hypotheses, H1a and H1b, are supported.

H1a: The HIC algorithm outperformed the link-based method for both the Web forums ($p < 0.01$).

H1b: The HIC algorithm outperformed the similarity-based method for both the Web forums ($p < 0.01$).

5.2.5 Results and Discussion

The HIC algorithm performed better than both the link-based and similarity-based methods for our test bed. The F-measure was 8–15% higher than the other two techniques. Such improved performance was consistent across all threads in our test bed. The enhanced accuracy of the HIC algorithm was attributable to the incorporation of both system and linguistic features and its ability to handle various forms of CMC noise. For the LNSG forum, lexical relations were more commonly used as interaction cues, resulting in the improved performance of the similarity method over the link-based method on this forum. The LNSG forum members were less likely to utilize system features, which are heavily relied upon by the link-based method.

6 Conclusions

In this study, we applied interactional coherence analysis to Web forums. We developed a hybrid approach that uses both CMC system features and linguistic features for constructing interaction patterns from Web discourse. The results show that our approach outperformed traditional link-based and similarity-based methods due to the use of a robust set of interaction features. In the future, we will work on analyzing user roles in Web forums based on interaction networks generated by the HIC algorithm. We are also interested in identifying interaction across different forums so that we can understand the information dissemination patterns across multiple forums, and in exploring the effectiveness of using thread-level interaction networks to identify important threads in Web forums. Another attractive direction is to apply our techniques to other CMC modes such as blogs and chat room discussions. Blogs have very similar system features to Web forums, including headers and quotations. Bloggers also share usage idiosyncrasies with Web forum posters, such as typos and misspellings. Chat rooms, however, usually do not have system features, and the chat postings are often too short to provide useful lexical information. By applying our algorithm to these two types of datasets, we may be able to identify the potential differences in their interactional coherence.

Acknowledgments This research was funded in part by the following grant: NSF Information and Data Management, "SGER: Multilingual Online Stylometric Authorship Identification: An Exploratory Study," August 2006–August 2007.

References

Abbasi, A. and Chen, H. (2006). Visualizing Authorship for Identification. In the 4th IEEE Symposium on Intelligence and Security Informatics (ISI 2006).

Adamson, G. W. and Boreham, J. (1974). The Use of an Association Measure Based on Character Structure to Identify Semantically Related Pairs of Words and Document Titles. Information Storage and Retrieval, (10), 253–260.

Bagga, A., and Baldwin, B. (1998). Entity-Based Cross-Document Coreferencing Using the Vector Space Model. In Proceedings of the 17th International Conference on Computational Linguistics.

Barcellini, F., Detienne, F., Burkhardt, J. and Sack, W. (2005). A Study of Online Discussions in an Open-Source Software Community: Reconstructing Thematic Coherence and Argumentation from Quotation Practices. In Proceedings of the Communities and Technologies Conference.

Barzilay, R. and Elhadad, M. (1997). Using Lexical Chains for Text Summarization. In Proceedings of the ACL Workshop on Intelligent Scalable Text Summarization, 10–17.

Beaugrande, R.A. and Dressler, W.U. (1996) Introduction to Text Linguistics. Longman., New York, 84–112.

Burris, V., Smith, E., and Strahm, A. (2000). White Supremacist Networks on the Internet. Sociological Focus, (33:2), 215–235.

Chen, H. (2005). Introduction to the Special Topic Issue: Intelligence and Security Informatics. Journal of the American Society for Information Science and Technology, 56(3), 217–220.

Choi, F. Y. Y. (2000). Advances in Domain Independent Linear Text Segmentation. In Proceedings of the Meeting of the North American Chapter of the Association for Computational Linguistics (ANLP-NAACL-00), 26–33.

Commer, D. and Peterson L. (1986). Conversation-based Mail. In TOCS 4(4), ACM Press, 299–319.

De Roeck, A. N. and Al-Fares, W. (2000). A Morphologically Sensitive Clustering Algorithm for Identifying Arabic Roots. In Proceedings of ACL-2000 (ACL, 2000), Hong Kong.

Donath, J., Karahalio, K. and Viegas, F. (1999). Visualizing Conversation. In Proceedings of the 32nd Conference on Computer-Human Interaction (CHI' 02), Chicago, USA.

Donath, J. (2002). A Semantic Approach to Visualizing Online Conversations. Communications of the ACM, 45(4), 45–49.

Eklundh, K. S. (1998). To Quote or Not to Quote: Setting the Context for Computer-Mediated Dialogues. Technical report TRITA-NA-P9807, IPLab-144, Royal Institute of Technology, Stockholm.

Eklundh, K. S. and Rodriguez, H. (2004). Coherence and Interactivity in Text-based Group Discussions around Web Documents. In Proceedings of the 37th Annual Hawaii International Conference on System Sciences.

Fiore, A. T., Tiernan, S. L., and Smith, M. A. (2002). Observed Behavior and Perceived Value of Authors in Usenet Newsgroups: Bridging the Gap. Proceeding of the SIGCHI Conference on Human Factors in Computing Systems: Changing Our World, Changing Ourselves, 323–330.

Hale, C. (1996). Wired Style: Principles of English, Usage in the Digital Age. HardWired, San Francisco.

Halliday, M. A. and Hasan, R. (1976). Cohesion in English. Longman, London.

Hayne, S. C., Pollard, C. E., and Rice, R. E. (2003). Identification of Comment Authorship in Anonymous Group Support Systems. Journal of Management Information Systems, Volume 20, Number 1, Summer 2003, 301–329.

Hearst, M. A. (1994). Multi-paragraph Segmentation of Expository Text. Proceedings of the ACL'94, Las Cruces, NM.

Herring, S. C. and Nix, C. (1997). Is "Serious Chat" an Oxymoron? Academic vs. Social Uses of Internet Reply Chat. Presented at the American Association of Applied Linguistics, Orlando, FL.

Herring, S. C. (1999). Interactional Coherence in CMC. In Proceeding of the 32nd Hawaii International Conference on System Science.

Khan, F. M. (2002). Mining Chat-room Conversations for Social and Semantic Interactions. http://www.cse.lehigh.edu/techreports/2002/LU-CSE-02–011.pdf.

Khan, M., Klavans, J. L., and Mckeown, K. R. (1998). Linear Segmentation and Segment Significance. In Proceedings of the 6th International Workshop of Very Large Corpora (WVLC-6), 197–205, Montreal, Quebec, Canada.

Kjell, B., Woods, W. A., and Frieder, O. (1994). Discrimination of Authorship Using Visualization. Information Processing and Management, (30:1), 141–150.

Lewis, D. D. and Knowles, K. A. (1997). Threading Electronic Mail: A Preliminary Study. Information Processing and Management, (33:2), 209–217.

McDonald, D., Chen, H., Su, H., and Marshall, B. (2004). Extracting Gene Pathway Relations using a Hybrid Grammar: The Arizona Relation Parser. Bioinformatics, (20:18), 3370–3378.

Meho, L. (2006). E-Mail Interviewing in Qualitative Research: A Methodological Discussion. Journal of the American Society for Information Science and Technology, 57(10), 1284–1295.

Miller, G. A., Ed. (1990). WordNet: An On-line Lexical Database. International Journal of Lexicography, 3(4), 235–312.

Morris, J. (1988). Lexical Cohesion, the Thesaurus, and the Structure of Text. Technical Report CSRI 219, Computer System Research Institute, University of Toronto.

Nahnsen, T., Uzuner, O., and Katz, B. (2005). Lexical Chains and Sliding Locality Windows in Content-based Text Similarity Detection. CSAIL Memo, AIM-2005–017.

Nash, M. C. (2005). Cohesion and Reference in English Chatroom Discourse. In Proceedings of the 38th Annual Hawaii International Conference on System Sciences (HICSS'05).

Nasukawa, T. and Nagano, T. (2001) Text Analysis and Knowledge Mining System. IBM Systems Journal, (40:4), 967–984.

Newman, P. S. (2002). Exploring Discussion Lists: Steps and Directions. In Proceeding of the 2nd ACM/IEEE-CS Joint Conference on Digital Libraries, 126–134.

Osterlund, C. and Carlile, P. (2005) Relations in Practice: Sorting Through Practice Theories on Knowledge Sharing in Complex Organizations. The Information Society, (21), 91–107.

Paolillo, J. C. (2006). Conversational Codeswitching on Usenet and Internet Relay Chat. Computer-Mediated Conversation, S. Herring (Ed.).

Radford, M. L. (2006). Encountering Virtual Users: A Qualitative Investigation of Interpersonal Communication in Chat Reference. Journal of the American Society for Information Science and Technology, 57(8), 1046–1059.

Resnik, P. (1995). Disambiguating Noun Groupings with Respect to WordNet Senses. In Proceedings of the Third Workshop on Very Large Corpora.

Sack, W. (2001). Conversation Map: An Interface for Very Large-Scale Conversations. Journal of Management Information Systems, (17:3), 73–92.

Salton, G. and McGill, M. J. (1986). Introduction to Modern Information Retrieval. McGraw-Hill, Inc., New York, NY.

Schafer, J. (2002). Spinning the Web of Hate: Web-based Hate Propagation by Extremist Organizations. Journal of Criminal Justice and Popular Culture, (9:2), 69–88.

Smith, M. A. and Fiore, A. T. (2001). Visualization Components for Persistent Conversations. In Proceedings of the SIGCHI Conference on Human Factors in Computing Systems, 136–143.

Soon, W. M., Ng, H. T., and Lim D. C. Y. (2001). A Machine Learning Approach to Coreference Resolution of Noun Phrases. Computational Linguistics, (27), 521–544.

Spiegel, D. (2001). Coterie: A Visualization of the conversational dynamics within IRC. Online. Master's Dissertation, http://alumni.media.mit.edu/~spiegel/thesis/Thesis.pdf.

Voorhees, E. M. (1993). Using WordNet to Disambiguate Word Senses for Text Retrieval. Annual ACM Conference on Research and Development in Information Retrieval, In Proceedings of the 16th Annual International ACM SIGIR Conference on Research and Development in Information Retrieval, Pittsburgh, Pennsylvania, United States, 171–180.

Walther, J. B., Anderson, J. F., and Park, D. W. (1994). Interpersonal Effects in Computer-Mediated Interaction: A Meta-Analysis of Social and Antisocial Communication. Communication Research, Vol. 21, No. 4, 460–487.

Wasko, M, M. and Faraj, S. (2005). Why Should I Share? Examining Social Capital and Knowledge Contribution in Electronic Networks of Practice. MIS Quarterly (29:1), 35–57.

Xiong, R., Smith, M. A., and Drucker, S. M. (1998). Visualizations of Collaborative Information for End-Users. Technical Report MSRTR-98–52, Microsoft Research.

Yee, K. P. (2002). Zest: Discussion Mapping for Mailing Lists. In CSCW 2002 Conference Supplement, ACM Press, 123–126.

Chapter 8
Dark Web Attribute System

1 Introduction

The weekly news coverage of excerpts from messages and videos produced and webcast by terrorists/extremists has shown that terrorists and extremists have become exploiters of the Internet beyond routine communication operations. The Internet has dramatically increased their ability to influence the outside world (Arquilla and Ronfeldt, 1993). Several virtues of the Internet, such as ease of access, anonymity of posting, huge audience, and lack of regulations, have enabled terrorists to directly speak to millions of people – both supporters and adversaries – with little chance of being detected. As posited by Jenkins (2004), through operating their own web sites and online forums, terrorists have effectively created their own "terrorist news network."

Terrorist/extremist organizations have generated thousands of web sites that support psychological warfare, fundraising, recruitment, coordination, and distribution of propaganda materials. From those terrorist/extremist web sites, supporters can download multimedia training materials, buy games, T-shirts, and music CDs and access forums and chat services such as PalTalk (Elison, 2000; Tekwani, 2002; Bowers, 2004; Muriel, 2004; Weimann, 2004). Some web sites such as those associated with the jihad terrorist/extremist movement are extremely dynamic in that they emerge overnight, frequently modify their contents, and then swiftly "disappear" by changing their URLs which are later announced via online forums (Weimann, 2004). They are often hosted on free web space servers or by unsecured and poorly maintained commercial servers. Such web sites are technically supported by those who are Internet savvy to provide sophisticated propaganda images and videos via proxy servers to mask ownerships (Armstrong and Forde, 2003; El Deeb, 2004). The level of technical sophistication of the Islamic terrorist/extremist organizations' web sites has increased according to Katz, who monitors Islamic fundamentalist Internet activities (Internet Haganah, 2005). The rapid proliferation and increased sophistication of web sites and online forums run by terrorist/extremist organizations are indications of the growing popularity of the Internet in terrorism campaigns. They also indicate that there is a vast pool of sympathizers that such

H. Chen, *Dark Web: Exploring and Data Mining the Dark Side of the Web*,
Integrated Series in Information Systems 30, DOI 10.1007/978-1-4614-1557-2_8,

organizations have attracted, with some applying their IT expertise as contributions to the cause (Jesdanun, 2004).

Although this alternate side of the Internet, referred to as the "Dark Web," has received extensive government and media attention, there is a dearth of empirical studies that examine the sophistication of terrorist/extremist organizations' web sites and how they support strategic and tactical information operations. Therefore, some basic questions about terrorist/extremist organizations' Internet usage remain unanswered. For example, what are the major Internet technologies that they have used on their web sites? How sophisticated and effective are the technologies in terms of supporting communications and propaganda activities?

In this chapter, we explore an integrated approach for collecting and monitoring terrorist-created web contents and propose a systematic content analysis approach to enable quantitative assessment of the technical sophistication of terrorist/extremist organizations' Internet usage. The rest of this chapter is organized as follows: In Sect. 2, we briefly review previous works on terrorists' use of the Internet. In Sect. 3, we present our research questions and the proposed methodologies to study those questions. In Sect. 4, we describe the findings obtained from a case study of the analysis of technical sophistication, content richness, and web interactivity features of major Middle Eastern terrorist/extremist organizations' web sites and a benchmark comparison of Middle Eastern terrorist/extremist web sites and web sites from the US government. In the last section, we provide conclusions and discuss the future directions of this research.

2 Literature Review

2.1 *Terrorism and the Internet*

Previous research showed that terrorists/extremists mainly utilize the Internet to enhance their information operations surrounding propaganda, communication, and psychological warfare (Thomas, 2003; Denning, 2004; Weimann, 2004). To achieve their goals, terrorists/extremists often need to maintain a certain level of publicity for their causes and activities to attract more supporters. Prior to the Internet era, terrorists/extremists maintained publicity mainly by catching the attention of traditional media such as television, radio, or print media. This was difficult for them because terrorists/extremists often could not meet the editorial selection criteria of those public media (Weimann, 2004). With the Internet, terrorists/extremists can bypass the requirements of traditional media and directly reach hundreds of millions of people globally, 24/7.

Terrorist/extremist groups have sought to replicate or supplement the communication, fundraising, propaganda, recruitment, and training functions on the Internet by building web sites with massive and dynamic online libraries of speeches, training manuals, and multimedia resources that are hyperlinked to other sites that share similar beliefs (Coll and Glasser, 2005; Weimann, 2004). The web sites are designed to communicate with diverse global audiences of members, sympathizers, media, enemies, and the public (Weimann, 2004). Table 8.1 summarizes terrorist/extremist groups' objectives and tasks that are supported by web sites.

2.2 *Existing Dark Web Studies*

In recent years, there have been studies of how terrorists/extremists use the web to facilitate their activities (Zhou et al., 2005; Chen et al., 2004; ISTS, 2004; Thomas, 2003; Tsfati and Weimann, 2002; Weimann, 2004). For example, researchers at the Institute for Security Technology Studies (ISTS) have analyzed dozens of terrorist/extremist organizations' web sites and identified five categories of terrorists' use of the web: propaganda, recruitment and training, fundraising, communications, and targeting. These usage categories are supported by other studies such as those by Thomas (2003), Katz at SITE Institute (2004), and Weimann (2004).

Since the late 1990s, several organizations, such as SITE Institute, the Anti-Terrorism Coalition (ATC), and the Middle East Media Research Institute (MEMRI), started to monitor contents from selected terrorist/extremist web sites for research and intelligence purposes. Tsfati and Weimann (2002) studied the content types and target audiences of terrorist/extremist organizations' web sites by analyzing the content of 29 Middle Eastern web sites. Table 8.2 lists some of the organizations that capture and analyze terrorists/extremists' web sites (and the collection start dates) grouped into three functional categories: archive, research center, and vigilante community.

Except for the Artificial Intelligence (AI) Lab, none of the enumerated organizations seem to use automated methodologies for both collection building and analysis of the web sites. Due to the enormous size and the dynamic nature of the web, the manual collection and analysis approaches have limited the comprehensiveness of their analyses. Furthermore, none of the studies have provided empirical evidence of the levels of technical sophistication or compared terrorist/extremist organizations' cyber capabilities with those of mainstream organizations. Since technical knowledge required to maintain web sites provides an indication of terrorist/extremist organizations' technology adoption strategies (Jackson, 2001), we believe it is important to analyze the technologies required to maintain terrorist/extremists' web sites from the perspectives of technical sophistication, content richness, and web interactivity.

2.3 *Dark Web Collection Building*

The first step toward studying the terrorist/extremist web presence is to capture terrorist web sites and store them in a repository for further analysis. Web collection

Table 8.1 How web sites support objectives of terrorist/extremist groups

Terrorist/extremists' objectives	Tasks supported by web sites	Web features (Preece, 2000)
Enhance communication (Becker, 2005; Weimann, 2004)	• Composing, sending, and receiving messages • Searching for messages, information, and people • One-to-one and one-to-many communications • Maintaining anonymity	• Synchronous (chat, video conferencing, MUDs, and MOOs) and asynchronous (e-mail, bulletin board, forum, and usenet newsgroup) • GUI • Help function • Feedback form • Login • E-mail address for webmaster and organization contact
Increase fundraising (Weimann, 2004)	• Publicizing need for funds • Providing options for collecting funds	• Payment instruction and facility • E-commerce application • Hyperlinks to other resources
Diffuse propaganda (Weimann, 2004)	• Posting resources in multiple languages • Providing links to forums, videos, and other groups' web sites • Using web sites as an online clearinghouse for statements from leaders	• Content management • Hyperlinks • Directory for documents • Navigation support • Search, browsable index • Free web site hosting • Accessible
Increase publicity (Coll and Glasser, 2005; Jenkins, 2004)	• Advertising groups' events, martyrs, history, and ideologies • Providing groups' interpretation of the news	• Downloadable files • Animated and flashy banner, logo, and slogan • Clickable maps • Information resources (e.g., international news)
Overcome obstacles from law enforcement and military (Coll and Glasser, 2005; Kelley, 2001)	• Send encrypted messages via e-mail, forums, or post on web sites • Move web sites to different servers so that they are protected	• Anonymous e-mail accounts • Password-protected or encrypted services • Downloadable encryption software • E-mail security • Stenography
Provide recruitment and training (Weimann, 2004)	• Hosting martyrs' stories, speeches, and multimedia that are used for recruitment • Using flashy logos, banners, and cartoons to appeal to sympathizers with specialized skills and similar views • Build massive and dynamic online libraries of training resources	• Interactive services (e.g., games, cartoons, and maps) • Online registration process • Directory • Multimedia (e.g., videos, audios, and images) • FAQs, alerts • Virtual community

Table 8.2 Organizations that capture and analyze terrorists' web sites

Organization	Description	Access
Archives		
1. Internet Archive (IA)	1996 – Collect open access HTML pages (every 2 months)	Via http://www.archive.org
Research centers		
2. Anti-Terrorism Coalition (ATC)	2003 – Jihad watch. Has 448 terrorist web sites and forums	Via http://www.atcoalition.net
3. Artificial Intelligence (AI) Lab, University of Arizona	2003 – Spidering (every 2 months) to collect terrorist web sites. Has thousands of web sites: US domestic, Latin America, and middle eastern web sites	Via test bed portal called Dark Web Portal (ai.arizona.edu)
4. MEMRI	2003 – Jihad and terrorism studies project	Access reports via http://www.memri.org
5. SITE Institute	2003 – Capture web sites every 24 h. Extensive collection of thousands of files	Access reports and fee-based intelligence services http://siteinstitute.org
6. Weimann (University of Haifa, Israel)	1998 – Capture web sites daily. extensive collection of thousands of files	Closed collection
Vigilante Community		
7. Internet Haganah	2001 – Confronting the global Jihad project. Has hundreds of links to web sites	Provides snapshots of terrorist web sites http://haganah.us

building is the process of gathering and organizing unstructured information from pages and data on the web. Previous studies have suggested three types of approaches to collecting web contents in specific domains: manual approach, automatic approach, and semiautomatic approach.

In order to build the September 11 and Election 2002 Web Archives (Schneider et al., 2003), the Library of Congress collected seed URLs for a given theme. The seeds and their close neighbors (distance 1) are then downloaded. The limitation of such a manual approach is that it is time consuming and inefficient.

Albertsen (2003) used an automatic approach in the "Paradigma" project. The goal of Paradigma is to archive Norwegian legal deposit documents on the web. It employed a focused web crawler (Kleinberg et al., 1998), an automatic program that discovers and downloads web sites in particular domains by following web links found in the HTML pages of a starting set of web pages. Metadata was then extracted and used to rank the web sites in terms of relevance. The automatic approach is more efficient than the manual approach; however, due to the limitations of current focused crawling techniques, automatic approaches often introduce noise (off-topic web pages) into the collection.

The "Political Communications Web Archiving" group employed a semiautomatic approach to collecting domain-specific web sites (Reilly et al., 2003). Domain experts provided seed URLs as well as typologies for constructing metadata that can be used in the crawling process. Their project's goal is to develop a methodology for constructing an archive of broad-spectrum political communications over the web. We believe that the semiautomatic approach is most suitable for collecting terrorist/extremist web sites because it combines the high accuracy and high efficiency of manual and automatic approaches.

2.4 *Dark Web Content Analysis*

In order to reach an understanding of the various facets of terrorist/extremist web usage and communications, a systematic analysis of the web sites' content is required. Researchers in the terrorism domain have used observation and content analysis to analyze web site data. In Bunt's (2003) overview of jihadi movements' presence on the web, he described the reaction of the global Muslim community to the content of jihadi terrorist web sites. His assessment of the influence such content had on Muslims and Westerners was based on a qualitative analysis of message contents extracted from Taliban and al-Qaeda web sites. Tsfati and Weimann (2002) conducted a content analysis of the characteristics of terrorist groups' communications. They said that the small size of their collection and the descriptive nature of their research questions made a quantitative analysis infeasible.

Demchak et al. (2000) provided a well-defined methodology for analyzing communicative content in government web sites. Their work focused on measuring "openness" of government web sites. To achieve this goal, they developed a Web Site Attribute System (WAES) tool that is basically composed of a set of high-level attributes such as transparency and interactivity. Each high-level attribute is associated with a second layer of attributes at a more refined level of granularity. For example, the increase of "operational information" and "responses" on a given web page can induce an increase in the openness level of a government web site. This WAES system is an example of a well-structured and systematic content analysis methodology.

Demchak et al.'s work provides guidance for this chapter. However, the "openness" attributes used in their work were designed specifically for e-government studies. We surveyed research in e-commerce, e-government, and e-education domains and identified several sets of attributes that could be used to study the technical advancement and effectiveness of terrorists/extremists' use of the Internet.

Palmer and David's (1998) study identified a set of 15 attributes (called "technical characteristics" in the original work) to evaluate two aspects of e-commerce web sites: technical sophistication and media richness. More specifically, the technical sophistication attributes measure the level of advancement of the techniques used in the design of web sites, e.g., "use of HTML frames," "use of Java scripts," etc. The media richness attributes measure how well the web sites use multimedia to deliver information to their users, e.g., "hyperlinks," "images," "video/audio files," etc.

Another set of attributes called web interactivity has been widely adopted by researchers in e-government and e-education domains to evaluate how well web sites facilitate the communication among web site owners and users. Two organizations, the United Nations Online Network in Public Administration and Finance (UNPAN; www.unpan.org) and the European Commission's IST program (www.cordis.lu/ist/), have conducted large-scale studies to evaluate the interactivity of government web sites of major countries in the world. The web interactivity attributes can be summarized into three categories: one-to-one-level interactivity, community-level interactivity, and transaction-level interactivity.

The one-to-one-level interactivity attributes measure how well the web sites support individual users to give feedback to the web site owners (e.g., provide e-mail contact, provide guest book functions, etc.). The community-level interactivity attributes measure how well the web sites support the two-way interaction between site owners and multiple users (e.g., use of forums, online chat rooms, etc.). The transaction-level interactivity measures how well users are allowed to finish tasks electronically on the web sites (e.g., online purchasing, online donation, etc.).

Chou's (2003) study proposed a detailed four-level framework to analyze e-education web sites' level of advancement and effectiveness. Attributes in the first level (called learner-interface interaction) of Chou's framework are very similar to the technical sophistication attributes used in Palmer and David's (1998) study. Attributes in the other three levels (learner-content interaction, learner-instructor interaction, and learner-learner interaction) of Chou's framework are similar to the three-level web interactivity attributes used in the e-government evaluation projects as mentioned above.

To date, no study has employed the technical sophistication, media richness, and web interactivity attributes as well as the WAES framework in the terrorism domain. We believe that these web content analysis metrics can be applied in terrorist/extremist web site analysis to deepen our understanding of the terrorists' tactical use of the web.

3 Proposed Methodology: Dark Web Collection and Analysis

The research questions postulated in this chapter are:

1. What design features and attributes are necessary to build a highly relevant and comprehensive Dark Web collection for intelligence and analysis purposes?
2. For terrorist/extremist web sites, what are the levels of technical sophistication in their system design?
3. For terrorist/extremist web sites, what are the levels of richness in their online content?
4. For terrorist/extremist web sites, what are the levels of web interactivity to support individual, community, and transaction interactions?

To study the research questions, we propose a Dark Web analysis tool which contains several components: a systematic procedure for collecting and monitoring Dark Web contents and a Dark Web Attribute System to enable quantitative analysis of Dark Web content (see Fig. 8.1).

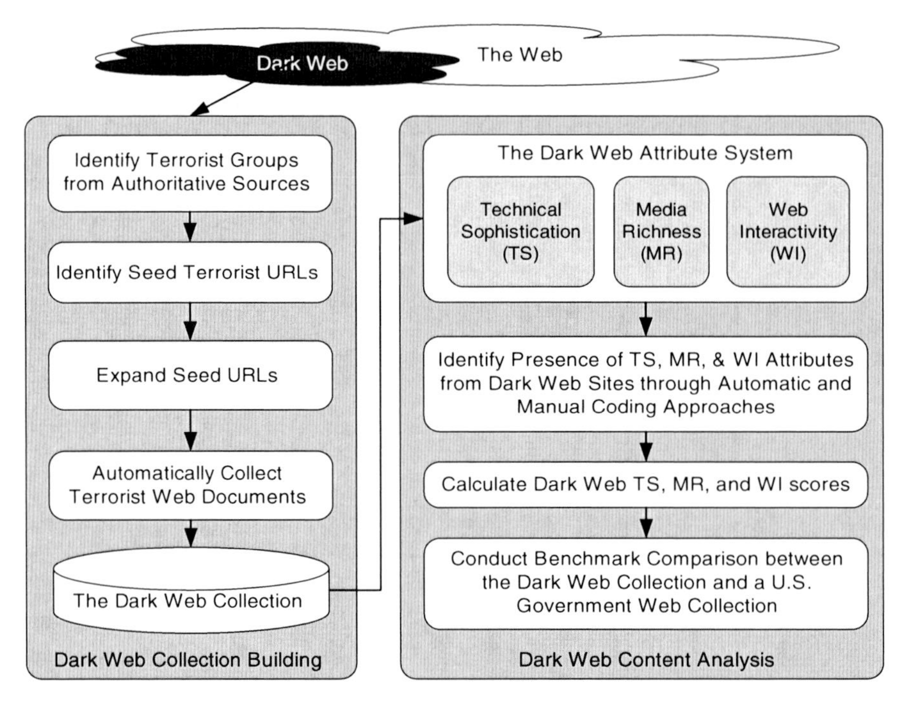

Fig. 8.1 The Dark Web collection-building and content analysis framework

3.1 Dark Web Collection Building

The first step toward studying terrorists' tactical use of the web is to build a high-quality Dark Web collection. To ensure the quality of our collection, based on our review of web collection-building methodologies, we propose to use a semiautomated approach to collecting Dark Web contents (Reid et al., 2004). Our collection-building approach contains the following steps (see Fig. 8.2 for graphical depiction):

1. *Identify terrorist/extremist groups*: Defining terrorism is complicated by the fact that people almost never define themselves as terrorists, and the use of the label by others often has political overtones. We start the collection-building process by identifying the groups that are considered by authoritative sources as terrorist/extremist groups. The sources include government agency reports (e.g., US State Department reports, FBI reports, government reports from United Kingdom, Australia, Japan, and P. R. China, etc.), authoritative organization reports (e.g., Counter-Terrorism Committee of the UN Security Council, US Committee for A Free Lebanon, etc.), and studies published by terrorism research centers such as the Anti-Terrorism Coalition (ATC), the Middle East Media Research Institute (MEMRI), Dartmouth College, etc. Information such as terrorist group names, leaders' names, and terrorist jargon is identified from the sources to create a terrorism keyword lexicon for use in the next step.

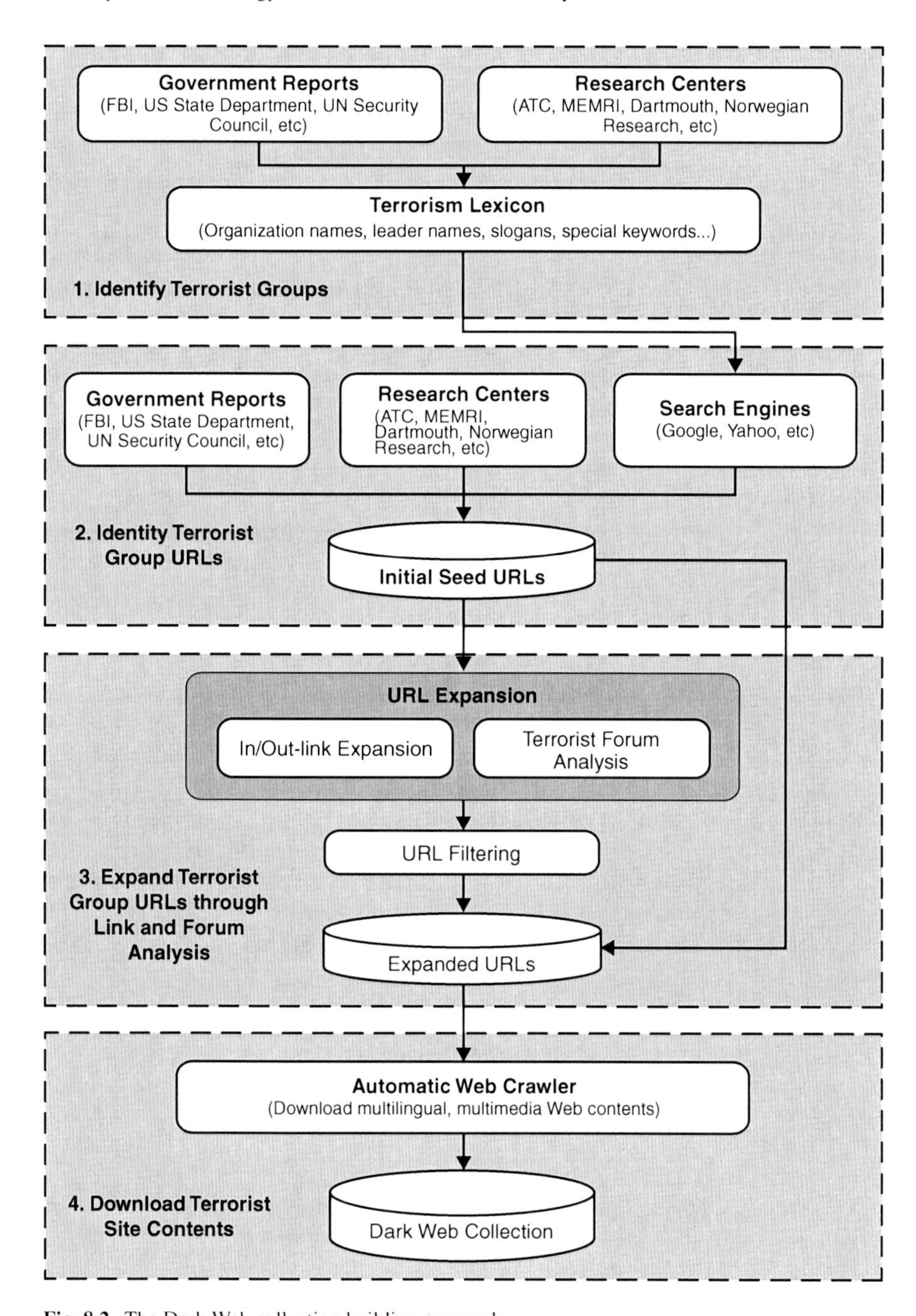

Fig. 8.2 The Dark Web collection building approach

2. *Identify terrorist/extremist group URLs*: We manually identify a set of seed terrorist group URLs from two sources. First, terrorist group URLs can be directly identified from the authoritative sources and literature used in the first step. Second, terrorist group URLs can be identified by using the terrorism keyword lexicon to query major search engines on the web. The identified set of terrorist group URLs will serve as the seed URLs for the next step.

   Expand terrorist/extremist URL sets through link and forum analysis: After identifying the seed URLs, out-links and in-links of the seed URLs were automatically extracted using link analysis programs. The out-links are extracted from the HTML contents of "favorite link" pages under the seed web sites. The in-links are extracted from Google in-link search service through Google API. Automatic out-link and in-link expansion is an effective way to expand the scope of our collection. We also have language experts who browse the contents of terrorist-supporting forums and extract the terrorist/extremist URLs posted by terrorist supporters. Because bogus or unrelated web sites can make their way into our collection through the expansion, we have developed a robust filtering process based on evidence and clues from the web sites.
3. Aside from sites which explicitly identify themselves as the official sites of a terrorist organization or one of its members, a web site that contains even minor praise of or adopts ideologies espoused by a terrorist group is included in our collection.
4. *Download terrorist/extremist web site contents*: Once the terrorist/extremist web sites are identified, a program is used to automatically download all their contents. Unlike the tools used in previous studies, our program was designed to download not only the textual files (e.g., HTML, TXT, PDF, etc.) but also multimedia files (e.g., images, video, audio, etc.) and dynamically generated web files (e.g., PHP, ASP, JSP, etc.). Moreover, because terrorist organizations set up forums within their web sites whose contents are of special value to research communities, our program also can automatically log into the forums and download the dynamic forum contents. The automatic downloading method allows us to effectively build Dark Web collections with millions of documents. This greatly increases the comprehensiveness of our Dark Web study.

To keep the Dark Web collection comprehensive and up-to-date, steps 2 to 4 are periodically repeated. Collections built using such a recursive procedure can also provide information about the evolution and diffusion of the Dark Web.

3.2 The Dark Web Attribute System (DWAS)

Instead of using observation-based qualitative analysis approaches (Thomas, 2003), we propose a systematic approach to enable the quantitative study of terrorist/extremist groups' use of the web. The proposed Dark Web Attribute System is similar to the WAES framework in Demchak et al.'s study (2000). However, instead of the openness attributes used in WAES, our framework focuses on the attributes that could help us better understand the level of advancement and effectiveness of terrorists' web usage, namely, technical sophistication attributes, content richness

Table 8.3(a) Technical sophistication attributes

TS attributes		Weights
Basic HTML techniques	Use of lists	1
	Use of tables	2
	Use of frames	2
	Use of forms	1.5
Embedded multimedia	Use of background image	1
	Use of background music	2
	Use of stream audio/video	3.5
Advanced HTML	Use of DHTML/SHTML	2.5
	Use of predefined script functions	2
	Use of self-defined script functions	4.5
Dynamic web programming	Use of CGI	2.5
	Use of PHP	4.5
	Use of JSP/ASP	5.5

Table 8.3(b) Content richness attributes

CR Attributes	Scores
Hyperlink	# Hyperlinks
File/software download	# Downloadable documents
Image	# Images
Video/audio file	# Video/audio files

Table 8.3(c) Web interactivity attributes

WI attributes	Weights
One-to-one interactivity	
E-mail feedback	1.75
E-mail list	2.25
Contact address	1.25
Feedback form	2.75
Guest book	1.5
Community-level interactivity	
Private message	4.25
Online forum	4.25
Chat room	4.75
Transaction-level interactivity	
Online shop	4
Online payment	4
Online application form	4

attributes (an extension of the traditional media richness attributes), and web interactivity attributes. Based on previous literatures in e-commerce (Palmer and David, 1998), e-government (Demchak et al. 2000), and e-education domains (Chou, 2003), we selected 13 technical sophistication attributes, 5 content richness attributes, and 11 web interactivity attributes for our DWAS framework. A list of these attributes is summarized in Tables 8.3a to 8.3c.

1. *Technical sophistication (TS) attributes:* The technical sophistication attributes can be grouped into four categories as shown in Table 8.3a. The first category of four attributes, called the basic HTML technique attributes, measures how well the basic HTML layout techniques (i.e., lists, tables, frames, and forms) are applied in web sites to organize web contents. The second category, called the embedded media attributes, measures how well the web sites deliver their information to the user in multimedia formats such as images, animations, and audio/video clips. The third category of three attributes, called the advanced HTML attributes, measures how well advanced HTML techniques, such as DHTML and SHTML, and predefined and self-defined script functions (e.g., JavaScript, VBScript, etc.) are applied to implement security and dynamic functionalities. The last category, called the dynamic web programming attributes, measures how well dynamic web programming languages such as PHP, ASP, and JSP are utilized to implement dynamic interaction functionalities such as user login, online request or application, and online transaction processing. The four technical sophistication attributes and associated subattributes are present in most of the Dark Web sites we collected.

   The presence of different attributes indicates different levels of technical sophistication. For example, a web site which uses JSP techniques should be considered more technically sophisticated than a site which only uses static HTML. Different weights should be assigned to the attributes to reflect the differences (Chou, 2003). We determined the weights based on web experts' opinions collected through an e-mail survey. Surveys were sent to webmasters and network administrators of several web sites belonging to the University of Arizona, and they were encouraged to forward the survey to their webmaster colleagues. In the survey, we asked the experts to give each of our attributes a weight of 1–10 (1 is the least advanced/sophisticated). Six experts sent their responses back to us. For each attribute, the average weight assigned by the experts was used in the final framework. Among the six experts, two are webmasters of academia web sites, two are webmasters of commercial web sites, one is a web developer in a commercial company, and the last one is a professor teaching web development courses in a university. On average, they have 7 years of professional experience in web technology. To ensure the reliability of the weights, we conducted a reliability test on the experts' answers. The reliability score (Cronbach's alpha) calculated for the experts' answers was 0.89 which was well above the 0.70 required for acceptable scale reliability (Nunnally, 1978). The TS attributes and their weights are summarized in Table 8.3a.
2. *Content richness (CR) attributes:* In traditional media richness studies, researchers only focused on the variety of media used to deliver information (Trevino et al., 1987; Palmer and Griffith, 1998). However, to have a deep understanding of the richness of Dark Web contents, we would like to measure not only the variety of the media but also the amount of information delivered by each type of media. In this chapter, we expand the media richness concept by taking the volume of information into consideration. More specifically, as shown in Table 8.3b, we calculated the average number of four types of web elements as the indication of Dark Web content richness: hyperlinks, downloadable documents, images, and video/audio files.

3. *Web interactivity (WI) attributes:* For the web interactivity attributes (see Table 8.3c), we followed the standard built by the UNPAN and the European Commission's IST program as well as Chou's (2003) work to group the attributes into three levels: the one-to-one-level interactivity, the community-level interactivity, and the transaction-level interactivity. The one-to-one-level interactivity contains five attributes (i.e., e-mail feedback, e-mail list, contact address, feedback form, and guest book) that provide basic one-to-one communication channels for Dark Web users to contact the terrorist web site owners (see Table 8.3c). The community-level interactivity contains three attributes (i.e., private message, online forum, and chat room) that allow Dark Web site owners and users to engage in synchronized many-to-many communications with each other. The transaction-level interactivity contains three attributes (i.e., online shop, online payment, and online application form) that allow Dark Web users to complete tasks such as donating to terrorist/extremist groups, applying for group membership, etc. The presence of these attributes in the Dark Web sites indicates how well terrorists/extremists utilize Internet technology to facilitate their communication with their supporters.

Similar to the TS attributes, different weights should be assigned to the WI attributes to indicate their different levels of support on communications. We asked web experts to assign weights of 1 to 10 to the WI attributes in the same e-mail survey where the TS attributes' weights were determined. The WI attributes and their weights are summarized in Table 8.3c.

We developed strategies to efficiently and accurately identify the presence of the DWAS attributes from Dark Web sites. The TS and CR attributes are marked by HTML tags in page contents or file extension names in the page URL strings. For example, an HTML tag "<image>" indicates that an image is inserted into the page content. A URL string ending with ".jsp" indicates that the page utilizes JSP technology. We developed programs to automatically analyze Dark Web page contents and URL strings to extract the presence of the TS and CR attributes. Since there are no clear indications or rules that a program could follow to identify WI attributes from Dark Web contents with a high degree of accuracy, we developed a set of coding schemes to allow human coders to identify their presence in Dark Web sites. Technical sophistication, content richness, and web interactivity scores are calculated for each web site based on the presence of the attributes to indicate how advanced and effective the site is in terms of supporting terrorist/extremist groups' communications and interactions.

4 Case Study: Understanding Middle Eastern Terrorist Groups

To test our proposed approach, we conducted a case study to collect and analyze the web presence of major Middle Eastern terrorist groups. We also conducted a benchmark comparison between the terrorist/extremist web sites and US federal and state government web sites to evaluate the terrorist/extremist organizations' online capabilities. The terrorist/extremist groups we studied mainly include Islamic terrorist groups rooted in Middle Eastern countries, for example, al-Qaeda, Palestinian Islamic Jihad, Hamas, etc. These terrorist/extremist groups are the focus of most current counterterrorism studies. We chose US government web sites as benchmarks because government web sites and terrorist/extremist web sites have common overall objectives – to inform the public about their goals, programs, and strategies. To achieve this objective, similar web features must be implemented in both government and terrorist/extremist web sites. Furthermore, the US government was ranked the top in the world by the CyPRG group (http://www.cyprg.arizona.edu/) in terms of web technical sophistication and interactivity. With the US government web sites as high-standard benchmarks, we can better understand the terrorist/extremist web sites' levels of technical advancement and effectiveness.

4.1 Building Dark Web Research Test Bed

Following the collection-building procedure discussed in Sect. 3.1, we created a Middle Eastern terrorist/extremist web site collection and a US government web site collection as the test beds for this study.

The Middle Eastern terrorist/extremist web collection was created in June of 2004. We identified 36 Middle Eastern terrorist/extremist groups from authoritative sources mentioned in Sect. 3.1. Based on the information of these terrorist/extremist groups, we constructed a lexicon of Middle Eastern terrorism keywords with the help of Arabic language experts. Examples of relevant keywords include terrorist leaders' names such as "الشيخ المجاهد بن لادن" ("Sheikh Mujahid bin Laden"), terrorist groups' names such as "خلق ايران" ("Khalq Iran"), and special words used by terrorists/extremists such as "حرب صليبية" ("Crusader's War") and "الكفار" ("Infidels"). This lexicon was used to query major search engines for identification and retrieval of terrorist/extremist groups' URLs. The URLs identified from the search engines, together with the terrorist/extremist URLs listed in the terrorism literature and reports, served as seed URLs for the out-link and in-link expansion process. We performed a one-level-deep in-link expansion using Google's in-link search tool and a one-level-deep out-link expansion. After carefully filtering the expansion results, we obtained the URLs of 86 Middle Eastern terrorist/extremist web sites. Using SpidersRUs, a digital library building toolkit developed by our group, we collected about 222,000 multimedia web documents from the identified terrorist/extremist web sites.

Table 8.4 summarizes the detailed file-type breakdown of the terrorist/extremist collection; 179,223 out of the total 222,687 documents in the terrorist/extremist collection are indexable files.

These are textual files such HTML files, plain text files, PDF/Word documents, and dynamic files generated by web applications (e.g., ASP, JSP, etc.). Interestingly, the majority of indexable files (130,972 files out of 179,223 total files) in the terrorist/

Table 8.4 Middle eastern terrorist/extremist web collection file types

Terrorist/extremist collection	# Files	Volume (bytes)
Grand total	222,687	12,362,050,865
Indexable files total	179,223	4,854,971,043
HTML files	*44,334*	*1,137,725,685*
Word files	*278*	*16,371,586*
PDF files	*3,145*	*542,061,545*
Dynamic files	*130,972*	*3,106,537,495*
Text files	*390*	*45,982,886*
PowerPoint files	*6*	*6,087,168*
XML files	*98*	*204,678*
Multimedia files total	35,164	5,915,442,276
Image files	*31,691*	*525,986,847*
Audio files	*2,554*	*3,750,390,404*
Video files	*919*	*1,230,046,468*
Archive files	1,281	483,138,149
Nonstandard files	7,019	1,108,499,397

extremist collection are dynamic files. We conducted a preliminary analysis on the contents of these dynamic files and found that most dynamic files were forum postings. This indicates that online forums play an important role in terrorists/extremists' web usage. Other than indexable files, multimedia files also make a significant presence in the terrorist/extremist collection. While the quantity of multimedia files is not as large as the indexable files, multimedia files are the largest category in the collection in terms of their volume. This indicates heavy use of multimedia technologies in terrorist/extremist web sites. The last two categories, archive files (1,281 files) and nonstandard files (7,019 files), made up less than 5% of the collection. Archive files are compressed file packages such as .zip files and .rar files. They could be password protected. Nonstandard files are files that cannot be recognized by the Windows operating system. These files may be of special interest to terrorism researchers and experts because they could be encrypted information created by terrorists/extremists. Further analysis is needed to study the contents of these two types of files.

The benchmark US government web collection was built in July of 2004. All 92 federal and state government URLs under Yahoo!'s "Government" category were selected as seed URLs. Around 277,000 web documents were automatically collected from these government web sites using the SpidersRUs toolkit. The detailed file type breakdown of the US government web collection is summarized in Table 8.5. The file-type distribution of the government collection is similar to the terrorist/extremist collection. Indexable files (221,684 files) are the largest category, the majority of which are dynamic files (145,590 files). However, in the government collection, we did not find as many forum postings as in the terrorist/extremist collection. Many dynamic files in the government collection are articles dynamically retrieved from large-document databases at users' requests. Multimedia files also have a significant presence in the government collection, indicating heavy multimedia usage in government web sites.

Table 8.5 US government web collection file types

US government collection	# Files	Volume (bytes)
Grand total	277,274	19,341,345,384
Indexable files total	221,684	6,502,288,302
HTML files	*71,518*	*2,632,912,620*
Word files	*298*	*210,906,045*
PDF files	*841*	*663,293,376*
Dynamic files	*145,590*	*2,071,734,849*
Text files	*2,878*	*555,403,447*
Excel files	*4*	*98,560*
PowerPoint files	*5*	*725,017*
XML files	*554*	*367,214,389*
Multimedia files total	49,582	10,835,029,216
Image files	*45,707*	*850,011,712*
Audio files	*3,429*	*8,153,419,931*
Video files	*449*	*1,831,597,573*
Archive files	538	286,312,990
Nonstandard files	5,471	1,717,714,876

4.2 Collection Analysis and Benchmark Comparison

Following the DWAS approach, the presence of technical sophistication and media richness attributes was automatically extracted from the collections using programs. The presence of web interactivity attributes was extracted from each web site by language experts based on the coding scheme in DWAS. Because of the time limitation, language experts examined only the top two levels of web pages in each web site. For each web site in the two collections, three scores (technical sophistication, content richness, and web interactivity) were calculated based on the presence of the attributes and their corresponding weights in DWAS. Statistical analysis was conducted to compare the advancement/effectiveness scores achieved by the terrorist/extremist collection and the US government collection.

4.2.1 Benchmark Comparison Results: Technical Sophistication

The technical sophistication comparison results are shown in Table 8.6. The results showed that:

- The US government web sites are significantly more advanced than the terrorist web sites in terms of basic HTML techniques ($p<0.0001$). Government agencies paid a great deal of attention to the design of their web sites, and they used many of the HTML features to organize their web contents. Terrorists/extremists, on the other hand, did not organize the contents on their web sites very well.
- The US government web sites are significantly more advanced than the terrorist web sites in terms of utilizing dynamic web programming languages ($p=0.0066$). Most government web sites employed web programming technologies (e.g., PHP,

Table 8.6 Technical sophistication comparison results

TS attributes	Weighted average score		t-Test result
	US	*Terrorists*	
Basic HTML techniques	**0.9130434**	0.710526	$\mathbf{p < 0.0001}$**
Embedded multimedia	0.565217	**0.833333**	$\mathbf{p = 0.0027}$**
Advanced HTML	1.789855	1.771929	$p = 0.139$
Dynamic web programming	**2.159420**	1.407894	$\mathbf{p = 0.0066}$**
Average	1.356884	1.180921	$p = 0.06$

**Significant level is at 0.05

ASP, JSP, etc..) to implement functionalities such as user login, online application, online purchase, etc. Few terrorist/extremist web sites implemented such dynamic functionalities.

- There is no significant difference between the terrorist web sites and the US government web sites in terms of applying advanced HTML techniques at a significant level of 0.05 ($p = 0.139$).
- The terrorist web sites have a significantly higher level of embedded media usage than the US government web sites ($p = 0.0027$). This unique characteristic of terrorist/extremist web sites is discussed in detail below.
- When taking all four sets of attributes into consideration, there is no significant difference between the technical sophistication of the Middle Eastern terrorist web sites and the US government web sites at a significant level of 0.05 ($p = 0.06$).

The extensive use of media in terrorist/extremist groups' web sites is of special interest. While the terrorist/extremist groups are not as good as the US government in terms of organizing their web pages into clear layouts or implementing dynamic web functionalities, they employed a significantly higher level of embedded multimedia techniques, especially images and audio/video clips, to catch the interest of their target audience. In the terrorist/extremist groups' collection, 46% of the web sites embedded audio/video clips into their pages, while only 29% of the US government web sites provided audio/video clips.

Multimedia content is more attractive and tends to leave a stronger impression on people than pure textual content. For example, the militant Islamic group Hamas foments a violent resistance to their "enemies" by disseminating graphic posters on their web sites (see Fig. 8.3). Moreover, terrorists often post images, audio, or video clips from their leaders or martyrs to boost the spirits of their members and supporters. For example, Osama bin-Laden's portrait appears on homepages of many Middle Eastern terrorist/extremist web sites. Recently, posters of the Iraqi terrorist leader Abu Mus'ab Zarqawi, who is suspected to be responsible for the beheading of several Western hostages, can also be found in Middle Eastern terrorist web sites (see Fig. 8.4). These posters explicitly mention that Abu Mus'ab Zarqawi is a "beheader" and praise his brutal killing of innocents as a way to protect Iraq. Terrorists/extremists also post images and audio/video clips of their "martyrdom operations" as a way to demonstrate their resolve to fight their enemies and inspire

Fig. 8.3 A Hamas poster inviting men to join their military struggle. The text on the poster says, "Have you fought for the sake of God? You say no. Then you should have your mouth shot" (Source: http://www.palestine-info.com)

Fig. 8.4 A poster depicting terrorist leader in Iraq, Abu Mus'ab Zarqawi. The text on the poster says, "Emir Zarqawi, may God save him. Eagle of Iraq, volcano of jihad, and the beheader" (Source: http://www.islamic-f.net/vb/)

Table 8.7 Content richness comparison results

CR attributes	Average counts per sites		t-Test result
	US	Terrorists	
Hyperlink	**3513.254654**	3172.658483	$p<0.0001$**
Downloaded documents	**400.9674532**	151.868427	$p=0.0103$**
Image	582.352456	540.0484563	$p=0.466$
Video/audio file	**91.55434783**	50.9736828	$p<0.0001$**

**Significant level is at 0.05

their supporters. Many movie clips of several suicide bombing attacks in Iraq were posted by terrorists in one of the terrorist online forums (http://wwwlb.dm.net.lb/ubb/Forum4/) to show off their "triumph over the US invaders." The "Fighting Islamic Group" posted a set of detailed documentations with pictures describing their assassination attempt of Libyan president Mu'amar Khadafi and praising the "heroism" of their members.

The multimedia content posted on terrorist/extremist web sites is not only for terrorist supporters but for enemies. For example, the video clip of American Nicholas Berg being beheaded was spread to the public from a Malaysian terrorist web site. The video of the final minutes of another American hostage, Robert Jacobs, was first posted on Middle Eastern militant group web sites. We also found that an Iraqi terrorist/extremist group posted pictures of executed "traitors" on their web sites, warning other Iraqi people not to cooperate with the US Forces. Materials of such nature are usually considered to be too shocking to televise by most TV news producers. However, through the Internet, terrorists/extremists have successfully spread these gruesome materials to as many people as possible, especially in the West where Internet use is more common.

4.2.2 Benchmark Comparison Results: Content Richness

The content richness comparison results are summarized in Table 8.7. The results showed that:

- The US government web sites provided significantly more hyperlinks ($p<0.0001$), downloadable documents ($p=0.0103$), and video/audio clips ($p<0.0001$) than the terrorist/extremist web sites.
- The US government web sites provided more images than the terrorist/extremist web sites, but the difference is not significant at a significant level of 0.05.
- Overall, the terrorist/extremist web sites are not as good as the US government web sites in terms of content richness ($p<0.0001$) because the volume of contents in terrorist/extremist web sites is often smaller than US government web sites.

The content richness comparison results are not contradictory with the technical sophistication comparison results. The content richness results showed that the US government web sites provide a larger volume of multimedia content; while the technical sophistication results indicated that a higher percentage of terrorist/extremist groups' web sites provide multimedia contents. The terrorist/extremist web sites also utilize more advanced technology to deliver their multimedia contents.

One possible explanation for the smaller volume of multimedia content provided by the terrorist/extremist groups' web sites is the lower capacity and instability of terrorists/extremists' web servers. Unlike the US government web sites which are usually hosted on dedicated web servers, many of the terrorist/extremist groups' web sites in our collection are hosted on web servers provided by free public ISPs such as GeoCities. The public web servers usually have restrictions on the size and bandwidth of the web sites they host. The restrictions would limit terrorist/extremist groups' ability to host multimedia information on their web sites. The instability of the terrorist/extremist groups' web sites also makes it difficult for them to host multimedia information. Many web sites frequently move their web contents to other web servers because their old sites were shut down by ISPs or hacked. While textual web pages can be quickly and easily duplicated to the new servers, multimedia documents are more difficult to transfer and more prone to loss because of their larger size.

Nevertheless, terrorist/extremist groups still manage to host a considerable amount of downloadable documents and multimedia information on their web sites. These media cover a wide variety of topics ranging from propaganda campaigns to tutorials of weapon operations and guerilla tactics. For example, the web site of extremist cleric Sheikh Hamed Al Ali (see Fig. 8.5) hosts a list of audio clips consisting of preaching in the Salafi ideology and political issues. The Anbaar Iraqi terrorist/extremist group's web site (see Fig. 8.6) provides a collection of songs and hymns praising the "Holy war" that they are conducting.

4.2.3 Benchmark Comparison Results: Web Interactivity

Table 8.8 summarizes the web interactivity comparison results. The results showed that:

- In terms of supporting one-to-one-level interactivity, the US government agencies are doing significantly better than terrorist/extremist web sites by providing their contact information (e.g., e-mail, mail address, etc.) on their sites ($p=0.024$). Because of their covert nature, terrorist/extremist groups seldom disclose their contact information on their web sites.

In terms of supporting community-level interactivity, terrorist/extremist web sites are doing significantly better than government web sites.

- By providing online forums and chat rooms ($p=0.0025$). Few government agencies provided such online forum and chat room support on their web sites.
- Our experts did not identify transaction-level interactivity in terrorist/extremist web sites, although such interactivity might be hidden in their sites.

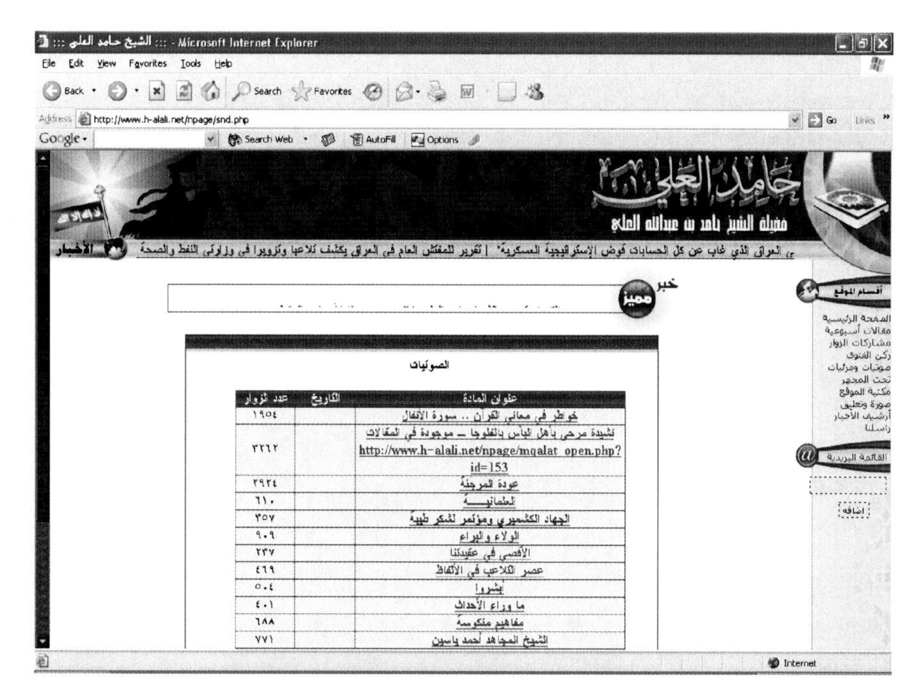

Fig. 8.5 A list of audio clips from the web site of extremist cleric Sheikh Hamed Al Ali which consists of preaching in the Salafi ideology and political issues (Source: http://www.h-alali.net)

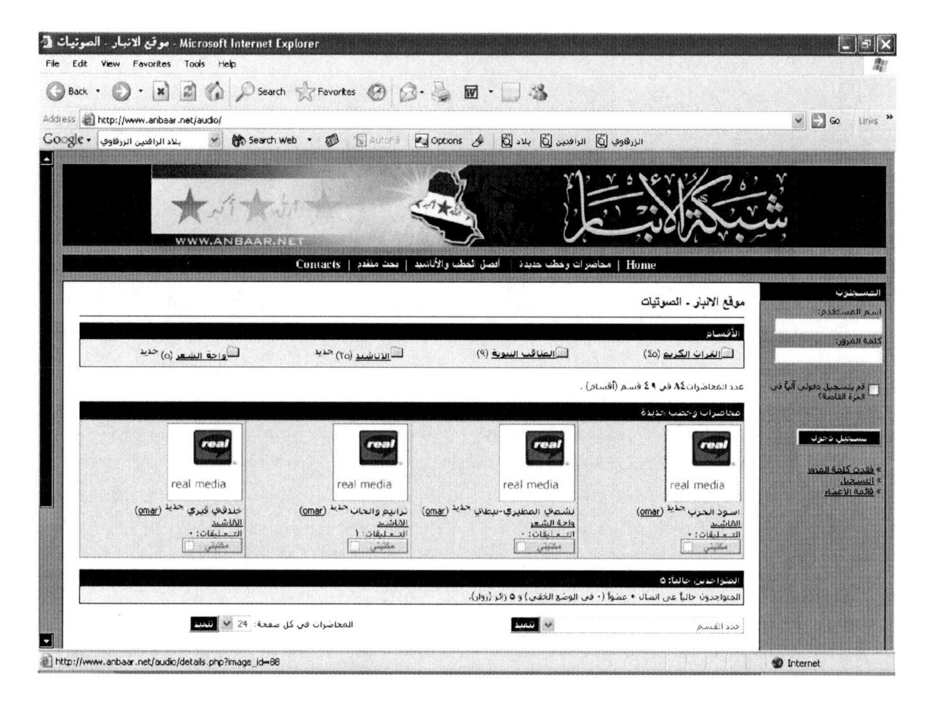

Fig. 8.6 "Holy war" songs and hymns presented on Anbaar Iraqi terrorist/extremist group's web site audio section (Source: http://www.anbaar.net/audio)

Table 8.8 Web interactivity comparison results

WI attributes	Weighted average score US	Terrorist	t-Test result
One-to-one	**0.342857**	0.292169	$p=0.024$**
Community	0.028571	**0.168675**	$p=0.0025$**
Transaction	0.3	*Not presented*	
Average (transaction not included)	0.185714	0.230422	$p=0.056$

**Significant level is at 0.05

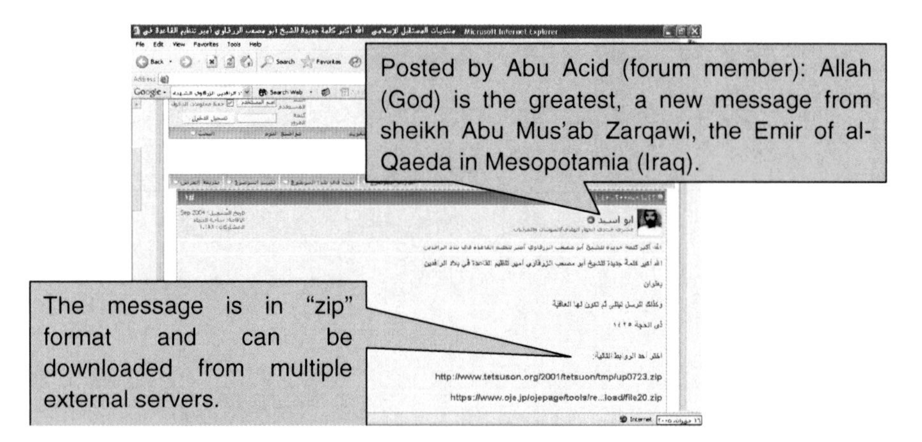

Fig. 8.7 Discussion forums are used to share important messages from terrorist leaders among the members of the terrorist groups and their supporters (Source: www.islamic-f.net)

- Taking both one-to-one and community-level interactivity into consideration, we did not find significant difference between the terrorist/extremist web sites and the US government web sites ($p=0.056$) at a significant level of 0.05.

Several previous studies implied that terrorists are relying on Internet-based communication tools such as online chat rooms and forums to facilitate their daily communication, command and control, and even operation planning and coordination (Zhou et al., 2005; Whine, 1999; FBIS, 1995). Our results further confirmed these observations. The Middle Eastern terrorist/extremist groups are very active in terms of hosting and maintaining online forums and bulletin boards. Among the largest terrorist-supporting forums that we have been monitoring, www.shawati.com has 31,894 registered forum members and 418,196 posts; www.kuwaitchat.net has 11,531 registered members and 624,694 posts. Not all of the forum members are terrorists or extremists. Many of them are just supporters or sympathizers. Members of these large forums participate in daily discussions, express their support of the terrorist groups, and reinforce each other's beliefs in the terrorist/extremist groups' courses. They sometimes can get messages directly from active members of terrorist/extremist groups. For example, messages from the Iraqi terrorist leader Abu Mus'ab Zarqawi can often be found at the online forum www.islamic-f.net (see Fig. 8.7). These dynamic forums provide snapshots of terrorist/extremist groups' activities, communications, ideologies, relationships, and evolutionary developments.

5 Conclusions and Future Directions

In this chapter, we proposed a systematic procedure to collect Dark Web contents and a Dark Web Attribute System (DWAS) to enable quantitative analysis of terrorists' tactical use of the Internet. The automatic collection building and content analysis components used in the proposed methodology allow the efficient collection and analysis of thousands of Dark Web documents. This enables our Dark Web study to achieve a higher level of comprehensiveness than previous manual approaches. Furthermore, the DWAS is a systematic content analysis tool that, we believe, brings more insights into terrorist/extremist groups' Internet usage than previous observation-based studies provided.

Using the proposed collection-building procedure and framework, we built a high-quality Middle Eastern terrorist/extremist group web collection and benchmarked it against the US government web site collection. The results showed that terrorist/extremist groups adopted levels of web technologies similar to that of US government agencies. Moreover, terrorists/extremists had a strong emphasis on multimedia usage, and their web sites employed significantly more sophisticated multimedia technologies than government web sites. We also found that terrorists/extremists seem to be as effective as the US government agencies in terms of supporting communication and interaction using web technologies. More specifically, terrorists/extremists make heavy use of web forums to facilitate their communication and coordination.

Our study provides insights for policy makers to better apply counterterrorism measures on the web. Our results showed that Internet technologies, especially forums and chat rooms, have become a major means for terrorists/extremists to reach out to a broad audience. They have invested a significant amount of effort and technical expertise into building their web infrastructure. Security and law enforcement experts should pay more attention to terrorists/extremists' online communication. We identified very high levels of communicative activities in terrorist/extremist forums in our collection. Some documents in our collection were not readable using conventional applications. Some of these documents might contain hidden information from terrorists/extremists. Monitoring and deciphering such hidden messages could help disrupt terrorist/extremist communication and prevent terrorism attacks. Furthermore, we believe that the proposed Dark Web research methodology could contribute to the terrorism research domain. The richness of the Dark Web contents calls for more studies being devoted to this domain to help enrich our understanding of terrorists/extremists' Internet usage, online propaganda campaigns, and their psychological warfare strategies.

We have several future research directions to pursue. First, we plan to experiment with better data analysis methods and collaborate with more terrorism/extremism domain experts to better analyze and interpret our study results. For example, for the content richness comparisons, we would like to conduct a more detailed study to compare the richness of terrorist/extremist web sites to government web sites based on the percentage of each type of media in the overall contents. We also plan to conduct a cross-comparison which takes both the TS and WI attributes into consideration to gain more insight about the correlation of these attributes. Second, we plan to cooperate with web technology experts to further improve the DWAS by incorporating additional attributes and adjusting the relevant weights. Third, we plan to expand the scope of our study by conducting a comparative analysis of terrorist/extremist groups' web sites across different regions of the world. We also plan to conduct a time series analysis study on the Dark Web to analyze the evolution and diffusion of terrorist/extremist groups' web presence. Last but not least, we also plan to explore more advanced machine-learning techniques to detect the technology and media usage patterns in terrorist/extremist web sites to gain more insights into terrorists/extremists' technology usage.

Acknowledgments This research has been supported in part by the following grants: (1) NSF, "COPLINK Center: Information and Knowledge Management for Law Enforcement," July 2000–September 2005; (2) NSF/ITR, "COPLINK Center for Intelligence and Security Informatics Research – A Crime Data Mining Approach to Developing Border Safe Research," September 2003–August 2005; and (3) DHS/CNRI, "BorderSafe Initiative," October 2003–March 2005.

We would like to thank Dr. Joshua Sinai, formerly at the Department of Homeland Security; al-Qaeda expert, Dr. Marc Sageman; Dr. Chip Ellis, from the Memorial Institute for the Prevention of Terrorism; and all the other anonymous domain experts for their insightful comments and suggestions on our project. We would also like to thank all members of the Artificial Intelligence Lab at the University of Arizona who have contributed to the project, in particular, Homa Atabakhsh, Cathy Larson, Chun-Ju Tseng, and Shing Ka Wu.

References

Armstrong, H. L. and Forde, P. J., "Internet Anonymity Practices in Computer Crime," Information Management and Computer Security, 11(5), pp. 209–215, 2003.

Arquilla, J. and Ronfeldt, D., "Cyber War Is Coming!" Comparative Strategy, 12(2), 1993.

Becker, A. "Technology and Terror: the New Modus Operandi," Frontline, 2005, available at http://www.pbs.org/wgbh/pages/frontline/shows/front/special/tech.html.

Bowers, F., "Terrorists Spread their Messages Online," Christian Science Monitor, July 28, 2004, available at http://www.csmonitor.com/2004/0728/p03s01-usgn.htm.

Chen, H., Qin, J., Reid, E., Chung, W., Zhou, Y., Xi, W., Lai, G., Bonillas, A. A. and Sageman, M., "The Dark Web Portal: Collecting and Analyzing the Presence of Domestic and International Terrorist Groups on the Web," In Proc. of International IEEE Conference on Intelligent Transportation Systems, 2004.

Chou, C., "Interactivity and Interactive Functions in Web-based Learning Systems: A Technical Framework for Designers," British Journal of Educational Technology, 34(3), pp. 265–279, 2003.

Coll, S. and Glasser, S. B., "Terrorists Turn to the Web as Base of Operations," Washington Post, Aug 7, 2005.

Demchak, C., Friis, C., and La Porte, T. M., "Webbing Governance: National Differences in Constructing the Face of Public Organizations," Handbook of Public Information Systems. G. D. Garson. NYC, Marcel Dekker, 2000.

Denning, D. E., "Information Operations and Terrorism," Journal of Information Warfare (draft), 2004, available at http://www.jinfowar.com.

Elison, W., "Netwar: Studying Rebels on the Internet," The Social Studies, 91, pp. 127–131, 2000.

FBIS, "Arab Afghans Said to Launch Worldwide Terrorist War," Paris al-Watan al-'Arabi, FBIS-TOT-96–010-L, pp.22–24, December 1, 1995.

Hillman, D. C. A., Willis, D. J., and Gunawardena, C. N., "Learner-interface Interaction in Distance Education: An Extension of Contemporary Models and Strategies for Practitioners," The American Journal of Distance Education, 8(2), pp. 30–42, 1994.

ISTS, "Examining the Cyber Capabilities of Islamic Terrorist Groups," Report, Institute for Security Technology Studies, 2004. http://www.ists.dartmouth.edu/.

Internet Haganah, Internet Haganah report, 2005, available at http://en.wikipedia.org/ wiki/ Internet_Haganah.

Jackson, Brian J., Technology Acquisition by Terrorist Groups: Threat Assessment Informed by Lessons from Private Sector Technology Adoption, Studies in Conflict and Terrorism, vol. 24, pp. 183–213, 2001.

Jenkins, B. M., "World Becomes the Hostage of Media-Savvy Terrorists: Commentary," USA Today, August 22, 2004. http://www.rand.org/.

Jesdanun, A., "WWW: Terror's Channel of Choice," CBS News, June 20, 2004.

Kelley, J., "Terror Groups Hide Behind Encryption," USA Today, Feb 5, 2001, available at http://www.usatoday.com/tech/news/2001-02-05-binladen.htm.

Muriel, D., "Terror Moves to the Virtual World," CNN News, April 8, 2004, available at http://edition.cnn.com/2004/TECH/04/08/internet.terror/.

Nunnally, J., Psychometric Theory. McGraw Hill, New York, 1978.

Palmer, J. W. and Griffith, D. A., "An Emerging Model of Web Site Design for Marketing," Communications of the ACM, 41(3), pp. 45–51, 1998.

Preece, J., Online Communities: Designing Usability, Supporting Sociability., New York City, Wiley, 2000.

Reid, E., Qin, J., Chung, W., Xu, J., Zhou, Y., Schumaker, R., Sageman, M., and Chen H., "Terrorism Knowledge Discovery Project: a Knowledge Discovery Approach to Addressing the Threats of Terrorism," in Proc. of 2nd Symposium on Intelligence and Security Informatics, ISI 2004, Tucson, Arizona, 2004.

Schneider, S. M., Foot, K., Kimpton, M., and Jones, G., "Building Thematic Web Collections: Challenges and Experiences from the September 11 Web Archive and the Election 2002 Web Archive," in Proc. of the 3rd ECDL Workshop on Web Archives, Trondheim, Norway, August 2003.

Tekwani, S., "Cyberterrorism: Threat and Response," Institute of Defence and Strategic Studies, in Proc. of the Workshop on the New Dimensions of Terrorism, 21–22, Singapore, 2002.

Thomas, T. L., "Al Qaeda and the Internet: The Danger of 'Cyberplanning,'" Parameters, Spring 2003, pp. 112–23, available at http://carlisle-www.army.mil/usawc/ Parameters/03spring/ thomas.htm.

Tsfati, Y. and Weimann, G., "www.terrorism.com: Terror on the Internet," Studies in Conflict and Terrorism, 25, pp. 317–332, 2002.

Trevino, L. K., Lengel, R. H., and Daft, R. L., "Media Symbolism, Media Richness, and Media Choice in Organizations: A Symbolic Interactionist Perspective," Communication Research, 14(5), pp. 553–574, 1987.

Weimann, G., "www.terror.net: How Modern Terrorism Uses the Internet," Special Report, U.S. Institute of Peace, 2004, Available at http://www.usip.org/pubs/ specialreports/sr116.pdf.

Whine, M., "Cyberspace: A New Medium for Communication, Command and Control by Extremists," 1999, available at http://www.ict.org.il/articles/ cyberspace.htm.

Zhou, Y., Reid, E., Qin, J., Chen, H., and Lai, G., "U.S. Domestic Extremist Groups on the Web: Link and Content Analysis," IEEE Intelligent Systems Special Issue on Homeland Security, 20(5), pp. 44–51, 2005.

Chapter 9
Authorship Analysis

1 Introduction

Analysis of web content is becoming increasingly important due to amplified communication via Internet sources such as e-mail, web sites, and forums. The anonymous nature of these channels makes them an ideal source of contact for militant groups and terrorist organizations. Furthermore, the global nature of criminal activity necessitates the exploration of online communication in a multilingual context. Application of authorship analysis techniques across multilingual web media is of dire importance for assisting in the identification and prevention of potential criminal activity with national security implications.

Specifically, Arabic has garnered greater attention in recent years for sociopolitical reasons and ties between Middle Eastern groups and terrorism; however, there has been an absence of studies aimed at applying authorship techniques across the language. The morphological challenges pertaining to Arabic pose several critical problems for authorship identification, which could be partially responsible for the lack of previous research relating to Arabic.

We modified an existing framework for the application of authorship analysis of online messages and applied it to Arabic and English web forum messages associated with extremist groups. Special multilingual components were developed for the extraction and identification of Arabic messages. These components were geared toward addressing the unique characteristics of the language. Furthermore, a complex message extraction component was incorporated in order to allow the use of a more comprehensive set of features tailored specifically toward online messages.

H. Chen, *Dark Web: Exploring and Data Mining the Dark Side of the Web*,
Integrated Series in Information Systems 30, DOI 10.1007/978-1-4614-1557-2_9,

2 Literature Review: Authorship Analysis

Authorship analysis is the process of evaluating writing characteristics in order to make inferences about authorship. It is rooted in the linguistic area known as stylometry, which is defined as the statistical analysis of literary style. Two major categories of authorship analysis are authorship identification and authorship characterization.

Authorship identification deals with attributing authorship of unidentified writing based on stylistic similarities between the author's known works and the unidentified piece. In contrast, a*uthorship characterization* attempts to formulate an author profile by making inferences about gender, education, and cultural background based on writing style. Authorship identification deals with classification problems, whereas authorship characterization is used for clustering.

We are primarily concerned with the application of authorship identification to English and Arabic online messages. In order to apply authorship identification successfully, we had to begin by determining the relevant features and techniques to utilize. Unfortunately, the lack of consensus in previous authorship analysis literature coupled with the application of identification methodologies to a new language made this task an arduous endeavor.

2.1 Writing Style Features

Writing style features are characteristics that can be derived from a message in order to facilitate authorship attribution. Numerous types of features have been used in previous studies including n-grams and the frequency of spelling and grammatical errors; however, four categories used extensively are lexical, syntactic, structural, and content-specific features.

Lexical features can be broken into two categories: word-based and character-based. *Word-based* lexical features include total number of words, words per sentence, word-length distribution, vocabulary richness, etc. *Character-based* lexical features include total number of characters, characters per sentence, characters per word, and the usage frequency of individual letters.

Syntax refers to the patterns used for the formation of sentences. This category of features is comprised of the tools used to structure sentences, such as punctuation and function words. Examples of function words are "while" and "upon." Usage patterns of function words can be effective features for authorship identification. For example, the difference between using the word "thus" or "hence" may appear subtle but can constitute a significant stylistic difference.

Structural features deal with the organization and layout of the text. This set of features has been shown to be particularly important for online messages (De Vel et al. 2001). Previously, structural features used have focused on *word structure* such as the use of greetings and signatures, or the number of paragraphs and average paragraph length. While these features are important discriminators, online messages also provide additional information such as fonts, images, and links, which are not captured. Although these characteristics are not writing style features per se, they provide important insight into the authorship characteristics of the writer. The use of various font sizes and colors requires a conscientious effort, hence making it a style marker. Similarly, the ability to embed images and icons or provide links to different types of web sites can be a reflection of technical prowess. Evaluation of technical characteristics measured in terms of the use of images, hyperlinks, and audio or video media is not a novel idea; it has been applied to web sites (Palmer and Griffith 1998). Thus, we propose a new subcategory of structural features called *technical structure* which encompasses font, hyperlink, and embedded image characteristics.

Content-specific features are words that are important within a specific topic domain. An example of content-specific words for a discussion on computers might be "RAM" and "laptop." The rationale for content-specific words is similar to that of other word usage features but at a finer level of granularity.

2.2 Analysis Techniques

The two most commonly used analytical techniques for authorship attribution are statistical and machine learning approaches. Many multivariate statistical approaches, such as principal component analysis, have been shown to provide a high level of accuracy. However, some of the pitfalls associated with statistical approaches include the need for more stringent models and assumptions.

Drastic increases in computational power have caused the emergence of machine learning techniques such as support vector machines (SVM), neural networks, and decision trees. These techniques have gained wider acceptance in authorship analysis studies in recent years (Zheng et al. 2003). Machine learning approaches provide greater scalability in terms of number of features that can be handled and are less susceptible to noisier data as compared to statistical techniques. These benefits are ideal for identification of online messages which entails classification of more authors and a large feature set.

2.3 Online Messages

Online messages pose several problems for authorship identification as compared to conventional forms of writing, with the biggest concern being message length. Writing style markers are far less visible for messages shorter than a few hundred words, with identification becoming cumbersome or even impossible in such situations. The problem is further amplified by the larger pool of potential authors in online attribution situations.

Additional difficulties associated with online messages center around the casual style of online communication. E-mail and forum postings tend to be less formal than traditional writing, resulting in more misspellings, unorthodox structure, increased use of abbreviations, and improper use of punctuation. As a consequence, application of authorship identification to web content is intrinsically faced with a quagmire of noisy data.

Despite all the challenges, the unique structural characteristics of online messages may also provide helpful discriminators for identification. Greetings, signatures, quotes, links, and the use of contact information (phone, e-mail) can offer significant insight for authorship identification. As previously elaborated, this set of features can be further enhanced with the inclusion of technical structure features.

2.4 Multilingual Issues

Applying authorship identification across different languages is becoming increasingly important due to the rapid proliferation of the Internet and the ensuing threats that are created. Online analysis of Arabic is especially important due to the emergence of terrorist organizations such as al-Qaeda. Nevertheless, there has been a lack of multilingual research with the exception of a few studies done on Greek and Chinese (Peng et al. 2003; Stamatatos et al. 2001; Zheng et al. 2003). The language dimension can create enormous challenges for authorship identification since previously applied features and techniques were designed for English. For example, the lack of word segmentation in Chinese makes word-based lexical features (e.g., number of words in a sentence) difficult to extract. Additionally, the larger volume of words in Chinese makes vocabulary richness measures less effective. Similarly, Arabic also poses some unique challenges with respect to the structural and stylistic properties of the language.

3 Arabic Language Characteristics

Arabic is a Semitic language belonging to the group of Afro-Asian languages. Semitic languages have several characteristics that can cause difficulties for authorship analysis, including properties such as inflection, diacritics, word length, and elongation.

3.1 Inflection

Inflection is the derivation of stem words from a root. There are approximately 5,000 roots in Arabic with each root being a 3–5-letter consonant combination (Beesley 1996). Stems are created by adding affixes (e.g., prefixes) to the root. Over 85% of Arabic words are derived from roots, and words with common roots are semantically related (Al-Fedaghi and Al-Anzi 1989). The orthographical and morphological properties of Arabic result in a great deal of lexical variation since words can take on numerous forms (Larkey and Connell 2001). Inflection creates feature extraction problems due to the larger number of possible words, impacting vocabulary richness measures (Zheng et al. 2003).

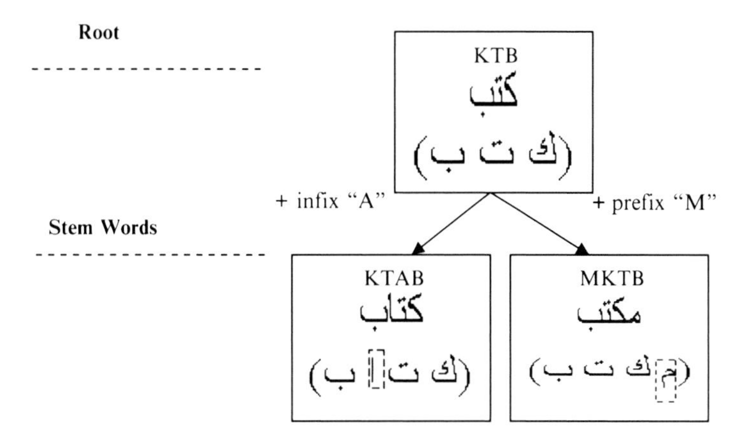

Fig. 9.1 Inflection example

Figure 9.1 shows an inflection example demonstrating the derivation of two words (KTAB, MKTB) from the root KTB. For the root and stems, the top row shows the word written using English alphabet characters, and the second row shows the word written in Arabic. Since Arabic letters are joined, making it difficult for non-Arabic readers to decipher individual letters, the third row shows the decomposed Arabic word in parentheses. The words KTAB (book) and MKTB (desk) are created with the addition of the infix "A" and the prefix "M," respectively.

3.2 *Diacritics*

Diacritics are markings above or below letters used to indicate special phonetic values. An example of diacritics in English would be the little mark found on top of the letter "e" in the word "résumé." These markings alter the pronunciation and meaning of the word. Arabic uses diacritics in every word to represent short vowels, consonant lengths, and relationships between words; however, they are rarely used in online communication. Although readers can use the sentence semantics to decipher proper meaning, this is not feasible for an automated extraction program. For example, the words "resume" and "résumé" would look identical to a computer without diacritics. The lack of diacritics can significantly impact the effectiveness of word-usage-based features such as function words. For example, in Arabic, it is impossible to differentiate between the words "who" and "from" without diacritics.

3.3 *Word Length and Elongation*

The shorter length of Arabic words as compared to English words may result in a reduction in the impact of many lexical features. Word-length features are less effective since Arabic word-length distributions have a smaller range. While the use

Table 9.1 Elongation example

Elongated	English	Arabic	Word length
No	MZKR	مذكر	4
Yes	M----ZKR	مـــذكر	8

of lengthier words in English can sometimes be associated with greater writing complexity, this assumption does not hold true for Arabic. Additionally, Arabic words are sometimes elongated for purely stylistic reasons using a special character that resembles a dash ("–"). Since Arabic characters are combined during writing, elongation is possible by lengthening the joins between letters. Although elongation provides an important authorship style marker, it can also create problems as illustrated by Table 9.1. The word MZKR ("remind") is extended with the addition of four dashes between the "M" and the "Z" (denoted by a faint oval) resulting in doubling of the word size.

Elongation can significantly inflate the values of word-length features. Handling elongation in terms of feature extraction is an important issue that must be resolved.

4 Research Questions and Research Design

We designed a series of experiments to test the efficacy of authorship identification techniques in an online setting. The objective of our experiments was to address numerous questions including:

- Will authorship analysis techniques be applicable in identifying authors in Arabic?
- What are the effects of using different types of features in identifying authors?
- Which classification techniques are appropriate for authorship analysis?
- How does identification performance differ between English and Arabic?
- What are the important feature differences between the English and Arabic groups and language models?

4.1 Test Bed

Our test bed consisted of English and Arabic datasets extracted from web forum messages. In both instances, 20 messages were extracted for each of 20 authors, resulting in a total of 400 messages per language. The English dataset had an average message length of 76.6 words, while the Arabic average message length was 580.69 words. The English messages were derived from a forum of a chapter of the White Knights of the Ku Klux Klan. The content associated with this group revolved around political, racial, and religious issues. Members commonly used profanities and advocated the use of violence against groups whom they detested. In addition to general anger and animosity, there were also disturbing references to specific members of society. In some instances, complete contact information including address was provided for these targeted individuals.

The Arabic dataset was extracted from forum messages associated with an extremist group called the Al-Aqsa Martyrs. These messages had a strong anti-American slant with members providing lengthy arguments in favor of their vantage point. There was an abundant use of embedded images and links relating to the war in Iraq and treatment of al-Qaeda prisoners. The image content was extremely graphic in nature and intended to be used as supporting material for the authors' central arguments. Much like the English messages, authors in the Arabic forum advocated the infliction of physical harm upon groups whom they disliked.

4.2 *Analysis Techniques*

In this chapter, we adopted two machine learning classifiers: C4.5 and SVM. C4.5 is a powerful decision tree-based classifier that has been shown to rival other techniques in terms of performance. SVM was developed on the premise of the structural risk minimization principle derived from computational learning theory and has gained popularity in the past decade.

Both techniques have been applied in previous authorship analysis research (Zheng et al. 2003). We incorporated SVM for its classification power and robustness. SVM is able to handle a large number of input values with great ease due to its capacity for dealing with noisy data. C4.5 was used for its analytical and explanatory potential. Decision trees provide an effective way to assess key differences between the English and Arabic feature sets.

4.3 *Addressing Arabic Characteristics*

In order to create an effective Arabic feature set, we had to address the morphological and orthographical properties of the language. Overcoming the diacritics problem would require the use of a semantic tagger. Since no feasible solutions exist, we decided to focus our attention toward the other challenges, namely, inflection, elongation, and word length.

4.3.1 Inflection

Word roots have been shown to provide superior performance to normal Arabic words in information retrieval. As a result of heavy inflection in Arabic, root indexing outperforms word indexing on both precision and recall (Hmeidi et al. 1997).

We complemented our feature set by tracking usage frequencies of a select set of word roots. The use of word roots was intended to help compensate for the loss in effectiveness of vocabulary richness measures. Tracking root frequencies required matching words to their appropriate roots which could be accomplished using a clustering algorithm.

De Roeck and Fares (2000) created a clustering algorithm specifically designed for Arabic, consisting of five steps. However, this algorithm is meant to compare words against other words as opposed to roots. Since we are comparing words against a list of roots (an easier task), not all parts of the algorithm are necessary. We adapted the algorithm by using three of the five steps, including blank insertion, cross, and Jaccard's similarity score equation.

Root frequencies were extracted by calculating similarity scores for each word against a dictionary containing over 4,500 roots. Words were assigned to the root with the highest similarity score, and the usage frequency of the selected root was incremented. An important issue was determining the number of roots to include in the final feature set. A trial-and-error approach was used since such methods have been used in other multilingual authorship studies due to a lack of previous research (Stamatatos et al. 2001). In order to determine the number of roots to include, between 0 and 500 of the most frequently occurring roots were added to the complete Arabic feature set. The classification power of these roots was tested using SVM as the classifier. The optimal number (50 roots) was integrated into the feature set.

4.3.2 Word Length and Elongation

Arabic words tend to be shorter than English words, with lengthier words (longer than ten characters) less common in Arabic. However, elongation of Arabic words can distort word-length distributions by artificially inflating them. The use of elongation is an important authorship style marker, and hence, the occurrence and degree (extent of stretching) of elongation should be tracked. However, a filter should be embedded into the feature extractor to remove elongation once it has been tracked in order to allow for precise capturing of word length.

4.4 Feature Sets

The English feature set was adapted from previous online authorship studies (De Vel et al. 2001; Zheng et al. 2003) and was composed of 301 features including 87 lexical, 158 syntactic, 45 structural, and 11 content-specific features. The major difference from prior studies was that our feature set was enhanced with the inclusion of technical structure features. The technical structure features fell into four categories: font color, font size, and the use of embedded images and hyperlinks.

Inspection of the datasets revealed that there were 15 different font colors used in the English messages and over 120 in the Arabic! A closer look showed that

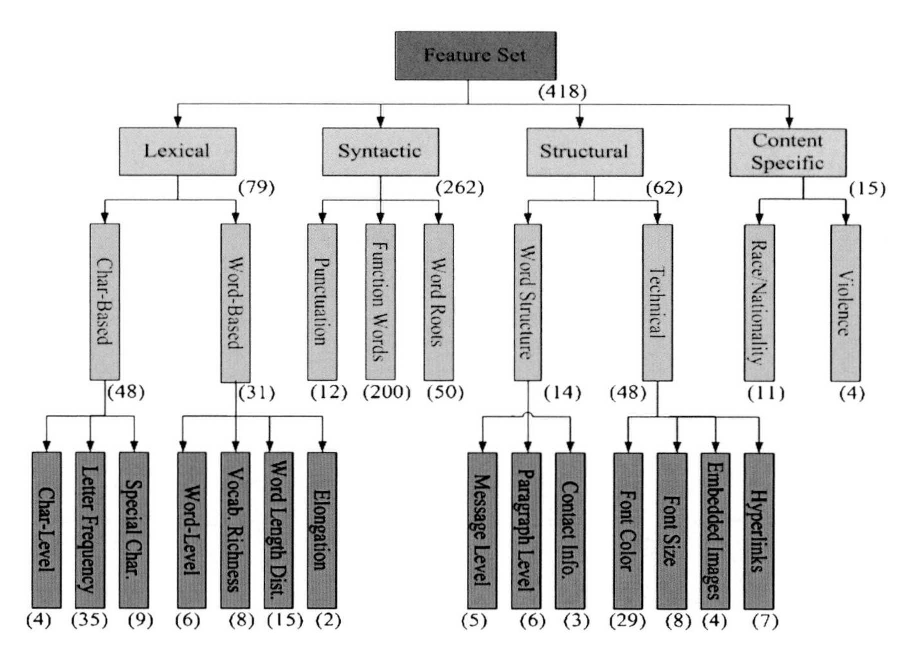

Fig. 9.2 Arabic feature set

Table 9.2 Key differences between English and Arabic feature sets

Feature type	Feature	English	Arabic
Lexical	Short word count	<=3	<=2
	Word-length distribution	1–20	1–15
	Elongation	N/A	2
Syntactic	Function words	150	200
	Word roots	N/A	50
Structural	Technical structure	31	48

many of the Arabic font colors were minor modifications of standard colors resulting in an inflated count. Since most of these modified colors were seldom used, we felt that they should not be included in the feature set in order to avoid overfitting. The consolidated color count consisted of 12 colors for English and 29 for Arabic. Other technical structure features consisted of 8 font size, 4 embedded image, and 7 hyperlink features.

The Arabic feature set, shown in Fig. 9.2, was modeled after the English feature set. It was composed of 418 features, including 79 lexical, 262 syntactic, 62 structural, and 15 content-specific features.

The differences between the English and Arabic feature sets are highlighted by Table 9.2. In order to compensate for the lack of diacritics and inflection, a larger number of function words and 50 word roots were used. A smaller word-length distribution and short word threshold were also included in the Arabic dataset.

5 Authorship Identification Procedure

The complete online authorship identification process consisted of three main steps: collection, extraction, and experimentation. Figure 9.3 shows the complete process design for Arabic authorship identification.

5.1 Collection and Extraction

Web forums were identified using spidering programs that crawled through the Internet searching for "Dark Web" material, which is content involving potentially dangerous or criminal activity that may be of interest for cybercrime and homeland security-related issues. Once the forums were recognized, collection programs stored the messages in text and HTML format. Extraction programs then derived

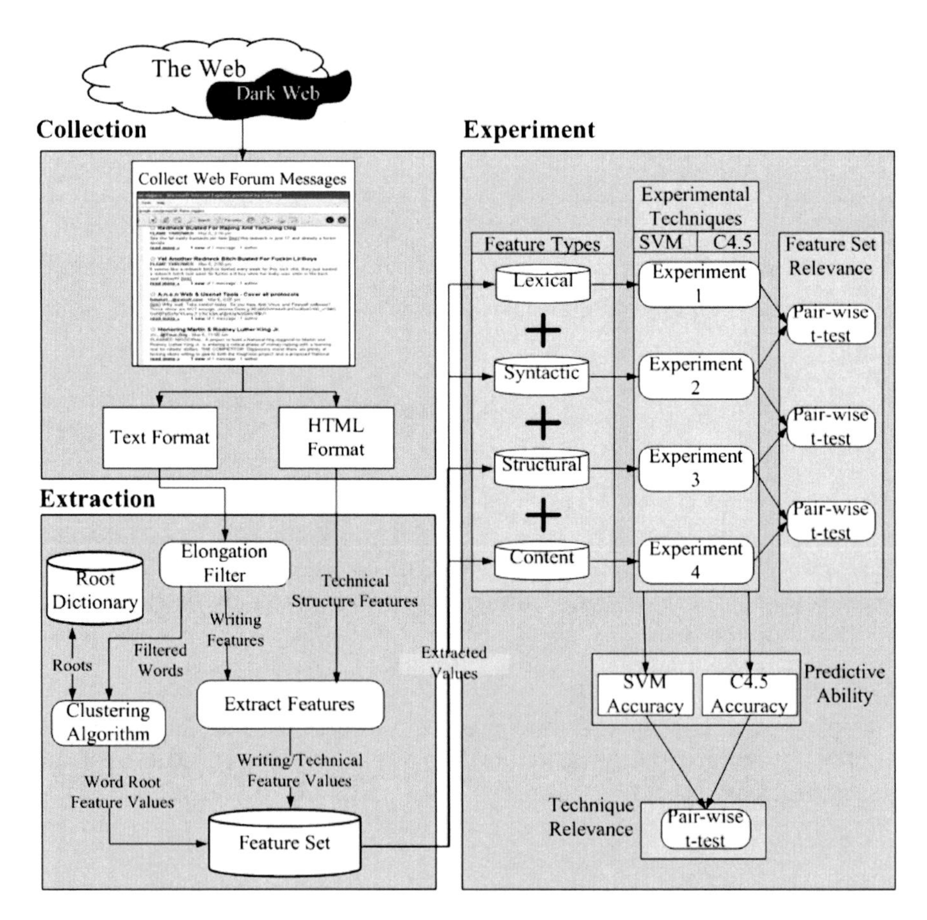

Fig. 9.3 Authorship identification procedure for Arabic

writing style characteristics identified in the feature sets from each message. The Arabic feature extractor was a bit more complex than the English one, owing to the need to account for elongation and inflection. An elongation filter, clustering algorithm, and root dictionary were integrated into the Arabic extraction process.

5.2 *Experiment*

Once the feature values were extracted, they were categorized into four feature sets. The first feature set (F1) consisted of lexical features, while the second (F2) encompassed lexical and syntactic features. Structural features were added to the first two groups in the third feature set (F3), and content-specific features were inserted with the other three categories in the fourth set (F4), which consequently contained all features (lexical, syntactic, structural, and content specific). Such a stepwise increment of features was utilized due to our perceptions concerning the order of importance of feature categories. Studies have shown that lexical and syntactic features are the most important categories and hence form the foundation for structural and content-specific features. We applied this concept in our design for testing the relevance of feature categories for online English and Arabic messages. For the experiment, we created 30 randomly selected samples of five authors which were used in all experiments. Each sample of five authors was evaluated using all 20 messages per author and 30-fold cross-validation with C4.5 and SVM. The overall accuracy was the average precision (# correctly identified/total messages) across the 30 samples. The feature type and classification accuracies were evaluated using pairwise *t* tests across the samples (n = 30).

6 Results and Discussion

Authorship identification accuracy results for the comparison of the different feature types and techniques are summarized in Table 9.3. The overall accuracies were exceptional, especially considering the difficult nature of the task and in comparison to previous authorship studies. Perhaps most surprising was the relatively small drop in performance across languages. In both datasets, the accuracy kept increasing with the addition of more feature types. The maximum accuracy was achieved with the use of SVM and all features for English and Arabic.

Table 9.3 Accuracy for different feature sets across techniques

	English dataset		Arabic dataset	
Features	C4.5 (%)	SVM (%)	C4.5 (%)	SVM (%)
F1	85.76	88.00	61.27	87.77
F1 + F2	87.23	90.77	65.40	91.00
F1 + F2 + F3	88.30	96.50	71.71	94.23
F1 + F2 + F3 + F4	90.10	97.00	71.93	94.83

Table 9.4 *P* values of pairwise *t* tests on accuracy using different feature types

Features	C4.5	SVM
t Test results for English dataset n = 30		
F1 vs. F1 + F2	0.000***	0.000***
F1 + F2 vs. F1 + F2 + F3	0.000***	0.000***
F1 + F2 + F3 vs. F1 + F2 + F3 + F4	0.000***	0.1628
t Test results for Arabic dataset n = 30		
F1 vs. F1 + F2	0.000***	0.000***
F1 + F2 vs. F1 + F2 + F3	0.000***	0.000***
F1 + F2 + F3 vs. F1 + F2 + F3 + F4	0.1216	0.0224**

**Significant with alpha = 0.05
***Significant with alpha = 0.01

Table 9.5 *P* values of pairwise *t* tests on accuracy using different classification techniques

Technique/features	F1	F1 + F2	F1 + F2 + F3	F1 + F2 + F3 + F4
t Test results for English dataset n = 30				
C4.5 vs. SVM	0.000***	0.000***	0.000***	0.000***
t Test results for Arabic dataset n = 30				
C4.5 vs. SVM	0.000***	0.000***	0.000***	0.000***

***Significant with alpha = 0.01

6.1 Comparison of Feature Types

All feature categories improved classification accuracy in the stepwise analysis of features. Pairwise *t* tests were conducted to show the statistical significance of the additional feature types added. The results shown in Table 9.4 indicate that all feature types significantly improved accuracy for Arabic and English, except for content-specific words. This category of features was statistically insignificant in two situations ($P = 0.1628$, $P = 0.1216$) and significant at a lower alpha level in a third instance ($P = 0.0224$). The weaker performance of content-specific features could be attributable to their less prominent representation in the feature set in terms of number of features. There were only 11 and 15 content-specific features used in the English and Arabic feature sets, respectively. This number is far less than all other categories of features. Overall, the impact of the different feature types for Arabic was consistent with English results.

6.2 Comparison of Classification Techniques

Table 9.5 reveals that SVM significantly outperformed the decision tree classifier in all cases. This is consistent with previous studies that have shown SVM to be better equipped to handle larger feature sets and noisier data (characteristics associated with online authorship identification). The difference in accuracy between classifiers across Arabic was far greater than English: SVM outperformed C4.5 by over 20% on all feature set combinations.

7 Analysis of English and Arabic Group Models

We evaluated the important features for the two group forums based on decision tree analysis and overall feature usage. The analysis highlighted some of the key differences between the language models and revealed some interesting trends pertaining to the English and Arabic groups.

7.1 Decision Tree Analysis

The C4.5 decision tree can be used as an effective analytical tool due to its descriptive nature. Decision trees can be visualized to look at the effect of individual features, since trees choose the features with the highest discriminatory power, measured in terms of entropy reduction. We analyzed the C4.5 trees for the English and Arabic group models and extracted a list of the important features based on decision tree outputs.

Table 9.6 highlights the key differences between the English and Arabic models, based on the decision tree evaluations. For a particular group of features, the "Used" column indicates the number of features used, while the "Total" column refers to the total number of that feature type in the feature set. The "% Used" column indicates the percentage of that feature group incorporated by the decision tree and provides a good basis for comparing the KKK and Al-Aqsa Martyrs feature usage.

The specific features integrated into the feature set for Arabic played an important role based on the decision tree analysis. Both elongation features and nearly half the word roots were deemed as vital attributes based on the C4.5 output, indicating that these are important Arabic characteristics that should be adopted in future studies. Furthermore, as expected, word length played a more critical role in the English KKK messages (40%) as compared to Arabic Al-Aqsa Martyrs messages (20%).

Table 9.6 Summary of key features based on evaluation of decision trees

	English			Arabic		
Features	Used	Total	% Used	Used	Total	% Used
Elongation	N/A	N/A	N/A	2	2	100
Word length	8	20	40	3	15	20
Punctuation	4	8	50	7	12	58.33
Function words	31	150	20.67	62	200	31
Root words	N/A	N/A	N/A	22	50	44
Word structure	8	14	57.14	8	14	57.14
Technical structure	12	31	38.71	32	48	66.67
Content-specific	3	11	27.27	3	15	20

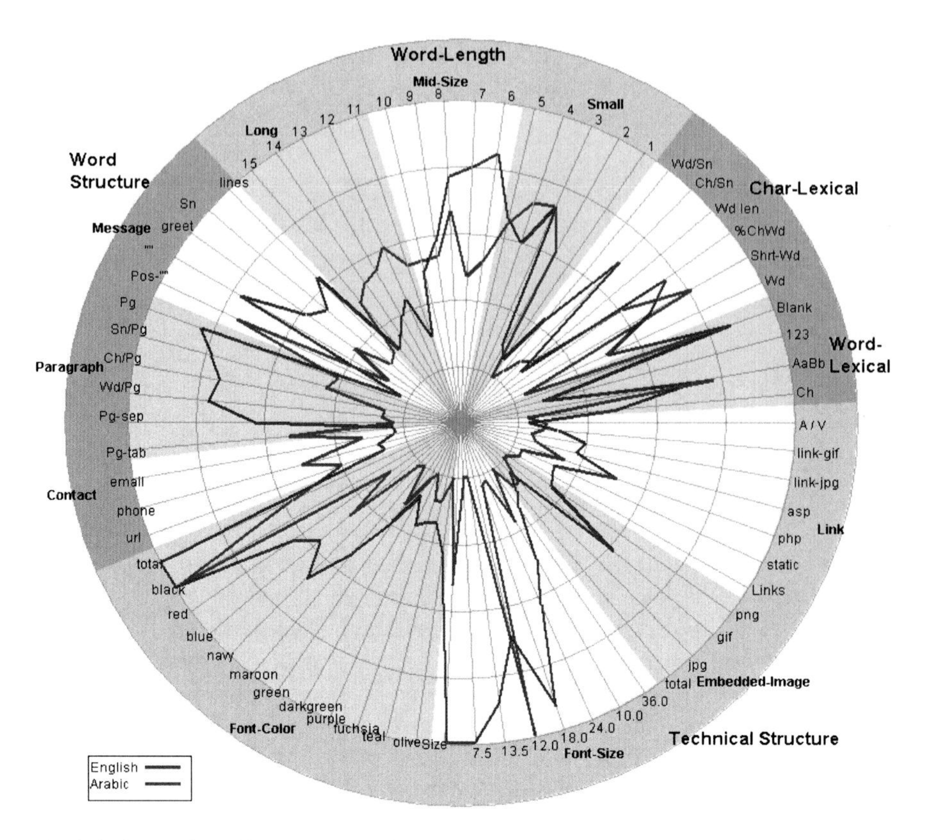

Fig. 9.4 Comparison of group authorship characteristics

The importance of punctuation, function words, and word-based structural features was fairly consistent across both languages. This suggests that syntactical and structural features are fairly robust feature categories that can be applied across languages. The largest disparity in terms of feature importance was in the technical structure category. The use of font size, color, hyperlinks, and embedded images was more important in classifying messages from the Al-Aqsa Martyrs. The prevalence in usage of technical structure features for the Arabic group did not come as a complete surprise; however, the amount of such features used by the decision tree (66.7%) was beyond our expectations.

7.2 *Feature Usage Analysis*

In order to provide a more in-depth analysis of the differences between the KKK and Al-Aqsa Martyrs messages, a graph consisting of writing attributes common to the two groups was constructed. The visualization consisted of only lexical and structural features, since these feature groups are mostly language independent. Figure 9.4 shows the average usage by language for each of these attributes.

The values were normalized to a 0–1 scale in order to facilitate more accurate comparisons. Five major feature groups within the lexical and structural categories were identified, consisting of character lexical, word lexical, word length, word structure, and technical structure. These groups were further decomposed into subgroups (e.g., paragraph structure) represented in either light gray or white. In addition to demonstrating obvious linguistic dissimilarities, the comparison revealed several interesting subtleties which may be attributable to group or cultural differences.

7.2.1 Word/Character Lexical

The word- and character-level lexical features showed that the Al-Aqsa messages tended to be considerably longer than the KKK messages. In addition to overall length, sentence lengths were longer as well for the Al-Aqsa Martyrs messages.

7.2.2 Word Length

Based on our data, midsized Arabic words in the 6–10-letter range were far more prevalent than English words. However, longer Arabic words (greater than length ten) were less common. This is consistent with previous research suggesting that Arabic has a narrower word-length distribution than English.

7.2.3 Word Structure

Overall, the Al-Aqsa messages had a more formal structure, with greater use of greetings, more sentences, more paragraphs, and lengthier paragraphs. Unsurprisingly, author contact information was not provided very often, but the KKK authors more commonly supplied e-mail addresses and phone numbers. Typically, the addresses and phone numbers provided belonged to groups/individuals disliked by the author.

7.2.4 Technical Structure

Al-Aqsa messages used a plethora of font colors and sizes, often using them as tools to emphasize a certain point. Red, blue, and navy were used almost as much as black. This was in sharp contrast to the KKK messages, where black 10–12-point fonts were a fixture, with the exception of the occasional deviation to green or blue.

The Al-Aqsa messages had a far higher frequency of embedded images than the KKK messages (approximately 20 times more). The images were either photos or graphics represented by JPEGs and GIFs. The majority of the disparity in the use of embedded images was with respect to GIF and PNG file usage. The Al-Aqsa Martyrs forum messages frequently used GIFs to represent slogans and logos while there were no signs of this in the KKK messages.

The Al-Aqsa group's messages also had far greater links to static, dynamic, and image pages. Links to multimedia files existed in both forums; however, such direct links were not very common. Some multimedia links were via web sites, thus classified as web page links by the parser.

7.2.5 Inferences

Both forums consisted of messages that stated opinions and beliefs. However, the structure and dynamics of the two groups were quite different. The KKK forum messages were shorter and more conversational, implying greater familiarity between members. The Al-Aqsa group messages were more structured and formal and had a stronger persuasive inclination. The authors appeared to be making a concerted effort to state and justify their position by using a systematic and thorough writing approach. Bulleted points, paragraphs with headings, and generally longer message lengths, supported by embedded images and links, were the standard structural theme for the Al-Aqsa messages.

8 Conclusions and Future Directions

In this research, we successfully applied authorship identification techniques for the classification of English and Arabic extremist group forum messages. In order to accomplish this task, we used techniques and features to overcome the challenges realized based on the linguistic properties of Arabic. All feature types incorporated (lexical, syntactic, structural, and content specific) showed significant discriminating power for Arabic and English, resulting in exceptional classification accuracy.

With an established set of features and techniques for multilingual authorship analysis, we have several potential future directions. One of the limitations of current authorship identification methodologies is the number of authors that it can be applied to. In order to truly address the online anonymity problem, the techniques would require significant upward scalability to help discriminate between hundreds of potential authors. The development of more complex methodologies for differentiating between a larger set of authors is an important future endeavor. We also plan a more comprehensive analysis of English and Arabic extremist group authorship tendencies in order to distinguish group-level differences from linguistic disparities inherent between English and Arabic. For example, do the "persuasive" tendencies observed regarding the Al-Aqsa Martyrs messages have broader applicability to other extremist Arabic groups? Furthermore, what role does geographic proximity and time play on group and individual authorship characteristics? Evaluation of these questions could prove to be an interesting venture.

Acknowledgments This research was supported by the following grant: NSF, ITR-0326348, 2003–2005, "ITR: COPLINK Center for Intelligence and Security Informatics Research – A Crime Data Mining Approach to Developing Border Safe Research." The authors also express their gratitude for the research assistance provided by fellow members of the Dark Web Project team in the Artificial Intelligence Lab, including Jialun Qin, Yilu Zhou, Greg Lai, and a couple of team members who wish to remain anonymous.

References

Al-Fedaghi, S. S. and Al-Anzi, F. (1989) A new algorithm to generate Arabic root-pattern forms. Proceedings of the 11th National Computer Conference (KFUPM, 1989), Dhahran, Saudi Arabia.

Beesley, K.B. (1996). Arabic Finite-State Morphological Analysis and Generation. Proceedings of COLING-96, 89–94.

De Roeck, A. N. and Al-Fares, W. (2000). A morphologically sensitive clustering algorithm for identifying Arabic roots. In Proceedings ACL-2000 (ACL, 2000), Hong Kong, 2000.

De Vel, O., Anderson, A., Corney, M., and Mohay, G. (2001). Mining E-mail content for author identification forensics. SIGMOD Record, 30(4), 55–64.

Hmeidi, I., Kanaan, G. and Evens, M. (1997). Design and Implementation of Automatic Indexing for Information Retrieval with Arabic Documents. Journal of the American Society for Information Science, 48(10), 867–881.

Larkey, L. S. and Connell, M. E. (2001). Arabic information retrieval at UMass in TREC-10 (TREC 2001), Gaithersburg, Maryland, (NIST 2001).

Palmer, J.W. and Griffith, D.A. (1998). An Emerging Model of Web Site Design for Marketing. Communications of the ACM, 41(3), 44–51.

Peng, F., Schuurmans, D., Keselj, V., and Wang, S. (2003). Automated authorship attribution with character level language models. Paper presented at the 10th Conference of the European Chapter of the Association for Computational Linguistics (EACL 2003).

Stamatatos, E., Fakotakis, N., and Kokkinakis, G. (2001). Computer-based Authorship Attribution without Lexical Measures. Computers and the Humanities, 35(2), 193–214.

Zheng, R., Qin, Y., Huang, Z., and Chen, H. (2003). Authorship Analysis in Cybercrime Investigation. In Proceedings of the first NSF/NIJ Symposium, ISI 2003, Tucson, AZ, USA.

Chapter 10
Sentiment Analysis

1 Introduction

Analysis of Web content is becoming increasingly important due to augmented communication via computer-mediated communication (CMC) Internet sources such as e-mail, Web sites, forums, and chat rooms. The numerous benefits of the Internet and CMC have been coupled with the realization of some vices, including cybercrime. In addition to misuse in the form of deception, identity theft, and the sales and distribution of pirated software, the Internet has also become a popular communication medium and haven for extremist and hate groups. This problematic facet of the Internet is often referred to as the Dark Web (Chen 2006).

Stormfront, what many consider to be the first hate group Web site (Kaplan and Weinberg 1998), was created around 1996. Since then, researchers and hate watch organizations have begun to focus their attention toward studying and monitoring such online groups (Leets 2001). Despite the increased focus on analysis of such group's Web content, there has been limited evaluation of forum postings, with the majority of studies focusing on Web sites. Burris et al. (2000) acknowledged that there was a need to evaluate forum and chat room discussion content. Schafer (2002) also stated that it was unclear as to how much and what kind of forum activity was going on with respect to hateful cyber activist groups. Due to the lack of understanding and current ambiguity associated with the content of such groups' forum postings, analysis of extremist group forum archives is an important endeavor.

Sentiment analysis attempts to identify and analyze opinions and emotions. Hearst (1992) and Wiebe (1994) originally proposed the idea of mining direction-based text, i.e., text containing opinions, sentiments, affects, and biases. Traditional forms of content analysis, such as topical analysis, may not be effective for forums. Nigam and Hurst (2004) found that only 3% of Usenet sentences contained topical information. In contrast, Web discourse is rich in sentiment-related information (Subasic and Huettner 2001). Consequently, in recent years, sentiment analysis has been applied to various forms of Web-based discourse (Agrawal et al. 2003;

H. Chen, *Dark Web: Exploring and Data Mining the Dark Side of the Web*, Integrated Series in Information Systems 30, DOI 10.1007/978-1-4614-1557-2_10,

Efron 2004). Application to extremist group forums can provide insight into significant discussions and trends.

In this study, we propose the application of sentiment analysis techniques to hate/extremist group's forum postings. Our analysis encompasses classification of sentiments on two forums: a US supremacist and Middle Eastern extremist group. The remainder of this chapter is organized as follows. Section 2 presents a review of current research on sentiment classification. Section 3 describes research gaps and questions, while Sect. 4 presents our research design. Section 5 describes the EWGA algorithm and our proposed feature set. Section 6 presents experiments used to evaluate the effectiveness of the proposed approach and discussion of the results. Section 7 concludes with closing remarks and future directions.

2 Related Work

Extremist groups often use the Internet to promote hatred and violence (Glaser et al. 2002). The Internet offers a ubiquitous, quick, inexpensive, and anonymous means of communication for such groups (Crilley 2001). Zhou et al. (2005) did an in-depth analysis of US hate group Web sites and found significant evidence of fundraising, propaganda, and recruitment-related content. Abbasi and Chen (2005) also corroborated signs of Web usage as a medium for propaganda by US supremacist and Middle Eastern extremist groups. These findings provide insight into extremist group Web usage tendencies; however, there has been little analysis of Web forums. Burris et al. (2000) acknowledged the need to evaluate forum and chat room discussion content. Schafer (2002) was also unclear as to how much and what kind of forum activity was going on with respect to extremist groups. Automated analysis of Web forums can be an arduous endeavor due to the large volumes of noisy information contained in CMC archives. Consequently, previous studies have predominantly incorporated manual or semiautomated methods (Zhou et al. 2005). Manual examination of thousands of messages can be an extremely tedious effort when applied across thousands of forum postings. With increasing usage of CMC, the need for automated text classification and analysis techniques has grown in recent years. While numerous forms of text classification exist, we focus primarily on sentiment analysis for two reasons. First, Web discourse is rich in opinion and emotion-related content. Second, analysis of this type of text is highly relevant to propaganda usage on the Web since directional/opinionated text plays an important role in influencing people's perceptions and decision making (Picard 1997).

2.1 Sentiment Classification

Sentiment analysis is concerned with analysis of direction-based text, i.e., text containing opinions and emotions. We focus on sentiment classification studies which attempt to determine whether a text is objective or subjective, or whether a subjective

Table 10.1 A taxonomy of sentiment polarity classification

Tasks		
Category	Description	Label
Classes	Positive/negative sentiments or objective/subjective texts	C1
Level	Document- or sentence-/phrase-level classification	C2
Source	Whether source/target of sentiment is known or extracted	C3
Features		
Category	Examples	Label
Syntactic	Word/POS tag n-grams, phrase patterns, punctuation	F1
Semantic	Polarity tags, appraisal groups, semantic orientation	F2
Link-based	Web links, send/reply patterns, and document citations	F3
Stylistic	Lexical and structural measures of style	F4
Techniques		
Category	Examples	Label
Machine learning	Techniques such as SVM, naïve Bayes, etc.	T1
Link analysis	Citation analysis and message send/reply patterns	T2
Similarity score	Phrase pattern matching, frequency counts, etc.	T3
Domains		
Category	Description	Label
Reviews	Product, movie, and music reviews	D1
Web discourse	Web forums and blogs	D2
News articles	Online news articles and Web pages	D3

text contains positive or negative sentiments. Sentiment classification has several important characteristics including the various tasks, features, techniques, and application domains. These are summarized in the taxonomy presented in Table 10.1.

We are concerned with classifying sentiments in extremist group forums. Based on the proposed taxonomy, Table 10.2 shows selected previous studies dealing with sentiment classification. We discuss the taxonomy and related studies in detail below.

2.2 *Sentiment Analysis Tasks*

There have been several sentiment polarity classification tasks. Three important characteristics of the various sentiment polarity classification tasks are the classes, classification levels, and assumptions about sentiment source and target (topic). The common two-class problem involves classifying sentiments as positive or negative (Pang et al. 2002; Turney 2002). Additional variations include classifying messages as opinionated/subjective or factual/objective (Wiebe et al. 2001, 2004). A closely related problem is affect classification which attempts to classify emotions instead of sentiments. Examples of affect classes include happiness, sadness, anger, horror, etc. (Subasic and Huettner 2001; Grefenstette et al. 2004; Mishne 2005).

Sentiment polarity classification can be conducted at the document, sentence, or phrase (part of sentence) level. Document-level polarity categorization attempts to

Table 10.2 Selected previous studies in sentiment polarity classification

Study	Features				Reduce features	Techniques			Domains			Number of languages
	F1	F2	F3	F4	Yes/No	T1	T2	T3	D1	D2	D3	1-n
Subasic and Huettner 2001	✓	✓			No			✓			✓	1
Tong 2001	✓	✓			No			✓	✓			1
Morinaga et al. 2002	✓				Yes			✓	✓			1
Pang et al. 2002	✓				No	✓			✓			1
Turney 2002	✓	✓			No			✓	✓			1
Agrawal et al. 2003	✓		✓		No	✓	✓			✓		1
Dave et al. 2003	✓				No	✓		✓	✓			1
Nasukawa and Yi 2003	✓	✓			No			✓	✓			1
Riloff et al. 2003		✓		✓	No	✓					✓	1
Yi et al. 2003	✓	✓			Yes			✓	✓		✓	1
Yu and Hatzivassiloglou 2003	✓	✓			No	✓		✓			✓	1
Beineke et al. 2004		✓			No	✓		✓	✓			1
Efron 2004	✓		✓		No	✓	✓			✓		1
Fei et al. 2004		✓			No			✓	✓			1
Gamon 2004	✓			✓	Yes	✓			✓			1
Grefenstette et al. 2004	✓	✓			No			✓		✓		1
Hu and Liu 2004	✓	✓			No			✓	✓			1
Kanayama et al. 2004	✓	✓			No			✓	✓			1
Kim and Hovy 2004		✓			No			✓		✓		1
Pang and Lee 2004	✓	✓			No	✓		✓	✓			1
Mullen and Collier 2004	✓	✓			No	✓			✓			1
Nigam and Hurst 2004	✓	✓			No	✓				✓		1
Wiebe et al. 2004	✓			✓	Yes	✓		✓		✓	✓	1
Liu et al. 2005	✓	✓			No			✓	✓			1
Mishne 2005	✓	✓		✓	No	✓				✓		1
Whitelaw et al. 2005	✓	✓			No	✓			✓			1
Wilson et al. 2005	✓	✓			No	✓					✓	1

classify sentiments in movie reviews, news articles, or Web forum postings (Wiebe et al. 2001; Pang et al. 2002; Mullen and Collier 2004; Pang and Lee 2004; Whitelaw et al. 2005). Sentence-level polarity categorization attempts to classify positive and negative sentiments for each sentence (Yi et al. 2003; Mullen and Collier 2004; Pang and Lee 2004) or whether a sentence is subjective or objective (Riloff et al. 2003). There has also been work on phrase-level categorization in order to capture multiple sentiments that may be present within a single sentence (Wilson et al. 2005).

In addition to sentiment classes and categorization levels, different assumptions have also been made about the sentiment sources and targets (Yi et al. 2003). In this study, we focus on document-level sentiment polarity categorization (i.e., distinguishing positive and negative sentiment texts). However, we also review related sentence-level and subjectivity classification studies due to the relevance of the features and techniques utilized and the application domains.

2.3 Sentiment Analysis Features

There are four feature categories that have been used in previous sentiment analysis studies. These include syntactic, semantic, link-based, and stylistic features. Along with semantic features, syntactic attributes are the most commonly used set of features for sentiment analysis. These include word n-grams (Pang et al. 2002; Gamon 2004), part-of-speech (POS) tags (Pang et al. 2002; Yi et al. 2003; Gamon 2004), and punctuation. Additional syntactic features include phrase patterns, which make use of POS tag n-gram patterns (Nasukawa and Yi 2003; Yi et al. 2003; Fei et al. 2004). They noted that phrase patterns such as "n + aj" (noun followed by positive adjective) typically represented positive sentiment orientation while "n + dj" (noun followed by negative adjective) often expressed negative sentiment (Fei et al. 2004). Wiebe et al. (2004) used collocations, where certain parts of fixed word n-grams were replaced with general word tags, thereby also creating n-gram phrase patterns. For example, the pattern "U-adj as-prep" would be used to signify all bigrams containing a unique (once occurring) adjective followed by the preposition "as." Whitelaw et al. (2005) used a set of modifier features (e.g., very, mostly, not); the presence of these features transformed appraisal attributes for lexicon items.

Semantic features incorporate manual/semiautomatic or fully automatic annotation techniques to add polarity or affect intensity-related scores to words and phrases. Hatzivassiloglou and McKeown (1997) proposed a semantic orientation (SO) method later extended by Turney (2002) that uses a mutual information calculation to automatically compute the SO score for each word/phrase. The score is computed by taking the mutual information between a phrase and the word "excellent" and subtracting the mutual information between the same phrase and the word "poor." In addition to pointwise mutual information, the SO approach was later also evaluated using latent semantic analysis (Turney and Littman 2003).

Manually or semiautomatically generated sentiment lexicons (e.g., Tong 2001; Fei et al. 2004; Wilson et al. 2005) typically use an initial set of automatically generated terms which are manually filtered and coded with polarity and intensity information. The user-defined tags are incorporated to indicate whether certain phrases convey positive or negative sentiment. Riloff et al. (2003) used semiautomatic lexicon generation tools to construct sets of strong subjectivity, weak subjectivity, and objective nouns. Their approach outperformed the use of other features, including bag-of-words, for classification of objective versus subjective English documents. Appraisal groups (Whitelaw et al. 2005) are another effective method for annotating semantics to words/phrases. Initial term lists are generated using WordNet, which are then filtered manually to construct the lexicon. Developed based on appraisal theory (Martin and White 2005), each expression is manually classified into various appraisal classes. These classes include attitude, orientation, graduation, and polarity of phrases. Whitelaw et al. (2005) were able to get very good accuracy using appraisal groups on a movie review corpus, outperforming several previous studies (e.g., Mullen and Collier 2004), the automated mutual-information-based approach (Turney 2002), as well as the use of syntactic features

(Pang et al. 2002). Manually crafted lexicons have also been used for affect analysis. Subasic and Huettner (2001) used affect lexicons along with fuzzy semantic typing for affect analysis of news articles and movie reviews. Abbasi and Chen (2007, 2008) used manually constructed affect lexicons for analysis of hate and violence in extremist Web forums.

Other semantic attributes include contextual features representing the semantic orientation of surrounding text, which have been useful for sentence-level sentiment classification. Riloff et al. (2003) utilized semantic features that considered the subjectivity and objectivity of text surrounding a sentence. Their attributes measured the level of subjective and objective clues in the sentence prior to and following the sentence of interest. Pang and Lee (2004) also leveraged coherence in discourse by considering the level of subjectivity of sentences in close proximity to the sentence of interest.

Link-based features use link/citation analysis to determine sentiments for Web artifacts and documents. Efron (2004) found that opinion Web pages heavily linking to each other often shared similar sentiments. Agrawal et al. (2003) observed the exact opposite for Usenet newsgroups discussing issues such as abortion and gun control. They noticed that forum replies tended to be antagonistic. Due to the limited usage of link-based features, it is unclear how effective they may be for sentiment classification. Furthermore, unlike Web pages and Usenet, other forums may not have a clear message link structure, and some forums are serial (no threads).

Stylistic attributes include lexical and structural attributes incorporated in numerous prior stylometric/authorship studies (e.g., De Vel et al. 2001; Zheng et al. 2006). However, lexical and structural style markers have seen limited usage in sentiment analysis research. Wiebe et al. (2004) used hapax legomena (unique/once occurring words) effectively for subjectivity and opinion discrimination. They observed a noticeably higher presence of unique words in subjective texts as compared to objective documents across a Wall Street Journal corpus and noted "Apparently, people are creative when they are being opinionated" (p. 286). Gamon (2004) used lexical features such as sentence length for sentiment classification of feedback surveys. Mishne (2005) used lexical style markers such as words per message and words per sentence for affect analysis of Web blogs. While it is unclear whether stylistic features are effective sentiment discriminators for movie/product reviews, style markers have been shown to be highly prevalent in Web discourse (Abbasi and Chen 2005; Zheng et al. 2006; Schler et al. 2006).

2.4 Sentiment Classification Techniques

Previously used techniques for sentiment classification can be classified into three categories. These include machine learning algorithms, link analysis methods, and score-based approaches.

Many studies have used machine learning algorithms, with support vector machines (SVM) and naïve Bayes (NB) being the most commonly used. SVM has been used extensively for movie reviews (Pang et al. 2002; Pang and Lee 2004; Whitelaw et al. 2005) while naïve Bayes has been applied to reviews and Web discourse (Pang et al. 2002; Pang and Lee 2004; Efron 2004). In comparisons, SVM has outperformed other classifiers such as NB (Pang et al. 2002). While SVM has become a dominant technique for text classification, other algorithms such as Winnow (Nigam and Hurst 2004) and AdaBoost (Wilson et al. 2005) have also been used in previous sentiment classification studies.

Studies using link-based features and metrics for sentiment classification have often used link analysis. Efron (2004) used cocitation analysis for sentiment classification of Web site opinions while Agrawal et al. (2003) used message reply link structures to classify sentiments in Usenet newsgroups. An obvious limitation of link analysis methods is that they are not effective where link structure is not clear or links are sparse (Efron 2004).

Score-based methods are typically used in conjunction with semantic features. These techniques generally classify message sentiments based on the total sum of comprised positive or negative sentiment features. Phrase pattern matching (Nasukawa and Yi 2003; Yi et al. 2003; Fei et al. 2004) requires checking text for manually created polarized phrase tags (positive and negative). Positive phrases are assigned a plus one while negative phrases are assigned a minus one. All messages with a positive sum are assigned to the positive sentiment while negative messages are assigned to the negative sentiment class. The semantic orientation approach (Hatzivassiloglou and McKeown 1997; Turney 2002) uses a similar method to score the automatically generated polarized phrase tags. Score-based methods have also been used for affect analysis where the affect features present within a message/ document are scored based on their degree of intensity for a particular emotion class (Subasic and Huettner 2001).

2.5 Sentiment Analysis Domains

Sentiment analysis has been applied to numerous domains including reviews, Web discourse, and news articles and documents. Reviews include movie, product, and music reviews (Morinaga et al. 2002; Pang et al. 2002; Turney 2002). Sentiment analysis of movie reviews is considered to be very challenging since movie reviewers often present lengthy plot summaries and also use complex literary devices such as rhetoric and sarcasm. Product reviews are also fairly complex since a single review can feature positive and negative sentiments about particular facets of the product.

Web discourse sentiment analysis includes evaluation of Web forums, newsgroups, and blogs. These studies assess sentiments about specific issues/topics. Sentiment topics include abortion, gun control, and politics (Agrawal et al. 2003;

Efron 2004). Robinson (2005) evaluated sentiments about 9/11 in three forums in the United States, Brazil, and France. Wiebe et al. (2004) performed subjectivity classification of Usenet newsgroup postings.

Sentiment analysis has also been applied to news articles (Yi et al. 2003; Wilson et al. 2005). Henley et al. (2002) analyzed newspaper articles for biases pertaining to violence-related reports. They found that there was a significant difference between the manner in which the Washington Post and the San Francisco Chronicle reported news stories relating to antigay attacks, with the reporting style reflecting newspaper sentiments. Wiebe et al. (2004) classified objective and subjective news articles in a Wall Street Journal corpus.

Some general conclusions can be drawn from Table 10.2 and the literature review. Most studies have used syntactic and semantic features. There has also been little use of feature reduction/selection techniques which may improve classification accuracy. In addition, most previous studies have focused on English data, predominantly in the review domain.

3 Research Gaps and Questions

Based on our review of previous literature and conclusions, we have identified several important research gaps. First, there has been limited previous sentiment analysis work on Web forums, and most studies have focused on a sentiment classification of a single language. Second, there has been almost no usage of stylistic feature categories. Finally, little emphasis has been placed on feature reduction/selection techniques.

3.1 Web Forums in Multiple Languages

Most previous sentiment classification of Web discourse has focused on Usenet and financial forums. Applying such methods to extremist forums is important in order to develop a viable set of features for assessing the presence of propaganda, anger, and hate in these online communities. Furthermore, there has been little evaluation on non-English content, with the exception of Kanayama et al. (2004) performing sentiment classification on Japanese text. Even in that study, machine translation software was used to convert the text to English. Thus, multiple language features have not been used for sentiment classification. The globalized nature of the Internet necessitates more sentiment analysis across languages.

3.2 Stylistic Features

Previous work has focused on syntactic and semantic features. There has been little use of stylistic features such as word-length distributions, vocabulary richness measures, character- and word-level lexical features, and special character frequencies.

Gamon (2004) and Pang et al. (2002) pointed out that many important features may not seem intuitively obvious at first. Thus, while prior emphasis has been on adjectives, stylistic features may uncover latent patterns that can improve classification performance of sentiments. This may be especially true for Web forum discourse, which is rich in stylistic variation (Abbasi and Chen 2005; Zheng et al. 2006). Stylistic features have also been shown to be highly prevalent in other forms of computer-mediated communication, including Web blogs (Herring and Paolillo 2006).

3.3 Feature Reduction for Sentiment Classification

Different automated and manual approaches have been used to craft sentiment classification feature sets. Little emphasis has been given to feature subset selection techniques. Gamon (2004) and Yi et al. (2003) used log likelihood to select important attributes from a large initial feature space. Wiebe et al. (2004) evaluated the effectiveness of various potential subjective elements (PSEs) for subjectivity classification based on their occurrence distribution across classes. However, many powerful techniques have not been explored. Feature reduction/selection techniques have two important benefits (Li et al. 2006). They can potentially improve classification accuracy and also provide greater insight into important class attributes, resulting in a better understanding of sentiment arguments and characteristics (Guyon and Elisseeff 2003). Using feature reduction, Gamon (2004) was able to improve accuracy and narrow in on a key feature subset of sentiment discriminators.

3.4 Research Questions

We propose the following research questions.

1. Can sentiment analysis be applied to Web forums in multiple languages?
2. Can stylistic features provide further sentiment insight and classification power?
3. How can feature selection improve classification accuracy and identify key sentiment attributes?

4 Research Design

In order to address these questions, we propose the use of a sentiment classification feature set consisting of syntactic and stylistic features. Furthermore, utilization of feature selection techniques such as genetic algorithms (Holland 1975) and information gain (Shannon 1948; Quinlan 1986) is also included to improve classification accuracy and gain insight into the important features for each sentiment class. Based on the prevalence of stylistic variation in Web discourse, we believe that lexical and

Table 10.3 Text classification studies using GA, IG, and SVM weights

Technique	Task	Study
GA	Stylometric analysis	Li et al. 2006
IG	Topic classification	Efron et al., 2003
	Stylometric analysis	Juola and Baayen, 2005
		Koppel and Schler 2003
		Abbasi and Chen 2006
SVM weights	Topic classification	Mladenic et al. 2004
	Gender categorization	Koppel et al. 2002

structural style markers can improve the ability to classify Web forum sentiments. Integrated stylistic features include attributes such as word-length distributions, vocabulary richness measures, letter usage frequencies, use of greetings, presence of requoted content, use of URLs, etc.

We also propose the use of an entropy weighted genetic algorithm (EWGA) that incorporates the information gain (IG) heuristic with a genetic algorithm (GA) to improve feature selection performance. GA is an evolutionary computing search method (Holland 1975) that has been used in numerous feature selection applications (Siedlecki and Sklansky 1989; Yang and Honavar 1998; Li et al. 2006; 2007). Oliveira et al. (2002) successfully applied GA to feature selection for handwritten digit recognition. Vafaie and Imam (1994) showed that GA outperformed other heuristics such as greedy search for image recognition feature selection. Like most random search feature selection methods (Dash and Liu 1997), it uses a wrapper model where the performance accuracy is used as the evaluation criterion to improve the feature subset in future generations.

In contrast, IG is a heuristic based on information theory (Shannon 1948). It uses a filter model for ranking features which makes it computationally more efficient than GA. IG has outperformed numerous feature selection techniques in head-to-head comparisons (Forman 2003). Since our experiments will use the SVM classifier, we also plan to compare the proposed EWGA technique against the use of SVM weights for feature selection. In this method, the SVM weights are used to iteratively reduce the feature space, thereby improving performance (Koppel et al. 2002). SVM weights have been shown to be effective for text categorization (Koppel et al. 2002; Mladenic et al. 2004) and gene selection for cancer classification (Guyon et al. 2002). GA, IG, and SVM weights have been used in several previous text classification studies as shown in Table 10.3. A review of feature selection for text classification can be found in Sebastiani (2002).

A consequence of using an optimal search method such as GA in a wrapper model is that convergence toward an ideal solution can be slow when dealing with very large solution spaces. However, as previous researchers have argued, feature selection is considered an "offline" task that does not need to be repeated constantly (Jain and Zongker 1997). This is why wrapper-based techniques using genetic algorithms have been used for gene selection with feature spaces consisting of tens of thousands of genes (Li et al. 2007). Furthermore, hybrid GAs have previously been used for product design optimization (Alexouda and Papparrizos 2001; Balakrishnan et al. 2004) and scheduling problems (Levine 1996) to facilitate improved accuracy and convergence efficiency (Balakrishnan et al. 2004). We developed the EWGA hybrid GA that utilizes the information gain (IG) heuristic with the intention of improving feature selection quality. More algorithmic details are provided in the next section.

5 System Design

We propose the following system design (shown in Fig. 10.1). Our design has two major steps: extracting an initial set of features and performing feature selection. These steps are used to carry out sentiment classification of forum messages.

5.1 Feature Extraction

We incorporated syntactic and stylistic features in our sentiment classification attribute set. These features are more generic and applicable across languages. For instance, syntactic, lexical, and structural features have been successfully used in

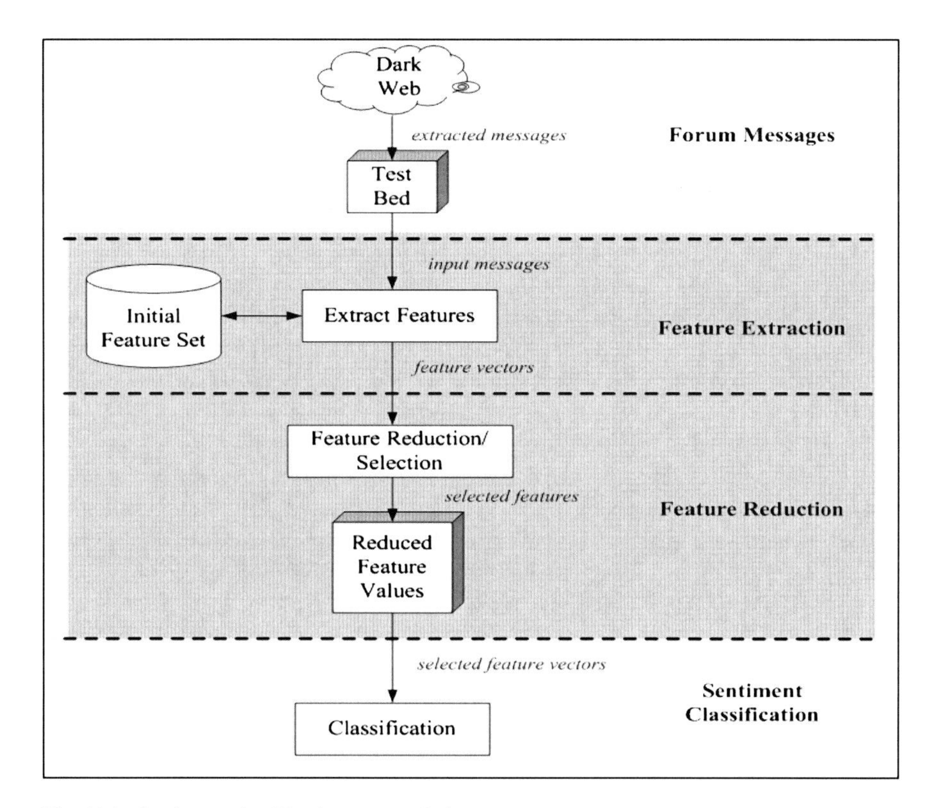

Fig. 10.1 Sentiment classification system design

stylometric analysis studies applied to English, Chinese (Peng et al. 2003; Zheng et al. 2006), Greek (Stamatatos et al. 2003), and Arabic (Abbasi and Chen 2005; 2006). Link-based features were not included since our messages were not in sequential order (insufficient cross-message references). These types of features are only effective where the test bed consists of entire threads of messages and message-referencing information is available. Semantic features were not used since these attributes are heavily context-dependent (Pang et al. 2002). Such features are topic and language specific. For example, the set of positive polarity words describing a good movie may not be applicable to discussions about racism. Unlike stylistic and syntactic features, semantic features such as manually crafted lexicons incorporate an inherent feature selection element via the human involvement. Such human involvement makes semantic features (e.g., lexicons and dictionaries) very powerful for sentiment analysis. Lexicon developers will only include features that are considered to be important and weight these features based on their significance, thereby reducing the need for feature selection. For example, Whitelaw et al. (2005) used WordNet to construct an initial set of features, which were manually filtered and weighted to create the lexicon. Unfortunately, the language specificity of semantic features is particularly problematic for application to the Dark Web, which contains text in dozens of languages (Chen 2006). We hope to overcome the lack of semantic features by incorporating feature selection methods intended to isolate the important subset of stylistic and syntactic features and remove noise.

5.1.1 Determining Size of Initial Feature Set

Our initial feature set consisted of 14 different feature categories which included POS tag n-grams (for English), word roots (for Arabic), word n-grams, and punctuation for syntactic features. Style markers included word- and character-level lexical features, word-length distributions, special characters, letters, character n-grams, structural features, vocabulary richness measures, digit n-grams, and function words. The word-length distribution includes the frequency of 1–20 letter words. Word-level lexical features include total words per document, average word length, average number of words per sentence, average number of words per paragraph, total number of short words (i.e., ones less than four letters), etc. Character-level lexical features include total characters per document, average number of characters per sentence, average number of characters per paragraph, percentage of all characters that are in words, and the percentage of alphabetic, digit, and space characters. Vocabulary richness features include the total number of unique words used, hapax legomena (number of once occurring words), dis legomena (number of twice occurring words), and various previously defined statistical measures of richness such as Yule's K, Honore's R, Sichel's S, Simpson's D, and Brunet's W measures. The structural features encompass the total number of lines, sentences, and paragraphs, as well as whether the document has a greeting or a signature. Additional structural attributes include whether there is a separation between paragraphs, whether the paragraphs

Table 10.4 English and Arabic feature sets

Category	Feature group	English	Arabic	Examples
Syntactic	POS n-grams	Varies	–	Frequency of part-of-speech tags (e.g., NP_VB)
	Word roots	–	Varies	Frequency of roots (e.g., slm, ktb)
	Word n-grams	Varies	Varies	Word n-grams (e.g., senior editor, editor in chief)
	Punctuation	8	12	Occurrence of punctuation marks (e.g., !,;, :, and ?)
Stylistic	Letter n-grams	26	36	Frequency of letters (e.g., a, b, c, etc.)
	Char. n-grams	Varies	Varies	Character n-grams (e.g., abo, out, ut, ab, etc.)
	Word lexical	8	8	Total words,% char. per word
	Char. lexical	8	8	Total char.,% char. per message
	Word length	20	20	Frequency distribution of 1–20 letter words
	Vocab. richness	8	8	Richness (e.g., hapax legomena and Yule's K)
	Special char.	20	21	Occurrence of special char. (e.g., @, #, $,%, ^, &, *, and +)
	Digit n-grams	Varies	Varies	Frequency of digits (e.g., 100, 17, and 5)
	Structural	14	14	Has greeting, has URL, requoted content, etc.
	Function words	250	200	Frequency of function words (e.g., of, for, and to)

are indented, the presence and position of quoted and forwarded content, and whether the document includes e-mail, URL, and telephone contact information. Further descriptions of the lexical vocabulary richness and structural attributes can be found in De Vel et al. (2001), Zheng et al. (2006), and Abbasi and Chen (2005). The Arabic function words were Arabic words translated from the English function word list, as done in previous research (e.g., Chen and Gey 2002). Only words were considered; for convenience, no affixes were included.

Many feature categories are predefined in terms of the number of potential features. For example, there are only a certain number of possible punctuation and stylistic lexical features (e.g., words per sentence, words per paragraph, etc.). In contrast, there are countless potential n-gram-based features. Consequently, some shallow selection criterion is typically incorporated to reduce the feature space for n-grams. A common approach is to select features with a minimum usage frequency (Mitra et al. 1997; Jiang et al. 2004). We used a minimum frequency threshold of 10 for n-gram-based features. Less common features are sparse and likely to cause overfitting. In addition, we only used unigrams, bigrams, and trigrams as these higher level n-grams tend to be redundant. Using only up to trigrams has been shown to be effective for stylometric analysis (Kjell et al. 1994) and sentiment classification (Pang et al. 2002; Wiebe et al. 2004). Based on this criterion for n-gram features, Table 10.4 shows the English and Arabic feature sets.

5.1.2 Feature Extraction Component

Due to the challenging morphological characteristics of Arabic, our attribute extraction process features a component for tracking elongation as well as a root extraction algorithm (illustrated in Fig. 10.2).

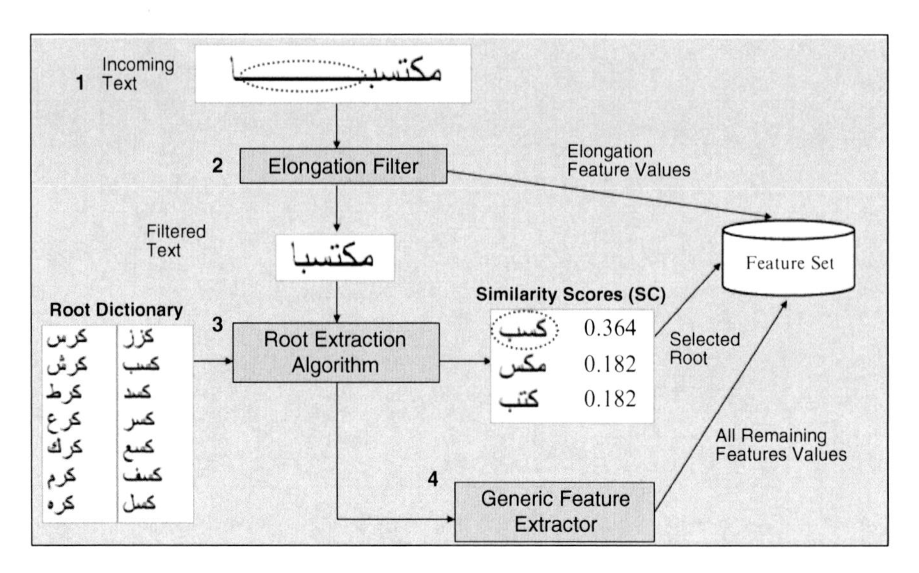

Fig. 10.2 Arabic extraction component

Elongation is the process of using a dash-like "kashida" character for stylistic word stretching (shown in step 1 in Fig. 10.2). The use of elongation is very prevalent in Arabic Web forum discourse (Abbasi and Chen 2005). In addition to tracking the presence and extent of elongation, we filter out these "kashida" characters in order to ensure reliable extraction of the remaining features (step 2 in Fig. 10.2). The filtered words are then passed through a root extraction algorithm (Abbasi and Chen 2005) that compares each word against a root dictionary to determine the appropriate word-root match (step 3). Root frequencies are tracked in order to account for the highly inflective nature of Arabic which reduces the effectiveness of standard bag-of-words features. The remaining stylistic and syntactic features are then extracted in a similar manner for English and Arabic (step 4).

5.2 *Feature Selection: Entropy Weighted Genetic Algorithm (EWGA)*

Most previous hybrid GA variations combine GA with other search heuristics such as beam search, where the beam search output is used as part of the initial GA population (Alexouda and Papparrizos 2001; Balakrishnan et al. 2004). Additional hybridizations include modification of the GA's crossover (Aggarwal et al. 1997) and mutation operators (Balakrishnan et al. 2004). The entropy weighted genetic algorithm (EWGA)

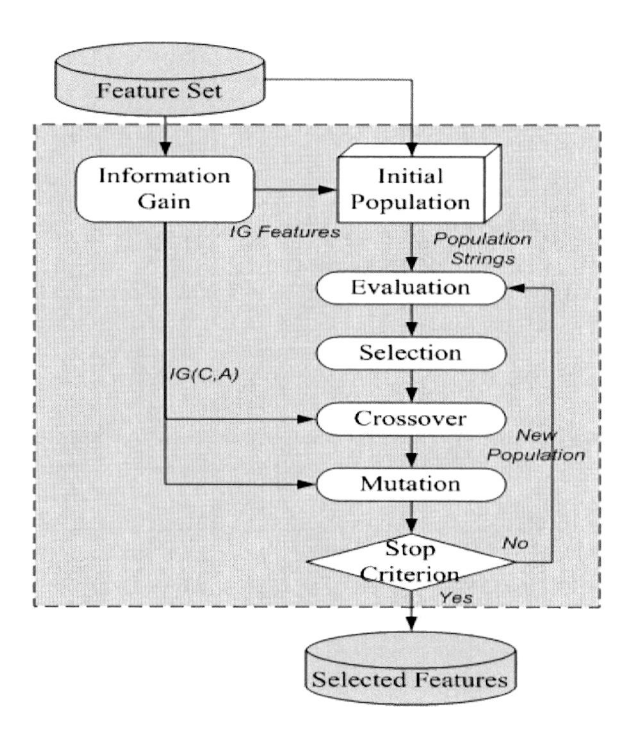

Fig. 10.3 EWGA illustration

uses the information gain (IG) heuristic to weight the various sentiment attributes. These weights are then incorporated into the GA's initial population and crossover and mutation operators. The major steps for the EWGA are as follows:

EWGA Steps

1. Derive feature weights using IG.
2. Include IG selected features as part of initial GA solution population.
3. Evaluate and select solutions based on fitness function.
4. Crossover solution pairs at point that maximizes total IG difference between the two solutions.
5. Mutate solutions based on feature IG weights.
6. Repeat steps 3–5 until stopping criterion is satisfied.

Figure 10.3 shows an illustration of the EWGA process. A detailed description of the IG, initial population, evaluation and selection, crossover, and mutation steps is presented in Fig. 10.3.

5.2.1 Information Gain

For information gain (IG), we used the Shannon entropy measure (Shannon 1948) in which:

$$\mathrm{IG}(\mathrm{C,A}) = \mathrm{H}(\mathrm{C}) - \mathrm{H}(\mathrm{C} \mid \mathrm{A})$$

where:

$IG(C, A)$	information gain for feature A;
$H(C) = -\sum_{i=1}^{n} p(C = i) \log_2 p(C = i)$	entropy across sentiment classes C;
$H(C \mid A) = -\sum_{i=1}^{n} p(C = i \mid A) \log_2 p(C = i \mid A)$	specific feature conditional entropy;
n	total number of sentiment classes.

If the number of positive and negative sentiment messages is equal, $H(C)$ is 1. Furthermore, the information gain for each attribute A will vary along the range 0–1 with higher values indicating greater information gain. All features with an information gain greater than 0.0025 (i.e., $IG(C,A) > 0.0025$) are selected. The use of such a threshold is consistent with prior work using IG for text feature selection (Yang and Pederson 1997).

5.2.2 Solution Structure and Initial Population

We represent each solution in the population using a binary string of length equal to the total number of features, with each binary string character representing a single feature. Specifically, 1 represents a selected feature while 0 represents a discarded one. For example, a solution string representing five candidate features, "10011," means that the first, fourth, and fifth features are selected, while the other two are discarded (Li et al. 2006). In the standard GA, the initial population of n strings is randomly generated. In the EWGA, n-1 solution strings are randomly generated while the IG solution features are used as the final solution string in the initial population.

5.2.3 Evaluation and Selection

We use the classification accuracy as the fitness function used to evaluate the quality of each solution. Hence, for each genome in the population, tenfold cross-validation with SVM is used to assess the fitness of that particular solution. Solutions for the next iteration are selected probabilistically with better solutions having a higher probability of selection. While several population replacement strategies exist, we use the generational replacement method originally defined by Holland (1975) in which the entire population is replaced every generation. Other replacement alternatives include steady-state methods where only a fraction of the population is replaced every iteration, while the majority is passed over to the next generation (Levine 1996). Generational replacement is used in order to maintain solution diversity and prevent premature convergence attributable to the IG seed solution dominating the other solutions (Bentley 1990; Aggarwal et al. 1997; Balakrishnan et al. 2004).

5.2.4 Crossover

From the n solution strings in the population (i.e., $n/2$ pairs), certain adjacent string pairs are randomly selected for crossover based on a crossover probability P_c. In the standard GA, we use single-point crossover by selecting a pair of strings and swapping substrings at a randomly determined crossover point x.

S = 010010	→ $x = 3$ →	S = 010 | 010	→	S = 010100
T = 110100		T = 110 | 100		T = 110010

The IG heuristic is utilized in the EWGA crossover procedure in order to improve the quality of the newly generated solutions. Given a pair of solution strings S and T, the EWGA crossover method selects a crossover point x that maximizes the difference in cumulative information gain across strings S and T. Such an approach is intended to create a more diverse solution population: those with heavier concentrations of features with higher IG values and those with fewer IG features. The crossover point selection procedure can be formulated as follows:

$$\arg\max_{x} \left| \sum_{A=1}^{x} IG(C,A)(S_A - T_A) + \sum_{A=x}^{m} IG(C,A)(T_A - S_A) \right|$$

where: $IG(C,A)$ information gain for feature A; S_A Ath character in solution string S; T_A Ath character in solution string T; m total number of features; x crossover point in solution pair S and T, where $1 < x < m$.

Maximizing the IG differential between solution pairs in the crossover process allows the creation of potentially better solutions. Solutions with higher IG contain attributes that may have greater discriminatory potential while the lower IG solutions help maintain the diversity balance in the solution population. Such balance is important to avoid premature convergence of solution populations toward local maxima (Aggarwal et al. 1997).

5.2.5 Mutation

The traditional GA mutation operator randomly mutates individual feature characters in a solution string based on a mutation probability constant P_m. The EWGA mutation operator factors the attribute information gain into the mutation probability as shown below. This is done in order to improve the likelihood of inclusion into the solution string for features with higher information gain while decreasing the probability of features with lower information gain. Our mutation operator sets the probability of a bit to mutate from 0 to 1 based on the feature's information gain, whereas the probability to mutate from 1 to 0 is set to the value one minus the feature's information gain. Balakrishnan et al. (2004) demonstrated the potential for modified mutation operators that favored features with higher weights in their hybrid genetic algorithm geared toward product design optimization.

$$P_m(A) = \begin{cases} B[IG(C,A)], \textit{if } S_A = 0 \\ B[1 - IG(C,A)], \textit{if } S_A = 1 \end{cases}$$

where $P_m(A)$ probability of mutation for feature A; $IG(C,A)$ information gain for feature A; S_A Ath character in solution string S;B constant in the range 0–1.

5.3 *Classification*

Because our research focus is on sentiment feature extraction and selection, in all experiments, SVM is used with tenfold cross-validation and bootstrapping to classify sentiments. We chose SVM in our experiments because it has outperformed other machine learning algorithms for various text classification tasks (Pang et al. 2002; Abbasi and Chen 2005; Zheng et al. 2006). We use a linear kernel with the sequential minimal optimization (SMO) algorithm (Platt 1999) included in the Weka data mining package (Witten and Frank 2005).

6 System Evaluation

Experiments were conducted on English and Arabic Web forums. The overall accuracy was the average classification accuracy across all tenfold where the classification accuracy was computed as follows:

$$\text{Classification Accuracy} = \frac{\text{Number of Correctly Classified Documents}}{\text{Total Number of Documents}}$$

In addition to tenfold cross-validation, bootstrapping was used to randomly select 50 samples for statistical testing, as done in previous research (e.g., Whitelaw et al. 2005). For each sample, we used 5% of the instances for testing and the other 95% for training. Pairwise t tests were performed on the bootstrap values to assess statistical significance.

We conducted two experiments to evaluate the effectiveness of our features as well as feature selection methods for sentiment classification of messages from English and Arabic extremist Web forums. SVM was run using tenfold cross-validation, with 900 messages used for training and 100 for testing in each fold. Bootstrapping was performed by randomly selecting 50 messages for testing and the remaining 950 for training, 50 times. In experiment 1a, we evaluated the effectiveness of syntactic and stylistic features. Experiment 1b focused on evaluating the effectiveness of feature selection for sentiment analysis across English and Arabic forums.

6.1 Test Bed

Our test bed consists of messages from two major extremist forums (one US and one Middle Eastern) collected as part of the Dark Web project (Chen 2006). This project involves spidering the Web and collecting Web sites and forums relating to hate and extremist groups. The initial list of group URLs is collected from authoritative sources such as government agencies and the United Nations. These URLs are then used to gather additional relevant forums and Web sites.

The US forum www.nazi.org is an English forum that belongs to the Libertarian National Socialist Green Party (LNSG). This is an Aryan supremacist group that gained notoriety when a forum member was involved in a school shooting in 2004. The Middle Eastern forum www.la7odood.com is a major Arabic-speaking partisan forum discussing the war in Iraq and support for the insurgency. The forum's content includes numerous al-Qaeda speeches and beheading videos.

We randomly selected 1,000 polar messages from each forum, which were manually tagged. The polarized messages represented those in favor of (agonists) and against (antagonists) a particular topic. The number of messages used is consistent with previous classification studies (Pang et al. 2002). In accordance with previous sentiment classification experiments, a maximum of 30 messages was used from any single author. This was done in order to ensure that sentiments were being classified as opposed to authors. For the US forum, we selected messages relating to racial issues. Agonistic sentiment messages were considered to be those in favor of racial diversity. In contrast, antagonistic sentiment messages had content denouncing racial diversity, integration, interracial marriages, and race mixing. For the Middle Eastern forum, we selected messages relating to the insurgency in Iraq. Agonistic messages were considered to be those opposed to the insurgency. These messages had positive sentiments about the Iraqi government and US troops in Iraq. Antagonistic sentiment messages were those in favor of the insurgents and against

Table 10.5 Characteristics of English and Arabic test bed

Forum	Messages	Authors	Average length (char.)	Data range
US	1,000	114	854	3/2004–9/2005
Middle Eastern	1,000	126	1,126	11/2005–3/2006

Table 10.6 Characteristics of English and Arabic test bed

US forum				
Features	Accuracy (%)	Bootstrap (%)	Standard dev.	Number features
Stylistic	71.40	71.07	3.324	867
Syntactic	87.00	87.13	2.439	12,014
Stylistic + syntactic	**90.63**	**90.59**	2.042	12,881
Middle Eastern forum				
Features	Accuracy (%)	Bootstrap (%)	Standard dev.	Number features
Stylistic	80.20	80.01	4.145	1,166
Syntactic	85.42	85.23	2.457	12,645
Stylistic + syntactic	**90.81**	**90.69**	2.093	13,811

the current Iraqi government and US forces. These messages had negative sentiments about the Iraqi government and US troops. The occurrence of messages with opposing sentiments is attributable to the presence of agitators (also referred to as trolls) and debaters in these forums (Donath 1999; Herring et al. 2002; Viegas and Smith 2004). Thus, while the majority of the forum membership may have negative sentiments about a topic, a subset has opposing sentiment polarity. For the sake of simplicity, from here on, we will refer to agonistic messages as "positive" and antagonistic messages as "negative" as these terms are more commonly used to represent the two sides in most previous sentiment analysis research. Here, we use the terms positive and negative as indicators of semantic orientation with respect to the specific topic; however, the "positive" messages may also contain sentiments about other topics (which may be positive or negative) as described by Wiebe et al. (2005). This is similar to the document-level annotations used for product and movie reviews (Pang et al. 2002; Yi et al. 2003). Using two human annotators, 500 positive (agonistic) and 500 negative (antagonistic) sentiment messages were incorporated from each forum. Both annotators/coders were bilingual, fluent in English and Arabic. The message annotation task by the independent coders had a kappa (k) value of 0.90 for English and 0.88 for Arabic, which is considered to be reliable, suggesting sufficient intercoder reliability. Table 10.5 shows some summary statistics for our English and Arabic Web forum test bed.

6.2 Experiment 1a: Evaluation of Features

In our first experiment, we repeated the feature set tests previously performed on the movie review dataset. The three permutations of stylistic and syntactic features were used. Table 10.6 shows the results for the three feature sets across the US and Middle Eastern forum message datasets.

Table 10.7 P values for pairwise t tests on accuracy (n = 50)

Features/test bed	US	Middle Eastern
Sty. vs. syn.	<0.0001*	<0.0001*
Sty. vs. syn + sty.	<0.0001*	<0.0001*
Syn. vs. syn + sty	<0.0001*	<0.0001*

**P*-values significant at alpha = 0.05

The best classification accuracy results using SVM were achieved when using both syntactic and stylistic features. The combined feature set statistically outperformed the use of only syntactic or stylistic features across both datasets. The increase was more prevalent in the Middle Eastern forum messages, where the use of stylistic and syntactic features resulted in a 5% improvement in accuracy over the use of syntactic features alone. Surprisingly, stylistic features alone were able to attain over 80% accuracy for the Middle Eastern messages, nearly a 9% improvement in the effectiveness of these features as compared to the English forum messages. This finding is consistent with previous stylometric analysis studies that have also found significant stylistic usage in Middle Eastern forums, including heavy usage of fonts, colors, elongation, numbers, and punctuation (Abbasi and Chen 2005).

Table 10.7 shows the pairwise t tests conducted on the bootstrap samples to evaluate the statistical significance of the improved results using stylistic and syntactic features. As expected, syntactic features outperformed stylistic features when both were used alone. However, using both feature categories significantly outperformed the use of either category individually. The results suggest that stylistic features are prevalent and important in Web discourse, even when applied to sentiment classification.

6.3 *Experiment 1b: Evaluation of Feature Selection Techniques*

This experiment was concerned with evaluating the effectiveness of feature selection for sentiment classification of Web forums. The same experimental settings as experiment 1a were used for all techniques. Table 10.8 shows the results for the four feature reduction methods and the number feature selection baseline applied across the US and Middle Eastern forum messages. All four feature selection techniques improved the classification accuracy over the baseline. The EWGA had the best performance across both test beds in terms of overall accuracy, resulting in a 3–4% improvement in accuracy over the number feature selection baseline. Furthermore, the EWGA was also the most efficient in terms of the number of features used, improving accuracy while utilizing a smaller subset of the initial feature sets. EWGA-based feature selection was able to identify a more concise set of key features that was 50–70% smaller than IG and SVM weights (SVMW) and 75–90% smaller than the baseline. GA also used a smaller number of features; however, the use of EWGA resulted in considerably improved accuracy.

Table 10.8 Experiment 1b results

US forum				
Technique	Tenfold CV (%)	Bootstrap (%)	Standard dev.	Number features
Base	90.61	90.56	1.831	12,881
IG	92.22	92.10	1.612	1,057
GA	91.83	91.64	1.396	511
SVMW	92.33	92.28	1.512	1,000
EWGA	**94.72**	**94.94**	1.671	502
Middle Eastern forum				
Technique	Tenfold CV (%)	Bootstrap (%)	Standard dev.	Number features
Base	90.79	90.57	1.932	13,811
IG	93.41	93.38	1.665	1,045
GA	92.14	92.24	1.438	462
SVMW	93.28	93.26	1.337	1,000
EWGA	**93.62**	**93.84**	2.831	338

Table 10.9 P values for pairwise t tests on accuracy (n = 50)

Technique/test bed	US	Middle Eastern
Base vs. IG	**<0.0001***	**<0.0001***
Base vs. GA	**0.0001***	**0.0134***
Base vs. EWGA	**<0.0001***	**<0.0001***
Base vs. SVMW	**<0.0001***	**<0.0001***
IG vs. GA	**0.0356***	0.0685
IG vs. EWGA	**<0.0001***	0.2783
IG vs. SVMW	0.2934	0.4130
GA vs. EWGA	**<0.0001***	**0.0456***
GA vs. SVMW	**0.0279***	0.0728
SVMW vs. EWGA	**<0.0001***	0.2025

**P*-values significant at alpha = 0.05

Table 10.9 shows the pairwise t tests conducted on the bootstrap values to evaluate the statistical significance of the improved results using feature selection.

EWGA outperformed the baseline and GA for both datasets significantly. In addition, EWGA provided significantly better performance than IG and SVMW on the English Web forum messages. EWGA also outperformed IG and SVMW on the Middle Eastern forum dataset, though the improved performance was not statistically significant.

6.4 Results Discussion

Fig. 10.4 shows the selection accuracy and number of features selected (out of over 12,800 potential features) for the US forum using EWGA as compared to GA across the 200 iterations (average of tenfold). The Middle Eastern forum graphs looked

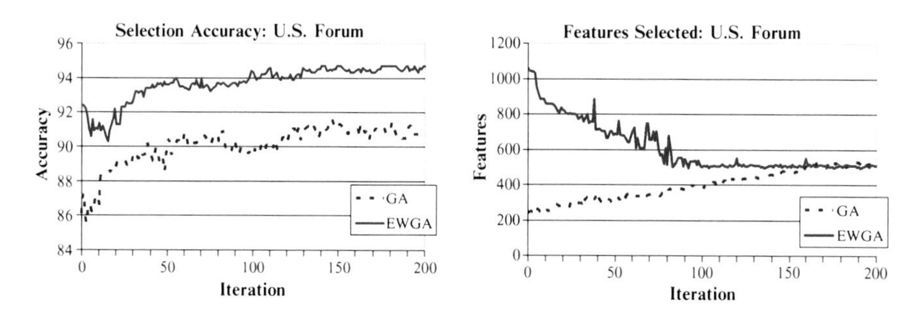

Fig. 10.4 US forum results using EWGA and GA

similar to the US forum and were hence not included. The EWGA accuracy declines initially despite being seeded with the IG solution. This is due to the use of generation replacement which prevents the IG solution from dominating the other solutions and creating a stagnant solution population. As intended, the IG solution features are gradually disseminated to the remaining solutions in the population until the new solutions begin to improve in accuracy around the 20th iteration. Overall, the EWGA is able to converge on an improved solution while only using half of the features originally transferred from IG. It is interesting to note that EWGA and GA both converge to a similar number of features when applied to the US forum; however, the EWGA is better able to isolate the more effective sentiment discriminators.

6.4.1 Analysis of Key Sentiment Features

We chose to analyze the EWGA features since they provided the highest performance with the most concise set of features. Thus, the EWGA-selected features are likely to be the most significant discriminators with the least redundancy. Figure 10.5 shows the number of each feature category selected by the EWGA for the English and Arabic feature set. As expected, more syntactic features (POS tags, n-grams, word roots) were used since considerably more of these features were included.

While Fig. 10.5 shows the number of features selected by the EWGA for each feature category, Fig. 10.6 shows the percentage of the overall number of features in each category that were selected. For example, the EWGA selected 12 structural features from the US (English) feature set; however, this represents 86% percent of the structural features, as shown in Fig. 10.6.

Looking at theusage percentage, stylistic features were more efficient than word n-grams and POS tags/roots. Many of the stylistic feature groups had over 40% usage whereas syntactic features rarely had such high usage with the exception of punctuation. For the US feature set, some categories such as word length, vocabulary richness, special characters, and structural features had well over 80% representation in the final feature subset. Comparing across regions, US features had higher usage rates than the Middle Eastern feature set. Approximately 10% of the Middle Eastern features were used by the EWGA versus 25% of the US attributes.

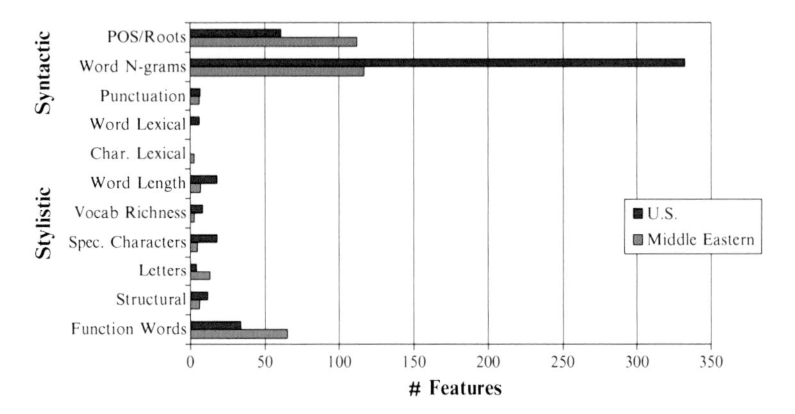

Fig. 10.5 Key feature usage frequencies by category

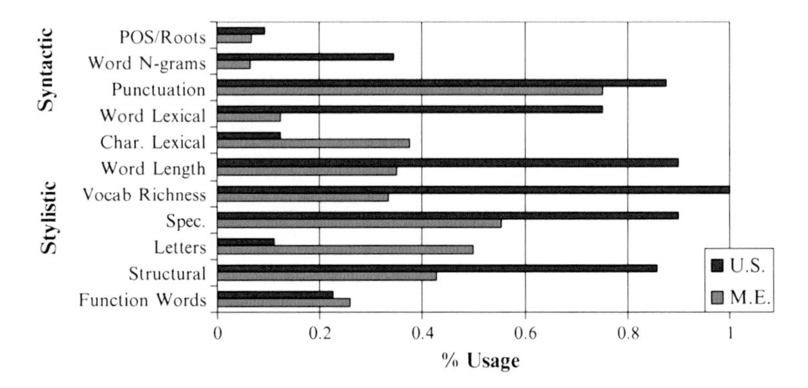

Fig. 10.6 Key feature usage percentage by category

6.4.2 Key Stylistic Features

Figure 10.7 shows some of the important stylistic features for the US forum. The diagram to the left shows the normalized average feature usage across all positive and negative sentiment messages. The table to the right shows the description for each feature as well as its IG and SVM weight.

The positive sentiment messages (agonists, in favor of racial diversity) tend to be considerably shorter (feat. 1), containing a few long sentences. These messages also feature heavier usage of conjunctive function words such as "however," "therefore," and "nevertheless" (feat. 6–8). In contrast, the negative sentiment messages are nearly twice as long and contain lots of digits (feat. 5) and special characters (feat. 2–4). Higher digit usage in the negative messages is due to references to news articles used to stereotype. Article snippets begin with a date, resulting in the higher digit count. The negative messages also feature shorter sentences. The stylistic feature usage statistics suggest that the positive sentiment messages follow more of a

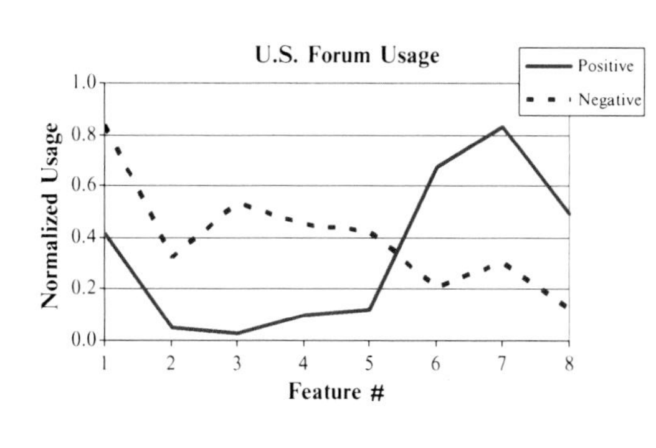

Feature	IG	SVM
total char.	0.027	0.243
$	0.029	0.130
&	0.017	0.141
{	0.012	0.126
digit count	0.015	0.316
therefore	0.021	-0.104
however	0.017	-0.120
nevertheless	0.014	-0.119

Fig. 10.7 Key stylistic features for US forum

debating style with shorter, well-structured arguments. In contrast, the negative sentiment messages tend to contain greater signs of emotion. The following verbal joust between two members in the US forum exemplifies the stylistic differences across sentiment classes. It should be noted that some of the content in the messages has been sanitized for vulgar word usage; however, the stylistic tendencies that are meant to be illustrated remain unchanged.

Negative

You are a total%#$*@ idiot!!! You walk around thinking you're doing humanity a favor, sympathizing with such barbaric slime. They use your sympathy as an excuse to fail. They are a burden to us all!!! Your opinion means nothing.

Positive

Neither does yours. But at least my opinion is an educated and informed one backed by well-reasoned arguments and careful skepticism about my assumptions. Race is nothing more than a social classification. What have you done for society that allows you to deem others a burden?

Figure 10.8 shows some of the important stylistic features for the Middle Eastern forum.

There are a few interesting similarities between the US and Middle Eastern forum feature usage tendencies across sentiment lines. The positive sentiment messages in the Middle Eastern forum (agonists, opposed to the insurgency) also tend to be considerably shorter than the negative sentiment messages in terms of total number of characters (feat. 2). Additionally, like their US forum counterparts, the negative Arabic messages contain heavy digit usage attributable to news article snippets (feat. 5). The negative sentiment messages make greater use of stylistic word stretching (elongation) which is done in order to emphasize key words (feat. 3). Consequently, the negative messages include greater use of words longer than 10 characters (feat. 4) while the positive messages are more likely to use shorter words, less than 4 characters in length (feat. 1). The negative sentiment messages also have higher vocabulary richness (feat. 6–9, various vocabulary richness formulas).

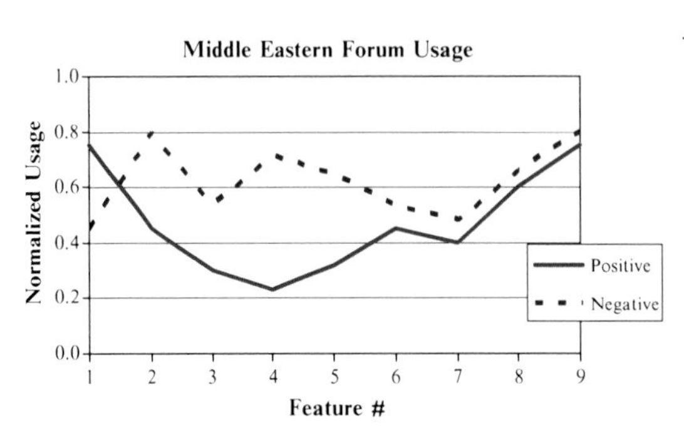

Feature	IG	SVM
short words	0.026	-0.020
total char.	0.039	0.210
elongation	0.037	0.319
long words	0.035	0.362
digits	0.014	0.086
Simpson	0.029	0.135
Yule	0.023	0.137
Brunet	0.027	0.109
Honore	0.024	0.140

Fig. 10.8 Key stylistic features for Middle Eastern forum

Table 10.10 Key n-grams for various sentiment classes

US forum		Middle Eastern forum	
Positive (agonist)	Negative (antagonist)	Positive (agonist)	Negative (antagonist)
Racist terms:	Racist terms:	Racist Shia terms:	Racist Sunni terms:
"Racism"	"Criminals"	"Terrorists"—"ارهابيين"	"Freedom fighters"—"مجاهدين"
"Subhuman racist"	"Whites"	"Shia"—"شيعة"	"Martyrdom"—"استشهاد"
"Anti-Semitism"	"Americans"	"Shiite"—"شيعي"	"Zarqawi"—"الزرقاوي"
"Ignorant slime"	"Get a job"		"Sunni"—"سني"
	"lmwao"		"American"—"امريكي"
	"Odin's rage"		"Iraq"—"امريكي"
	"Urban jungle"		"International forces"—"دولية قوات"

6.4.3 Key Syntactic Features

Table 10.10 shows the key word n-grams for each sentiment class selected by the EWGA. Many of the terms and phrases were racist content that was not included in the table but rather represented using a description label. Items in quotes indicate actual terms (e.g., "criminals") while nonquoted items signify term descriptions (e.g., racist terms). For the Middle Eastern forum, sentiments seem to be drawn along sectarian lines. In contrast, US forum sentiments are not clearly separated along racial lines. While the majority of the negative sentiments toward racial issues are generated by white supremacists, many of the positive sentiments are also presented by those with the same self-proclaimed racial affiliations. This reduced the amount of racial name calling across sentiments in the US forums, resulting in the need for considerably larger numbers of n-grams to effectively discern sentiment classes. Consequently, the number of n-grams used for the US feature set (332) is nearly threefold those used for the Middle Eastern sentiment classification (117).

7 Conclusions and Future Directions

In this study, we applied sentiment classification methodologies to English and Arabic Web forum postings. In addition to syntactic features, a wide array of English and Arabic stylistic attributes including lexical, structural, and function word style markers were included. We also developed the entropy weighted genetic algorithm (EWGA) for efficient feature selection in order to improve accuracy and identify key features for each sentiment class. EWGA significantly outperformed the number feature selection baseline and GA on all test beds. It also outperformed IG and SVMW on all three datasets (statistically significant for the movie review and US forum datasets) while isolating a smaller subset of key features. EWGA demonstrated the utility of these key features in terms of classification performance and for content analysis. Analysis of EWGA-selected stylistic and syntactic features allowed greater insight into writing style and content differences across sentiment classes in the two Web forums. Our approach of using stylistic and syntactic features in conjunction with the EWGA feature selection method achieved a high level of accuracy, suggesting that these features and techniques may be used in the future to perform sentiment classification and content analysis of Web forum discourse. Applying sentiment analysis to Web forums is an important endeavor, and the current accuracy is promising for effective analysis of forum conversation sentiments. Such analysis can help provide a better understanding of extremist group usage of the Web for information and propaganda dissemination.

In the future, we would like to evaluate the effectiveness of the proposed sentiment classification features and techniques for other tasks such as sentence- and phrase-level sentiment classification. We also intend to apply the technique to other sentiment domains (e.g., news articles and product reviews). Moreover, we believe the suggested feature selection technique may also be appropriate for other forms of text categorization and plan to apply our technique to topic, style, and genre classification. We also plan to investigate the effectiveness of other forms of GA hybridization, such as using the SVM weights instead of the IG heuristic.

References

Abbasi, A., and Chen, H. 2005. Identification and comparison of extremist-group web forum messages using authorship analysis, *IEEE Intelligent Systems 20*, 5, 67–75.

Abbasi, A., and Chen, H. 2006. Visualizing authorship for identification, In *Proceedings of the 4th IEEE International Conference on Intelligence and Security Informatics*, San Diego, CA, 60–71.

Abbasi, A., and Chen, H. 2007. Affect intensity analysis of Dark Web forums, In *Proceedings of the 5th IEEE International Conference on Intelligence and Security Informatics*, New Brunswick, NJ, 282–288.

Abbasi, A., and Chen, H. 2008. Analysis of affect intensities in extremist group forums, In *Terrorism Informatics*, (Eds.) H. Chen, E. Reid, H. Chen, J. Sinai, A. Silke, B. Ganor, Springer-Verlag.

Alexouda, G., and Papparrizos, K. 2001. A genetic algorithm approach to the product line design problem using the seller's return criterion: An extensive comparative computational study, *European Journal of Operational Research 134*, 165–178.

Aggarwal, C.C., Orlin, J., and Tai, R.P. 1997. Optimized crossover for the independent set problem, *Operations Research 45*, 2, 226–234.

Agrawal, R., Rajagopalan, S., Srikant, R. and Xu, Y. 2003. Mining newsgroups using networks arising from social behavior, In *Proceedings of the 12th International World Wide Web Conference*, 529–535.

Balakrishnan, P.V., Gupta, R., and Jacob, V.S. 2004. Development of hybrid genetic algorithms for product line designs, *IEEE Transactions on Systems, Man, and Cybernetics 34*, 1, 468–483.

Beineke, P., Hastie, T., and Vaithyanathan, S. 2004. The sentimental factor: Improving review classification via human-provided information, In *Proceedings of the 42nd Annual Meeting of the Association for Computational Linguistics*, 263.

Burris, V., Smith, E. and Strahm, A. 2000. White supremacist networks on the Internet, *Sociological Focus 33*, 2, 215–235.

Chen, A. and Gey, F. 2002. Building an Arabic stemmer for information retrieval, In *Proceedings of the 11th Text Retrieval Conference*, Gaithersburg, MD, 631–639.

Chen, H. 2006. *Intelligence and Security Informatics for International Security: Information Sharing and Data Mining*, London, Springer Press.

Crilley, K. 2001. Information warfare: New battle fields, terrorists, propaganda, and the Internet, *Aslib Proceedings 53*, 7, 250–264.

Dash, M. and Liu, H. 1997. Feature selection for classification, *Intelligent Data Analysis 1*, 131–156.

Dave, K. Lawrence, S. and Pennock, D.M. 2003. Mining the peanut gallery: Opinion extraction and semantic classification of product reviews, In *Proceedings of the 12th International Conference on the World Wide Web*, 519–528.

De Vel, O., Anderson, A., Corney, M., and Mohay, G. 2001. Mining e-mail content for author identification forensics, *ACM SIGMOD Record 30*, 4, 55–64.

Donath, J. 1999. Identity and deception in the virtual community, In Kollock, P., and Smith, M. (Eds.), *Communities in Cyberspace*, London: Routledge, 27–58.

Efron, M. 2004. Cultural orientations: Classifying subjective documents by cocitation analysis. In *Proceedings of the AAAI Fall Symposium Series on Style and Meaning in Language, Art, Music, and Design*, 41–48.

Efron, M., Marchionini, G., and Zhiang, J. 2003. Implications of the recursive representation problem for automatic concept identification in on-line government information, In *Proceedings of the ASIST SIG-CR Workshop*.

Fei, Z., Liu, J., and Wu, G. 2004. Sentiment classification using phrase patterns, *In Proceedings of the 4th IEEE International Conference on Computer Information Technology*, 1147–1152.

Forman, G. 2003. An extensive empirical study of feature selection metrics for text classification, *Journal of Machine Learning Research 3*, 1289–1305.

Gamon, M. 2004. Sentiment classification on customer feedback data: Noisy data, large feature vectors, and the role of linguistic analysis, In *Proceedings of the 20th International Conference on Computational Linguistics*, 841.

Glaser, J., Dixit, J., and Green, D. P. 2002. Studying hate crime with the Internet: What makes racists advocate racial violence? *Journal of Social Issues 58*, 1, 177–193.

Grefenstette, G.., Qu, Y., Shanahan, J. G.. and Evans, D. A. 2004. Coupling niche browsers and affect analysis for an opinion mining application, In *Proceedings of the 12th International Conference Recherche d'Information Assistee par Ordinateur*, 186–194.

Guyon, I., Weston, J., Barnhill, S., and Vapnik, V. 2002. Gene selection for cancer classification using support vector machines, *Machine Learning* 46, 389–422.

Guyon, I., and Elisseeff, A. 2003. An introduction to variable and feature selection, *Journal of Machine Learning Research 3*, 1157–1182.

Hatzivassiloglou, V. and McKeown, K. R. 1997. Predicting the semantic orientation of adjectives, In *Proceedings of the 35th Annual Meeting of the Association of Computational Linguistics*, 174–181.

Hearst, M. A. 1992. Direction-based text interpretation as an information access refinement. In P. Jacobs (Ed.), *Text-Based Intelligent Systems: Current Research and Practice in Information Extraction and Retrieval*. Mahwah, NJ, Lawrence Erlbaum Associates.

Henley, N. M., Miller, M. D., Beazley, J. A., Nguyen, D. N., Kaminsky, D., and Sanders, R. 2002. Frequency and specificity of referents to violence in news reports of anti-gay attacks, *Discourse and Society 13*, 1, 75–104.

Herring, S., Job-Sluder, K., Scheckler, R., and Barab, S. 2002. Searching for safety online: Managing "trolling" in a feminist forum, *The Information Society 18*, 5, 371–384.

Herring, S. and Paolillo, J. C. 2006. Gender and genre variations in weblogs, *Journal of Sociolinguistics*, *10*, 4, 439.

Holland, J. 1975. *Adaptation in natural and artificial systems*. Ann Arbor, University of Michigan Press.

Hu, M. and Liu, B. 2004. Mining and summarizing customer reviews. In *Proceedings of the ACM SIGKDD International Conference*, 168–177.

Jain, A. and Zongker, D. 1997. Feature selection: Evaluation, application, and small sample performance. *IEEE Transactions on Pattern Analysis and Machine Intelligence 19*, 2, 153–158.

Jiang, M., Jensen, E., Beitzel, S. and Argamon, S. 2004. Choosing the right bigrams for information retrieval, In *Proceedings of the Meeting of the International Federation of Classification Societies*.

Juola, P. and Baayen, H. 2005. A controlled-corpus experiment in authorship identification by cross-entropy, *Literary and Linguistic Computing 20*, 59–67.

Kanayama, H., Nasukawa, T., and Watanabe, H. 2004. Deeper sentiment analysis using machine translation technology, In *Proceedings of the 20th International Conference on Computational Linguistics*, 494–500.

Kaplan, J., and Weinberg, L. 1998. *The Emergence of a Euro-American Radical Right.*, New Brunswick, NJ, Rutgers University Press.

Kim, S. and Hovy, E. 2004. Determining the sentiment of opinions, In *Proceedings of the 20th International Conference on Computational Linguistics*, 1367–1373.

Kjell, B., Woods, W.A., and Frieder, O. 1994. Discrimination of authorship using visualization, *Information Processing and Management 30*, 1, 141–150.

Koppel, M., Argamon, S., and Shimoni, A.R. 2002. Automatically categorizing written texts by author gender, *Literary and Linguistic Computing 17*, 4, 401–412.

Koppel, M. and Schler, J. 2003. Exploiting stylistic idiosyncrasies for authorship attribution, In *Proceedings of the IJCAI Workshop on Computational Approaches to Style Analysis and Synthesis*, Acapulco, Mexico.

Levine, D. 1996. Application of a hybrid genetic algorithm to airline crew scheduling, *Computers and Operations Research 23*, 6, 547–558.

Leets, L. 2001. Responses to Internet hate sites: Is speech too free in cyberspace? *Communication Law and Policy 6*, 2, 287–317.

Li, J., Zheng, R., and Chen, H. 2006. From fingerprint to writeprint, *Communications of the ACM 49*, 4, 76–82.

Li, J. Su, H., Chen, H., and Futscher, B. 2007. Optimal search-based gene subset selection for gene array cancer classification, *IEEE Transactions on Information Technology in Biomedicine 11*, 4, 398–405.

Liu, B., Hu, M., and Cheng, J. 2005. Opinion observer: Analyzing and comparing opinions on the web, In *Proceedings of the 14th International World Wide Web Conference*, 342–351.

Martin, J. R. and White, P.R.R. 2005. *The Language of Evaluation: Appraisal in English*, London, Palgrave.

Mishne, G. 2005. Experiments with mood classification, In *Proceedings of the 1st Workshop on Stylistic Analysis of Text for Information Access*, Salvador, Brazil.

Mitra, M., Buckley, C., Singhal, A. and Cardie, C. 1997. An analysis of statistical and syntactic phrases, In *Proceedings of the 5th International Conference Recherche d'Information Assistee par Ordinateur*, Montreal, Canada, 200–214.

Mladenic, D., Brank, J., Grobelnik, M., and Milic-Frayling, N. 2004. Feature selection using linear classifier weights: Interaction with classification models, In *Proceedings of the 27th ACM SIGIR Conference on Research and Development in Information Retrieval*, Sheffield, UK, 234–241.

Morinaga, S., Yamanishi, K., Tateishi, K., and Fukushima, T. 2002. Mining product reputations on the web, In *Proceedings of the Eighth ACM SIGKDD International Conference on Knowledge Discovery and Data Mining*, Edmonton, Canada, 341–349.

Mullen, T., and Collier, N. 2004. Sentiment analysis using support vector machines with diverse information sources, In *Proceedings of the Empirical Methods in Natural Language Processing*, Barcelona, Spain, 412–418.

Nasukawa, T., and Yi, J. 2003. Sentiment analysis: Capturing favorability using natural language processing, In *Proceedings of the 2nd International Conference on Knowledge Capture*, Sanibel Island, Florida, 70–77.

Nigam, K., and Hurst, M. 2004. Towards a robust metric of opinion, In *Proceedings of the AAAI Spring Symposium on Exploring Attitude and Affect in Text.*

Oliveira, L.S., Sabourin, R., Bortolozzi, F., and Suen, C.Y. 2002. Feature selection using multi-objective genetic algorithms for handwritten digit recognition, In *Proceedings of the 16th International Conference on Pattern Recognition*, 568–571.

Pang, B., Lee, L., and Vaithyanathain, S. 2002. Thumbs up? Sentiment classification using machine learning techniques, In *Proceedings of the Conference on Empirical Methods in Natural Language Processing*, 79–86.

Pang, B., and Lee, L. 2004. A sentimental education: Sentimental analysis using subjectivity summarization based on minimum cuts, In *Proceedings of the 42nd Annual Meeting of the Association for Computational Linguistics*, 271–278.

Peng, F., Schuurmans, D., Keselj, V., and Wang, S. 2003. *Automated authorship attribution with character level language models. Paper presented at the 10th Conference of the European Chapter of the Association for Computational Linguistics (EACL 2003).*

Picard, R. W. 1997. *Affective Computing*, Cambridge, MA, MIT Press.

Platt, J. 1999. Fast training on SVMs using sequential minimal optimization, *In* Scholkopf, B., Burges, C., and Smola, A. (Ed.), *Advances in Kernel Methods: Support Vector Learning*, Cambridge, MA, MIT Press, 185–208.

Quinlan, J. R. 1986. Induction of decision trees, *Machine Learning 1*, 1, 81–106.

Riloff, E., Wiebe, J., and Wilson, T. 2003. Learning subjective nouns using extraction pattern bootstrapping, In *Proceedings of the Seventh Conference on Natural Language Learning Conference*, Edmonton, Canada, 25–32.

Robinson, L. 2005. Debating the events of September 11th: Discursive and interactional dynamics in three online for a, *Journal of Computer-Mediated Communication 10*, 4.

Schafer, J. 2002. Spinning the web of hate: Web-based hate propagation by extremist organizations, *Journal of Criminal Justice and Popular Culture 9*, 2, 69–88.

Schler, J., Koppel, M., Argamon, S., and Pennebaker, J. 2006. Effects of age and gender on blogging, In *Proceedings of the AAAI Spring Symposium Computational Approaches to Analyzing Weblogs*, Menlo Park, CA, 191–197.

Sebastiani, F. 2002. Machine learning in automated text categorization, *ACM Computing Surveys 34*, 1, 1–47.

Shannon, C. E. 1948. A mathematical theory of communication, *Bell System Technical Journal 27*, 4, 379–423.

Siedlecki, W. and Sklansky, J. 1989. A note on genetic algorithms for large-scale feature selection, *Pattern Recognition Letters 10*, 5, 335–347.

Stamatatos, E., Fakotakis, N., & Kokkinakis, G. 2001. Computer-based authorship attribution without lexical measures. *Computers and the Humanities 35*, 2, 193–214.

Subasic, P., and Huettner, A. 2001. Affect analysis of text using fuzzy semantic typing, *IEEE Transactions on Fuzzy Systems 9*, 4, 483–496.

Tong, R. 2001. An operational system for detecting and tracking opinions in on-line discussion, In *Proceedings of the ACM SIGIR Workshop on Operational Text Classification.* 1–6.

Turney, P. D. 2002. Thumbs up or thumbs down? Semantic orientation applied to unsupervised classification of reviews, In *Proceedings of the 40th Annual Meetings of the Association for Computational Linguistics*, Philadelphia, PA, 417–424.

Turney, P, D., and Littman, M, L. 2003. Measuring praise and criticism: Inference of semantic orientation from association, *ACM Transactions on Information Systems 21*, 4, 315–346.

Vafaie, H. and Imam, I. F. 1994. Feature selection methods: Genetic algorithms vs. greedy-like search, In *Proceedings of the International Conference on Fuzzy and Intelligent Control Systems*, 1994.

Viegas, F.B., and Smith, M. 2004. Newsgroup crowds and AuthorLines: Visualizing the activity of individuals in conversational cyberspaces, *In Proceedings of the 37th Hawaii International Conference on System Sciences*, Hawaii, USA.

Whitelaw, C., Garg, N., and Argamon, S. 2005. Using appraisal groups for sentiment analysis, In *Proceedings of the 14th ACM Conference on Information and Knowledge Management*, 625–631.

Wiebe, J. 1994. Tracking point of view in narrative, *Computational Linguistics 20*, 2, 233–287.

Wiebe, J., Wilson, T., and Bell, M. 2001. Identifying collocations for recognizing opinions, In *Proceedings of the ACL/EACL Workshop on Collocation*, Toulouse, France.

Wiebe, J., Wilson, T., Bruce, R., Bell, M., and Martin, M. 2004. Learning subjective language, *Computational Linguistics 30*, 3, 277–308.

Wiebe, J., Wilson, T., and Cardie, C. 2005. Annotating expressions of opinions and emotions in language, *Language Resources and Evaluation 1*, 2, 165–210.

Witten, I. H., and Frank, E. 2005. *Data Mining: Practical machine learning tools and techniques*, 2nd Edition,, San Francisco, CA, Morgan Kaufmann.

Wilson, T., Wiebe, J., and Hoffman, P. 2005. Recognizing contextual polarity in phrase-level sentiment analysis, In *Proceedings of the Human Language Technology Conference and Conference on Empirical Methods in Natural Language Processing*, British Columbia, Canada, 347–354.

Yang, Y. and Pederson, J. O. 1997. A comparative study on feature selection in text categorization, In *Proceedings of the 14th International Conference on Machine Learning*, 412–420.

Yang, J. and Honavar, V. 1998. Feature subset selection using a genetic algorithm, *IEEE Intelligent Systems 13*, 2, 44–49.

Yi, J., Nasukawa, T., Bunescu, R. and Niblack, W. 2003. Sentiment analyzer: Extracting sentiments about a given topic using natural language processing techniques, In *Proceedings of the 3 rd IEEE International Conference on Data Mining*, 427–434.

Yu, H. and Hatzivassiloglou, V. 2003. Towards answering opinion questions: Separating facts from opinions and identifying the polarity of opinion sentences, In *Proceedings of the Conference on Empirical Methods in Natural Language Processing*, 129–136.

Zheng, R., Li, J., Huang, Z., and Chen, H. 2006. A framework for authorship analysis of online messages: Writing-style features and techniques, *Journal of the American Society for Information Science and Technology 57*, 3, 378–393.

Zhou, Y., Reid, E., Qin, J., Chen, H., and Lai, G. 2005. U.S. extremist groups on the web: Link and content analysis, *IEEE Intelligent Systems 20*, 5, 44–51.

Chapter 11
Affect Analysis

1 Introduction

The need for enhanced information retrieval and knowledge discovery from computer-mediated communication (CMC) archives has been articulated by many individuals in recent years. One suggested information access refinement has been to mine directional text: text containing emotions and opinions (Hearst 1992; Wiebe 1994). Affects play an important role in influencing people's perceptions and decision making (Picard 1997). Analysis of sentiment and affects is particularly important for online discourse, where such information is often more pervasive than topical content (Subasic and Huettner 2001; Nigam and Hurst 2004). With the increased popularity of social computing, the presence and significance of affective text is likely to grow (Liu et al. 2003). There has been considerable recent work on sentiment analysis of online forums and product reviews (Turney and Littman 2003; Wiebe et al. 2004). However, research on analysis of affects (including emotions and moods) is still relatively sparse (Cho and Lee 2006). While recent studies have analyzed the presence of affects in blogs, online stories, chat dialog, transcripts, song lyrics, etc., it is unclear which features and techniques are most useful for affective computing of online texts. There is therefore a need to compare existing features for representing affective content as well as the techniques used for assigning emotive intensities.

In this chapter, we compare features and techniques for classification of affective intensities in online text. The features investigated include a large set of learned n-grams as well as automatically and manually generated affect lexicons used in prior research. We also propose a support vector regression correlation ensemble (SVRCE) method for text-based affect classification. SVRCE combines feature subset ensembles with affect correlation information for improved affect classification performance. Evaluation of the various feature representations and the proposed method in comparison with existing affect analysis techniques found that the use of SVRCE with n-grams is highly effective for affect classification of online forums, blogs, and stories.

H. Chen, *Dark Web: Exploring and Data Mining the Dark Side of the Web*, Integrated Series in Information Systems 30, DOI 10.1007/978-1-4614-1557-2_11,

The remainder of this chapter is organized as follows: Section 2 provides a review of related work on textual affect analysis. Section 3 outlines our research framework based on gaps and questions derived from the literature review. Section 4 presents an experimental evaluation of the various features and techniques incorporated in our framework. Section 5 features a brief case study illustrating how the proposed affect analysis methods can be applied to large CMC archives. Section 6 contains concluding remarks and describes future research directions.

2 Related Work

Affect analysis is concerned with the analysis of text containing emotions (Picard 1997; Subasic and Huettner 2001). Emotional intelligence, the ability to effectively recognize emotions automatically, is crucial for learning-preference-related information and determining the importance of particular content (Picard et al. 2001). Affect analysis is associated with sentiment analysis, which looks at the directionality of text, i.e., whether a text segment is positively or negatively oriented (Hearst 1992). However, there are two major differences between affect analysis and sentiment analysis. First, affect analysis involves a large number of potential emotions or affect classes (Subasic and Huettner 2001). These include happiness, sadness, anger, hate, violence, excitement, fear, etc. In contrast, sentiment analysis primarily deals with positive, negative, and neutral sentiment polarities. Second, while the sentiments associated with particular words or phrases are mutually exclusive, text segments can contain multiple affects (Subasic and Huettner 2001; Grefenstette et al. 2004b). For example, the sentence "I can't stand you!" has only a negative sentiment polarity but simultaneously contains hate and anger affects. Word-level examples include the verb form of "alarm," which can be attributed to fear, warning, and excitement affects (Subasic and Huettner 2001), and the adjective "gleeful," which can be assigned to the happiness and excitement affect classes (Grefenstette et al. 2004b). Additionally, certain affect classes may be correlated (Subasic and Huettner 2001). For instance, hate and anger often co-occur in text segments, resulting in a positive correlation. Similarly, happiness and sadness are opposing affects that are likely to have a negative correlation. In summary, affect analysis involves assigning text with emotive intensities across a set of mutually inclusive and possibly correlated affect classes. Important affect analysis characteristics include the features used to represent the presence of affects in text, techniques for assigning affective intensity scores, and the level of text granularity at which the analysis is performed. Table 11.1 presents a summary of the relevant prior studies based on these important affect analysis characteristics.

Based on the table, we can make several observations regarding the features and techniques used in previous affect analysis research.

1. Most prior research has used either manually generated lexicons, lexicons automatically created using WordNet or semantic orientation, or generic feature representations such as word and part-of-speech tag n-grams. It is unclear which of these feature representations is most effective for affect analysis.

Table 11.1 Related prior affect analysis studies

Study	Features	Technique(s)	Analysis level	Test bed and results
Donath et al. (1999)	Manual lexicon, punctuation	Posting scoring	Posting	Greek Usenet forums; visualization of anger intensities over time
Subasic and Huettner (2001)	Manual lexicon (fuzzy semantic typing)	Word scoring	Word	Movie reviews and news stories; visualization of 83 affects
Liu et al. (2003)	Language patterns derived from knowledge base	Sentence scoring	Sentence	User study on e-mail browser
Chuang and Wu (2004)	Manual lexicon	Support vector machine (SVM)	Sentence	Drama broadcast transcripts; 76.44% accuracy for 7 class experiments
Grefenstette et al. (2004a)	Manual lexicon, semantic orientation	Manual tagging, pointwise mutual information (PMI)	Word	Candidate affect words; scored intensities across 86 affects
Grefenstette et al. (2004b)	Manual lexicon	Word scoring	Word	Political web pages; scored text relating to certain topic
Read (2004)	Semantic orientation	Pointwise mutual information (PMI)	Sentence	Short stories; 47.14% accuracy for 2 class experiments
Ma et al. (2005)	Manual lexicon (WordNet-Affect database)	Word scoring	Sentence	Instant messaging chat data; no formal evaluation
Mishne (2005)	BOWs, POS tags, document length, emphasized words, semantic orientation, WordNet lexicon	Support vector machine (SVM)	Posting	LiveJournal blog postings; 60.25% accuracy for 2 class experiments
Cho and Lee (2006)	Manual lexicon, BOWs	Sentencez scoring, support vector machine (SVM)	Song	Korean song lyrics; 77.3% accuracy on 5 class experiments
Mishne and Rijke (2006)	Word n-grams	Pace regression	Posting	LiveJournal blog postings; average error of 52.53%, correlation coefficient of 0.827 for 2 class experiments
Wu et al. (2006)	Emotion generation and association rules	Separable mixture models	Posting	Student chat dialog; 80.98% accuracy for 3 class experiments

2. Techniques used for assigning affect intensities can be predominantly categorized into scoring methods and machine learning techniques. However, we are unaware of any prior work attempting to compare various techniques for affect classification.
3. Previous affect classification studies typically utilized between two and seven affect classes, applied at the word, sentence, or document levels. Despite the presence of multiple interrelated affects (Subasic and Huettner 2001; Grefenstette et al. 2004b), class correlation information was not leveraged for improved affect intensity assignment. Additionally, regression-based methods have seen limited usage despite their effectiveness in related application domains (Pang and Lee 2005; Schumaker and Chen 2006).
4. Prior studies mainly focused on a single application domain, such as movie reviews, web forums, blogs, chat dialog, song lyrics, stories, etc. Given the differences in the degree of interaction, language usage, and communication structure across these domains, it is unclear if an approach suitable for classifying story affects will be applicable on web forums and blogs. The features and techniques used in prior affect analysis research are expounded upon in the remainder of the section.

2.1 Features for Affect Analysis

The attributes used to represent affects can be classified into lexicon-based features and generic n-gram-based features. Considerable prior research has used manually or automatically generated lexicons. As previously stated, in affect lexicons, the same word/phrase can be assigned to multiple affect classes. The intensity score for an attribute is based on its degree of severity toward that particular affect class. Depending upon the semantic relation between affects, certain classes can have a positive or negative occurrence correlation (Subasic and Huettner 2001).

Many studies have incorporated manually developed affect lexicons. Subasic and Huettner (2001) used fuzzy semantic typing where each feature was assigned to multiple affect categories with varying intensity and centrality scores depending upon the word and usage context. For example, the word "rat" was assigned to the disloyalty, horror, and repulsion affect categories with intensity scores of 0.9, 0.6, and 0.7, respectively (on a 0.0–1.0 scale where 1.0 was highest). In order to compensate for word-sense ambiguity, their approach also assigned each word-affect pair a centrality score indicating the likelihood of the word being used for that particular affect class. For example, the word "rat" was assigned a centrality score of 0.3 for the disloyalty affect and 0.6 for the repulsion affect (also on a 0.0–1.0 scale) since the usage of "rat" to convey disloyalty is not as common. Thus, while "rat" was more intense for the disloyalty affect, it was also less central to this class. In Subasic and Huettner's (2001) approach, the intensity and centrality scores were both utilized for determining the affective composition of a text document. Although the accuracy for specific term affects may be inaccurate, the fuzzy logic approach is intended to capture the essence of a document's various affect intensities. A similar method for generating manual lexicons was employed in related work (Grefenstette et al. 2004a, b). Many other studies have also utilized manually constructed affect lexicons (Chuang and Wu 2004; Cho and Lee 2006). Donath et al. (1999) used a set of keywords relating to anger for analyzing Usenet forums. Ma et al. (2005) incorporated the WordNet-Affect database created by Valitutti et al. (2004). This database is comprised of manually assigned affect intensities for words found in the WordNet lexical resource (Fellbaum 1998). Liu et al. (2003) manually constructed sentence-level language patterns for identification of six affect classes, including happiness, sadness, anger, fear, etc.

Although manually created affect lexicons can provide powerful insight, their construction can be time consuming and tedious. As a result, many studies have explored the use of automated lexicon generation methods such as semantic orientation (Grefenstette et al. 2004a; Read 2004; Mishne 2005) and WordNet lexicons (Mishne 2005). These methods take a small set of manually generated seed/paradigm words which accurately reflect the particular affect class and use automated methods for lexicon expansion of candidate word scoring.

Based on the work of Turney and Littman (2003), the semantic orientation approach assesses the intensity of each word based on its frequency of co-occurrence with a set of core paradigm words reflective of that affect class (Grefenstette et al. 2004a). The occurrence frequencies for the paradigm words and candidate words are derived from search engines such as AltaVista (Grefenstette et al. 2004a; Read 2004; Mishne 2005) or Yahoo! (Mishne 2005). The number of paradigm words used for a particular affect class is generally five to seven (Grefenstette et al. 2004a; Read 2004). For example, the paradigm words for the *praise* affect may include "acclaim, praise, congratulations, homage, approval" (Grefenstette et al. 2004a), and additional lexicon items generated automatically using semantic orientation include the words "award, honor, extol." The semantic orientation approach is typically coupled with a pointwise mutual information (PMI) scoring mechanism for assigning candidate words intensity scores (Turney and Littman 2003). Traditional PMI assigns each word a score based on how often it occurs in proximity with positive and negative paradigm words; however, it has been modified to be applicable with affect classes (Read 2004; Grefenstette et al. 2004a). The affect analysis rendition of PMI proposed by Grefenstette et al. (2004a) is as follows:

$$\text{PMI Score}(word, Class) = \log_2\left(\frac{\prod_{cword \in Class} \text{hits}(word \text{ Near } cword)}{\prod_{cword \in Class} \log_2(\text{hits}(cword))}\right)$$

where *cword* is one of the paradigm words chosen for an affect class *Class* and *hits* is the number of pages found by Alta Vista.

Another automated affect lexicon generation method is WordNet lexicons. Originally proposed by Kim and Hovy (2004), this method is similar to semantic orientation. However, it uses WordNet to expand the seed words associated with a particular affect class by comparing each candidate word's synset with the seed word list (Mishne 2005). The intensity for a candidate word is proportional to the percentage of its synset also present in the seed word list for that particular affect class. Word scores are assigned using the following formula (Kim and Hovy 2004):

$$\text{WordNet Score}(word, Class) = P(Class)\frac{1}{count(c)}\sum_{i=1}^{n} count(syn_i, Class)$$

where *Class* is an affect class, syn_i is one of the n synonyms of *word*, and $P(Class)$ is the number of words in *Class* divided by the total number of words considered.

In addition to lexicon-based affect representations, studies have also used generic n-gram features. Mishne (2005) used bag-of-words (BOWs) and part-of-speech (POS) tags in combination with automatically generated lexicons, while Mishne and Rijke (2006) used word n-grams for affect analysis of blog postings. Cho and Lee (2006) used BOWs for classifying affects inherent in Korean song lyrics. N-grams have also been shown to be highly effective in the related area of sentiment classification (Wiebe et al. 2004; Abbasi et al. 2008), especially when combined with machine learning methods capable of learning n-gram patterns conveying opinions and emotions. While prior research has used various n-gram and lexicon representations, we are unaware of any work done to evaluate the effectiveness of various potential affect analysis features.

2.2 *Techniques for Assigning Affect Intensities*

Prior research has utilized scoring and machine learning methods for assigning affect intensities. Scoring-based methods, which are generally used in conjunction with lexicons, typically use the average intensity across lexicon items occurring in the text (i.e., word spotting) (Subasic and Huettner 2001; Liu et al. 2003; Cho and Lee 2006). Sentence-level averaging has also been performed in combination with the word-level PMI scores generated using semantic orientation (Turney and Littman 2003) as well as with WordNet lexicons (Kim and Hovy 2004). Studies that directly developed lexicons comprised of sentence patterns obviously do not use averaging (at least at the sentence level), but instead simply matching sentences with lexicon entries and assigning intensity scores accordingly (Liu et al. 2003; Cho and Lee 2006).

Machine learning techniques have also been used for assigning affect intensities. Many studies used support vector machine (SVM) for determining whether a text segment contained a particular affect class (Chuang and Wu 2004; Mishne 2005; Cho and Lee 2006). One shortcoming of using SVM is that it can only deal with discrete class labels, whereas affect intensities can vary along a continuum. Recent work has attempted to address this problem by using regression-based classifiers (Pang and Lee 2005). For example, Mishne and Rijke (2006) used word n-grams in unison with Pace regression (Witten and Frank 2005) for assigning affect intensities in LiveJournal blogs. Nevertheless, regression-based learning methods have seen limited usage despite their effectiveness in related application domains such as using news story text for stock price prediction (Schumaker and Chen 2006). Furthermore, although scoring and machine learning methods have been utilized for classifying affect intensities, there has been no research done to investigate the effectiveness of these methods.

3 Research Design

In this section, we highlight affect analysis research gaps based on our review of the related work. Research questions are then posed based on the relevant gaps identified. Finally, a research framework is presented in order to address these research questions, along with some research hypotheses. The framework encompasses various feature representations and techniques for assigning affective intensities to sentences.

3.1 Gaps and Questions

Prior research has utilized manually or automatically generated lexicons as well as generic n-gram features for representing affective content in text. Since most studies used a single feature category and did not compare different alternatives, it is unclear which emotive representation is most effective. Furthermore, prior research has used scoring-based techniques and machine learning methods such as SVM. Regression-based methods capable of assigning continuous intensity scores have not been explored in great detail, with the exception of Mishne and Rijke (2006). Leveraging the relationship between mutually inclusive affect classes in combination with powerful regression-based machine learning methods such as support vector regression (SVR) could be highly effective for accurate assignment of affect intensities. Additionally, most prior affect analysis research was applied to a single domain (e.g., blogs, stories, etc.). Application across multiple domains could lend greater validity to the effectiveness of affect analysis features and techniques. Based on these gaps, we present the following research questions:

- Which feature categories are best at accurately assigning affect intensities?
  - Can the use of an extended feature set enhance affect analysis performance over individual generic and lexicon-based feature categories?
- Can a regression ensemble that incorporates affect correlation information outperform existing machine learning and scoring-based methods?
- What impact will the application domain have on affect intensity assignment?

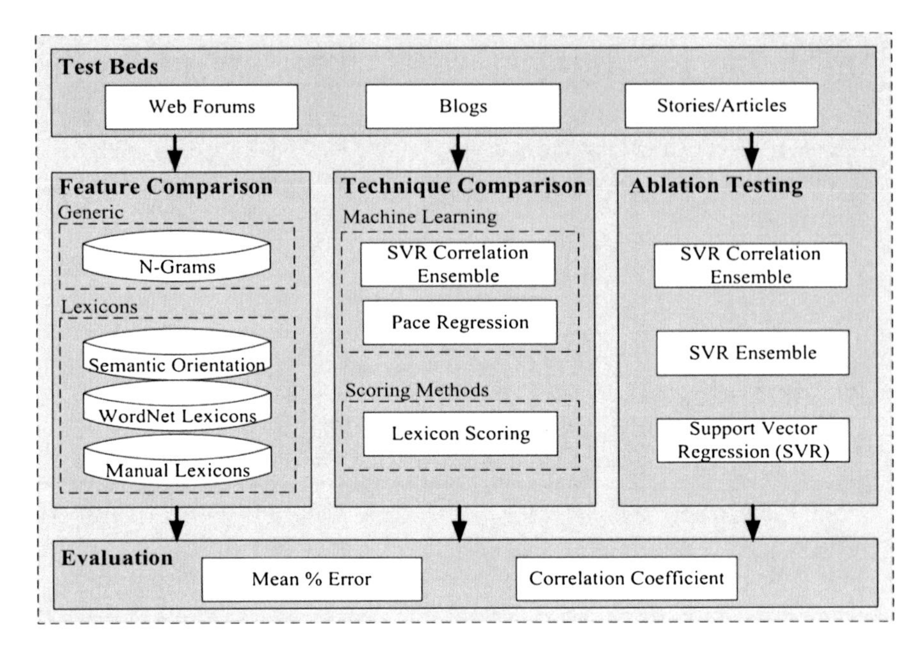

Fig. 11.1 Affect analysis research framework

3.2 Research Framework

Our research framework (shown in Fig. 11.1) relates to the features and techniques used for assigning affect intensity scores.

We intend to compare generic n-gram features with automatically and manually generated lexicons. We also plan to assess the effectiveness of using an extended feature set encompassing all these attributes in comparison with individual feature categories. With respect to affect analysis techniques, we propose a support vector regression (SVR) ensemble that considers affect correlation information when assigning emotive intensities to sentences. We intend to compare the SVR correlation ensemble (SVRCE) with other machine learning and scoring-based methods used in prior research. These include Pace regression (Witten and Frank 2005; Mishne and Rijke 2006), semantic orientation (Grefenstette et al. 2004a; Read 2004), WordNet (Kim and Hovy 2004), and manual lexicon scoring (Subasic and Huettner 2001).

We also plan to perform ablation testing to see how the different components of the proposed SVRCE method contribute to its overall performance. All testing will be performed on several test beds encompassing sentences derived from web forums, blogs, and stories. Features and techniques will be evaluated with respect to their percentage mean error and correlation coefficients in comparison with a human-annotated gold standard. Further details about the features, techniques, ablation testing, and our research hypotheses are presented below, while the test bed and evaluation metrics are discussed in greater detail in the ensuing evaluation section.

3.2.1 Affect Analysis Features

The n-gram feature set is comprised of word, character, and part-of-speech (POS) tag n-grams. For each n-gram category, we used up to trigrams only (i.e., unigrams, bigrams, and trigrams), as done in prior related research (Pang et al. 2002; Wiebe et al. 2004). Word n-grams, including unigrams (e.g., "LIKE"), bigrams (e.g., "I LIKE," "LIKE YOU"), and trigrams (e.g., "I LIKE YOU"), as well as POS tag n-grams (e.g., "NP VB," "JJ NP VB") have been used in prior affect analysis research (Mishne 2005). We also include character n-grams (e.g., "li," "ik," "ike"), which have been useful in related sentiment classification studies (Abbasi et al. 2008). In addition to standard word n-grams, we incorporate hapax legomena and dis legomena collocations (Wiebe et al. 2004). Such collocations replace once- (hapax legomena) and twice-occurring words (dis legomena) with "HAPAX" and "DIS" tags. Hence, the trigram "I hate Jim" would be replaced with "I hate HAPAX" provided "Jim" only occurs once in the corpus. The intuition behind such collocations is to remove sparsely occurring words with tags that will allow the extracted n-grams to be more generalizable, and hence, more useful (Wiebe et al. 2004). For instance, in the above example, the fact that the writer hates is more important from an affect analysis perspective than the specific person the hate is directed toward.

The lexicons employed are comprised of automated lexicons derived using semantic orientation and WordNet models as previously done by Grefenstette et al. (2004a) and Mishne (2005). We selected seven paradigm words for each affect class for input into the semantic orientation algorithm, as described in Sect. 2.1. For the WordNet models, sets of up to 50 words were used as the seeds, following the guidelines described by Kim and Hovy (2004).

Our feature set also consists of a manually crafted word-level lexicon. The lexicon is comprised of over 1,000 affect words for several emotive classes (e.g., happiness, sadness, anger, hate, violence, etc.). Each word is assigned an intensity and ambiguity score between 0 and 1. The intensities are assigned based on the word's degree of severity or valence for its particular affect category (with 1 being highest). This approach is consistent with the intensity score assignment methods incorporated in previous studies that utilized manually crafted lexicons (Donath et al. 1999; Subasic and Huettner 2001; Grefenstette et al. 2004b; Chuang and Wu 2004). Each affect feature is also assigned an ambiguity score. The ambiguity score is the probability of an instance of the feature having semantic congruence with the affect class represented by that feature. The ambiguity score for each feature is determined by taking a sample set of instances of the feature's occurrence and coding each occurrence as to whether the term usage is relevant to its affect. A maximum of 20 samples was used per term. Using more instances would be exhaustive, and we observed that the size used was sufficient to accurately capture the probability of an affect being relevant. The ambiguity score for each word can be computed as the number of correctly appearing instances divided by the total number of instances sampled for that word. Hence, an ambiguity value of one suggests that the term always appears in the appropriate affective connotation. The intensity and ambiguity assignment was done by two independent coders. Each coder initially assigned values without

Table 11.2 Manual lexicon examples for the violence affect

Term	Intensity	Ambiguity	Weight
Hit	0.210	0.800	0.168
Beat	0.400	0.667	0.267
Stab	0.575	1.000	0.575
Hang	0.800	0.650	0.520
Kill	0.850	0.950	0.808
Lynch	1.000	1.000	1.000

consulting the other. The coders then consulted one another in order to resolve tagging differences. The inter-coder reliability tests revealed a kappa statistic of 0.78 prior to coder discussions and 0.89 after discrepancy resolution. For situations where the disparity could not be resolved even after discussions, the two coders' values were averaged. Table 11.2 shows examples from the violent affect lexicon. The weight for each term is the product of its intensity and ambiguity value. This is the value assigned to each occurrence of the term in the text being analyzed. For example, "lynch" was considered more severe by the coders than "hang." Although the two terms represent similar actions, the more severe motivation behind "lynch" as compared to "hang" resulted in a higher intensity score. Furthermore, the word "lynch" was also less ambiguous, conveying only a single violent meaning in the samples analyzed by the coders during the disambiguation procedure.

3.2.2 Affect Analysis Techniques

Ensemble classifiers use multiple classifiers, with each built using different techniques, training instances, or feature subsets (Dietterich 2000). Particularly, the feature subset classifier approach has been shown to be effective for analysis of style and patterns. Stamatatos and Widmer (2002) used an SVM ensemble for music performer recognition. They used multiple SVMs, each trained using different feature subsets. Similarly, Cherkauer (1996) used a neural network ensemble for imagery analysis. Their ensemble consisted of 32 neural networks trained on eight different feature subsets. The intuition behind using a feature ensemble is that it allows each classifier to act as an "expert" on its particular subset of features (Cherkauer 1996; Stamatatos and Widmer 2002), thereby improving performance over simply using a single classifier. We propose the use of a support vector regression ensemble that incorporates the relationship between various affect classes in order to enhance affect classification performance. Our ensemble includes multiple SVR models, each trained using a subset of features most effective for differentiating emotive intensities for a single affect class. We use the information gain (IG) heuristic to select the features for each SVR classifier. Since affect intensities are continuous, discretization must be performed before IG can be applied. We use 5 and 10 class bins (e.g., an intensity value of 0.15 would be placed into class 1 of 5 and 2 of 10 using 5 and 10 class bins).

Let $M=\{1,2,\ldots,m\}$ denote the set of training instances.
The SVR correlation ensemble intensity score for instance i and affect class c can be computed as follows:

$$\mathrm{SVRCE}_c(i) = \mathrm{SVR}_c(i) + \sum_{a=1}^{n}\left(\mathrm{Corr}(c,a)^2(\mathrm{SVR}_a(i) - \mathrm{SVR}_c(i))K\right)$$

Where:

$\mathrm{SVR}_c(i)$ is the prediction for instance i for affect class c using an SVR model trainde on M;
the feature subset for SVR_c is selected using the IG heuristic;
c and a are part of the set of n affect classes being investigated, and $c \neq a$;
$K = 1$ if $\mathrm{Corr}(c,a) > 0$, $K = -1$ otherwise;
$\mathrm{Corr}(c,a)$ is the correlation coefficient for affect classes c and a across the m training instances as follows:

$$\mathrm{Corr}(c,a) = \frac{\sum_{x=1}^{m}(c_x - \bar{c})(a_x - \bar{a})}{\sqrt{\sum_{x=1}^{m}(c_x - \bar{c})^2 \sum_{x=1}^{m}(a_x - \bar{a})^2}}$$

For:

c_x and a_x are the actual intensity values for affects c and a assigned to $x \in M$;
$\bar{c}$ and $\bar{a}$ are the average intensity values for affects c and a across the m training instances.

Fig. 11.2 SVR correlation ensemble for assigning affect intensities

All features with an average information gain greater than a threshold t are selected, as done in prior research (Yang and Pederson 1997).

The support vector regression correlation ensemble (SVRCE) adjusts the affect intensity prediction for a particular sentence based on the predicted intensities of other affects. The amount of adjustment is proportional to the level of correlation between affect classes (i.e., the affect class being predicted and the ones being used to make the adjustment) as derived from the training data. The SVRCE formulation is shown in Fig. 11.2.

The rationale behind SVRCE is that in certain situations, a particular sentence may get misclassified by a trained model due to a lack of prior exposure to the affective cues inherent in its text. In such circumstances, leveraging the relationship between affect classes may help alleviate the magnitude of such erroneous classifications.

We intend to compare the proposed SVRCE method against machine learning and scoring-based methods used in prior affect analysis research. These include the Pace regression technique proposed by Witten and Frank (2005) which was used to analyze affect intensities in weblogs (Mishne and Rijke 2006) as well as the semantic orientation, WordNet model, and manual lexicon scoring approaches. In addition to comparing the proposed SVRCE against other affect analysis techniques, we also intend to perform ablation testing to better understand the impact different components of our proposed method have on classification performance. Since SVRCE uses correlation information and feature-subset-based ensembles, we plan to compare it against an SVR ensemble that does not use correlation information as well as an SVR trained using a single feature set for all affect classes. The hypotheses associated with our research framework are presented below.

3.3 Research Hypotheses

H1: Features
The use of learned generic n-gram features will outperform manually and automatically crafted affect lexicons. Additionally, using an extended feature set encompassing all features will outperform individual feature sets.

- H1a: N-Grams > manual lexicon, semantic orientation, WordNet models
- H1b: All features > n-grams, manual lexicons, semantic orientation, WordNet models

H2: Techniques
The proposed SVRCE method will outperform comparison techniques used in prior studies for affect analysis.

- H2: SVRCE > Pace regression, semantic orientation scores, WordNet model scores, manual lexicon scores

H3: Ablation Testing
The SVRCE method will outperform an SVR ensemble not using correlation information as well as SVR run using a single feature set. Furthermore, the SVR ensemble will also significantly outperform SVR run using a single feature set.

- H3a: SVRCE > SVR ensemble, SVR
- H3b: SVR ensemble > SVR

4 Evaluation

We conducted experiments to evaluate various affective feature representations along with different affect analysis techniques, including the proposed support vector regression correlation ensemble (SVRCE). The experiments were conducted on four test beds comprised of sentences taken from web forums, blogs, and short stories. This section encompasses a description of the test beds, experimental design, experimental results, and outcomes of the hypotheses testing.

4.1 Test Bed

Analyzing affect intensities across application domains is important in order to get a better sense of the effectiveness and generalizability of different features and techniques. As a result, our test bed consisted of sentences taken from two corpora (shown in Table 11.3). The first test bed was a set of supremacist web forums discussing issues relating to Nazi and socialist ideologies. The second was comprised of 1,000 sentences taken from a couple of Arabic language Middle Eastern forums discussing issues relating to the war in Iraq. Analysis of such forums is important to better understand cyber activism, social movements, and people's political sentiments.

Table 11.3 Test bed description

Test bed name	Source URL(s)	No. of sentences	Affect classes tagged	Inter-coder reliability
Supremacist Web forums (SF)	www.stormfront.org www.nazi.org	1,000	Violence, anger, hate, racism	0.89
Middle Eastern Web forums (MEF)	www.montada.com www.alfirdaws.com	1,000	Violence, anger, hate, racism	0.79*

*Kappa value from initial tagging

Two independent coders tagged the sentences for intensities across the four affect classes used for each test bed (shown in Table 11.3). Each sentence was tagged with an intensity score between 0 and 1 (with 1 being most intense) for each of the affects. The tagging followed the same format as the one used for the manual lexicon creation. Each coder initially assigned values without consulting the other. The coders then consulted one another in order to resolve tagging differences. For situations where the disparity could not be resolved even after discussion, the two coders' values were averaged. The inter-coder reliability kappa values shown in Table 11.3 are from after discrepancy resolution (prior to averaging). For the Middle Eastern forums, the coders were unable to meet to resolve coding differences. For this test bed, the kappa value shown is for the two coders' initial tagging.

4.2 Experimental Design

Based on our research framework and hypotheses presented in Sect. 3, three experiments were conducted. The first was intended to compare the performance of learned n-grams against manually and automatically crafted lexicons. We also investigated the effectiveness of an extended feature set comprised of n-grams and lexicons versus individual feature groups. The second experiment compared different affect analysis techniques, including the proposed SVRCE, Pace regression, and scoring methods. The final experiment pertained to ablation analysis of the major components of SVRCE, including the use of correlation information and an ensemble approach to affect classification. In order to allow statistical testing of results, we ran 50 bootstrap instances for each condition across all three experiments. In each bootstrap run, 95% of the sentences were randomly selected for training while the remaining 5% were used for testing (Argamon et al. 2007). The average results across the 50 bootstrap runs were reported for each experimental condition. Performance was evaluated using standard metrics for affect analysis, which include the mean percentage error and the correlation coefficient (Mishne and Rijke 2006):

$$\text{Mean \% Error} = \frac{100}{\text{n}} \sum |x - y| \quad \text{Corr}(X,Y) = \frac{\sum (x-\bar{x})(y-\bar{y})}{\sqrt{\sum (x-\bar{x})^2 \sum (y-\bar{y})^2}}$$

Table 11.4 Overall results for various feature sets

	Mean % error		Correlation coefficient	
Features/test bed	SF	MEF	SF	MEF
N-grams	**4.6360**	**3.8066**	**0.6627**	**0.7455**
SO	5.0725	4.4742	0.4558	0.5308
WNet	4.9646	4.5507	0.4952	0.5122
ML	4.9767	4.6147	0.5388	0.4121
All	4.8176	4.3522	0.6238	0.6036

where x and y are the actual and predicted intensity values for one of the n testing instances denoted by the vectors X and Y.

4.3 Experiment 1: Comparison of Feature Sets

In this experiment, we compared generic n-grams with semantic orientation (SO), WordNet model (WNet), and the manual lexicon (ML). We also constructed an extended feature set comprised of n-grams, SO, WNet, and ML (labeled "All"). All feature sets were evaluated using the support vector regression correlation ensemble (SVRCE). SVRCE was run using a linear kernel. N-grams were selected using the information gain heuristic applied at the affect level, as outlined in Sect. 3.2.2. The information gain was applied to the training data during each of the 50 bootstrap instances; these features were then used to train the SVRCE classifiers used on the testing data. This resulted in 16 n-gram feature subsets (one for each affect class across the four test beds) and a corresponding SVRCE model for each feature subset. SO and WNet were run using the formulas described in Sects. 2.1 and 3.2.2. For SO, WNet, and ML, the word-level scores were computed for each sentence, resulting in a vector of scores for each sentence. Since different paradigm/seed words were used for each affect across all four test beds, the lexicon methods also generated 16 sets of sentence vectors each. Consistent with Mishne (2005), these vectors were treated as features input into the SVRCE. For the "All" feature set, the lexicon sentence vectors were merged with the n-gram frequency vectors.

Table 11.4 shows the macrolevel experimental results for the mean percentage error and correlation coefficients across the five feature sets applied to the two test beds. The values shown were averaged across the four affect classes used within each test bed. The test bed labels correspond to the abbreviations presented in Table 11.3 under the column "Test bed name." The n-gram features appeared to have the best performance, with the lowest mean percentage error and highest correlation coefficient for all test beds. The automated (i.e., SO and WNet) and manual lexicons all had fairly similar performance, with mean errors typically in the 5–7% range and correlation coefficients between 0.2 and 0.5. As anticipated, the use of all features performed well, outperforming the use of individual lexicons. Surprisingly however, using all features (i.e., n-grams in conjunction with lexicons) did not outperform the use of n-grams alone. N-grams outperformed the extended feature set

Table 11.5 Results for experiment 2 (comparison of techniques)

	Mean % error		Correlation coefficient	
Techniques/test bed	SF	MEF	SF	MEF
SO	8.6590	14.8759	0.4673	0.2530
WNet	5.9899	8.6639	0.5837	0.5224
ML	6.7270	8.3860	0.5500	0.5251
SVRCE	**4.6360**	**3.8066**	**0.6627**	**0.7455**
Pace	6.3038	5.8473	0.5692	0.6124

by as much as 0.5% and 0.14 on mean error and correlation coefficient, respectively. This suggests that the learned n-grams were able to effectively represent affective patterns in the text. Adding lexicon features introduced redundancy and, in some instances, noise. Further elaboration regarding the performance of n-grams in comparison with other feature sets is provided in the hypotheses testing section.

4.4 Experiment 2: Comparison of Techniques

The SVRCE method was compared against scoring and machine learning methods used in prior studies. The comparison techniques included Pace regression (Mishne and Rijke 2006), WordNet (WNet) scores (Kim and Hovy 2004; Mishne 2005), the pointwise mutual information scores from the semantic orientation (SO) approach, and the scores from our manual lexicon (ML). For SO, WNet, and ML, the average word-level intensities were used as the sentence-level scores as done in prior affect analysis research (Subasic and Huettner 2001; Grefenstette et al. 2004a; Read 2004; Cho and Lee 2006). SVRCE and Pace regression were both run using the n-gram features. N-grams were used since they had the best performance in experiment 1. Both techniques (i.e., SVRCE and Pace) were run using identical features, with each using 16 feature subsets selected using the information gain heuristic as described in experiment 1. Any scores outside the 0–1 range were adjusted to fit the possible range of intensities (this was done in order to avoid inflated errors stemming from values well outside the feasible range).

Table 11.5 shows the macrolevel experimental results for the mean percentage error and correlation coefficients across the five techniques. The SVRCE method had the best performance, with the lowest mean percentage error and highest correlation coefficient for all four test beds. Pace regression, WordNet (WNet) models, and the manual lexicon (ML) scoring methods were all in the middle, while the semantic orientation scoring method had the worst performance. The results are consistent with prior research that has also observed large differences between the word-level scores assigned using WNet and SO (Mishne 2005). The machine learning methods (SVRCE and Pace) both fared well with respect to their correlation coefficients. Pace also performed well on the supremacist and Middle Eastern forums in terms of mean percentage error, but not on the blogs test bed (LJ).

Table 11.6 Results for experiment 3 (ablation testing)

	Mean % error		Correlation coefficient	
Techniques/test bed	SF	MEF	SF	MEF
SVRCE	**4.6360**	**3.8066**	**0.6627**	**0.7455**
SVRE	5.0776	4.0667	0.5990	0.7231
SVR	5.7676	5.0460	0.5631	0.5757

4.5 Experiment 3: Ablation Testing

Ablation testing was performed to evaluate the effectiveness of the different SVRCE components. The SVRCE was compared against a support vector regression ensemble (SVRE) that does not utilize correlation information, as well as a support vector regression classifier using only a single feature set (SVR). The SVR was trained using a single feature set (for each test bed) selected by using all n-grams occurring at least five times in the corpus (Jiang et al. 2004). The SVRE and SVRCE were both run using information gain on the training data to select the 16 feature subsets most representative of each affect class. The experiment was intended to evaluate the two core components of SVRCE: (1) its use of feature ensembles to better represent affective content; (2) the use of correlation information for enhanced affect classification. Table 11.6 shows the macrolevel results for the mean percentage error and correlation coefficients for SVRCE, SVRE, and SVR.

The SVRCE method had the best performance, with the lowest mean percentage error and highest correlation coefficient for all test beds. SVRCE marginally outperformed SVRE, while both techniques outperformed SVR. The results suggest that use of feature ensembles and correlation information are both useful for classifying affective intensities.

4.6 Hypotheses Results

We conducted pairwise t tests on the 50 bootstrap runs for all three experiments. Given the large number of comparison conditions, a Bonferroni correction was performed to avoid spurious positive results. All P values less than 0.0005 were considered significant at alpha = 0.01.

4.6.1 H1: Feature Comparison

Table 11.7 shows the results for pairwise t tests conducted to compare the effectiveness of the extended and n-gram feature sets with other feature categories.

N-grams and the extended feature set both significantly outperformed the lexicon-based representations on all test beds with respect to mean error and correlation (all P values < 0.0001). Surprisingly, the extended feature set did not outperform n-grams.

Table 11.7 P values for pairwise t tests (n = 50) on feature comparisons

	Mean % error		Correlation coefficient	
Comparison features	SF	MEF	SF	MEF
All vs. n-gram	<0.0001*	<0.0001*	<0.0001*	<0.0001*
All vs. SO	<0.0001	<0.0001	<0.0001	<0.0001
All vs. WNet	<0.0001	<0.0001	<0.0001	<0.0001
All vs. ML	<0.0001	<0.0001	<0.0001	<0.0001
N-gram vs. SO	<0.0001	<0.0001	<0.0001	<0.0001
N-gram vs. WNet	<0.0001	<0.0001	<0.0001	<0.0001
N-gram vs. ML	<0.0001	<0.0001	<0.0001	<0.0001

*Result contradicts hypothesis

Table 11.8 Sample learned n-grams and lexicon items for hate affect

Learned n-grams		
Category	N-gram	Lexicon items
Character n-grams	uck, ck, fuc	awful, stupid, terrible, sicken, s**t, f**k
Word n-grams	terribly, suck, the stupid, the s**t, the f**k	
Hapax and dis legomena collocations	HAPAX so awful	
POS tag n-grams	PERSON_SG, WEEKDAY_NNP, TIME_SG	

Table 11.9 P values for pairwise t tests (n = 50) on technique comparisons

	Mean % error		Correlation coefficient	
Comparison features	SF	MEF	SF	MEF
SVRCE vs. SO	<0.0001	<0.0001	<0.0001	<0.0001
SVRCE vs. WNet	<0.0001	<0.0001	<0.0001	<0.0001
SVRCE vs. ML	<0.0001	<0.0001	<0.0001	<0.0001
SVRCE vs. Pace	<0.0001	<0.0001	<0.0001	<0.0001

In contrast, the n-gram feature set significantly outperformed the use of all features (n-grams plus the three lexicons), with all P values significant at alpha = 0.01.

Table 11.8 provides examples of learned n-grams taken from the LiveJournal test bed for the hate affect.

It also shows some related hateful items from the manual lexicon. The n-grams were able to learn many of the concepts conveyed in the lexicon. Furthermore, the n-grams were able to provide better context for some features and also learn deeper patterns in several instances. For example, the hate in LiveJournal blogs is often directed toward specific people and frequently involves places and times. This pattern is captured by the POS tag n-grams. In contrast, word lexicons cannot accurately represent such complex patterns. The example illustrates how the n-gram features learned were more effective than the lexicons employed in this study.

Table 11.10 P values for pairwise t tests (n = 50) on ablation testing

	Mean % error		Correlation coefficient	
Comparison features	SF	MEF	SF	MEF
SVRCE vs. SVRE	<0.0001	<0.0001	<0.0001	0.0013
SVRCE vs. SVR	<0.0001	<0.0001	<0.0001	<0.0001
SVRE vs. SVR	<0.0001	<0.0001	<0.0001	<0.0001

4.6.2 H2: Technique Comparison

As shown in Table 11.9, the SVRCE method significantly outperformed all four comparison techniques on mean percentage error and correlation coefficient across all four test beds. All P values were less than 0.0005 and therefore significant at alpha = 0.01. The results indicate that the SVRCE method's use of ensembles of learned n-gram features combined with affect correlation information allows the classifier to assign affect intensities with greater effectiveness than comparison approaches used in prior research.

4.6.3 H3: Ablation Tests

Table 11.10 shows the P values for pairwise t tests conducted to assess the contribution of the major components of the SVRCE method. The results of SVRCE versus SVRE revealed that the use of correlation information significantly enhanced performance on most test beds (significant for three out of four test beds on mean error and correlation). The results were not significant for mean error on the LiveJournal blog test bed (P value = 0.3452) as well as for correlation on the Middle Eastern forum dataset (P value = 0.0013). Both SVRCE and SVRE also significantly outperformed SVR, indicating that the use of feature ensembles is effective for classifying affect intensities (all P values less than 0.0001, significant at alpha = 0.01).

5 Case Study: Al-Firdaws vs. Montada

Many prior studies have used brief case studies to illustrate the utility of their proposed affect analysis methods (Subasic and Huettner 2001; Mishne and Rijke 2006). In order to demonstrate the usefulness of the SVRCE method coupled with a rich set of learned n-grams, we analyzed the affective intensities in two popular Middle Eastern web forums: www.alfirdaws.org/vb and www.montada.com. Analysis of affects in such forums is important for sociopolitical reasons and to better our understanding of social phenomena in online communities. Al-Firdaws is considered a more extreme forum by domain experts, with considerable content dedicated to support the Iraqi insurgency and al-Qaeda. In contrast, Montada is a general discussion forum with content and discussion pertaining to various social matters. We hypothesized that our SVRCE

Table 11.11 Summary statistics for the two web forums collected

Forum	No. of authors	No. of threads	No. of messages	No. of sentences	Duration
Al-Firdaws	2,189	14,526	39,775	244,917	January 2005–July 2007
Montada	31,692	114,965	869,264	2,052,511	September 2000–July 2007

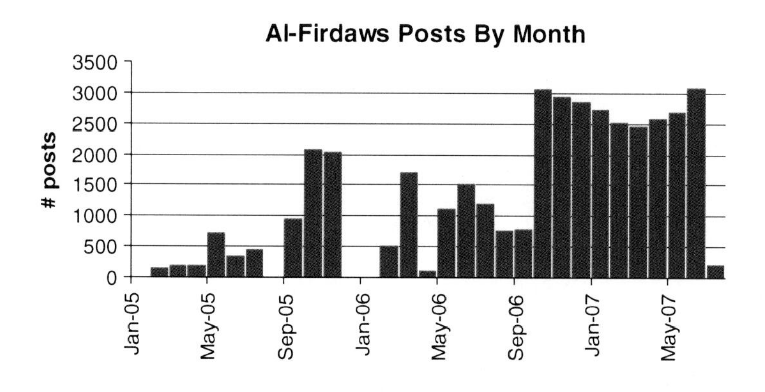

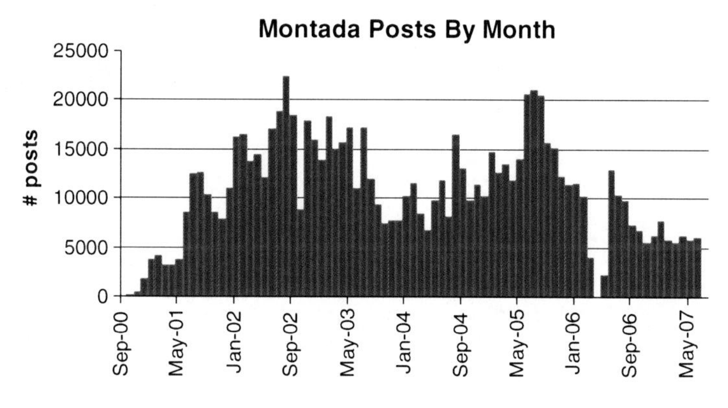

Fig. 11.3 Posting frequency for the two web forums

method would be able to effectively depict the likely intensity differences between these two web forums for appropriate affect classes.

We used spidering programs to collect the content in both web forums. Table 11.11 shows summary statistics for the content collected from the two forums. The Montada forum was considerably larger, with over 31,000 authors and a large number of threads and postings, partially because it had been around for approximately 7 years. Al-Firdaws was a relatively newer forum, beginning in 2005. Due to the nature of its content and time duration of existence, this forum had fewer authors and postings.

Figure 11.3 shows the number of posts for each month the forums have been active. Montada was very active in 2002 and 2005, with over 20,000 posts in some months,

Table 11.12 Affect intensities per posting across two web forums

Intensity	Forum	Violence	Anger	Hate	Racism
Average per message	Al-Firdaws	**0.084**	**0.018**	**0.037**	**0.032**
	Montada	0.027	0.012	0.010	0.014
Total per message	Al-Firdaws	**0.523**	**0.127**	**0.178**	**0.191**
	Montada	0.246	0.105	0.092	0.134

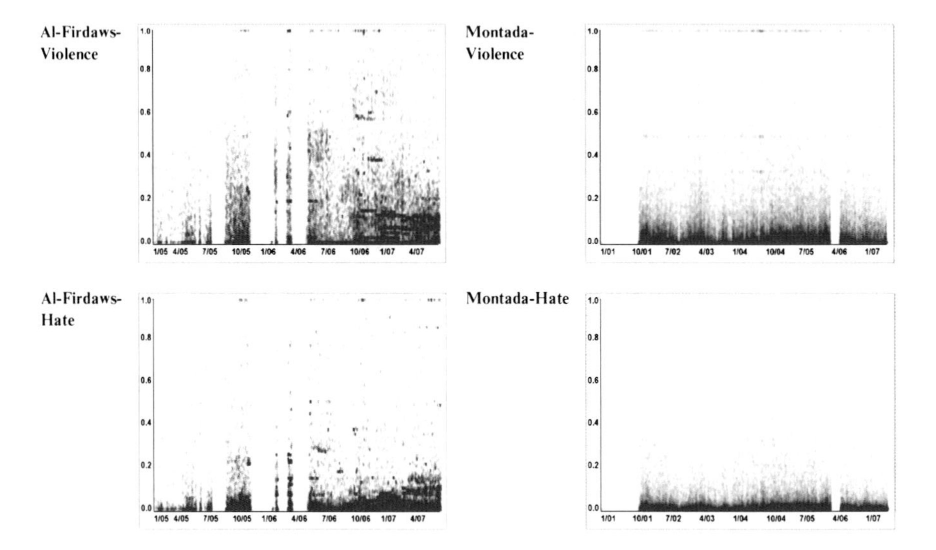

Fig. 11.4 Temporal view of intensities in two web forums

yet appears to be in a down phase in 2007 (similar to 2004). Al-Firdaws consistently had between 2,500 and 3,000 posts per month since the second half of 2006.

The SVRCE classifier was employed in conjunction with the n-gram feature set to analyze affect intensities in the two web forums. Analysis was performed on violence, hate, racism, and anger affects. We computed the average posting level intensities (averaged across all sentences in a posting) as well as the total intensity per post (the summation of sentence intensities in each posting). The analysis was performed on all postings in each forum (approximately 900,000 postings and 2.3 million sentences). As shown in Table 11.12, the Al-Firdaws forum had considerably higher affect intensities for all four affect classes, usually 2–3 times greater than Montada.

Figure 11.4 depicts the average message violence and hate intensities over time for all postings in each of the two web forums. The x-axis indicates time, while the y-axis shows the intensities (on a scale of 0–1). Each point represents a single message; areas with greater message concentrations are darker. The blank periods in the diagrams correspond to periods of posting inactivity in forums (see Fig. 11.3 for correspondence). Based on the diagram, we can see that Al-Firdaws has considerably higher violence and also greater hate intensity across time. Al-Firdaws also appears to have increasing violence intensity in 2007 (based on the concentration of postings), possibly attributable to the increased activity in this forum. In contrast, violence and hate intensities are consistently low in Montada. The results generated using SVRCE and n-gram features are consistent with existing knowledge regarding these two forums. The case study illustrates how the proposed features and techniques can be successfully applied toward affect analysis of computer-mediated communication text.

6 Conclusions

In this chapter, we evaluated various features and techniques for affect analysis of online texts. In addition, the support vector regression correlation ensemble (SVRCE) was proposed. This method leverages an ensemble of SVR classifiers with each constructed for a separate affect class. The ensemble of predictions combined with the correlation between affect classes is leveraged for enhanced affect classification performance. Experimental results on test beds derived from online forums, blogs, and stories revealed that the proposed method outperformed existing affect analysis techniques. The results also suggested that learned n-grams can improve affect classification performance in comparison with lexicon-based representations. However, combining n-gram and lexicon features did not improve performance due to increased amounts of noise and redundancy in the extended feature set. A case study was also performed to illustrate how the proposed features and techniques can be applied to large cyber communities in order to reveal affective tendencies inherent in these communities' discourse. To the best of our knowledge, the experiments conducted in this study are the first to evaluate features and techniques for affect analysis. Furthermore, we are also unaware of prior research applied to such a vast array of domains and test beds.

We believe this chapter provides an important stepping stone for future work intended to further enhance the feature representations and techniques used for classifying affects. Based on this work, we have identified several future research directions. We intend to apply the techniques across a larger set of affect classes (e.g., 10–12 affects per test bed). We are also interested in exploring additional feature representations, such as the use of richer learned n-grams (e.g., semantic collocations, variable n-gram patterns, etc.). We also plan to evaluate the effectiveness of real-world knowledge bases such as those employed by Liu et al. (2003).

References

Abbasi, A., Chen, H., and Salem, A. "Sentiment analysis in multiple languages: Feature selection for opinion classification in web forums," *ACM Trans. on Information Systems*, vol. 26(3), 2008.

Argamon, S., Whitelaw, C., Chase, P., Hota, S.R., Garg, N., and Levitan, S. "Stylistic text classification using functional lexical features," *Journal of the American Society for Information Science and Technology*, vol. 58(6), pp. 802–822, 2007.

Chuang, Z. and Wu, C. "Multi-modal emotion recognition from speech and text," *Computational Linguistics and Chinese Language Processing*, vol. 9(2), pp. 45–62, 2004.

Cherkauer, K.J., "Human expert-level performance on a scientific image analysis task by a system using combined artificial neural networks," In Chan, P. (Ed.), *Working Notes of the AAAI Workshop on Integrating Multiple Learned Models*, pp. 15–21, 1996.

Cho, Y.H. and Lee, K.J. "Automatic affect recognition using natural language processing techniques and manually built affect lexicon," *IEICE Tran. Information Systems*, vol. E89(12), pp. 2964–2971, 2006.

Dietterich, T. G. "Ensemble methods in machine learning," In *Proc. of the 1st Int'l Workshop on Multiple Classifier Systems*, pp. 1–15, 2000.

Donath, J., Karahalio, K. and Viegas, F. "Visualizing conversation," In *Proc. of the 32nd Conf. on Computer-Human Interaction*, Chicago, USA, 1999.

Fellbaum, C. (Ed.). *WordNet: An Electronic Lexical Database*, Cambridge, MA. MIT Press, 1998.

Grefenstette, G., Qu, Y., Evans, D. A., and Shanahan, J. G. "Validating the coverage of lexical resources for affect analysis and automatically classifying new words along semantic axes," In Yan Qu, James Shanahan, and Janyce Wiebe, (Eds.), *Exploring Attitude and Affect in Text: Theories and Applications*, AAAI-2004 Spring Symposium Series, pp. 71–78, 2004a.

Grefenstette, G., Qu, Y., Shanahan, J. G., and Evans, D. A. "Coupling niche browsers and affect analysis for an opinion mining application," In *Proc. of the 12th Int'l Conf. Recherche d'Information Assistee par Ordinateur*, pp. 186–194, 2004b.

Hearst, M. A. "Direction-based text interpretation as an information access refinement," In P. Jacobs (Ed.), *Text-Based Intelligent Systems: Current Research and Practice in Information Extraction and Retrieval*. Mahwah, NJ: Lawrence Erlbaum Associates, 1992.

Jiang, M., Jensen, E., Beitzel, S., and Argamon, S. "Choosing the right bigrams for information retrieval" *In Proc. of the Meeting of the Int'l Federation of Classification Societies*, 2004.

Kim, S. and Hovy, E. "Determining the sentiment of opinions," In *Proc. of the 20th Int'l Conf. on Computational Linguistics*, pp. 1367–1373, 2004.

Liu, H., Lieberman, H., Selker, T. "A model of textual affect sensing using real-world knowledge." In *Proc. of the 8th Int'l Conf. on Intelligent User Interfaces*, Miami, Fl., 2003.

Ma, C., Prendinger, H., and Ishizuka, M. "Emotion estimation and reasoning based on affective textual interaction," In *Proc. of the First Int'l Conf. on Affective Computing and Intelligent Interaction*, pp. 622–628, 2005.

Mishne, G. "Experiments with mood classification," In *Proc. of Stylistic Analysis of Text for Information Access Workshop*, 2005.

Mishne, G. and Rijke, M. de. "Capturing global mood levels using blog posts," In *Proc. of the AAAI 2006 Spring Symposium on Computational Approaches to Analysing Weblogs*, 2006.

Nigam, K. and Hurst, M. "Towards a robust metric of opinion," In *Proc. of the AAAI Spring Symposium on Exploring Attitude and Affect in Text*, 2004.

Pang, B., Lee, L., and Vaithyanathain, S. "Thumbs up? Sentiment classification using machine learning techniques," In *Proc. of the Empirical Methods in Natural Language Processing*, pp. 79–86, 2002.

Pang, B. and Lee, L. "Seeing stars: Exploiting class relationships for sentiment categorization with respect to rating scales," In *Proc. of the Annual Meeting on Association for Computational Linguistics*, pp. 115–124, 2005.

Picard, R.W. *Affective Computing*. Cambridge, MA. MIT Press, 1997.

Picard, R.W., Vyzas, E., and Healey, J. "Toward machine emotional intelligence: analysis of affective physiological state," *IEEE Tran. Pattern Analysis and Machine Intelligence*, vol. 23(10), pp. 1179–1191, 2001.

Read, J. "Recognizing affect in text using point-wise mutual information," *Master's Thesis*, 2004.

Schumaker, R. and Chen, H. "Textual analysis of stock market prediction using financial news articles," *Americas Conference on Information Systems (AMCIS-2006)*, Acapulco, Mexico, 2006.

Stamatatos, E. and Widmer, G. "Music performer recognition using an ensemble of simple classifiers," In *Proc. of the 15th European Conf. on Artificial Intelligence*, Lyon France, 2002.

Subasic, P. and Huettner, A. "Affect analysis of text using fuzzy semantic typing," *IEEE Tran. Fuzzy Systems*, vol. 9(4), pp. 483–496, 2001.

Turney, P.D. and Littman, M.L. "Measuring praise and criticism: inference of semantic orientation from association," *ACM Trans. Information Systems*, vol. 21(4), pp. 315–346, 2003.

Valitutti, A., Strapparava, C., and Stock, O. "Developing affective lexical resources," *PsychNology Journal*, vol. 2(1), pp. 61–83, 2004.

Wiebe, J. "Tracking point of view in narrative," *Computational Linguistics*, vol. 20(2), pp. 233–287, 1994.

Wiebe, J., Wilson, T., Bruce, R., Bell, M., and Martin, M. "Learning subjective language," *Computational Linguistics*, vol. 30(3), pp. 277–308, 2004.

Witten, H. and Frank, E. *Data mining: Practical machine learning tools and techniques*, Morgan Kauffman, 2nd Edition, 2005.

Wu, C., Chuang, Z., and Lin, Y. "Emotion recognition from text using semantic labels and separable mixture models," *ACM Trans. Asian Language Information Processing*, vol. 5(2), pp. 165–182, 2006.

Yang, Y. and Pederson, J.O. "A comparative study on feature selection in text categorization," In *Proc. of the 14th Int'l Conference on Machine Learning*, pp. 412–420, 1997.

Chapter 12
CyberGate Visualization

1 Introduction

Computer-mediated communication (CMC) has seen tremendous growth due to the fast propagation of the Internet. Text-based modes of CMC include e-mail, listservs, forums, chat, and the World Wide Web (Herring 2002). These modes of CMC continue to have a profound impact on organizations due to their quick and ubiquitous nature. Electronic communication methods have redefined the fabric of organizational culture and interaction. With the persistent evolution of communication processes (Welck 1987) and constant advancements in technology, such metamorphoses are likely to continue. One such trend has been the increased use of online communities to support business operations (Wenger and Snyder 2000). Business online communities provide invaluable mechanisms for various forms of interaction including knowledge dissemination, transfer of goods and services, and product/service reviews (Cothrel 2000). Electronic communities (Wenger and Snyder 2000) and networks (Wasko and Faraj 2005) of practice allow companies to tap into the wealth of information and expertise available across corporate boundaries. Internet marketplaces enable the efficient transfer of goods and services in cyberspace while consumer rating forums have also emerged, providing product reviews that can be equally useful to potential customers and marketing departments (Turney and Littman 2003; Pang et al. 2002).

In spite of the numerous benefits of electronic communication, it is not without its pitfalls. Two characteristics of computer-mediated communication have proven to be particularly problematic: online anonymity and the enormity of data present in cyber communities. These vices undermine the numerous benefits associated with CMC and online business communities (Davenport 2002).

The anonymous nature of the Internet has resulted in several trust-related issues including online deception (Nissenbaum 1996; Friedman et al. 2000). In addition to deceit, cyberspace is crawling with agitators (also referred to as "trolls") that attempt to peeve community members and disrupt online discourse (Donath 1999).

H. Chen, *Dark Web: Exploring and Data Mining the Dark Side of the Web*, Integrated Series in Information Systems 30, DOI 10.1007/978-1-4614-1557-2_12,

Newsgroups and knowledge exchange communities also suffer from lurkers that free ride off of others (Smith 2002; Wasko and Faraj 2005). Collectively, these concerns can cast serious doubts onto the quality of information exchanged in such online communities (Davenport 2002; Viegas and Smith 2004; Wasko and Faraj 2005). Cyber communities also contain large volumes of information, including various communication modes, topics, threads, messages, and authors. CMC environments feature very large-scale conversations involving thousands of people and messages (Sack 2000; Herring 2002). The enormous information quantities make such places noisy and difficult to navigate (Viegas and Smith 2004).

Hence, there is a need for techniques to evaluate, summarize, and present computer-mediated communication content. Many believe that the solution is to develop systems that support navigation and knowledge discovery (Wellman 2001). Such CMC information systems can enhance informational transparency which benefits community participants and observers (Sack 2000; Kelly et al. 2002). Erickson and Kellogg (2000) argued that tools supporting social translucence in cyberspace would improve participant accountability. Smith (1999) suggested that methods for providing social accounting data could be mutually beneficial to online community members and researchers. Consequently, numerous CMC information systems have been developed in order to address these needs (Xiong and Donath 1999; Fiore and Smith 2002; Viegas et al. 2004; Viegas and Smith 2004). These techniques generally visualize data provided in the message headers such as interaction- (send/reply structure) and activity-based (posting patterns) information. Little support is provided for analysis of textual information contained in the message bodies. In the instances where text analysis is provided, simple feature representations such as those used in information retrieval systems (Sack 2000; Whitelaw and Patrick 2004) are utilized, e.g., bag-of-words (Mladenic 1999).

In addition to topical information, online discourse is also rich in social cues including emotion, opinion, style, and genre (Yates and Orlikowski 1992, 2002; Henri 1992; Hara et al. 2000). There is a need for improved CMC system content analysis capabilities (Paccagnella 1997) based on a richer textual representation. This requires the incorporation of a complex set of features, techniques, and visual representations to facilitate enhanced text analysis. Unfortunately, text analysis features and visualization techniques are not well defined. Consequently, there is a need for a set of design guidelines for CMC systems supporting text analysis (Sack 2000).

This chapter proposes a design framework for the creation of CMC systems that can provide improved text analysis capabilities by incorporating a richer set of textual information types. Our framework addresses several important issues from the text mining literature en route to the development of a set of guidelines geared toward CMC text analysis. We then develop the CyberGate system based on our design framework. CyberGate includes the Writeprint and Ink Blot techniques that can be used for analysis and categorization of CMC text.

The remainder of this chapter is organized as follows. Section 2 provides a review of CMC systems developed to support content analysis. This section outlines the need for systems to/that support CMC textual analysis. Section 3 reviews some of the relevant text mining issues related to CMC text analysis. We then present a design framework for CMC text analysis in Sect. 4 that takes into consideration the text mining concerns outlined in the previous section. Section 5 uses the framework to develop CyberGate: a CMC text analysis system that features the Writeprint and Ink Blot techniques. In Sect. 6, we present two sample cases of how CyberGate can be used to help with Dark Web forum content and author attribution. The case studies involve two major extremist forums: Clearguidance.com and Ummah.com. We conclude this chapter in Sect. 7.

2 Background

2.1 CMC Content

Several dimensions have been proposed for CMC content analysis (Henri 1992; Hara et al. 2000). These include participation/activity, interaction, social cues, topics, emotions, roles, linguistic variation, question types, response complexity, etc. (Henri 1992; Rourke et al. 2001). The information utilized for CMC content analysis can be broadly categorized as either structural or textual, as shown in Fig. 12.1. Structural features of CMC content include all attributes based on communication topology. These features are extracted from message headers, without any use of the message body (Sack 2000). Structural features support activity and interaction analysis. Posting activity–related features include number of posts, number of initial messages, number of replies, number of responses to a particular author's posts, etc. (Fiore et al. 2002). These features can be used to represent an author's social accounting metrics (Smith 2002). Activity/participation-based attributes and analysis can provide insight into different roles such as debaters, experts, etc. (Zhu and Chen 2001;

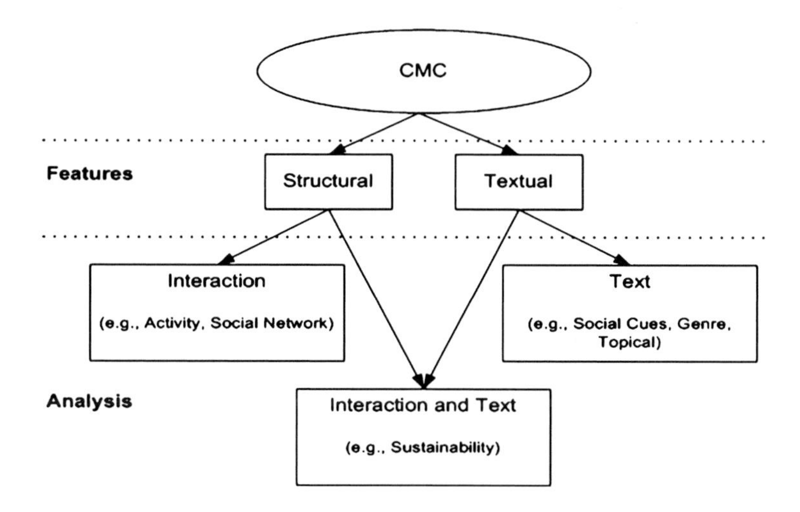

Fig. 12.1 CMC content analysis model

Viegas and Smith 2004). While activity analysis looks at individuals independent of their fellow participants, interaction analysis is specifically concerned with contact-based/interpersonal information. This type of analysis typically involves the construction of social networks based on who is talking to whom (Sack 2000; Smith and Fiore 2001). Such analysis can help identify key members and relationships by utilizing social network analysis metrics such as centrality and link densities. Interaction analysis can also provide insight into important roles such as major/minor discussants, reply sources, and reply sinks (Smith and Fiore 2001).

Textual features include all attributes derived from the message body. Although the informational richness of CMC text was previously questioned (Daft and Lengel 1986), numerous studies have since demonstrated the richness of CMC content (Yates and Orlikowski 1992, 2002; Lee 1994; Panteli 2002). Thus, in additional to topical information and events (e.g., Allan et al. 1998), textual online discourse contains several types of information including emotions (Picard 1997; Subasic and Huettner 2001), opinions (Hearst 1992), style (Abbasi and Chen 2006; Zheng et al. 2006), and genres (Santini 2004). These additional forms of information can support several forms of analysis including social cues (Spears and Lea 1992, 1994), power cues (Panteli 2002), and genre analysis (Yates and Orlikowski 1992, 2002). Social cues include textual elements not related to formal content or subject matter (Henri 1992). Hara (2000) provided examples of social cues that included self-introductions, expressions of feeling, greetings, signatures, jokes, use of symbolic icons, and compliments. Text also contains evidence of power cues, where the style of messages differs depending upon whether the interaction is between participants at the same or different levels within an organization (Panteli 2002). Genre theory identified types of writing based on purpose and form (e.g., memos, meetings, reports, etc.) and demonstrated how different genres served as a source of organizing structures and communicative norms (Yates and Orlikowski 1992; Yates et al. 1999).

2.2 CMC Systems

CMC systems can be categorized into two categories based on functionality: those that support the communication process and those that support analysis of communication content (Sack 2000). While it is certainly possible for a single system to support both functions (e.g., Erickson and Kellogg 2000), we focus on only the analysis functionalities provided by these systems due to its relevance to CMC content analysis. Table 12.1 provides a review of previous CMC systems based on the type of analysis features included.

A plethora of CMC systems have been developed to support structural features. Several tools visualize posting activity patterns, such as Loom (Donath et al. 1999) and Authorlines (Viegas and Smith 2004). PeopleGarden and Communication Garden both use garden metaphors with flower glyphs to display author and thread activity (Xiong and Donath 1999; Zhu and Chen 2001). The number of petals, number of thorns, petal colors, and stem lengths are used to represent activity features such

Table 12.1 Previous CMC systems

System name	References	Feature types		Feature descriptions
		Struct.	Text	
Chat Circles	Donath et al. (1999)	√	√	Length, headers
Loom	Donath et al. (1999)	√	√	Terms, punctuation, headers
PeopleGarden	Xiong and Donath (1999)	√		Headers
Babble	Erickson and Kellogg (2000)	√		Headers
Conversation Map	Sack (2000)	√	√	Semantic, headers
Communication Garden	Zhu and Chen (2001)	√	√	Noun phrases, headers
Coterie	Donath (2002)	√		Headers
Newsgroup Treemaps	Fiore and Smith (2002)	√		Headers
PostHistory	Viegas et al. (2004)	√		Headers
Social Network Fragments	Viegas et al. (2004)	√		Headers
Authorlines	Viegas and Smith (2004)	√		Headers
Newsgroup Crowds	Viegas and Smith (2004)	√		Headers

as total number of posts and the number of threads an author has been active in. Babble (Erickson and Kellogg 2000) and Coterie (Donath 2002) are both geared toward showing activity patterns in persistent conversation. In these systems, all participants are displayed in a two-dimensional space. More active authors are shown in the center, while participants with fewer postings gradually shift toward the outside. The visual effect represents a good method for identifying active participants versus lurkers and free riders (Donath 2002). Chat Circles (Donath et al. 1999) presents recently posted messages as bubbles that fade over time as newer content is displayed. In contrast to systems providing activity-based functions, systems displaying interaction information have also been developed. Conversation Map visualizes social networks based on send/reply patterns (Sack 2000). Netscan displays message and author interactions (Smith and Fiore 2001; Smith 2002), while Loom shows thread-level interaction structures (Donath et al. 1999).

Previous CMC systems offer limited support for text features. Loom (Donath et al. 1999) shows some content patterns based on message moods. The moods are assigned based on the occurrence of certain terms and punctuation in the message text. Chat Circles (Donath et al. 1999) displays messages based on body text length. Conversation Map (Sack 2000) and Communication Garden (Zhu and Chen 2001) provide more in-depth topical analysis. Conversation Map uses computational linguistics to build semantic networks for discussion topics, while Communication Garden performs topic categorization based on noun phrases. Overall, the features used in CMC systems are insufficient to effectively capture textual content in online discourse (Sack 2000; Whitelaw and Patrick 2004). Text systems are a related class of systems that are used for information retrieval. However, information retrieval (IR) systems are more concerned with information access than analysis (Hearst 1999). Mladenic (1999) presented a review of 29 IR systems, all of which used bag-of-words to represent textual features. Thus, text-based information systems (TBIS) for supporting in-depth text analysis of CMC content remain nonexistent.

2.3 Need for CMC Systems Supporting Text Analysis

Numerous CMC researchers and analysts have stated the need for tools to support computer-mediated communication text analysis. Textual features are important yet often overlooked in e-mail analysis (Panteli 2002). Cothrel (2000) stated that discussion content is an essential dimension of online community success measurement, yet proper definition and measurement remains elusive. Hara et al. (2000) noted that there has been limited CMC content analysis since manual methods are time consuming. For instance, features such as use of greetings and signatures, which may be important power cues, can easily be captured using stylistic feature extractors (Zheng et al. 2006). Manual extraction of textual features can be time consuming and typically results in smaller data samples used for experimentation. Paccagnella (1997) also suggested that computer programs to support CMC text analysis would be helpful, yet do not exist. He noted numerous ways in which text analysis systems could benefit CMC content analysis. Based on his recommendations, six of the pertinent tasks that CMC text analysis systems could support are listed below:

1. *Data linking*: connecting relevant data segments to each other, forming categories, clusters, or networks of information.
2. *Content analysis*: counting frequencies, sequences, or locations of words and phrases.
3. *Data display*: placing selected or reduced data in a condensed, organized format, such as a matrix or network, for inspection.
4. *Conclusion-drawing and verification*: aiding the analyst to interpret displayed data and to test or confirm findings.
5. *Theory-building*: developing systematic, conceptually coherent explanations of findings; testing hypotheses.
6. *Graphic mapping*: creating diagrams that depict findings or theories.

Given the need for CMC text analysis and lack of systems that address this need, an important and obvious question arises. Why do most CMC systems support structural information but not textual content features? There are three major differences between structural and textual features that are likely responsible for the disparity between the numbers of systems representing these feature types. These three factors are feature definitions, extraction, and presentation. Structural features are well defined, easy to extract, and easy to visualize. Activity- (Fiore et al. 2002) and interaction-based features and metrics have been well defined in the sociology literature. Furthermore, posting activity features and interaction features (e.g., network metrics) can each easily be extracted from message headers.

The extracted features are typically visualized using bar chart variants for activity frequency (e.g., Xiong and Donath 1999; Zhu and Chen 2001; Viegas and Smith 2004) and networks for interaction (e.g., Donath et al. 1999; Smith and Fiore 2001). In contrast, rich textual features are not well defined, difficult to extract, and harder to present to end users. Text categorization requires a complex set of subjective features (Donath et al. 1999). For example, over 1,000 features have been used for analyzing style, with no consensus (Rudman 1998). Additionally, text feature extraction can be challenging due to high levels of noise in online discourse text (Knight 1999; Nasukawa and Nagano 2001). Finally, text presentation requires the use of multiple presentation views (Losiewicz et al. 2000) since standard visualization techniques may not apply to text (Keim 2002). Consequently, different techniques have been developed to support various facets of text visualization (Wise 1999; Miller et al. 1998; Rohrer et al. 1998; Huang et al. 2005) with no ideal solution.

In light of these challenges, Sack (2000) argues for a new CMC system design philosophy that incorporates automatic text analysis techniques. He states "…it is necessary to formulate a complementary design philosophy for CMC systems in which the point is to help participants and observers spot emerging groups and changing patterns of communication…" (p. 86). Design guidelines are needed because of the lack of previous tools for CMC textual analysis, complexity of text analysis tasks, and the fact that guidelines for appropriate features, feature selection, and presentation styles are not well defined. Hence, a design framework for text-based information systems must address the aforementioned text mining issues. A review of each of these issues is presented in the following section.

3 CMC Text Mining

Text mining is concerned with the process of extracting interesting information and knowledge from unstructured/free text. In order to develop a set of design guidelines for text-based information systems, we first present a review of the relevant text mining elements pertaining to text analysis systems, which include: tasks, information types, features, feature selection methods, and visualization.

3.1 Tasks

Categorization and analysis are two important text mining tasks (Lewis 1992; Tan 1999).

Categorization refers to the assignment of text to predefined categories based on content (Dumais et al. 1998; Chen 2001). This includes classification and clustering operations. Analysis is concerned with the trends, patterns, and comparisons derived from text (Hearst 1999; Nasukawa and Nagano 2001). Categorization tasks use textual feature occurrence frequencies, similarities, and variations for classification or clustering, while analysis tasks use that information for identification and assessment of trends and patterns. For CMC text analysis, categorization and analysis are both important. Categorization enables the classification of messages by different information types, while analysis functionalities provide support for analyzing patterns and trends across various dimensions including forums, threads, authors, messages, and time. The categorization and analysis tasks correspond to the data-linking and content analysis functions that Paccagnella (1997) suggested CMC systems should support.

3.2 Information Types

Systemic Functional Linguistic Theory states that language has three kinds of meaning: ideational, interpersonal, and textual (Halliday 1994). *Ideational* means that language consists of ideas. *Textual* indicates that language has organization and structure. *Interpersonal* refers to the fact that language is a medium of exchange. The interpersonal dimension of language is effectively covered by social/interaction networks (Sack 2000). However, text-based information systems should incorporate a wide range of ideational and textual information. Examples of ideational information types found in text include topics, events, opinions, and emotions.

Topical information is the most commonly represented form of text information. It is supported by all information retrieval systems (Mladenic 1999). Common features used to represent topics include bag-of-words, noun phrases, and named entities. For example, HelpfulMed (Chen et al. 2003) creates topical categorization maps based on the noun phrases extracted from the content of medical document corpora. In contrast, events are specific topics/incidents with a temporal dimension. While "hurricane" is a topic, "Hurricane Katrina" is an event. Event detection/tracking has garnered significant attention in recent years. Although event detection and tracking are important areas of text analysis, they continue to present challenges, and effective features and techniques remain elusive (Allan et al. 1998).

Additional forms of ideational information include opinions and emotions. Opinions include sentiments about a particular topic, such as agonistic, neutral (no opinion), or antagonistic (Hearst 1992). Popular applications of opinion-related information include mining movie and product review sites (Turney and Littman 2003). Text is also rich in emotional information (Picard 1997). Emotions encompassed in online communication include various affects such as happiness, horror, anger, etc. (Subasic and Huettner 2001).

Language also contains textual features such as organization and structure. Examples of textual information types include style, genres, and vernaculars. Style includes numerous stylistic attributes, including vocabulary richness, word choice, and punctuation usage (Argamon et al. 2003; Abbasi and Chen 2006). Example styles include formal (use of greetings, structured sentences, paragraphs), informal (no sentences, no greetings, erratic punctuation usage), etc. Style is based on the literary choices an author makes, which can be a reflection of his/her context

Table 12.2 Text information types

Information type	Examples	Analysis types	References
Ideational	Topics	Topical analysis	Mladenic (1999), Chen et al. (2003)
	Events	Event detection	Allan et al. (1998)
	Opinions	Sentiment analysis	Hearst (1992), Turney and Littman (2003)
	Emotions	Affect analysis	Picard (1997), Subasic and Huettner (2001)
Textual	Style	Authorship analysis Deception detection Power cues	Argamon et al. (2003), Abbasi and Chen (2006), Zhou et al. (2004), Panteli (2002)
	Genres	Genre analysis	Yates and Orlikowski (1992, 2002), Santini (2004)
	Metaphors/ vernaculars	Semantic networks	Sack (2000)

(who, what, when, why, where) and personal background (education, gender, etc.). Stylistic information is utilized in numerous forms of analysis. Authorship analysis attempts to identify and characterize individuals based on their writing style (Argamon et al. 2003; Abbasi and Chen 2006), deception detection attempts to determine if an individual's writing is deceitful (Zhou et al. 2004), while power cue identification looks at the stylistic differences in writing between superiors and subordinates in organizational settings (Panteli 2002). Genres are classes of writing. Genres found in an organizational communication setting include inquiries, informational messages, news articles, memos, resumes, reports, interviews, etc. (Yates and Orlikowski 1992; Santini 2004). Genres have a profound impact on the structure and organization of text in computer-mediated communication. Table 12.2 shows example for each information type and their corresponding analysis applications.

3.3 *Features*

Linguistic features can be classified into two broad categories: language resources and processing resources (Cunningham 2002). Both categories are often used in conjunction to complement each other. Language resources are data-only resources such as lexicons, thesauruses, word lists (e.g., pronouns), etc. Such self-standing features exist independent of the context and provide powerful discriminatory potential. However, language resource construction is typically manual, and features may be less generalizable across information types (Pang et al. 2002).

Processing resources require programs/algorithms for computation. For instance, parts of speech, n-grams, statistical features (e.g., vocabulary richness), and bag-of-words are all examples of processing resources since they require processing operations before being used. The majority of these features are context-dependent, meaning that they change according to the text corpus. However, the extraction procedures/definitions remain constant, making processing resources highly generalizable across information types. Consequently, processing resource features such as bag-of-words, parts of speech, and n-grams are used to represent numerous information types including topics, events, sentiments, emotions, style, and genres (Pang et al. 2002; Argamon et al. 2003; Santini 2004). Using language and processing resources in conjunction can improve text categorization and analysis capabilities since processing resources provide breadth while language resources offer depth.

3.4 Feature Selection

Given the complexities of language and text content, a large number of potential linguistic features are used for categorization and analysis. However, only a subset of these features may be relevant or useful for a particular instance (e.g., an author, document, message, etc.). Three types of feature selection techniques have been identified in previous research (Guyon and Elisseeff 2003), all of which have also been used in text mining. These include ranking, projection, and subset selection methods. Ranking techniques are those that rank/sort attributes based on some heuristic (Duch et al. 2004; Hearst 1999). Examples of ranking techniques include information gain, chi-squared, and Pearson's correlation coefficient (Forman 2003). Projection methods are transformation-based techniques that utilize dimensionality reduction (Huber 1985; Huang et al. 2005). These include techniques such as principal component analysis (PCA), multidimensional scaling (MDS), and self-organizing map (Chen et al. 2003; Huang et al. 2005). Subset selection techniques select a subset of the original attributes. These techniques typically use search strategies, including exhaustive and random search, to consider different feature combinations that comprise subsets of the original feature set (Dash and Liu 1997). Each of the three feature selection techniques has its advantages and disadvantages.

Ranking methods have been used in several text categorization and analysis studies. Efron et al. (2004) used information gain to select bag-of-words for topic categorization. Abbasi and Chen (2005) used decision tree models to select key features in order to analyze the important stylistic differences between web forum authors. Pang et al. (2002) used minimum frequency thresholds to filter out n-grams for sentiment classification. Ranking methods offer greater explanatory potential since they preserve the original feature set and simply sort the features based on some heuristic (Seo and Shneiderman 2002). These techniques also offer simplicity and scalability; however, they typically only consider an individual feature's predictive power (Guyon and Elisseeff 2003; Li et al. 2006). Thus, important feature interactions may be lost.

Projection methods have been used to transform text feature spaces into two- or three-dimensional projections for authorship and topical categorization (Allan et al. 2001; Chen et al. 2003; Abbasi and Chen 2006). Projection methods are highly robust against noise, which makes them very useful for text analysis since they can

Table 12.3 Feature selection methods applied to text

Selection method	Example	Analysis type	References
Ranking	Information gain	Topical	Efron et al. (2004)
	Decision tree model	Authorship	Abbasi and Chen (2005)
	Minimum frequency	Sentiment	Pang et al. (2002)
Projection	Principal component analysis	Authorship	Abbasi and Chen (2006)
	Multidimensional scaling	Topical	Allan et al. (2001)
	Self-organizing map	Topical	Chen et al. (2003)
Subset selection	Genetic algorithm	Authorship	Li et al. (2006)

uncover important underlying patterns (Abbasi and Chen 2006). However, the use of transformation from the original features to projections results in reduced explanatory potential (Seo and Shneiderman 2002). Projection methods can describe important high-level patterns and trends but have difficulty explaining details about specific features.

Subset selection methods provide a high level of power, often outperforming other techniques in terms of predictive abilities (Dash and Liu 1997). For example, Li et al. (2006) demonstrated the effectiveness of a genetic algorithm for feature subset selection for authorship analysis. Subset selection techniques consider feature interactions (unlike many ranking methods). However, subset selection techniques can be computationally inefficient. Such search-based methods are often considered to be a "brute force" approach since they simply try numerous feature combinations (Guyon and Elisseeff 2003). This is a big concern for systems where feature selection operations are likely to see heavy usage (such as text-based systems). In addition, the large number of potential features in text analysis can further decrease the efficiency of subset selection methods since it results in even larger search spaces with greater potential feature combinations. Table 12.3 shows example selection methods that have been applied to text mining and the type of analysis performed. Ranking and projection methods have seen greater use due to their simplicity/efficiency and propensity to handle noise, respectively.

3.5 Visualization

Text visualization is challenging since text cannot easily be described by numbers (Keim 2002). It requires the use of multiple views, representing different data types (Losiewicz et al. 2000), with varying dimensionalities. Wise (1999) noted that text analysis should "...provide a basis for altered visualization of the information for different users and purposes...why should we preconceive that there is only one 'correct' visualization of text information in a document corpus?" (p. 1230). For instance, text itself is one-dimensional, textual features are multidimensional (Huang et al. 2005), and the relation between features and the text they represent (e.g., structural, semantic) is often established using 2D–3D text overlay (e.g., Cunningham 2002). Thus, while several types of text visualizations have been developed, we focus on numeric techniques that visualize text feature statistics and text techniques that visualize features as they occur in the text since these two view types provide important complementary functionality.

Multidimensional views are often used to visualize text feature statistics. Such statistics, including frequency, variance, and similarity, provide important insight and summarization yet abstract away important meaning from the underlying, nonnumeric content they are intended to represent. These techniques can tell us what features are important, but not how or why. Text overlay techniques serve two important functions. They allow users to see exactly how and where the features occur within their proper context. They are also important in order to allow users to assess the quality of feature extraction and representation (Losiewicz et al. 2000) due to the high levels of noise in text (Knight 1999; Nasukawa and Nagano 2001). Thus, it is important to incorporate multidimensional presentation formats that can summarize feature statistics as well as text formats that can bridge the gap between feature statistics and their actual occurrences in text.

Several multidimensional techniques have been used for text visualization. These techniques, including graphs/plots and reduced dimensionality views, are used to display feature occurrence statistics and patterns. Several graphs and plots have been used including radar charts, parallel coordinates, and scatter plot matrices. Subasic and Huettner (2001) and Abbasi and Chen (2005) used radar charts to view affect and stylistic feature occurrences, respectively. Huang et al. (2005) used parallel coordinates and scatter plot matrices to view topical features extracted from biomedical text. Reduced dimensionality visualizations decrease the feature space to show essential patterns. These techniques typically are used in conjunction with projection-based feature selection techniques such as PCA and MDS to create a two- or three-dimensional view. Examples include Writeprints (Abbasi and Chen 2006), ThemeRiver© (Havre et al. 2002), Text Blobs (Rohrer et al. 1998), and Themescapes™ (Wise 1999).

Text overlay combines text with feature occurrence patterns to provide greater insight. Examples include the Stereoscopic Document View (Miller et al. 1998) and text annotation (Cunningham 2002). The Stereoscopic Document View in Topic Islands™ uses wavelet transformations to show the key topical patterns within a document, superimposed onto the text (Miller et al. 1998). Text annotation simply highlights the feature occurrences in the text (Cunningham 2002).

4 A Design Framework for CMC Text Analysis

Design is a product and a process (Walls et al. 1992; Hevner et al. 2004). The design product is the set of requirements and necessary design features that should guide the IT artifact construction. The design process is the steps and procedures taken to develop the artifact. Information systems development typically follows an iterative process of building and evaluating (March and Smith 1995), which is analogous to

Table 12.4 Components of an ISDT design product

Design product	
1. Kernel theories	Theories from natural and social sciences governing design requirements
2. Meta-requirements	Describes a class of goals to which theory applies
3. Meta-design	Describes a class of artifacts hypothesized to meet meta-requirements
4. Testable hypotheses	Used to test whether meta-design satisfies meta-requirements

the generate/test cycle proposed by Simon (1996). Such an approach is particularly important in design situations involving a complex or poorly defined set of user requirements (Markus et al. 2002). We believe that the ambiguities associated with CMC text analysis component alternatives also warrant the use of such a design process. Thus, we focus on the design product elements of Walls et al.'s (1992) model, which are presented in Table 12.4.

Using Walls et al.'s guidelines, we propose a design framework for CMC text analysis. Systemic Functional Linguistic Theory illustrates the need for incorporating different information types for text analysis, such as ideational and textual information. However, capturing multiple information types for representational richness can be challenging. For instance, topical information has a well-established set of features that are typically used (i.e., bag-of-words); however, as previously mentioned, over 1,000 features have been used for style alone (Rudman 1998). Opinion has also seen the use of lexical, syntactic, lexicon, and structural features. Thus, improving representational richness may result in increased complexity with respect to the myriad of potential features, selection methods, and visualization techniques that can be utilized.

When dealing with numerous methods with varying strengths and weaknesses, methodological triangulation (Denzin 1970) is useful for overcoming the problems that may stem from the overt dependency on any one method (Balakrishnan and Jacob 1995). In the social sciences, methodological triangulation suggests that researchers should use divergent methods for measuring and analyzing constructs (Campbell and Fiske 1959). Balakrishnan and Jacob (1995) extended the concept of methodological triangulation to be applicable to information systems design guidelines. They designed and developed a Decision Support System (DSS) that incorporated multiple complementary search strategies to support product design optimization. Their DSS utilized exhaustive search methods which provided high performance yet low efficiency and random search methods which provided varying levels of performance and improved efficiency. Balakrishnan and Jacob (1995) argued that methodological triangulation can improve performance and user confidence, resulting in greater system usage.

In our review of the key CMC text mining characteristics, we described the strengths and weakness of different types of features and selection and presentation methods. We also noted how different categories of these characteristics can complement each other for improved analysis and categorization capabilities. For instance, the use of language and processing resources may be beneficial as compared to simply using a single feature type. Similarly, ranking and projection methods for feature selection can complement each other by facilitating analysis of overview (projection methods) and specific feature details (ranking methods).

4.1 Meta-requirements

Using Systemic Functional Linguistic Theory and methodological triangulation as our kernel guiding principles, we propose several meta-requirements based on our review of the key text mining elements. The primary objective of our meta-requirements is to improve the representational richness of text analysis systems over the standard bag-of-words features. We wish to couple this richer feature set with the appropriate selection and visualization techniques to allow support for CMC text analysis:

1. Support for various text mining tasks.
2. Support in-depth content analysis across various information types for improved representational richness.
3. Support large sets of complementary and contrasting textual features.
4. Feature selection methods to support selection of attributes that best capture the underlying content based on user needs.
5. Visualization techniques to support presentation of overview and details, make comparisons, and assess similarity.

4.2 Meta-design

Based on our meta-requirements and review of the important categories for each element, we propose the following meta-design:

1. Include categorization and analysis functionality (from Requirement 1).
2. Incorporate information types to capture ideational and textual language meaning (from Requirement 2).
3. Language and processing resources-based textual features (from Requirement 3).
4. Ranking and projection-based feature selection methods (from Requirement 4).
5. Basic, multidimensional, and text overlay visualization techniques (from Requirement 5).

Based on our meta-design, we present a framework for the design of text-based information systems to support computer-mediated communication content analysis (shown in Fig. 12.2). While the framework is applicable to all forms of text, we focus specifically on computer-mediated communication due to its high level of interaction and informational richness.

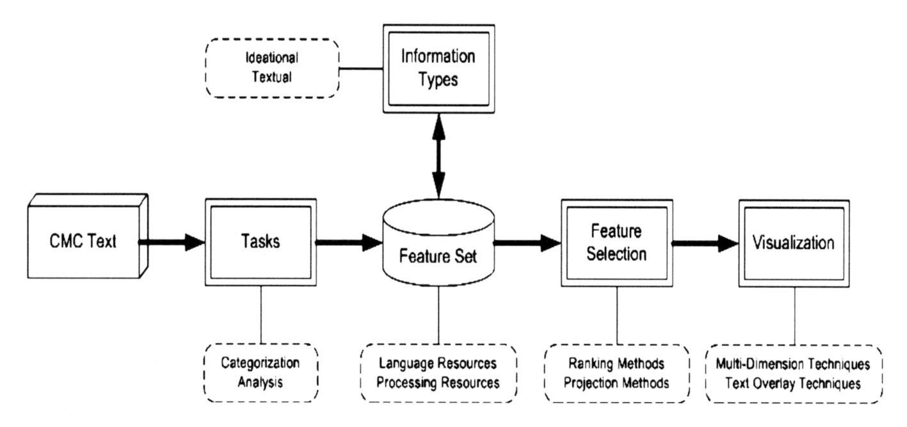

Fig. 12.2 A design framework for CMC text analysis

5 System Design: The CyberGate System

Using our design framework as a guideline, we developed a text analysis system for CMC analysis called CyberGate. The system was developed using a cyclical design process involving several iterations of adding and testing system components. The testing phase included experiments for performance evaluation and feedback from CMC researchers and web analysts. In this section, we describe the system in its completed state. While the system supports several tasks, information types, features, feature selection methods, and visualization techniques, the two major components of CyberGate are Writeprints and Ink Blots. We first present an overview of the CyberGate system based on our design framework and then provide in-depth details of the Writeprint and Ink Blot techniques. Figure 12.3 shows an overview of the system design.

5.1 Information Types and Features

CyberGate supports several information types, including topics, sentiments, affects, style, and genres. Event information was not included since the current state of the art for event detection is insufficient for effectively capturing and representing events (Allan et al. 1998). In order to enable the capturing of such a breadth of information, several language and processing resources were included. These include language resources such as sentiment and affect lexicons, word lists, and the WordNet thesaurus (Fellbaum 1998). Embedded processing resources include n-grams, statistical features (Abbasi and Chen 2005; Zheng et al. 2006), parts of speech, noun phrases, and named entities (McDonald et al. 2004).

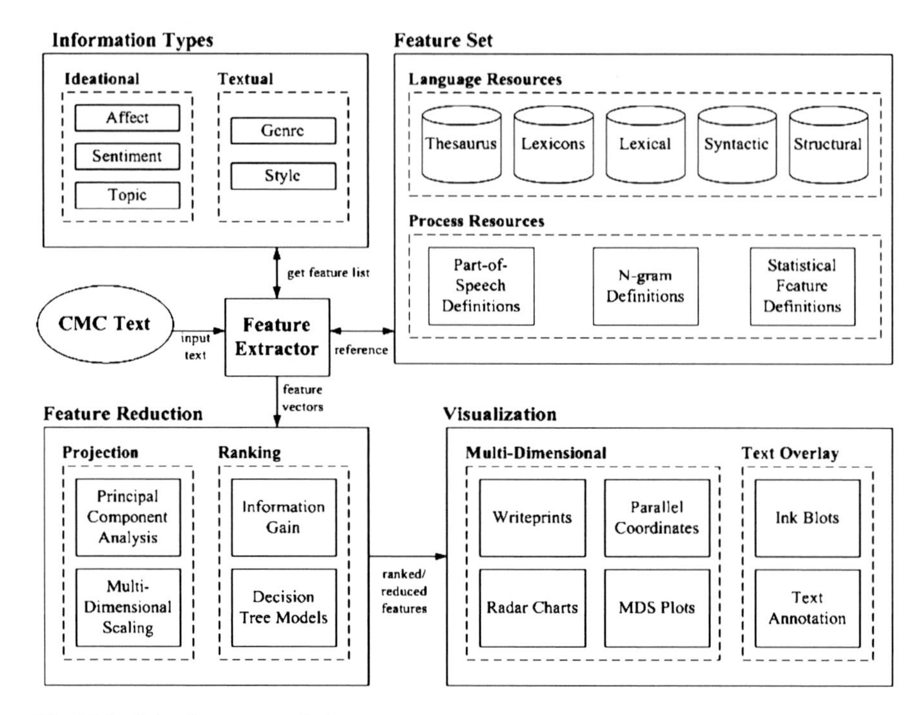

Fig. 12.3 CyberGate system design

a All features

No.	Description	Usage	Mean
0	MARK	0.726	0.054
1	COUNSELEN...	0.526	0.011
2	GENERAL	0.526	0.023
3	PRESIDENT	0.526	0.017
4	TAYLORVICE	0.526	0.011
5	WHOLESALE	0.526	0.041
6	SERVICES	0.368	0.035
7	TIME	0.284	0.165
8	FAX	0.274	0.038
9	SMITH	0.274	0.085
10	TEXAS	0.263	0.095
11	STREET	0.253	0.083
12	PARTY	0.221	0.04

b Projection

Ex	Ey
0.278	0.103
0.099	-0.355
0.019	-0.095
0.227	0.323
0.22	0.33
-0.15	0.06
0.152	0.052
0.247	0.118
0.078	-0.186
0.309	0.013
0.305	0.0040
0.0080	-0.201

c Ranking

Description	Weight
PANDA	60.71
DEVIL	28.85
LATERAL	18.9
KH	14.43
THROUGHPUT	12.14
TW	10.95
SUN	18.9
GAS	28.85
PIPELINE	18.9
EVI	21.86
ECT	5.93
HAT	1.29
ATE	1.0

Fig. 12.4 CyberGate feature selection examples: (**a**) shows the complete set of all features, (**b**) shows the top two dimensions of the PCA projections (Ex and Ey) and (**c**) shows the decision tree model rankings

5.2 *Feature Selection*

CyberGate uses both ranking and projection-based feature reduction methods. For feature ranking, it uses information gain (IG) and decision tree models (DTM). Both of these methods have been shown to be effective for textual feature selection (Forman 2003; Efron et al. 2004; Abbasi and Chen 2005). For projection, it uses PCA and MDS for lower dimension feature transformations. PCA and MDS have both been previously used for textual feature reduction (Abbasi and Chen 2006; Huang et al. 2005). Figure 12.4 shows examples of the feature selection techniques used in CyberGate. The table on the left (a) shows the complete set of features, while (b) shows the two-dimensional PCA projections (Ex and Ey) and (c) shows the decision tree model rankings where each feature is assigned a weight. Higher weights indicate greater feature rank.

5.3 Visualization

CyberGate includes multidimensional and text overlay–based visual representations. Multidimensional visualizations (shown in Fig. 12.5) include Writeprints (Fig. 12.5a) to show usage variation and parallel coordinates (Fig. 12.5b) to show feature similarities across messages or text windows (Abbasi and Chen 2006). Each circle in Writeprints denotes a single message or text window projected using principal component analysis (PCA). The blue polygonal lines in parallel coordinates also represent individual messages or text windows. The selected Writeprint point corresponds to the selected parallel coordinate's polygonal line. The intersection between a polygonal line and a vertical axis in parallel coordinates represents the occurrence frequency of that feature in that particular message. For example, the selected message in Fig. 12.5b has very high occurrence of feature #7 (occurs 21 times).

CyberGate also utilizes MDS plots (Fig. 12.5d) to show overall feature similarities, while radar charts (Fig. 12.5c) are used to compare feature occurrence statistics across authors. The radar chart shown compares the selected author against another author and the mean normalized usage frequencies for a particular set of features (which are numbered along the perimeter). The MDS plot in Fig. 12.5d shows features projected based on occurrence similarity for the bag-of-words features. We can see one large cluster and two smaller ones in addition to 3–4 features that are on their own. These features (e.g., "services") do not frequently co-occur with any of the three clusters.

The CyberGate system also features a couple of text overlay techniques (shown in Fig. 12.6). Text annotation simply highlights key features in the text (Cunningham 2002). Figure 12.6a shows the text annotation view in which the bag-of-words features are highlighted in blue, while the selected feature ("CounselEnron") is highlighted in red. Ink Blots (Fig. 12.6b) superimposes colored circles (blots) onto text for key features as identified by the particular underlying feature ranking method incorporated. The size of the blot indicates the feature rank/weight (based on the feature ranking technique). Features that are more unique to a particular author have higher weights than features that are equally common (less interesting) across authors. The color indicates the author's usage of the particular feature (red = high, blue = low, yellow = medium). The selected feature (again, "CounselEnron") is highlighted with a black circle. This particular feature is represented with large red blots indicating that the feature has a high weight (rarely used by others, unique to this author) and is frequently used by this author.

a Writeprints

Two dimensional PCA projections based on feature occurrences. Each circle denotes a single message. Selected message is highlighted in pink. Writeprints show feature usage/occurrence variation patterns. Greater variation results in more sporadic patterns.

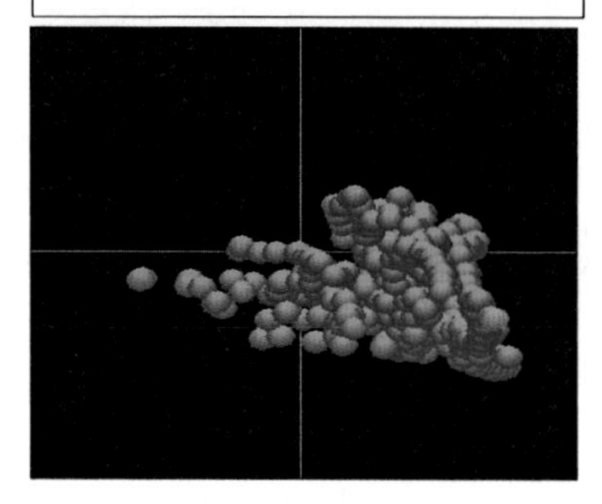

b Parallel Coordinates

Parallel vertical lines represent features. Bolded numbers are feature numbers (0-15). Smaller numbers above and below feature lines denote feature range. Blue polygonal lines represent messages. Selected message is highlighted in red. Selected feature is highlighted in pink (#2).

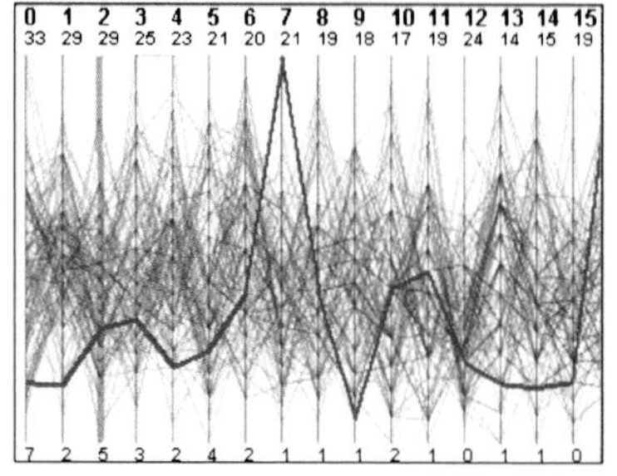

c Radar Charts

Chart shows normalized feature usage frequencies. Blue line represents author's average usage, red line indicates mean usage across all authors, and green line is another author (being compared against). The numbers represent feature numbers. Selected feature is highlighted (#6).

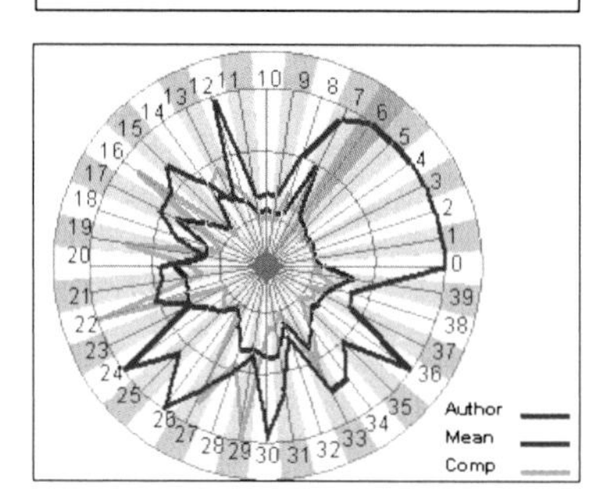

d MDS Plots

MDS algorithm used to project features into two-dimensional space based on occurrence similarity. Each circle denotes a feature. Closer features have higher co-occurrence. Labels represent feature descriptions. Selected feature is highlighted in pink (the term "services").

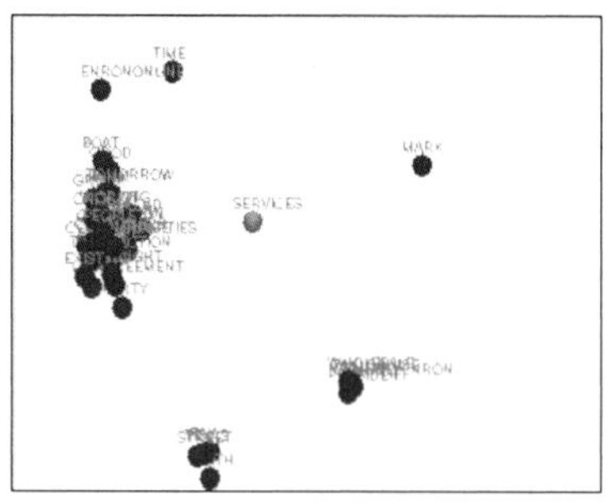

Fig. 12.5 Multidimensional views in CyberGate

a Text Annotation View

Feature occurrences are highlighted in blue. The selected bag-of-words feature is highlighted in red ("CounselEnron").

al CounselEnronWholesaleServices1400Smith Street -
EB3892HoustonTexas77008(713853-7459(713646-3490fax I
thought of another item to add to the list:Cruising World 2002
wall calendar -- available on Amazon.com for 11.95Mark
TaylorVicePresidentand GeneralCounselEnronWholesale
Services1400Smith Street - EB3892Houston, Texas
77008(713853-7459(713646-3490fax Hey Mike -- here are my de
Taylor 781322 Rutland St.Houston, TX 77008(713863-7190
(homemark.taylor@enron.comLet me know how I can help.Mark
TaylorVicePresidentand GeneralCounselEnronWholesale
Services1400Smith Street - EB3892Houston, Texas
77008(713853-7459(713646-3490fax If you need to reach me this
will be at my parent s home in Cincinnati: (513541-8649.Mark
TaylorVicePresidentand GeneralCounselEnronWholesale
Services1400Smith Street - EB3892Houston, Texas
77008(713853-7459(713646-3490fax In my view an assignment

b Ink Blot View

Colored circles (blots) superimposed onto feature occurrence locations in text. Blot size and color indicates feature importance and usage. Selected feature's blots are highlighted with black circles.

Fig. 12.6 Text views in CyberGate

5.4 *Writeprints and Ink Blots*

CyberGate includes the Writeprint and Ink Blot techniques, which are the core components driving the system's analysis and categorization functions. These techniques epitomize the essence of the proposed design framework: representational richness and methodological triangulation. With respect to representational richness, Writeprints and Ink Blots can incorporate a wide range of features representing various information types. Both techniques also utilize feature selection and visualization. Writeprints uses principal component analysis (PCA) with a sliding window algorithm to create two-dimensional plots that accentuate feature usage variation. Ink Blots uses decision tree models (DTM) to select features which are superimposed onto text to show usage frequencies as they occur within their textual structure. The techniques support methodological triangulation in the sense that Writeprints uses a projection method for feature selection along with a reduced dimensionality visual representation of multidimensional features. Ink Blots uses a ranking method for feature selection along with a text overlay visual presentation format. Writeprints is better suited to present a broad overview across a large number of features, while Ink Blots is more geared to show detailed examples of feature occurrences. Both techniques can be used for text categorization and analysis. Specific details of Writeprints and Ink Blots are presented below.

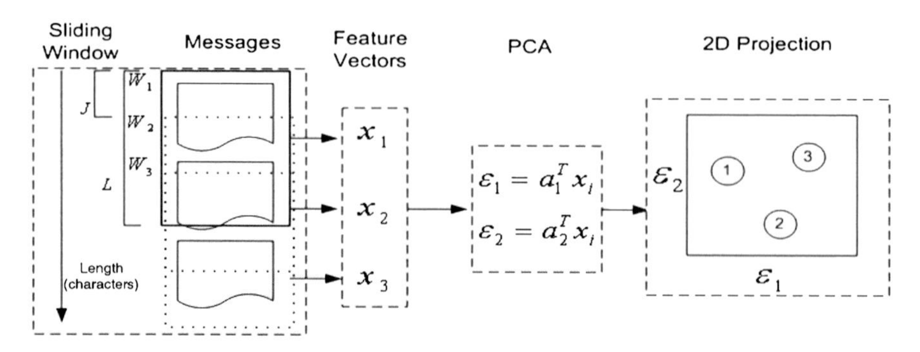

Fig. 12.7 Writeprint process illustration

5.4.1 Writeprints

The steps for the Writeprint technique are listed below.

Writeprint Technique Steps

1. Derive two primary eigenvectors (ones with the largest eigenvalues) from feature usage matrix.
2. Extract feature vectors for sliding window instance.
3. Compute window instance coordinates by multiplying window feature vectors with two eigenvectors.
4. Plot window instance points in two-dimensional space.
5. Repeat steps 2–4 for each window.

A sliding window of length L with a jump interval of J characters is run over the messages. The feature occurrence vector for each window is projected to a two-dimensional space by taking the product of the window feature vector and the two primary eigenvectors. The product of the window feature vector and the first eigenvector is used to get the x-axis coordinate (ε_1), while the product of the feature vector and the second eigenvector produces the y-axis coordinate (ε_2). Figure 12.7 illustrates the key steps in the Writeprints process.

Writeprints is geared toward showing occurrence variation patterns. These patterns can be used for text categorization of stylistic information or analysis across various types of ideational and textual information.

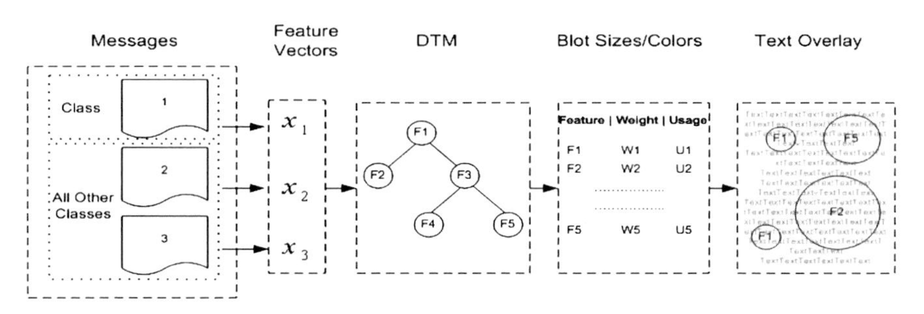

Fig. 12.8 Ink Blot process illustration

5.4.2 Ink Blots

The steps for the Ink Blot technique are listed below.

Ink Blot Technique Steps

1. Separate input text into two classes (one for class of interest, one class containing all remaining texts).
2. Extract feature vectors for messages.
3. Input vectors into DTM as binary class problem.
4. For each feature in computed decision tree, determine blot size and color based on DTM weight and feature usage.
5. Overlay feature blots onto their respective occurrences in text.
6. Repeat steps 1–5 for each class.

The Ink Blot process is shown in Fig. 12.8.

The Ink Blot technique identifies the most important features for a given class by using a binary class decision tree model (DTM) for feature ranking. In this case, a class can refer to an author, sentiment, emotion, topic, etc. The class of interest is input into the DTM along with a second class containing the text from all other classes. The DTM will determine the most important features that differentiate the class of interest from other classes, weighted by their level of entropy reduction. For each selected feature, the weights determined by the DTM are used to determine the attribute's blot size (higher weight = larger blot size). Blot colors are determined based on feature usage (red = high, yellow = medium, blue = low). A feature blot is assigned the color red only if this particular class uses it more than others. Similarly, a feature blot is assigned the color blue if it is never used by the class of interest. For instance, let us assume we have ten topics of interest for which we would like to identify the key blot features (number of classes equals ten). For each topic, we will generate a DTM comparing that topic against all others (one against all comparison) in order to determine the key features for that particular topic. The topic's key features are assigned weights and colors based on the DTM rankings and feature occurrence within that topic, respectively. The process is repeated for each topic (ten times, in this case). Finally, text overlay can be performed by superimposing a topic's blot features onto text by creating blots at every location within the text where a key feature occurs. The blot size and color are based on the class feature's weight and usage as previously mentioned.

Once each class's key features have been extracted, assigned weights, and assigned colors, they can be used for categorization and analysis. For categorization, superimposing a class's blots onto an unclassified text can provide insight into whether the text belongs to that particular class. Correct class-text matches should result in patterns featuring high levels of red and yellow (features that occur frequently in this class's texts) and a minimal amount of blue (features rarely or never occurring in this class's texts). Ink Blots can also be used to analyze how key class features occur and interact within a certain piece of text. When analyzing a class of text with its own blots, only red and yellow blots are likely to occur since a class's text can obviously not contain features it never uses (blue blots).

6 CyberGate Case Studies: Clearguidance.com and Ummah.com

In this section, we present two case studies demonstrating how CyberGate can be used for extremist forum content and author attribution and visualization.

6.1 Clearguidance.com

The Clearguidance.com forum has been reported to be the main forum that attracts members affiliated with the Toronto terror plot. At its peak, it had as many as 15,000 members worldwide. The site was removed in February 2004. Based on our extensive exploration and spidering work, we were able to recover some partial data which involves 269 authors and 877 messages in 135 threads, dated from September 2002 to February 2004. Approximately 2/3 of the 269 members specified a location country, with a majority based in the USA, United Kingdom, Canada, and the Middle East. The forum contains many messages relating to al-Qaeda and jihad and is intended to be an information resource for teenagers seeking "guidance." Many younger members inquired about things such as whether or not it is okay to support al-Qaeda, justification for jihad or suicide bombing, etc. Using message replies to connect forum communication, we were able to construct the member interaction network, as shown in Fig. 12.9. Blue nodes in the center indicate members with the greatest number of in-links and are considered the core forum "experts" and propagandists.

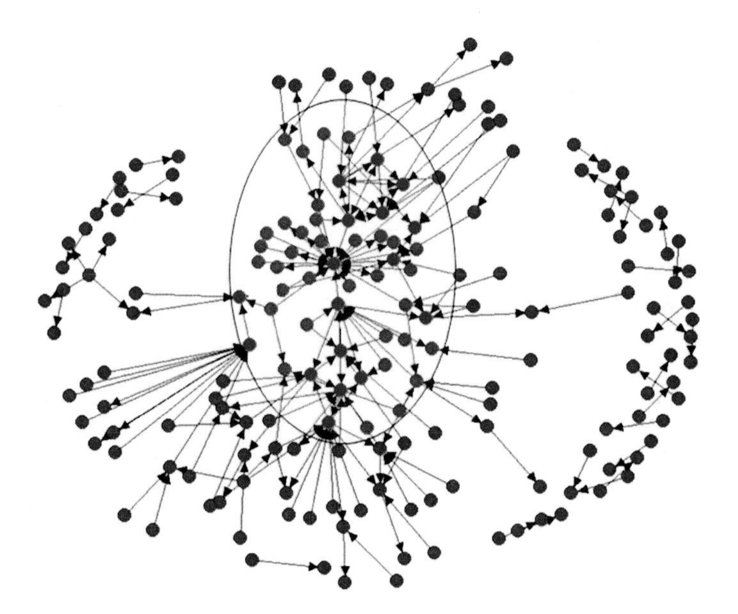

Fig. 12.9 Clearguidance.com member interaction network

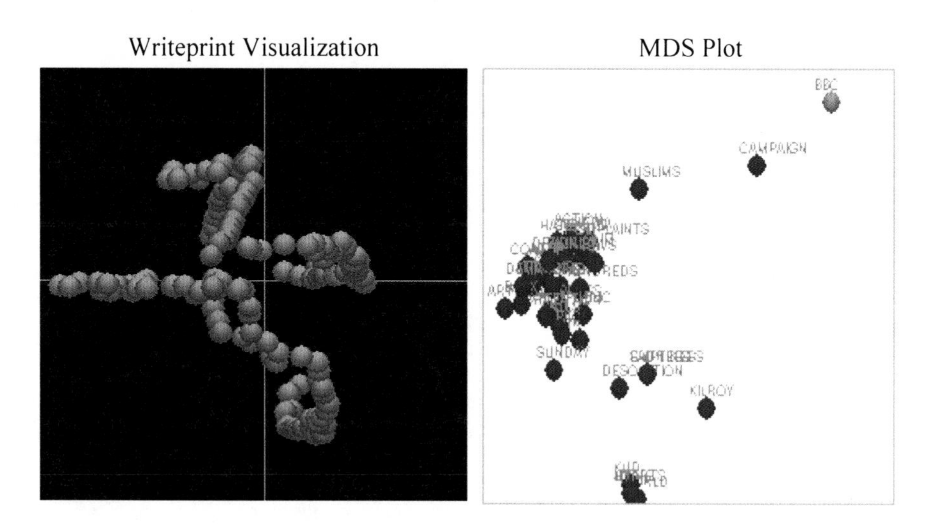

Fig. 12.10 CyberGate visualization of a typical forum author

Using CyberGate, we were able to analyze and visualize forum member topic interest. In Fig. 12.10, the Writeprint pattern suggests that this author has considerable topical variation, discussing several different things (with messages scattered all over the map). MDS-projected topics include BBC, Robert Kilroy, Muslim, Campaign, Chechnya, etc., again with little focus.

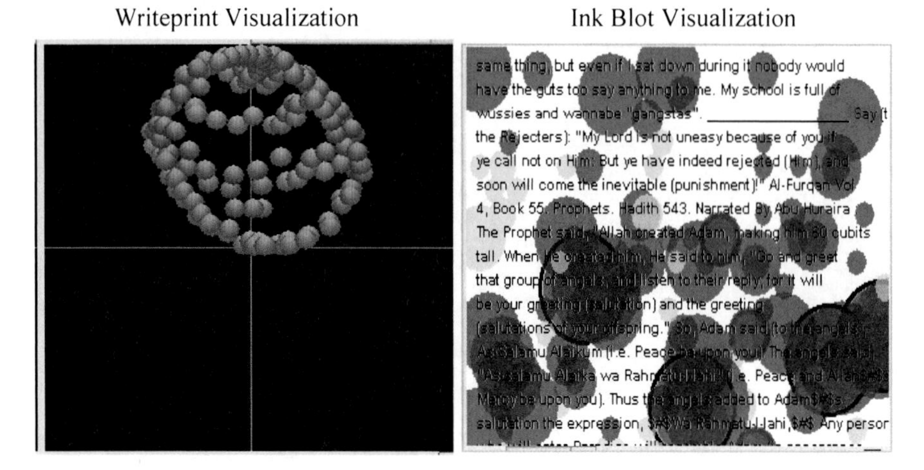

Fig. 12.11 CyberGate visualization of a religious expert

A different pattern emerges when we analyzed the communication content of a forum expert (members denoted as blue nodes in Fig. 12.9). As shown in Fig. 12.11, the series of overlapping circular patterns for bag-of-words features in Writeprint indicates that the author's discussion revolves around a closely related set of topics. Many large red blots indicate the presence of features unique to this author, e.g., angels, Allah, Adam, etc. This author attempts to use his religious "expertise" to justify jihad and violence. We also identify several other key members who act as history experts (topics on the conquest of Rome and Constantinople) or propagandists (topics on jihad, martyr, battle, war). Using CyberGate, we were able to identify a few members and their messages relating to acquiring al-Qaeda recruiting videos and shipping to Toronto, e.g., "IbnMardhiyah, how much would that be in CND$$ to ship to Toronto?…and, do they accept check or money order? Jazakallah. ~Allahu Akbar~!" (1–13–2003 02:15 PM).

6.2 Ummah.com

The Ummah is a large English language religious discussion forum which contains many religious experts with considerable knowledge. It also has a sizeable female presence. Unlike Clearguidance, this forum does not have significant extremist or violent content. Our collection included 2,338 members, with 81,976 messages in 3,787 threads. Figure 12.12 shows its member interaction network, which is significantly denser than the Clearguidance network. Again, core members are indicated in blue in the center of the network.

Fig. 12.12 Ummah.com member interaction network

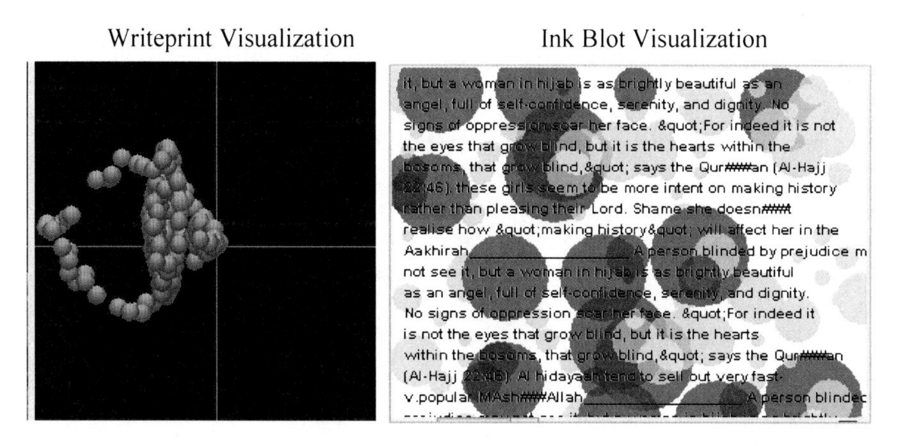

Fig. 12.13 CyberGate visualization of a women-issue expert

In Fig. 12.13, we show the authorship pattern of an expert specializing in women-related topics. This author has a circular bag-of-word variation pattern in Writeprint, reflective of topical consistency. It also has large red blots indicative of the presence of features unique to this author, e.g., hijab (head covering for Muslim women), serenity, dignity, bosoms, girl, etc. Similarly, Fig. 12.14 indicates the authorship pattern of a computer networking expert in Ummah.

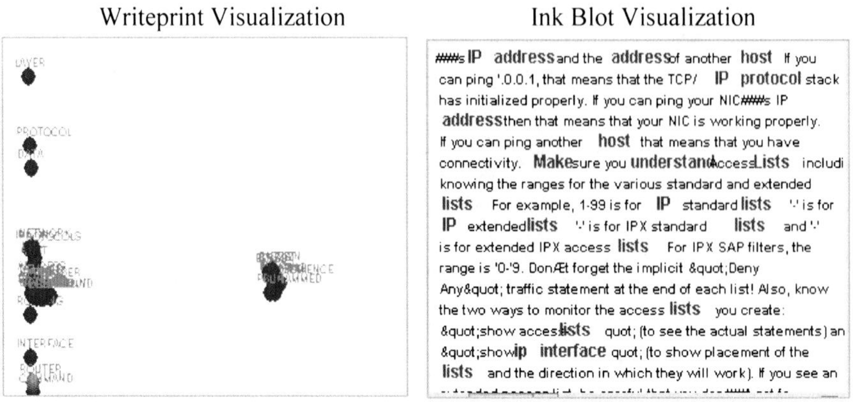

Writeprint Visualization

Ink Blot Visualization

Fig. 12.14 CyberGate visualization of a computer networking expert

7 Conclusions

Advanced text analysis systems can provide numerous benefits for CMC researchers, analysts, and participants. Despite the need for such systems that can support CMC text analysis, implementations have been scarce. This is attributable to the lack of design guidelines for the development of such systems. In this chapter, our major research contributions are twofold. First, we developed a framework for the categorization and analysis of computer-mediated communication text. Our framework is guided by the principles of representational richness, taken from Systemic Functional Linguistic Theory, and methodological triangulation. In order to develop our framework, we first reviewed several important CMC text mining issues including the relevant tasks, information types, feature selection methods, and visualization techniques. Second, we developed the CyberGate system, which features the Writeprint and Ink Blot techniques, to evaluate the efficacy of our design framework. Two Dark Web case studies were used to illustrate the ability of the CyberGate components for analysis and categorization of CMC text.

Our design framework and the resulting CyberGate system have several shortcomings. There are likely to be numerous additional information types and selection and visualization techniques that could have been considered but were omitted. An additional stage for clustering and categorization techniques could also have been added. Our intention and hope is that future research will improve upon our design framework, resulting in CMC systems with improved text analysis capabilities.

Acknowledgments This research has been supported in part by the following grant: National Science Foundation (NSF), "COPLINK Center: Social Network Analysis and Identity Deception Detection for Law Enforcement and Homeland Security," October 2004–September 2007.

References

Abbasi, A., and Chen, H. "Identification and Comparison of Extremist-Group Web Forum Messages using Authorship Analysis," *IEEE Intelligent Systems* (20:5), 2005, pp. 67–75.

Abbasi, A. and Chen, H. "Visualizing Authorship for Identification", In the 4th IEEE Symposium on Intelligence and Security Informatics (ISI 2006), San Diego, CA, 2006.

Allan, J. Carbonell, J, Doddington, G., Yamron, J. and Yang, Y "Topic detection and tracking pilot study: Final report," in *proceedings of the DARPA Broadcast News Transcription and Understanding Workshop,* 1998, pp. 194–218.

Allan, J., Leuski, A., Swan R. C., and Byrd, D. "Evaluating Combinations of Ranked Lists and Visualizations of Inter-Document Similarity," *Information Processing and Management*, (37:3), 2001, pp. 435–458.

Argamon, S., Saric, M., and Stein, S, S. "Style Mining of Electronic Messages for Multiple Authorship Discrimination: First Results," in *Proceedings of the ninth ACM SIGKDD International conference on Knowledge discovery and data mining,* 2003.

Balakrishnan, P. V., V. S. Jacob. "Triangulation in decision support systems: Algorithms for product design," *Decision Support Systems*, (14), 1995, pp. 313–327.

Campbell, D. T. and Fiske, D. W. "Convergent and Discriminant Validity by Multitrait-Multimethod Matrix," *Psychology Bulletin*, (56:2), 1959, pp. 81–105.

Chen, H. *Knowledge Management Systems. A Text Mining Perspective,* Knowledge Computing Corporation, 2001.

Chen, H., Lally, A.M., Zhu, B., and Chau, M. "HelpfulMed: Intelligent Searching for Medical Information over the Internet," *Journal of the American Society for Information Science and Technology* (54:7), 2003, pp. 683–694.

Cothrel, J, P. "Measuring the Success of an Online Community," *Strategy and Leadership* (20:2), 2000, pp. 17–21.

Cunningham, H. "GATE, a General Architecture for Text Engineering," *Computers and the Humanaties* (36), 2002, pp. 223–254.

Daft, R, L., and Lengel, R, H. "Organizational Information Requirements, Media Richness and Structural Design," *Management Science* (32:5), 1986, pp. 554–571.

Dash, M. and Liu, H. "Feature Selection for Classification," *Intelligent Data Analysis,* (1), 1997, pp. 131–156.

Davenport, D. "Anonymity on the Internet: Why the Price May Be too High," *Communications of the ACM*,(45:4), 2002, pp. 33–35.

Denzin, N. *The Research Act*, Aldine, Chicago, 1970.

Donath, J. "Identity and Deception in the Virtual Community," *In Communities in Cyberspace,* London, Routledge Press, 1999.

Donath, J., Karahalio, K. and Viegas, F. "Visualizing Conversation," in *Proceedings of the 32nd Conference on Computer-Human Interaction* (CHI' 02), Chicago, USA, 1999.

Donath, J. "A Semantic Approach to Visualizing Online Conversations," *Communications of the ACM*, 45(4), 2002, pp. 45–49.

Duch, W., Wieczorek, T., Biesiada, J., and Blachnik M. "Comparison of feature ranking methods based on information entropy," *Neural Networks*, 15, 2004.

Dumais, S., Platt, J., Heckerman, D. And Sahami, M. "Inductive Learning Algorithms and Representations for Text Categorization," *In Proceedings of the Seventh of ACM-CIKM,* 1998, pp. 148–155.

Efron, M., Marchionini, G., and Zhiang, J. "Implications of the Recursive Representation Problem for Automatic Concept Identification in On-Line Government Information," *In Proceedings of the ASIST SIG-CR Workshop*, 2004.

Erickson, T. and Kellogg, W. A. "Social Translucence: An Approach to Designing Systems that Support Social Processes," *ACM Transactions on Computer-Human Interaction* (7:1), 2000 pp. 59–83.

Fellbaum, C. *Wordnet: An Electronic Lexical Database*, The MIT Press, Cambridge, MA, 1998.

Fiore, A, T., and Smith, M, A. "Tree Map Visualizations of News Groups," *Poster Presented at IEEE Symposium on Information Visualization*, 2002, Boston, Massachusetts.
Forman, G. "An Extensive Empirical Study of Feature Selection Metrics for Text Classification," *The Journal of Machine Learning Research* (3), 2003, pp. 1289–1305.
Friedman, B., Kahn, P. H. and Howe, D. C. "Trust Online," *Communications of the ACM* (43:12), 2000, pp. 88–93.
Guyon, I., and Elisseef, A. "An Introduction to Variable and Feature Selection," *The Journal of Machine Learning Research* (3), 2003, pp. 1157–1182.
Halliday, M.A.K. *An Introduction to Functional Grammar*, 2nd (ed). London: Edward Arnold, 1994, p. 179.
Hearst, M. A. "Direction-Based Text Interpretation as an Information Access Refinement," In P. Jacobs (Ed.), *Text-Based Intelligent Systems: Current Research and Practice in Information Extraction and Retrieval*, Mahwah, NJ: Lawrence Erlbaum Associates, 1992.
Hearst, M. A. "Untangling Text Data Mining," in *Proceedings of the Association for Computational Linguistics*, 1999, pp. 3–10.
Hara, N., Bonk, C, J., and Angeli, C. "Content Analysis of Online Discussion In An Applied Educational Psychology Course," *Instructional Science* (28), 2000, pp. 115–152.
Havre, S., Hetzler, E., Whitney, P. and Nowell, L. "ThemeRiver: Visualizing Thematic Changes in Large Document Collections," *IEEE Transactions on Visualization and Computer Graphics*, (8:1), 2002, pp. 9–20.
Henri, F. "Computer Conferencing and Content Analysis," in *Collaborative Learning through Computer Conferencing: The Najaden papers*, A.R. Kaye, (ed), 1992, pp. 115–136.
Herring, S. C. "Computer-Mediated Communication on the Internet," *Annual Review of Information Science and Technology* (36:1), 2002, pp. 109–168.
Hevner, A, R., March, S, T., Park, J., and Ram, S. "Design Science in Information Systems Research," *MIS Quarterly* (28:1), 2004, pp. 75–105.
Huang, S., Ward, M, O., and Rundensteiner, E, A. "Exploration of Dimensionality Reduction For Text Visualization," in *Proceedings of The Third International Conference on Coordinated and Multiple Views in Exploratory Visualization* (CMV'05), 2005.
Huber, P. J. "Projection Pursuit," *Annals of Statistics*, (13:2), 1985, pp. 435–475.
Keim, D, A. "Information Visualization and Visual Data Mining," *IEEE Transactions on Visualization and Computer Graphics* (7:1), 2002, pp. 100–107.
Kelly, S. U., Sung, C., and Farnham, S. "Designing for Improved Social Responsibility, User Participation and Content in On-Line Communities," in *Proceedings of the Conference on Human Factors in Computing Systems* (CHI 2002), 2002.
Knight, K. "Mining Online Text," *Communications of the ACM* (42:11), 1999, pp. 58–61.
Lee, A, S. "Electronic Mail as a Medium of Rich Communication: An Empirical Investigation using Hermeneutic Interpretation," MIS Quarterly, 1994, pp. 143–157.
Lewis, D. "Text Representation for Intelligent Text Retrieval: A Classification Oriented View," In P. Jacobs (Ed.), *Text-Based Intelligent Systems: Current Research and Practice in Information Extraction and Retrieval*, Mahwah, NJ: Lawrence Erlbaum Associates, 1992.
Li, J., Zheng, R. and Chen, H. "From Fingerprint to Writeprint," *Communications of the ACM*, (49:4), 2006, pp. 76–82.
Losiewicz, P., Oard, D. and Kostoff, R. N. "Textual Data Mining to Support Science and Technology Management," *Journal of Intelligent Information Systems*, (15), 2000, pp. 99–119.
March, S. T. and Smith, G. "Design and Natural Science Research on Information Technology," *Decision Support Systems* (15:4), 1995, pp. 251–266.
Markus, M, L., Majchrzak, A., and Gasser, L. "A Design Theory for Systems That Support Emergent Knowledge Processes," *MIS Quarterly* (26:3), 2002, pp. 179–212.
McDonald, D., Chen, H., Hua S., and Marshall, B. "Extracting Gene Pathway Relations using a Hybrid Grammar: The Arizona Relation Parser," *Bioinformatics* (20:18), 2004, pp. 3370–3378.
Miller, N. E., Wong, P. C., Brewster, M. and Foote, H. "Topic Islands: A wavelet-based text visualization system," *in Proceedings of IEEE Visualization '98*, Research Triangle Park, NC, USA. 1998

Mladenic, D. "Text-Learning and Related Intelligent Agents: A Survey," *IEEE Intelligent Systems* (14:4), 1999, pp. 44–54.

Nasukawa, T. and Nagano, T. "Text Analysis and Knowledge Mining System," *IBM Systems Journal* (40:4), 2001, pp. 967–984.

Nissenbaum, H. "Accountability in a Computerized Society," *Science and Engineering Ethics* (2), 1996, pp. 25–42.

Paccagnella, L. "Getting the Seats of Your Pants Dirty: Strategies for Ethnographic Research on Virtual Communities," *Journal of Computer Mediated Communication* (3:1), 1997.

Pang, B., Lee, L., and Vaithyanathain, S. "Thumbs up? Sentiment classification using machine learning techniques", in *proceedings of the Empirical Methods in Natural Language Processing (EMNLP 2002)*, 2002.

Panteli, N. "Richness, Power Cues and Email Text," *Information and Management*, 2002, pp. 75–86.

Picard, R. W. *Affective Computing*, MIT Press, Cambridge, MA., 1997.

Rudman, J. "The state of authorship attribution studies: some problems and solutions," *Computers and the Humanities* (31), 1998, pp. 351–365.

Rohrer, R, M., Elbert, D, S., and Sibert, J, S. "The Shape of Shakespeare: Visualizing Text using Implicit Surfaces," in *Proceedings of the 1998 IEEE Symposium on Information Visualization* North Carolina, 1998, pp. 121–129.

Rourke, L., Anderson, T., Garrison, D. R., and Archer, W. "Methodological Issues in the Content Analysis of Computer Conference Transcripts," *International Journal of Artificial Intelligence in Education*, (12), 2001.

Sack, W. "Conversation Map: An Interface for Very Large-Scale Conversations," *Journal of Management Information Systems* (17:3), 2000, pp. 73–92.

Santini, M. "A Shallow Approach to Syntactic Feature Extraction for Genre Classification," in *Proceedings of the 7th Annual Colloquium for the UK Special Interest Group for Computational Linguistics (CLUK 04)*, 2004.

Seo, J., and Shneiderman, B. "Interactively Exploring Hierarchical Clustering Results," *IEEE Computer* (35:7), 2002, pp. 80–86.

Simon, H, A. *The Sciences of the Artificial*, 3rd (ED), *MIT Press*, Cambridge, MA, 1996.

Smith, M. A. "Invisible crowds in cyberspace: Mapping the Social Structure of Usenet," in M. Smith and P. Kollock (Eds.), *Communities in Cyberspace*, London, Routledge, 1999.

Smith, M, A., and Fiore, A, T. "Visualization Components for Persistent Conversations," *Proceedings of the SIGCHI conference on Human factors in computing systems*, Seattle, Washington, United States, 2001, pp. 136–143.

Smith, M. "Tools for Navigating Large social Cyberspaces," *Communications of ACM* (45:4), 2002, pp. 51–55.

Spears, R., and Lea, M. "Social Influence and the Influence of the Social in the Computer-Mediated Communication," in M. Lea (ED), *Contexts of Computer-Mediated Communication*, Hemel-Hempstead: Harvester Wheat sheaf, 1992, pp. 30–65.

Spears, R., and Lea, M. "Panacea or Panopticon? The Hidden Power in Computer-Mediated Communication." *Communication Research*, (4), 1994, pp. 427–459.

Subasic, P., and Huettner, A. "Affect Analysis of Text Using Fuzzy Semantic Typing," *IEEE Transactions on Fuzzy Systems* (9:4), 2001, pp. 483–496.

Tan, A. "Text Mining: The State of the Art and the Challenges," *In Proceedings of the PAKDD Workshop on Knowledge Discovery and Data Mining*, 1999.

Turney, P, D., and Littman, M, L. "Measuring Praise and Criticism: Inference of Semantic Orientation from Association," *ACM Transactions on Information Systems* (21:4), 2003, pp. 315–346.

Viegas, F.B., and Smith, M. "Newsgroup Crowds and AuthorLines: Visualizing the Activity of Individuals in Conversational Cyberspaces," in *Proceedings of the 37th Hawaii International Conference on System Sciences* (HICSS, 04), Hawaii, USA, 2004.

Walls, J, G., Widmeyer, G, R., and El Sawy, O, A. "Building an Information System Design Theory for Vigilant EIS," *Information Systems Research* (3:1), 1992, pp. 36–59.

Wasko, M, M., and Faraj, S. "Why Should I Share? Examining Social Capital and Knowledge Contribution in Electronic Networks of Practice," *MIS Quarterly* (29:1), 2005, pp. 35–57.

Welck, K. 1987 "Theorizing About Organizational Communication," in F. M. Jablin, L. L. Putnam, K. H. Roberts, and L. W. Porter (Eds.), *Handbook of Organizational Communication: An Interdisciplinary Perspective*, Newbury Park, CA, Sage, pp. 97–129.

Wellman, B. "Computers Networks as Social Networks," *Science* (293), 2001, pp. 2031–2034.

Wenger, E, C., and Snyder, W, M. "Communities of Practice: The Organizational Frontier," *Harvard Business Review*, 2000.

Whitelaw, C., and Patrick, J. "Selecting Systemic Features for Text Classification," in *Proceedings of AAAI Fall Symposium on Style and Meaning in Language, Art, and Music*, 2004.

Wise, J.A. "The Ecological Approach to Text Visualization," *Journal of the American Society for Information Science* (50:13), 1999, pp. 1224–1233.

Xiong, R., Donath, J., "PeopleGarden: Creating Data Portraits for Users," in *Proceedings of UIST* 1999.

Yates, J., and Orlikowski, W. J. "Genres of Organizational Communication: A Structurational approach to Studying Communication and Media," Academy of Management Review (17:2), 1992, pp. 299–326.

Yates, J., and Orlikowski, W. J. "Genre Systems: Structuring Interaction through Communicative Norms," *The Journal of Business Communication,* (39:1), 2002, pp. 13–35.

Yates, J., Orlikowski, W., and Okamura, K. "Explicit and Implicit Structuring of Genres in Electronic Communication: Reinforcement and Change of Social Interaction," *Organizational Science*, (10:1), 1999, pp. 83–103.

Zheng, R., Li, J., Chen, H., and Huang, Z. "A Framework of Authorship Identification of Online Messages: Writing-Style Features and Classification Techniques," *Journal of the American Society for Information Science and Technology (JASIST)*, 57(3), 2006, pp. 378–393.

Zhou, L., Burgoon, J. K., Nunamaker, J. F., and Twichell, D. "Automating Lingusitics-Based Cues for Deception Detection in Text-Based Asynchronous Computer-Mediated Communication," *Group Decision and Negotiation,* (13:1), 2004, pp. 81–106.

Zhu B. and Chen H. "Social Visualization for Computer-Mediated Communications: A Knowledge Management Perspective," in *Proceedings of the Eleventh Workshop on Information Technologies and Systems* 2001, Baton Rouge, LA, USA.

Chapter 13
Dark Web Forum Portal

1 Introduction

In recent years, there have been numerous studies from a variety of perspectives analyzing the Internet presence of hate and extremist groups. The use of the Internet by such groups has provoked terrorism research interest in various social sciences including psychology, sociology, criminology, and political science—computational scientists studying web mining and information extraction, and security analysts and others concerned with homeland and national policies and security.

Yet the web sites and forums of extremist and terrorist groups have long remained an underutilized resource due to their ephemeral nature and persistent access and analysis problems. They emerge quickly, often just as quickly disappearing or, in many cases, seeming to disappear by changing their uniform resource locators (URLs) but retaining much of the same content (Weimann 2004). Furthermore, some are hacked or closed by the ISPs. Thus, many researchers, students, analysts, and others face difficulties in identifying, collecting, and analyzing this content. Since terrorist and extremist groups are increasingly using the Internet to promulgate their agendas, it has become imperative to provide persistent access as well as user-friendly searching to this content. Given the sheer volume of sites, their dynamic and fugitive nature, different languages, and noise, it has become clear that systematic and automated procedures for identifying, collecting, and searching these sites must be provided.

Therefore, as reported in previous publications, the purpose of the Dark Web archive is to provide a research infrastructure for use by social scientists, computer and information scientists, policy and security analysts, and others (Zhang et al. 2009). The archive is currently comprised of 13 million postings from 29 international jihadist web forums. These forums collectively host 340,000 members whose discussions cover a wide range of sociopolitical, cultural, ideological, and religious topics. The forums collected for this project are in Arabic, English, French, German, and Russian and have been carefully selected with significant input from terrorism researchers, security and military educators, and other experts. The Arabic-language

H. Chen, *Dark Web: Exploring and Data Mining the Dark Side of the Web*,
Integrated Series in Information Systems 30, DOI 10.1007/978-1-4614-1557-2_13,

forums selected include major jihadist web sites, some of which are tracked by the CIA's Open Source Center. The English-language forums represent both extremist and more moderate groups in order to facilitate study of radicalization processes over time. Three French forums, and the single forums in German and Russian, provide representative content for extremist groups producing content in these languages. The content is updated regularly in order to remain fresh and relevant, and the infrastructure, described later in this chapter, includes tools for searching, browsing, translation, and analysis and visualization.

The Dark Web Forum Portal provides web-enabled access to these jihadist web forums (Zhang et al. 2009). In the following sections, this chapter will describe the significant extensions to the previous work, which included greatly increasing the scope of data collection; adding an incremental spidering component for regular data updates; enhancing the searching and browsing functions; enhancing multilingual machine translation for Arabic, French, German, and Russian; and adding advanced social network analysis.

2 Literature Review

2.1 Incremental Forum Spidering

Spiders (Cheong 1996) are defined as "software programs that traverse the World Wide Web information space by following hypertext links and retrieving web documents by standard HTTP protocol." There are six important characteristics (i.e., accessibility, collection type, content richness, URL ordering features, URL ordering techniques, and collection update procedure) of spider programs (Fu et al. 2010). Based on their suggestions, a functional spider program should be able to handle the registration requirement of targeted forums (accessibility), extract desired information from various types of collected data (collection type), filter out file types which are not of interest (content richness), sort queued URLs based on given heuristics (URL ordering features and techniques), and keep the collection up to date (collection update procedure).

The goal is to provide comprehensive Dark Web forum data from various sources in a timely manner. Therefore, accessibility and the collection update procedure are the two most critical issues. Accessibility problems can be solved by a human-assisted approach (Raghavan and Garcia-Molina 2001). For the collection update procedures, two well-adopted approaches are periodic and incremental spidering (Cho and Garcia-Molina 2000). The periodic approach respiders the entire collection of targeted forums in a fixed time interval. Due to the collection size of the Dark Web Forum Portal, spidering the entire collection would demand a great deal of time and make it difficult to keep the collection up to date. In addition, the spidering process generally starts from a list of given URLs and then attempts to download all the contents which can be linked from the starting URLs. Malicious links could potentially trap spiders in a loop.

Adopting the incremental spidering approach can minimize the problems mentioned above and is a good solution for keeping the collection up to date (Fu et al. 2010). Incremental spidering focuses on downloading only new content found in the targeted forums so that the spidering process can be done within a short time and even before the forums become aware of the spider. In addition, because incremental spidering only targets new contents, the structure of the targeted pages is very clear, and therefore, the malicious link problem can be avoided.

2.2 *Multilingual Translation*

The multilingual issue is critical in Dark Web research because much of jihadist content is written in various languages such as Arabic, German, French, etc. In order to process multilingual content, different methods have been explored to execute translation tasks, such as the machine translation–based approach, corpus-based approach, and dictionary-based approach (Zhou et al. 2005). The machine translation–based approach uses existing machine translation techniques to provide automatic translation. The corpus-based approach analyzes large document collections to construct a statistical translation model. In the dictionary-based approach, a bilingual dictionary is first constructed, and then, a translation result is obtained by looking up a given term in the dictionary. Google Translation is one of the most popular machine translation tools (http://code.google.com/apis/ajaxlanguage/ documentation/#Translation). The Google Translation Service provides translation functions for more than 80 languages. The language detection and language translation APIs it provides can be integrated into web pages using JavaScript. With this service, sentences in languages other than English can be translated to English automatically.

2.3 *Social Network Analysis on Web Forums*

Social network analysis (SNA) is a graph-based method to analyze the network structure of a group or population and its impact on social interactions (Liben-Nowel 2007). SNA has been widely used to study various real-world networks (Kossinets and Watts 2006). The social networks formed in illegal organizations are referred to as "dark networks" (Raab and Milward 2003). The dark networks can be either in the real world, such as the criminal networks studied by Hu et al. (2009), or in the virtual world, such as the social networks in web forums that are used by terrorists (Reid et al. 2004) to spread radical religious opinions, to organize terrorist activities, or to share knowledge about making weapons. Thus, terrorist web forums can be an important information resource for counterterrorism tasks (Coll and Glasser 2005).

In web forums, a thread is a collection of posts, displayed by time in ascending order. The first post is the thread starter, and the other threads are all replies to the thread starter. The social network in web forums is based on the thread structure and is usually referred to as a "reply network" (Zhang et al. 2007; Adamic et al. 2008). The reply network is a bipartite directed graph that consists of nodes and links. Every node represents a user in the web forum, and every link represents a reply-to relationship between two users, pointing from the author of the thread starter to the author of the reply.

Social network metrics are various measurements of nodes that can reflect the importance or connectivity of the node. There are several types of metrics that are commonly used in SNA for various purposes, such as centrality, cohesion, and reachability (Albert and Barabasi 2002). Centrality is a set of metrics that describes the importance of nodes based on their power of connecting the network, including betweenness, degree, and closeness. Cohesion of a social network describes how well a subset of nodes connects to each other in terms of forming a clique. Cohesion can be measured using a node's clustering coefficient, and a higher value indicates that if the node and the first neighbors of this node form a cluster, it is more likely that any two nodes in this cluster have a direct connection. Reachability describes the connectivity of two nodes, including metrics such as distance, which is the number of edges in the path that connects two nodes. In addition to these classic metrics, there are also other algorithms and metrics developed especially for estimating the importance in a web page network, such as PageRank scores and HITS scores.

3 Motivation and Research Questions

Since a large number of Dark Web forums exist, and they are heterogeneous and widely distributed, Dark Web forum data integration and retrieval are still obstacles for researchers to monitor Dark Web content (Albert and Barabasi 2002). The web sites and forums of extremist and terrorist groups have long remained an underutilized resource due to their ephemeral nature and persistent access and analysis problems. Thus, many researchers, students, analysts, and others face difficulties in identifying, collecting, and analyzing this content. A systematic and integrated approach to search, browse, and analyze multilingual terrorist/extremist forums is important and in demand. Few studies have incorporated social network analysis into a real-time, online Dark Web forum analysis system.

Based on the research gaps discussed above, we present the following research questions:

Q1: How can we conduct data updates regularly by adopting an incremental spidering process?

Q2: How can we identify different topics related to terrorist activities using the topic search component?

Q3: How can we identify the most influential members in the forums using the SNA analysis component?

Q4: How can the integrated system be used as an infrastructure to support the research of scientists from various disciplines (e.g., social science and computer science) who are interested in this content?

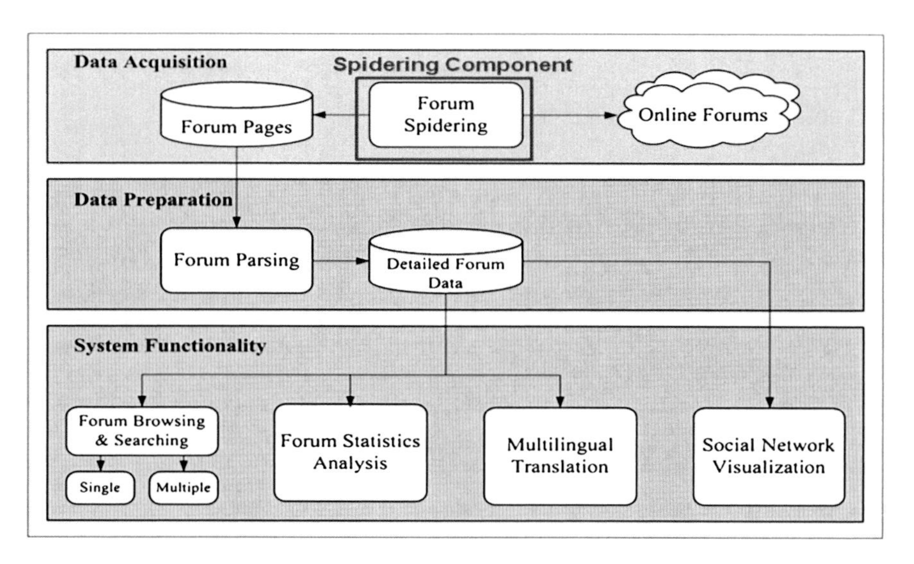

Fig. 13.1 System design of the Dark Web Forum Portal

4 System Design

As shown in Fig. 13.1, the Dark Web Forum Portal system contains three components: data acquisition, data preparation, and system functionality. The overall system design is similar to that reported in a previous publication (Zhang et al. 2009); a major difference is that since then, an incremental spidering component has been added to regularly update the collection. We detail each component in the following sections.

4.1 Data Acquisition

In this component, spidering programs are developed to collect the web pages from online forums that contain jihadist-related content identified by domain experts. The spidering component is composed of complete spidering and incremental spidering (Fig. 13.2). Complete spidering is applied to forums the first time they are added to our collection to collect all available postings, while incremental spidering is adopted if the forums already exist in the collection. Incremental spiders are designed to identify and collect postings posted after the last updating time of the forum so that only a small portion of forum data is collected, greatly improving the efficiency of the spidering process. To achieve this goal, an incremental spider is developed for each forum in the collection.

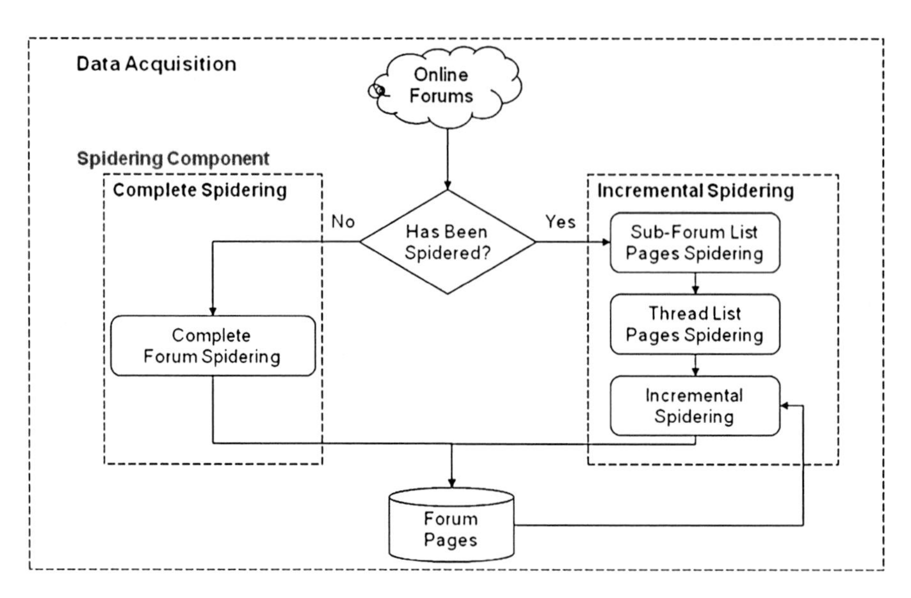

Fig. 13.2 Framework of the spidering component

The incremental spidering consists of three main steps: sub-forum list page spidering, thread list page spidering, and incremental spidering:

- *Sub-forum list page spidering*: Forums generally contain one or more sub-forums representing different discussion themes. In this step, incremental spiders first spider and parse sub-forum list pages of a forum and identify URLs of sub-forums.
- *Thread list page spidering*: Thread list pages contain the metadata of discussion threads (such as title, date of the last update, and author name) which are sorted by dates of the last update decreasingly. For each sub-forum, the incremental spider starts by downloading the first thread list page of the sub-forum, and dates of the last update of discussion thread are then extracted. Threads updated later than the date of the latest posting in the database are considered to be new threads, and their URLs are collected. If every thread listed in the first thread list page is a new thread, the spidering will move to the next thread page. Otherwise, the spidering of this sub-forum is complete.
- *Incremental spidering*: After collecting all the URLs of new threads, the incremental spider begins to download all of the postings within the new threads.

We conducted an experiment of incremental spidering using the Hanein Net forum (http://www.hanein.info/vb/), which is the most active forum identified by our domain experts. Because this forum is the most active one in our forum list, the experiment result can be considered as the upper bound of the time required for the incremental spidering. In the experiment, we collected postings from 11/1/2009 to 12/10/2009 (about 6 weeks). The experiment results show that during the time span, 3,504 threads and 29,016 postings were collected. The entire incremental spidering process was completed in 39 min. The result indicates that incremental spidering is a promising solution for keeping our forum data up to date.

4.2 Data Preparation

In this component, forum parsing programs are developed to extract the detailed forum data from the raw HTML web pages and store it in a local database. For each forum, the structured, detailed forum data extracted include thread names, main message bodies, member names, and post dates.

4.3 System Functionality

Different functions are developed and incorporated into the system as real-time services, including single and multiple forums, browsing and searching, forum statistics analysis, multilingual translation, and social network visualization. The Dark Web Forum Portal is implemented using Apache Tomcat, and the database is implemented using Microsoft SQL Server 2008. For forum statistics analysis, Java applet–based charts are created to show the trends based on the numbers of messages produced over time. The multilingual translation function is implemented using Google Translation Service, which can automatically detect non-English texts and translate them into English. The social network visualization function provides dynamic, user-interactive networks implemented using JUNG (http://jung.sourceforge.net/) to visualize the interactions among forum members.

5 Data Set: Dark Web Forums

Table 13.1 lists the forums incorporated into our system. Currently, the portal contains 29 jihadist forums, among which 17 are Arabic forums, 7 are English forums, 3 are French forums, and the other 2 are in German and Russian, respectively. The forums have been carefully selected with significant input from terrorism researchers, security and military educators, and other experts. The Arabic-language forums selected include major jihadist web sites, some of which include English-language sections. The English-language forums represent both extremist and more moderate groups. The French, German, and Russian forums provide representative content for extremist groups communicating in these languages and provide additional opportunity to evaluate multilingual translation. The total number of messages is about 13M; approximately 3M postings will be added annually through incremental spidering.

Table 13.1 Statistics of the Dark Web forum system functionality

Name	Language	Time span	Number of members	Number of threads	Number of messages
Al-Boraq	Arabic	01/08/2006–01/02/2010	3,503	52,322	223,648
Al-Fallujah	Arabic	09/19/2006–01/02/2010	5,853	74,899	547,712
Al-Firdaws[a]	Arabic	01/02/2005–02/06/2007	2,187	9,359	39,715
Midad al-Suyuf	Arabic	03/18/2006–1/02/2010	1,597	11,232	38,382
Alokab	Arabic	04/08/2005–12/31/2009	1,547	8,096	55,947
Al-Qimmah	Arabic	11/23/2007–01/02/2010	287	12,097	23,709
Alsayra	Arabic	04/05/2001–12/31/2009	66,705	147,598	1,227,207
Ansar	Arabic	11/07/2008–01/02/2010	1,224	12,041	46,928
At-tahadi	Arabic	04/14/2008–01/02/2010	313	2,599	5,406
Hanein Net	Arabic	11/27/2006–01/12/2010	2,837	96,239	821,478
Hawaa World	Arabic	01/01/2001–01/02/2010	113,579	40,501	2,251,553
Hadramout	Arabic	11/25/2000–12/29/2009	29,491	151,694	1,552,227
Ma'arik	Arabic	07/29/2007–01/03/2010	1,880	15,288	57,047
Al-Mujahidin	Arabic	11/09/2007–01/02/2010	4,259	29,980	140,930
Montada	Arabic	09/25/2000–12/29/2009	40,291	120,181	1,412,028
Ana al-Muslim	Arabic	10/08/1985–11/26/2009	12,215	179,791	1,343,370
Shumukh	Arabic	03/21/2007–01/02/2010	3,938	46,666	289,201
Ansar	English	12/08/2008–01/02/2010	377	11,133	29,056
Gawaher	English	10/24/2004–01/01/2010	6,790	210,656	569,709
Islamic Awakening	English	04/28/2004–12/31/2009	2,361	25,112	116,009
Islamic Network[a]	English	06/09/2004–05/07/2008	1,573	11,974	87,314
Islamic Web-Community	English	11/14/2000–12/31/2009	745	6,262	24,850
Turn To Islam	English	06/02/2006–01/01/2010	9,926	38,702	308,970
Ummah	English	04/01/2002–12/31/2009	14,349	71,218	1,192,583
Al Minha Dj	French	06/01/2008–01/04/2010	313	2,007	6,421
Forums d'aslama	French	10/06/2004–01/03/2010	2,665	20,468	131,559
Al-Mourabitoune	French	05/05/2002–03/27/2009	3,198	7,905	72,140
Ansar	German	02/27/2009–01/02/2010	62	726	1,645
KavkazChat	Russian	03/21/2003–01/03/2010	5,634	6,144	558,042
Total			339,699	1,422,890	13,174,786

[a]Forums marked with an asterisk are no longer active

6 System Functionality

As we described before, the system has four types of functions: single and multiple forum browsing and searching, forum statistics analysis, multilingual translation, and social network visualization. In this section, we describe our enhanced browsing and searching and social network visualization; for information about the other two functions, please refer to Zhang et al. 2009.

All Forums threads related to Topic: bomb, iraq, kill

This page shows all threads found which contain the search term.

Forum Name	Number of threads have been found
Forums in Arabic:	
Alboraq	0
AlFaloja	0
AlFirdaws	0
Almedad	0
Alokab	0
Alqimmah	0
Alsayra	0
AsAnsar	0
Atahadi	0
Hanein	0
Hawaa	0
Hdrmut	1
M3f	0
Majahden	0
Montada	0
Muslm	0
Shamikh	1
Forums in English:	
Ansar1	56
Gawaher	159
IslamicAwakening	3
IslamicNetwork	2
Myiwc	0
Ummah	5
TurnToIslam	0
Forums in French:	
Alminhadj	0
Aslama	0
Ribaat	0
Forums in German:	
DeAnsarnet	0
Forums in Russian:	
KavkazChat	0
IN ALL FORUMS	227

Fig. 13.3 Screenshot of the cross forum search based on keywords "bomb," "Iraq," and "kill" (AND operation)

6.1 Forum Browsing and Searching

6.1.1 Browsing and Searching for a Single Forum

The search function allows users to search message titles or bodies using multiple keywords. Users can choose the Boolean operations of the keywords to be either "AND" or "OR." Users can also express their search terms in English even when the forum is, for example, mainly Arabic. In that case, the search will return matches for both the English terms and the Arabic translations of those terms.

6.1.2 Browsing and Searching for Multiple Forums

In addition to browsing and searching information in a particular forum, our portal also supports multiple forum searching across all forums in the portal. For example, a total of 227 threads (Fig. 13.3) are retrieved across all forums that contain the keywords "bomb," "Iraq," and "kill" (AND operation) in the thread titles or message bodies.

Fig. 13.4 Screenshot of the detailed result for the forum "Gawaher" based on the cross forum search shown in Fig. 13.3

Among them, 159 are from the forum "Gawaher," 56 are from "Ansar1," 5 are from "Ummah," etc. "Gawaher" has more discussions on this topic than any of the other forums. Detailed searching results for each forum on these keywords can be found by clicking the row corresponding to a particular forum. Figure 13.4 shows the screenshot of the detailed result for the forum "Gawaher" based on the cross forum search in Fig. 13.3.

6.2 Social Network Visualization

The interface of the SNA function is shown in Fig. 13.5. It consists of three parts: the *search panel* (top box), *analysis panel* (middle box), and *visualization panel* (bottom box).

The search panel allows the user to choose three search criteria: forum, keyword, and time period. The threads that meet these search criteria are identified as "related threads" and are used to construct the social network. Any of the forums listed in the portal can be selected to perform SNA. The keywords are selected by the user, in

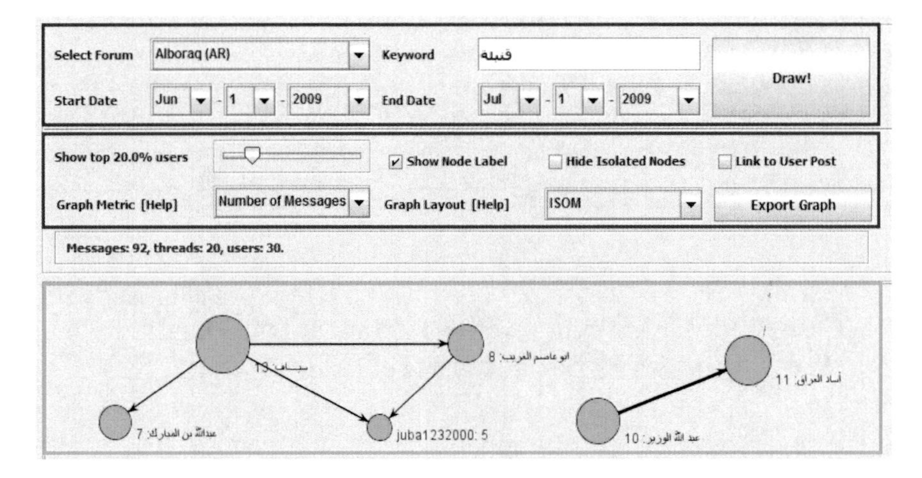

Fig. 13.5 SNA function interface

any language, separated with space or comma. Thread names, user names, and postings are searched using these keywords, and a thread is identified as a related thread if the thread name, or at least one posting, or at least one user name, contains any of the keywords. The start date and the end date are used to constrain the postings in the search result. When related threads are returned, the social network will be constructed based on the structure of these threads.

The analysis panel allows the user to select different metrics for SNA and to set the parameters for graph visualization. Every node in the social network has a set of attributes, including the screen name, the number of postings, and various social network metrics. After the social network is constructed, all nodes are ranked in descending order based on the number of postings. Since the resulting social network usually contains a large number of message authors, which makes the graph too crowded for analysis, the slide bar can be used to display only a portion of the top authors based on the ranking in order to make the graph easier to read. The label as well as the value of the selected metrics can be displayed beside each node by checking the corresponding box. An isolated node is defined as a node that has no connections to any other node. Removing isolated nodes is a useful function when too many cause noise in the graph. Checking the "Link to User Post" box will change the color of every node from green to red, and a click on any node will pop up a new window that shows all postings by this user during the selected time period.

The visualization panel displays the graph based on the settings in the analysis panel, with the thickness of the link proportionate to the intensity of interactions between two nodes. Any node can be dragged to any position in the panel, and all connected nodes and corresponding links will be highlighted when holding down the mouse button while moving the node. Different layouts are also provided for graph visualization. Four types of layout algorithms are integrated into the component, including static layout, circle layout, 3 force-based layouts (Fruchterman-Reingold, Kamada-Kawai, and Spring), and a self-organizing layout (ISOM). If users want to perform advanced analysis on the graph using other SNA tools such as UCINET, Pajek, and so on, clicking the "Export Graph" button allows the graph to be exported to a ".net" format file, which is the Pajek graph file format recognized by most SNA tools.

7 Case Study: Identifying Active Participants in Dark Web Forums Using Social Network Analysis

In this case study, we demonstrated a scenario on how to use the SNA component to identify active users on a particular topic of interest (in this case, "religious beliefs"). We chose Islamic Awakening for this case study. We searched "Muslim," "Islam," "Sharia," "Sunni," and "Shia" as keywords and generated the topic-based social network. The time period selected was 01/01/2009–12/31/2009 (1 year). As shown in Fig. 13.6, "Bint ul Islam," "Iloveislam," and "Abuhannah" were the three most active participants on these topics based on the "number of messages." We can show the SNA based on other different graph metrics such as betweenness, PageRank, in-degree, and out-degree.

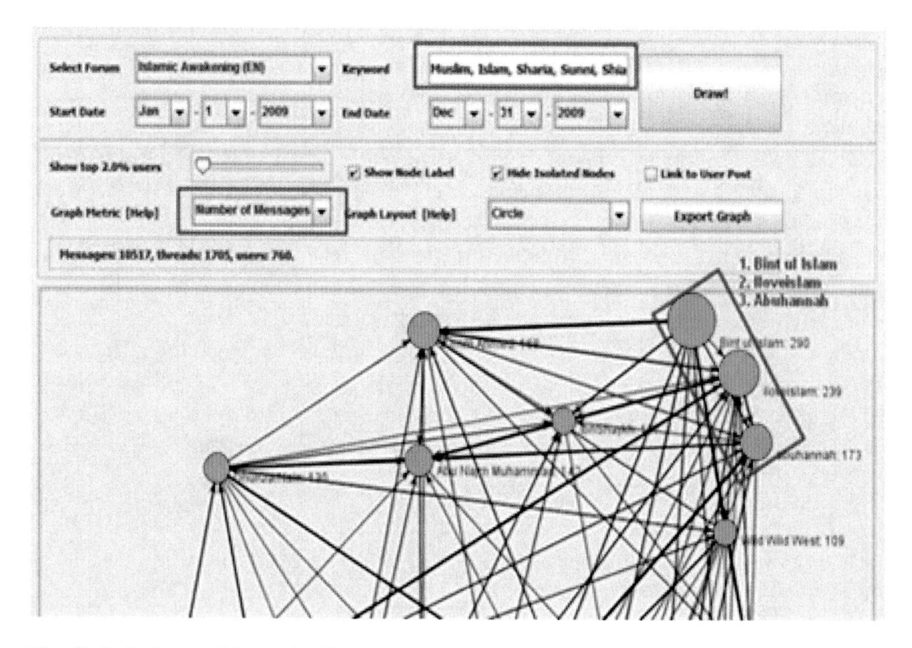

Fig. 13.6 Active participants in Islamic Awakening

By checking the "Link to User Post" box, the user can see the detailed messages. Example messages on this religious topic include the following:

> Sister, since you are a new Muslim, Allah will test you as He Says in the Qur'an: 'Do men think that they will be left alone on saying, "We believe," and that they will not be tested? We did test those before them, and Allah will certainly know those who are true from those who are false' [Al-Qur'an 29:2–3] You have to remain strong. ... by "iloveislam."
>
> Between The Past And The Future Imam Ibn ul Qayyim al Jawziyyah al-Fawaa'id, pp 151–152 Al-Istiqaamah, No. 2 Your life in the present moment is in between the past and the future. So what has preceded can be rectified by tawbah (repentance), nadam (regret) and istighfar (seeking Allah's forgiveness). ... by "Bint ul Islam."

8 Conclusions and Future Directions

In this chapter, we presented our work on developing a Dark Web collection and infrastructure for computational and social sciences. Significant extensions of our previous work included increasing the scope of our data collection; adding an incremental spidering component for regular data updates; enhancing the searching and browsing functions; enhancing multilingual machine translation for Arabic, French, German, and Russian; and adding advanced social network analysis. A case study is presented at the end. The Dark Web Forum Portal is an infrastructure which integrates heterogeneous forum data and will serve as a strong complement to the current databases, news reports, and other sources available to the research community.

Acknowledgments This work is supported by the NSF Computer and Network Systems (CNS) Program (CNS-0709338), September 2007–August 2010, and HDTRA1-09-1-0058, July 2009–July 2012. Any opinions, findings, and conclusions or recommendations expressed in this material are those of the author(s) and do not necessarily reflect the views of the National Science Foundation or DOD.

References

Adamic, L. A., Zhang, J., Bakshy, E., and Ackerman, M. S. (2008). "Knowledge Sharing and Yahoo Answers: Everyone Knows Something." In Proceeding of the 17th International Conference on World Wide Web (Beijing, China, April 21–25). WWW '08. ACM, New York, NY, pp. 665–674.

Albert, R. and Barabasi, A.-L. (2002), "Statistical Mechanics of Complex Networks," *Rev. Mod. Phys.*, Vol. 74, pp. 47–97.

Cheong, F.C. (1996). Internet Agents: Spiders, Wanderers, Brokers, and Bots. Indianapolis, IN: New Riders Publishing.

Cho, J. and Garcia-Molina, H. (2000). "The Evolution of the Web and Implications for an Incremental Crawler." In Proceedings of the 26th International Conference on Very Large Databases.

Coll, S. and Glasser, S.B. (2005). "Terrorists Turn to the Web as Base of Operations," *Washington Post*, August 7.

Fu, T.J., Abbasi, A., and Chen, H. (2010 online; forthcoming in print). "A Focused Crawler for Dark Web Forums," *Journal of the American Society for Information Science and Technology* (JASIST).

Hu, D., Kaza, S., and Chen, H. 2009. "Identifying Significant Facilitators of Dark Network Evolution," *JASIST*, Vol. 60, no. 4, pp. 655–665.

Kossinets, G. and Watts, D.J. (2006), "Empirical Analysis of an Evolving Social Network," *Science*, Vol. 311, pp. 88–90.

Liben-Nowel, D. (2007), "The Link-prediction Problem for Social Networks," *JASIST*, vol. 58, no. 7, pp. 1019–1031.

Raab, J. and Milward, H.B. (2003), "Dark Networks as Problems," *Journal of Public Administration Research and Theory*, Vol. 13, pp. 413–439.

Raghavan, S. and Garcia-Molina, H. (2001). "Crawling the Hidden Web." In Proceedings of the 27th International Conference on Very Large Databases.

Reid, E., Qin, J., Chung, W., Xu, J., Zhou, Y., Schumaker, R., Sageman, M., and Chen, H. (2004). "Terrorism Knowledge Discovery Project: A Knowledge Discovery Approach to Addressing the Threats of Terrorism." In Proceedings of the 2nd Symposium on Intelligence and Security Informatics (Tucson, June 10–11), pp. 125–145.

Weimann, G. (2004). "www.terror.net: How Modern Terrorism Uses the Internet." Special Report, United States Institute of Peace. Retrieved October 31, 2006. http://www.usip.org/pubs/specialreports/sr116.pdf.

Zhang, Y., Zeng, S., Fan, L., Dang, Y., Larson, C., and Chen, H. (2009). "Dark Web Forums Portal: Searching and Analyzing Jihadist Forums." In Proceedings of the IEEE International Intelligence and Security Informatics Conference (Dallas, Texas, June 8–11).

Zhou, Y., Qin, J., Chen, H., et al. (2005). "Multilingual Web Retrieval: An Experiment on a Multilingual Business Intelligence Portal." In Proceedings of the 38th Annual Hawaii International Conference on System Sciences (HICSS'2005).

Zhang, J., Ackerman, M. S., and Adamic, L. (2007). "Expertise Networks in Online Communities: Structure and Algorithms." In Proceedings of the 16th International Conference on World Wide Web (Banff, Alberta, Canada, May 08–12). WWW '07. ACM, New York, NY, pp. 221–230.

Part III
Dark Web Research: Case Studies

Chapter 14
Jihadi Video Analysis

1 Introduction

With the global expansion of jihadi (Holy war) movements (e.g., Egypt, Iraq, Spain, USA, UK), there has been an increase in radical Islamist and jihadi groups' use of the Internet. Some reports say that there are thousands of jihadist web sites that support the groups' community building (*ummah*) and distribute recruitment videos, strategy documents, speeches, and combat computer games.

Although these web sites provide an abundance of information, they are almost entirely in Arabic, tied to radical ideologies (Paz 2006), challenging to identify and capture (Chen et al. 2005), and part of the groups' communication strategies (Corman and Schiefelbein 2006). Videos produced by jihadi groups and their sympathizers are disseminated on the Internet, most notably in online discussion forums and dedicated jihadi web sites as well as free file hosting web sites. However, materials available on the Internet are also circulated as printed leaflets and videos within different countries (International Crisis Group 2006). For example, the videos are being sold in the local Iraqi market alongside pornography (Hendron 2006). They are also aired on Al-Zawraa TV which is a 24-h satellite station that airs video compilations of attacks on US forces in Iraq. This channel is viewed throughout the Middle East, North Africa, and parts of Europe (Coleman 2007).

These videos function as cultural screens for multiple enactments, viewings, and interpretations of accepted patterns, themes, and norms (e.g., suicide bombing, martyrdom) while perpetrating the development of shared understandings and evolving glossaries of radical visuals about their ideologies, goals, tactics, and mistakes. The use of recurring visuals and themes in jihadi web sites and multimedia was substantiated by a content analysis study of jihadi groups' Internet visual motifs (e.g., symbols, photographic images) conducted by the Combating Terrorism Center (Islamic Imagery Project 2006).

The volume of jihadi groups' multimedia artifacts disseminated over the web is vast (Weimann 2006). These artifacts are evanescent in nature, reflect cultural norms, and embed shared messages in them (Baran and Mathieu 2005). Consequently,

H. Chen, *Dark Web: Exploring and Data Mining the Dark Side of the Web*,
Integrated Series in Information Systems 30, DOI 10.1007/978-1-4614-1557-2_14,

the intelligence, law enforcement, and research communities spend substantial resources and efforts to identify, capture (harvest), monitor, translate, and analyze these video artifacts. However, there is an intellectual gap because there is still limited systematic and evidence-based research about the videos that can be used for comparative analysis and forecasting.

The purpose of this chapter is to provide an exploratory, evidence-based analysis of how jihadi extremist groups use videos to support their goals such as sharing ideologies and mobilization of potential recruits for perpetrating terrorist attacks. It describes how groups are using the videos to show their resolve, share messages, solicit funds, and support training. It uses the "Jihad Academy" video to provide an illustration of patterns associated with creating and distributing videos via the Internet. The illustration highlights the importance of conducting a content analysis of Arabic jihad videos. The content analysis is part of a systematic effort to apply automated methodologies to identify, harvest, classify, analyze, and visualize extremist groups' video artifacts usage. The content analysis involves the creation of a multimedia coding tool and coding scheme, as well as coding 60 Arabic videos to analyze the portrayed events and how the videos support the groups' goals and modus operandi.

2 Jihadi Groups' Videos

The jihadi groups' extensive use of the web, technical sophistication, and media savvy has been described in several studies (Becker 2005). Their web sites, blogs, and discussion forums provide hyperlinks to many video clips that vary in language (e.g., Arabic, English, French), size, format (e.g., wmv, ram, 3GP), level of technical sophistication (e.g., amateur, professionally produced), and purpose (e.g., document attacks, boost morale, commemorate martyrdom) (Salem et al. 2006).

The Afghani Mujahideen and, later, Chechen rebels pioneered the creation of videos that captured their operations (IntelCenter 2005). The idea behind this was that even if the attack against Russian soldiers was limited in scale, if the operation was filmed and then shown to the world, the impact would be greater. However, the Afghani Mujahideen and Chechen rebels never had the means and ability to disseminate their videos on a large scale. In contrast, over the last few years, the filming of attacks, the sophistication of video production, quantity, and speed of video dissemination on the web have become important operational strategies for jihadi extremist groups around the world, who demonstrated an ability to quickly adopt and adapt Internet technologies.

The Internet enables the groups to mobilize resources (e.g., communication, money, training, networks) to strengthen their movement (Duijvelaar 1996) and launch effective strategies to attain their goals. For jihadi groups, this supports three strategic communication goals: (1) legitimatize their movement by establishing its social and religious viability while engaging in violent acts; (2) propagate their visions, goals, and slogans by spreading messages to sympathizers in areas that they want to expand; and (3) intimidate their opponents (Corman and Schiefelbein 2006).

The jihadi professionally produce videos that are released through media outlets such as the Al-Sahab Institute for Media Production (video production arm of al-Qaeda) and appear frequently on the Al-Jazeera channel as well as the web (IntelCenter 2004). Videotaping extremist groups' operations resulted in a mimetic effect, similar to that of an "infectious idea." Its multiplier effect among jihadi extremist groups emboldened them to produce more videos documenting their brazen attacks on soft targets (especially the beheadings of defenseless civilians), which are then disseminated via the web.

Extremist groups, such as al-Qaeda and their avid sympathizers, have been incredibly successful in using videos to share messages (e.g., Osama bin-Laden's speeches) and provide training (IntelCenter 2005). The popular press focus on video reporting (especially the beheadings) has gotten global attention (Robertson 2006). This has heightened the importance of the videos and may have contributed to the increase of violence.

2.1 Dissemination of Extremist Groups' Videos

Some of the videos are mirrored hundreds of times at different web sites or forums within a matter of days (News from Russia 2006). The cyber gatekeepers provide global and sustainable access to selective videos in different formats and sizes based on the user's requirements. The storage and distribution of the videos involve using many file hosting service web sites (e.g., www.megaupload.com). For example, the Tracking Al Qaeda blog (Tracking Al Qaeda 2006) identified the Global Islamic Media Front (GIMF), which is affiliated with al-Qaeda in Iraq, as the producer of the "Jihad Academy" video that portrays the events of a single day in the life of the Mujahideen (warriors).

According to the Science Applications International Corporation (SAIC), a defense contractor that specializes in homeland security, the "Jihad Academy" video includes various jihadi attacks against the enemy and uses both English and Arabic languages, which suggests that it could be targeted toward a broad range of audiences including supporters, sympathizers, and enemies (GIMF releases new video 2006). The persuasive messages make the "Jihadi Academy" an excellent example of why videos are important resources that can support recruitment, propaganda, and collective mobilization of members and sympathizers (Salem et al. 2006).

Figure 14.1 shows the process of producing and distributing the "Jihad Academy" video, which contains several clips shot by Iraq jihadi groups such as Al-Jaysh al-Islami fil-'Iraq, Tandhim al-Qa'ida fi Bilad al-Rafidayn, and Jaysh Ansar al-Sunna (GIMF releases new video 2006). After the video is produced, copies in different formats are generated and widely disseminated in discussion forums and television outlets such as Al-Zawraa TV. The video has been posted on at least three discussion forums (la7odood.com, 3nabi.com, and almarsaa.net) within a relatively short period of time (Tracking Al Qaeda 2006). Each forum provides links to free file hosting web sites where the videos are made available.

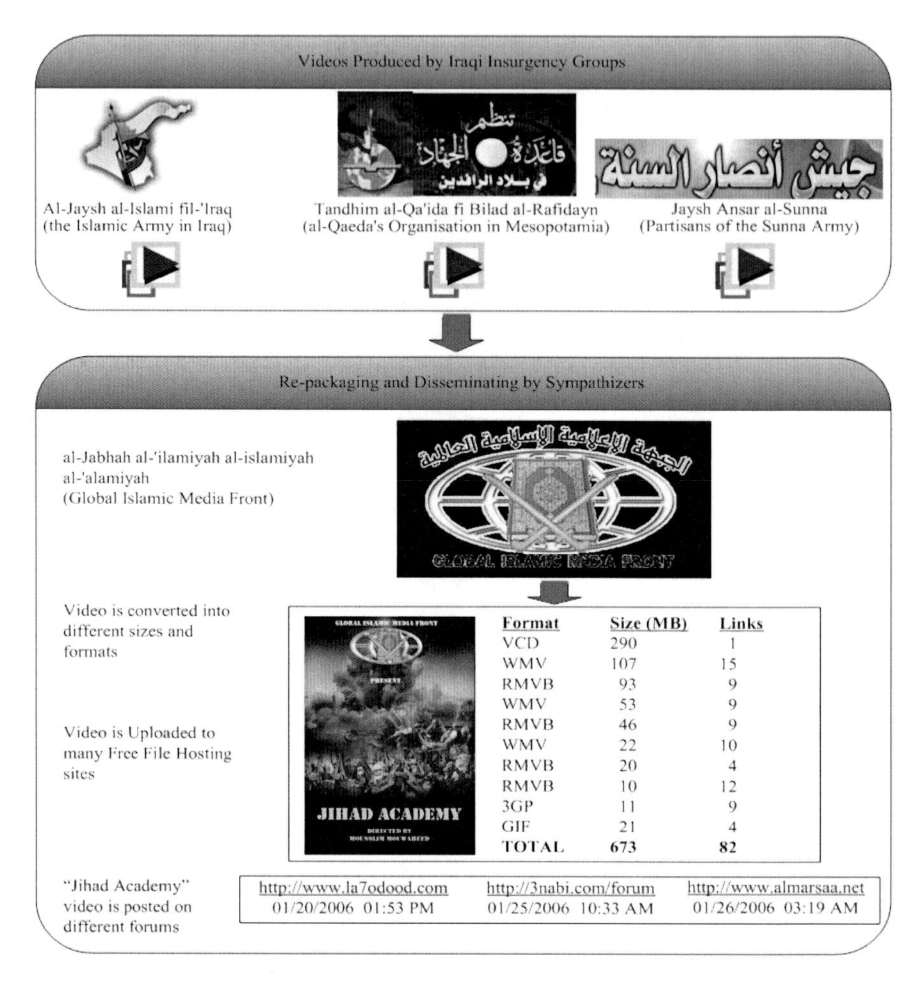

Fig. 14.1 Dissemination of videos in online discussion forums

3 Collections of Extremist Groups' Videos

The massive production and distribution of jihadi groups' videos have prompted organizations to identify, monitor, collect, translate, and analyze the videos. Table 14.1 identifies organizations that support the counterterrorism and law enforcement communities, such as the IntelCenter. Most of them monitor, collect, and analyze the videos and generate reports. The Artificial Intelligence (AI) Lab at the University of Arizona collects videos using a systematic web spidering approach

Table 14.1 Collections of jihadi groups' videos

Organizations	Collections	Number of videos	URLs
Counterterrorism organizations			
• Global terror alert (Director, E. Kohlmann)	Clearinghouse on international terrorism	134 titles	http://www.globalterroralert.com/archive.html
• IntelCenter (Director, B. Venzke)	Audio/video	60 volumes[a]	http://www.intelcenter.com
• Intelwire.com (Editor, J.M. Berger)	Jihad videos online archive	208 titles	http://intelfiles.com (partial list of videos)
• SITE Institute (Director, R. Katz)	Multimedia catalog	400 titles	http://www.siteinstitute.org (partial listing of videos)
Research centers			
• AI Lab, University of Arizona (Director, H. Chen)	Dark Web multimedia collection	706 titles	http://ai.eller.arizona.edu/ (closed research collection)
• MEMRI	Jihad and Terrorism Studies Project	23 titles	http://www.memri.org/jihad.html
Total (estimate)	1,471 titles		

Some are duplicate videos
[a]Not included in total number of titles

and performs research using content and link analysis (Zhou et al. 2006). The AI Lab's Dark Web video collection is intended for systematic research. The Dark Web is the alternate (covered and dark) side of the web used by extremist groups to spread their ideas (Chen et al. 2005).

The IntelCenter, SITE Institute, and MEMRI conduct high-level content analysis of the videos and code them based on several areas such as group, event, format, time, and language. The IntelCenter also categorizes their jihadi groups' video collections into seven types such as produced videos, which have the highest production values, and operational videos, which are short quick clips of attacks executed by a group (IntelCenter 2005). Table 14.2 presents categorizations of jihadi videos and examples.

Although there are few fine-grained content analysis schemes of jihadi videos, there are terrorism ontologies which provide concept classification of terrorist events in several areas: groups, targets, weapons, and regions (IntelCenter 2004).

Table 14.2 Categorization of jihadi extremist groups' videos

IntelCenter	IntelFiles	Objectives of videos and examples
1. Produced	1. Documentary and propaganda	Boost morale and psychological warfare (e.g., 19 Martyrs video, 2002)
2. Operational	2. Operations	Document attacks (e.g., Destruction of the destroyer, USS cole video, 2001)
3. Hostage	3. Direct terrorism	Document hostage attacks and/or executions (e.g., Ansar Al Sunna executes three Iraqi drivers video, 2005)
4. Statement	4. Communiqués	Spread messages, threats, etc. (e.g., July 7 Transit bombing statement video, 2005)
5. Tribute		Commemorate death of members (e.g., wills of the heroes video, 2003)
6. Internal training		Document training (e.g., Islamic extremist ops/training video)
7. Instructional		Provide instruction on skills (e.g., manufacture of the explosive belt for suicide bombing video, 2004) (SITE Institute 2004)

4 Content Analysis of Videos

Extremist groups use video to enable communication, deliver propaganda, and disperse their ideologies, tactics, and strategies. Researchers have identified several factors (e.g., the multiplier effect, the sophistication and ease of video production, low cost, compression options, and global dissemination via the Internet) that influence extremist groups' use of videos to support their terror campaigns. This study uses the resource mobilization (Duijvelaar 1996) framework to undertake a systematic content analysis of extremist groups' videos and to answer the following research questions:

- What types of video are produced by extremist groups?
- How are the videos used by the jihadi extremist groups?
- What modus operandi and production features are identified in extremist groups' videos?

From a resource mobilization perspective, the use of the videos is a rational choice for enhancing the groups' communications, propaganda, and training resources necessary to publicize, diffuse, and execute the campaigns. The content analysis process includes several steps described in Salem et al.'s study of jihadi groups' videos (Salem et al. 2006). The process includes the selection of the sample collection of videos, generation of a list of content categories and associated content features, assessment of coding reliability, design of a coding tool, coding the videos, and analysis of results.

Table 14.3 Dark Web video collection

Video type	Number of videos	Size (MB)	Play time (hh:mm:ss)
Documentary	291	2,376.91	35:15:31
Suicide attack	22	122.85	02:09:13
Beheading	70	294.95	04:44:03
Hostage taking	26	172.80	02:24:13
Tribute	13	128.69	02:49:40
Message	126	1,293.91	44:60:48
Propaganda	143	1,566.98	23:42:19
Instruction	1	16.72	00:08:24
Training	9	196.49	03:20:12
Newsletter	5	553.54	02:36:30
Total	706	6,723.84	122:06:53
Averages			
Average file size	Average playtime	Average bitrate	
9.5 MB	10:23	247.3 kbps	

4.1 Sample Collection

The collection development approach to identify and collect content from extremist groups' web sites is described in studies by the Artificial Intelligence (AI) Lab, University of Arizona (Reid et al. 2005). Table 14.3 provides a summary of the 706 multimedia files that were downloaded for the Dark Web multimedia collection (sixth batch) and categorized using the IntelCenter classification scheme.

From the Dark Web multimedia collection, the Arabic videos produced by insurgents in Iraq were identified. An arbitrary number of 60 videos were chosen for a randomly selected sample. They have a time span of 2 years, starting in January 2004. The sample videos are listed in Appendix A.

4.2 Coding the Videos

A multimedia coding tool (MCT) was designed to manage the coding process in a systematic and structured manner. MCT allows the user to create/edit the coding scheme, load the videos, play the videos, record observations, and generate reports. The content and technical features of each video were captured, classified, and stored in the MCT. For example, the group's name, video type, and other information were recorded as described in the coding scheme.

The coding scheme consists of *eight high-level classes* such as *general information* that are subdivided into 25 content categories (variables). Appendix B provides a list of the classes and content categories. The classes are (1) *general information* with content categories that include title, source, and type of video; (2) *date* with categories of reported and acquisition dates; (3) *production* with categories of languages, structure of video clip, and special effects; (4) *the group* with categories of group name and media agency; (5) *expressions* with categories of verbal and nonverbal; (6) *location* with categories of country and city; (7) *event* with categories of tactic and weapon; and (8) *nature of the target* with categories of types of victim. The scheme is based on the features of jihadi videos, terrorism ontologies, the IntelCenter's categorization, and terrorism incident databases such as the RAND-MIPT and the Institute for Counter-Terrorism (ICT) databases.

4.3 Inter-coder Reliability

The 60 videos were coded over a 3-week period by two domain analysts who speak Arabic. To deal with multiple responses for one content category (variable), the coders treated each possible content feature (response) as a separate variable. Due to the open-ended nature of content features, category reliability is measured using Holsti's formula for computing reliability (Holsti 1969). The percentage agreement between the two coders was higher than 0.80 for all content categories that were analyzed.

5 Types of Videos Produced

For the 60 jihadi groups' videos that were content analyzed, the average length was 6 min and 32 s. The video types, groups' modus operandi, production features, and the groups' video usage were analyzed to identify what types of videos are produced by extremist groups. The results identify two categories of videos (e.g., violent attacks and others) that are used to support the jihadist psychological warfare and mobilization strategies. Specific content such as the names of groups involved and the groups' modus operandi (e.g., tactics, targets, weapons) enable the extremist groups to (a) publicize their actions to diverse communities of supporters, sympathizers, media groups, and enemies; (b) claim responsibility; and (c) disseminate their messages globally to gain legitimacy for their causes. Cultural aspects, including production features (e.g., subtitles, logos), and verbal as well as nonverbal expressions (e.g., religious verses, kissing, hugging) are meticulously embedded in the content to help targeted audiences identify with the jihadi movement.

Table 14.4 provides the frequency count for jihadi groups' video types, which are grouped into two categories: violent attacks (e.g., documentary, suicide attack) and others (e.g., tribute, message such as leader statement). Appendix A provides a breakdown of the 60 sample Arabic videos by types. The violent attacks category has the largest number of videos with most videos classified as documentary.

5.1 Documentary Videos

The documentary (attack) videos are often filmed in real-time (show the attacks in action), instructive (take the viewer inside the planning and attack execution processes, including scenes of the different weapons such as rocket-propelled

Table 14.4 Video types

Video types	Frequency
Violent attacks	
Documentary	38
Suicide attack	4
Beheading	1
Hostage taking	5
Others	
Tribute	3
Message	6
Propaganda	2
Newsletter	1
Total	60

grenades and skills required for their operations), and low budget. They have limited promotional costs, as indicated by the low quality of some videos, and appeal to diverse audiences through the use of Arabic and English subtitles. The plots were simple (focus on few goals such as to destroy the enemy's tankers), versatile (can be used for meetings, training, fundraising, motivational sessions), persuasive (display actors' emotions and dedication), succinct (quickly present the materials in short videos), and targeted (producers have complete control over the message and sequence of events).

Documentary videos identify the name and sometimes the logo of the extremist groups, but rarely include a direct verbal message from the group. However, they are often accompanied by a wish for the success of the operation in the form of religious or semireligious phrases. For example, the "Road Side Bomb 1" video is only 12 s and is in Windows Media Video format (wmv). It shows a bombing in Dayali and identifies the group claiming responsibility as Al-Jabha al-Islamiya lil-Muqawama al-'Iraqiya (Appendix A #15).

The documentary videos often include improvised explosive devices (IEDs), artillery, and rocket attacks. In Fig. 14.2, the distribution of video types indicates the high number of documentary (63%) videos that are used by groups to document and claim responsibility for their attacks. Documentary videos include all types of attacks except suicide attack (7%), beheading (2%), and hostage taking (8%). According to the International Crisis Group, extremist groups in Iraq are waging a war of attrition by avoiding direct confrontation with coalition forces. They adopt hit-and-run tactics such as IED attacks, which constitute the bulk of the day-to-day operations.

In the sample, nine of the ten extremist groups produced documentary videos. Table 14.5 provides a breakdown of documentary videos by groups. For example, the Islamic Front of the Iraqi Resistance (Al-Jabha al-Islamiya lil-Muqawama al-'Iraqiya), an insurgency group in Iraq which was formed in 2004, has 11 videos in the sample, and all are documentary.

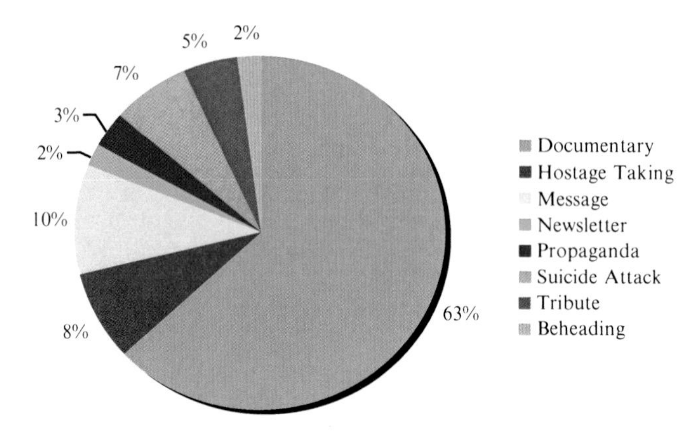

Fig. 14.2 Distribution of videos by types

Table 14.5 Breakdown of documentary videos by group

Number of videos	Group name
11	Al-Jabha al-Islamiya lil-Muqawama al-'Iraqiya (Islamic Front of the Iraqi Resistance)
5	Al-Jaysh al-Islami fil-'Iraq (Islamic Army in Iraq)
0	Al-Qiyada Al-Muwahada Lil-Mujahideen (Mujahideen Central Command)
1	Harakat al-Muqawama al-Islamiya fil-'Iraq (Islamic Resistance's Movement in Iraq)
1	Jaysh al-Iraq Al-Islami (Iraq Islamic Army)
2	Jaysh al-Jihad Al-Islami (Islamic Jihad Army)
4	Jaysh al-Mujahideen (Mujahideen's Army)
1	Jaysh al-Ta'ifa al-Mansoura (Victorious Group's Army)
2	Jaysh Ansar al-Sunna (Partisans of the Sunna Army)
3	Tandhim al-Qa'ida fi Bilad al-Rafidayn (al-Qaeda's Organisation in Mesopotamia)
8	Unclear/unknown

5.2 *Suicide Attack Videos*

In contrast to documentary videos, suicide bombing videos are in general more elaborate and show different stages of action. For example, a video of a suicide attack on a US base in Mosul, Iraq, illustrates a process associated with executing an individual attack (Appendix A #25).

In Fig. 14.3, scenes from the suicide attack video are used to illustrate the systematic approach of planning, preparation, execution, and outcome for a suicide bombing. In the same way, videos of beheadings and other types of executions follow a structure roughly consisting of first a message by the hostage, followed by a verdict or warning, and typically concluded with a grisly beheading or shooting of the hostage.

Fig. 14.3 The stages of an attack against a US base in Mosul, Iraq: (**a**) title and suicider name, (**b**) moral/religious justification, (**c**) planning, (**d**) farewell, (**e**) execution, and (**f**) aftermath

The category of video types entitled "Others" includes nonviolent activities such as tribute (5%), message (10%), propaganda (3%), and instruction, training, and newsletter (2%). The sample did not contain training or instructional material. It was observed that direct training and instructional content is in the form of text-based manuals.

6 How the Groups Used the Videos

Based on the analysis of video types, a matrix of jihadi groups' videos is used to describe how the videos are used by the extremist groups. The matrix classifies the videos into four basic types according to two usage dimensions: operational versus nonoperational and individual-oriented versus group-oriented. Figure 14.4 presents the schematic diagram.

In Fig. 14.4, a video can be classified as operational because it clearly displays a violent attack. Nonoperational videos center on showing nonviolent activities such as delivering a message or paying tribute to a fallen comrade. Although nonviolent activities may include threats, they are still considered as nonviolent acts. On the other hand, the other dimension involves several actors such as a group or a single individual. A suicide attack is a violent act (documentary) committed by a single individual. The focus of the documentary videos is the group as a whole (claiming responsibility), as opposed to the individual in the case of suicide attacks. Tributes and messages often focus on an individual, such as a martyr (shahid).

Most of the sample videos fall into quadrant 2 because they are group-oriented violent operations (e.g., bombings, beheadings). Table 14.6 shows to which quadrants the videos produced by various groups belong. Since most groups produced documentary videos, they seem to view this as the main and natural usage of videos.

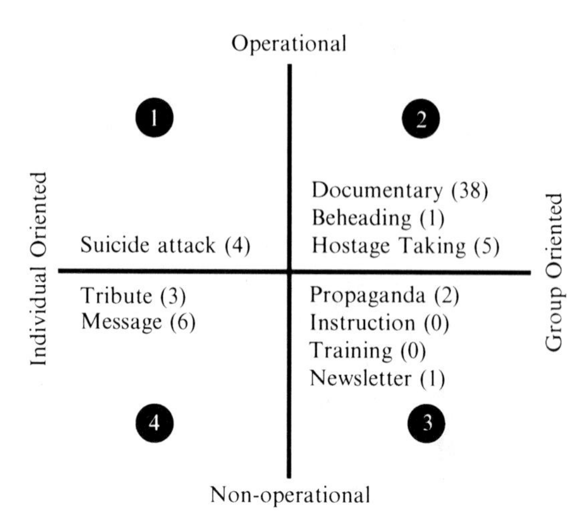

Fig. 14.4 Matrix of video types and usages (number of videos in parentheses)

Table 14.6 Breakdown of video matrix by group

Number of videos	Group name	Quadrant 1	2	3	4
11	*Al-Jabha al-Islamiya lil-Muqawama al-'Iraqiya*		X		
7	*Al-Jaysh al-Islami fil-'Iraq*		X		X
3	Al-Qiyada Al-Muwahada Lil-Mujahideen		X		X
1	Harakat al-Muqawama al-Islamiya fil-'Iraq		X		
1	Jaysh Al-Iraq Al-Islami		X		
3	Jaysh Al-Jihad Al-Islami		X		X
4	Jaysh al-Mujahideen		X		
2	Jaysh al-Ta'ifa al-Mansoura		X	X	
8	*Jaysh Ansar al-Sunna*	X	X	X	X
12	*Tandhim al-Qa'ida fi Bilad al-Rafidayn*	X	X		X
8	Unclear/unknown		X	X	
60	Total	4	44	3	9

In the sample, suicide attack videos were produced by only two groups: Jaysh Ansar al-Sunna and Al-Qiyada Al-Muwahada Lil-Mujahideen. Quadrant 3, nonoperational and group-oriented, has the second most videos. The main goals of the videos appear to be advertising their ideologies, legitimizing their actions, and indirectly communicating with supporters as well as sending threats to enemies.

6.1 Groups Identified

We identified ten unique groups that took credit for the videos. In eight videos, the groups did not identify themselves. In Table 14.6, the four most prominent groups (in terms of the number of videos) are in bold. According to the International Crisis Group (2006), they are also the most active groups of the Iraqi insurgency.

Among the groups identified, nine were involved in violent attacks. Extremist groups, such as Tandhim al-Qa'ida fi Bilad al-Rafidayn (previously led by al Zarqawi who was killed in a US operation in June 2006), produced the aforementioned Mosul suicide bombing depicted in Fig. 14.3. This video, in particular, supported organizational learning and provided mental models of the group members' dedication, closeness, emotional rituals, and skills as they executed the attacks. Scenes provide emotional and spiritual support because of the hugging, greeting, and praying together.

7 Groups' Modus Operandi and Production Features

The major targets identified in the videos are Western military vehicles. Table 14.7 provides a list of the types of targets identified.

Military vehicles constitute 56% of the total identified targets in the sample, while 20% of the identified targets are military bases. In this sample, there is a pattern of the emergence of specialization among extremist groups. For instance, Al-Jabha al-Islamiya lil-Muqawama al-'Iraqiya carried out ten roadside bombings and artillery attacks against military vehicles and bases. However, they did not conduct any beheadings. In contrast, Tandhim al-Qa'ida fi Bilad al-Rafidayn has executed several hostage takings. The International Crisis Group states that an informal division of labor and specialization is taking place within the Iraqi insurgency.

Most of the videos explicitly mentioned locations of the depicted attacks. The locations are often mentioned in the subtitles, and occasionally, the narrator provides the name of the location. Most attacks occurred in the Sunni triangle, which includes Balad, Dayali, Baghdad, Fallujah, and Abu-Ghraib. In addition, US military installations in Mosul are frequently attacked, as reported in the international media. Moreover, each extremist group operates in selected regions of the Sunni triangle. For instance, the Al-Jabha al-Islamiya lil-Muqawama al-'Iraqiya has operated mostly in the Salah al-Din, Diyala, and Baghdad governorates, while Tandhim al-Qa'ida fi Bilad al-Rafidayn operates mostly in Anbar, Baghdad, and Salah al-Din. Our results are compatible with media reports on the Iraqi insurgency.

IEDs are the most common type of weapons utilized. Preparing, implanting, and detonating the IED is often depicted. Mortar and rocket attacks are the second most

Table 14.7 Types of targets

Facility type	% Videos
Military vehicle	56
Military base	20
Unknown	13
Aircraft	7
Diplomatic	2
Transportation infrastructure	2

Fig. 14.5 RPG attack to disrupt a supply line

frequently observed weapon. Other weapons include assault rifles and rocket-propelled grenades (e.g., RPG) and, less frequently, suicide vehicles. These weapons are typically used in guerilla warfare. Figure 14.5 shows a member of an extremist group launching an RPG to disrupt a supply line (Appendix A #20).

The RPG is an inexpensive and single-shot weapon that is devastatingly effective yet easy to operate with very little training (Rocket propelled grenade 2006). The RPG and mortar attacks can impose considerable damage without conducting close-range engagements.

7.1 Production Features

A range of production quality patterns, from amateurish to professional, were identified. In addition, diverse special effects were identified, such as the use of subtitles (English or Arabic), the groups' logos, background hymns (with/without music), and excerpts of leaders' speeches. Fifty-five videos had special effects.

Figure 14.6 provides a screenshot of a video with English subtitles that shows bin-Laden giving a speech. Bin-Laden's messages are often directed toward a worldwide audience. Al-Qaeda's media agency "Al-Sahab" produced the bin-Laden interview excerpt in the sample (Appendix A #48). This agrees with reports from the SITE Institute and IntelCenter about al-Qaeda's use of a production company to plan and produce high-quality videos.

Ideologies and customs identified in the videos were consistent with real-world activities. For example, nasheed (hymns) in Tandhim al-Qa'ida fi Bilad al-Rafidayn videos were not accompanied by musical instruments, abiding by the strict stance on the use of such instruments in Salafi ideology.

In addition to being a central element of their psychological warfare, extremist groups in Iraq regard videos as an effective means for propagating their ideology. Jihadists, like other extremist groups, want to spread their movement to many places

Fig. 14.6 Video of bin-Laden with English subtitles

in the world (Corman and Schiefelbein 2006). To that end, they seek to legitimize the "Salafi Jihad Enterprise" as well as culturally identify with Muslims worldwide. The archetype is an elaborate audiovisual portrayal of the suffering inflicted by "Western occupiers" on Muslims in Iraq, followed by a selection of religious texts justifying violence. A majority of the videos referred to verses from the Koran justifying violence under certain circumstances. These observations have been reported previously by other researchers and columnists (Corman and Schiefelbein 2006).

Along the same lines, beheadings or gangster-style execution videos follow a common scenario. Whereby an extremist group member briefly introduces the action to follow, the hostage is then allowed to give his message and to answer questions, and finally, a verdict based on Salafi ideology is read and is immediately followed by the execution.

8 Conclusions

This chapter provided an exploratory analysis of 60 Arabic extremist groups' videos to identify the types of videos, groups' usage patterns including their modus operandi, and video production features. Forty-eight videos showed violent attacks; some included planning sessions with maps, diagrams, and logistical preparations. A matrix of jihadi groups' videos was proposed which classifies the videos based on usage dimensions: operational versus nonoperational and individual-oriented versus group-oriented activities.

By using videos, the jihadi extremist groups have at their disposal a potent media perfectly appropriate to our global communication avenues to spread extremist ideologies, radicalize sympathizers, recruit potential members, provide many learning opportunities, and tell/retell success stories. These galvanize an organizational saga that is then used to further legitimize the extremist groups' claims of victory. The organizational saga is emphasized when the viewers replay videos (reinforcement), store images and radical messages (e.g., usage of IEDs), hear expert commentary (suicide attack planning and execution), view interactions (social and emotional support with hugging) in the planning and execution (megacognitive event), and listen to devoted players in an operational environment (social event). This supports "borderless and informal" organizational learning and training because it becomes easier to acquire secondhand experience such as imitating violent attacks depicted in the documentary videos.

Although the intelligence, policy making, and research communities are monitoring, translating, and analyzing the videos for law enforcement investigation, troop training, planning, forecasting, and policy formulation, they must also exploit approaches for effective and efficient multimedia dissemination of credible counterarguments. Dissemination of credible counterarguments can help challenge the global diffusion of extremist ideology and offer alternative approaches. As described in the Critical Incident Analysis Group report (2007), the myths and disinformation propagated in the groups' videos can be countered by designing and disseminating credible messages from authentic and trusted sources at the grassroots level. The credible and multilingual messages should be in multimedia format (e.g., videos, games, audios) with graphic visuals to magnify the impact and be available via television, radio, and the Internet.

In addition to the counterarguments, the communities should further explore approaches for enhancing and strengthening people's understanding of diverse cultures and religions. Extremist groups' video production and dissemination have now anthromorphized into global, multicultural, virtual operations in which people collaboratively create, transform, duplicate, repackage, and distribute videos in various formats (e.g., wmv, vcd), sizes (e.g., 22 MB, 290 MB), languages (e.g., Arabic, English, French), and content options (e.g., downloadable file, streaming video). For an example, see the discussion of the "Jihad Academy" video.

The video content is also available via television, printed leaflets, cartoons, games, and audios. The ideas shared in the videos are viewed, interpreted, and acted upon by people from various cultural and religious backgrounds. Some may have limited diverse cultural and religious knowledge. This suggests that the communities must consider the use of cultural intelligence training and programs for enhancing people's capability to critically interpret and analyze violent messages generated by extremist groups.

A modest contribution of this exploratory study is the discernment of an informal division of labor and specialization among the extremist groups. The matrix we have developed in this study has helped disaggregate the groups, so their specializations (e.g., types of activities depicted in the videos, modus operandi, video production features) are made prominent. Clearly, the information content of these videos that are churned out in an exponentially increasing pattern does help reveal the various specializations that comprise the extremist groups. A chronoholistic approach to help further disaggregate the groups and their specialization is an advantage afforded by the videos.

The researchers, practitioners, and policy makers may perhaps get additional insight as to what counterterrorism strategies are effective and ineffective by understanding the groups' "growth" trajectories displayed in the videos. A clear understanding of the cultural intelligence at play within the various groups may help enhance our counterterrorism efforts. Video content analysis may provide us with important clues and information vital in recognizing how the extremist groups think, operate, and strategize. With funding, it should be expanded to include automatic extraction of structural (e.g., subtitles, images) and semantic content (e.g., weapons, target locations).

As with all research, this study has its limitations. Since it was limited to a sample of 60 Arabic video clips, future studies of this kind should endeavor to enlarge and broaden the sample and verify if similar results are found. Another limitation is the time span of 2 years. Further evidence-based research should also be done to provide additional insights into extremist groups' operations, organizational learning styles, and mobilization strategies.

Acknowledgments This research has been supported in part by NSF/ITR, "COPLINK Center for Intelligence and Security Informatics – A Crime Data Mining Approach to Developing Border Safe Research," EIA-0326348, September 2003–August 2006.

We would like to thank the staff of the Artificial Intelligence Lab at the University of Arizona who have contributed to this work, in particular Wei Xi, Homa Atabakhsh, Catherine Larson, Chun-Ju Tseng, and Shing Ka Wu. Finally, we also acknowledge the presence of native analysts on our team who, in addition to being proficient in translating Arabic text to English, are also able to place the content within the sociopolitical and cultural environmental context. The absence of this latter skill would have created serious gaps which could have derailed our analysis.

Appendix A List of Sample Videos

ID	File name	Time (m:s)	Title	Type
1	0405200501.rm	08:27	Abduction and execution of Jasim Mahdi	Beheading
2	01.(1).wmv	00:05	Hummer destruction in Taji	Documentary
3	0130200501.wmv	01:30	American embassy attack	Documentary
4	02.(1).wmv	00:49	Rocket attack on American base in Dayali	Documentary
5	0330200501.rm	00:15	Short clips	Documentary
6	03-320.wmv	00:55	Mortar attack on American base – Ad-Dalou'yah	Documentary
7	0416200502.wmv	00:14	2 IED attacks	Documentary
8	0504200501.wmv	03:20	Sniper attacks	Documentary
9	0504200502.wmv	04:35	The IED	Documentary
10	0504200503.wmv	08:25	Downing of a Bulgarian aircraft	Documentary
11	0513200501.wmv	11:05	Dedication to the pigs	Documentary
12	06-h.wmv	00:11	Double operation	Documentary
13	07-h.wmv	00:18	Attacking American base – Al-Khales	Documentary
14	08.wmv	00:43	IED's in Mushahada	Documentary
15	10.wmv	00:12	Road side bomb 1	Documentary

(continued)

ID	File name	Time (m:s)	Title	Type
16	11.wmv	00:58	Mortar attack on American base 1	Documentary
17	16.wmv	01:16	Attacking American base – Dyali	Documentary
18	18.wmv	00:18	Road side bomb 2	Documentary
19	abugraib.wmv	02:30	Variety of operation	Documentary
20	ahdath_fallujah.wmv	01:26	Supply line disruption	Documentary
21	almokawama3.wmv	00:36	Mortar attack on American base 2	Documentary
22	almokawama5.wmv	00:13	2 Attacks	Documentary
23	alsunnahcellphone.wmv	00:45	Attack on military vehicle	Documentary
24	amalyah.wmv	03:53	Mosul attack	Documentary
25	haifa-control.wmv	08:46	Haifa street battle excerpts	Documentary
26	ìê© ÿééáïâï¡.rm	01:45	Hummer Attack 1	Documentary
27	insurgent1.wmv	03:19	Collection of operations	Documentary
28	insurgent4.wmv	01:44	Gun battles	Documentary
29	Jihad Academy_ 93 Mb. rmvb	28:30	Jihad academy	Documentary
30	jonood.ram	01:08	Killing of seven American soldier	Documentary
31	labayk_fallujah.wmv	05:38	Labayk Fallujah	Documentary
32	mosul.wmv	36:46	Operations summary 1	Documentary
33	samara.(1).wmv	00:52	Attack on Iraqi interior minister vehicle	Documentary
34	shot2.wmv	00:58	Katibat Al-Ansar operation	Documentary
35	shot3.wmv	00:54	Mujahedeen operation in Abu-Ghraib	Documentary
36	taji_hit.rm	01:19	Taji rocket attack	Documentary
37	us_bomb_mosul.wmv	00:41	Mosul bomb	Documentary
38	us_humvee.wmv	00:50	Hummer attack 2	Documentary
39	walakinallahrama.wmv	40:48	Operations summary 2	Documentary
40	0330200502.wmv	05:01	Abducted and executed truck drivers	Hostage taking
41	0401200501.wmv	05:26	Capture and release of 16 workers	Hostage taking
42	0408200502.wmv	03:28	Execution of Iraqi policeman	Hostage taking
43	farisi.wmv	00:46	Iranian hostage	Hostage taking
44	muhafiz.rm	05:57	Muhafiz Abduction	Hostage taking
45	8-4-2005.rm	08:00	Jaish Islami Fi Al-Iraq Statement	Message
46	message-from-resistance. wmv	04:59	Communique NO. 6	Message
47	Movie.wmv	00:25	Usama clip	Message
48	obl09232004.mpeg	02:19	bin-Laden message	Message
49	rafidan4.wmv	05:54	Al-Jaish Al-Islami Fi Al-Iraq communique	Message
50	rafidan5.wmv	08:23	Dale C. Stoffel scandal 5	Message
51	serio0815.rm	03:56	Sawt Al-Khilafah (voice of the Caliphate)	Newsletter
52	1basha2er.wmv	44:11	Basha'r Al-Nassar	Propaganda
53	messages_fallujah.wmv	21:10	Fallujah volcano	Propaganda

(continued)

ID	File name	Time (m:s)	Title	Type
54	AlMuselmess.wmv	05:28	Abu Omar Al-Musli suicide attack	Suicide attack
55	hotel.wmv	02:50	Sodayr suicide attack	Suicide attack
56	suicidehand.wmv	02:51	Road side bomb – body parts	Suicide attack
57	voiture_kamikaze_1.wmv	04:06	First suicide attack British troops	Suicide attack
58	1rayat.wmv	31:38	Operations summary 3	Tribute
59	altawhiddocumentary.wmv	54:48	Sheikh Abu Anas Al-Shami (The Lion)	Tribute
60	shuhadas_alharamayn.wmv	06:47	Shuhada Alharamayn	Tribute

Appendix B Coding Scheme

Class	Content category	Content feature
General information	Title	Specify the reported title of the video
	Source	Specify the source web site/forum
	Batch number	Specify the AI batch number
	Type of video	Documentary, suicide attack, beheading, hostage taking, tribute, message, propaganda, instruction, training, newsletter
Date	Reported date	Specify the activity date
	Acquisition date	Date the video was obtained
Production	Language	Specify the language
	Special effects	Logo, English subtitles, Arabic subtitles, none
	Accompanying music/hymn	Hymn music, hymn + music, none
	Number of multiclips	Specify the number of multiclips
Extremist group	Extremist group name	Specify the reported group name
	Group media agency name	Specify the reported group media agency name
	Sub group name	Specify the reported group sub name
Expressions	Verbal	Religious verses, poetry, others
	Nonverbal	Kissing, hugging
	Reference to media	Arab media, Western media
Location	Country	Specify the event country
	City	Specify the reported city

(continued)

Class	Content category	Content feature
Event	Tactic	Suicide bomb, shooting, artillery attack, mortar attack, rocket/missile attack, knife attack, assassination, bombing, hijacking, hostage taking, vandalism, CBRN attack, threat, vehicle attack, aircraft downing, kidnapping, grenade attack, unclear/unknown
	Weapon	Improvised explosive devices (IEDs), mortar and rocket, bladed weapon, poison/biological agent, automobile/other vehicle, assault rifles, grenades, unclear/unknown
	Parts/stages	Threat and outcome, act and outcome, outcome, act being perpetrated, threat
Target	Victim type	Humanitarian/NGO, military personnel, religious figure, top government official, health care, diplomatic, civilian, businessman, government personnel, other, unknown
	Victim characteristics	Iranian, American, British, Iraqi, Canadian, French, Spanish, unknown
	Facility type	Military base, military vehicle, airports and airlines, business, government building, political party, paramilitary, transportation infrastructure, energy infrastructure, police facility, NGO, convoy, religious institutions, civilian vehicle, ship, hotel, unknown
	Facility characteristics	American, British, Iraqi, Bulgarian, unknown

References

Baran, David and Guidere, Mathieu. 2005. How to decode resistance propaganda Iraq: a message from the insurgents. Le Monde diplomatique, May.

Becker. A. 2005. Technology and terror: The new modus operandi. PBS, January 2005. [cited January 4 2006]. Available from http://www.pbs.org/wgbh/pages/frontline/shows/front/special/tech.html.

Chen, Hsinchun, et al. 2005. The Dark Web Portal project: collecting and analyzing the presence of terrorist groups on the Web. In Proceedings of the IEEE International Conference on Intelligence and Security Informatics (ISI 2005). Atlanta GA: Springer.

Coleman, Alistair. 2007. Iraq jihad TV' mocks coalition. BBC Monitoring, May 10, 2007. [cited August 10 2007]. Available from http://news.bbc.co.uk/2/hi/middle_east/6644103.stm.

Corman, Steven and Schiefelbein, Jill. 2006. Communication and media strategy in the jihadi war of ideas. In Consortium for Strategic Communication. Tempe AZ: Arizona State University. [cited April 12, 2006]. Available from http://www.asu.edu/clas/communication/about/csc/publications/jihad_comm_media.pdf.

Critical Incident Analysis Group (CIAG). 2007. NET worked radicalization: A counter-strategy. CIAG, Homeland Security Policy Institute. [cited July 19 2007]. Available from http://www.healthsystem.virginia.edu/internet/ciag/publications/NETworked-Radicalization_A-Counter-Strategy.pdf.

Duijvelaar, Christy. 1996. Beyond borders: East-east cooperation among NGOs in Central and Eastern Europe. Regional Environmental Center for Central and Eastern Europe. [Available December 2005]. Available from http://www.rec.org/REC/Publications/Beyond Borders/introSR.html.

GIMF releases a new video featuring a day in the life of a mujahid. 2006. SAIC, January 30, 2006. [cited August 4 2006]. Available from http://www.tecom.usmc.mil/caocl/OIF/IO_and_PA/insurgentacademy.doc. [Also cited January 23, 2006; available from http://trackingalqueda.blogspot.com/].

Hendron, John. 2006. Videos win support for Iraq insurgency, February 23, 2006. [cited August 4 2006]. Available from http://prairieweather.typepad.com/the_scribe/2006/03/22306npr_videos.html.

Holsti, Ole. 1969. Content analysis for the social sciences and humanities. Reading: Addison-Wesley.

IntelCenter. 2004. Al Qaeda videos and 3rd 9–11 anniversary v.1.0. Alexandria, VA: IntelCenter.

IntelCenter. 2005. Evolution of jihad video, v 1.0. Alexandria, VA: IntelCenter. [cited May 10 2006]. Available from http://www.intelcenter.com/EJV-PUB-v1–0.pdf.

IntelFiles, 2005. [cited August 10 2007]. Available from http://intelfiles.egoplex.com/.

International Crisis Group. 2006. In their own words: Reading the Iraqi insurgency. In Crisis Group Reports #50. Brussels, Belgium: International Crisis Group. [cited March 0 2006]. Available from http://www.crisisgroup.org/home/index.cfm?id=3953.

Islamic Imagery Project. 2006. West Point, N.Y.: Combating Terrorism Center. [cited December 12 2006]. Available from http://www.ctc.usma.edu/imagery.asp.

News from Russia. Terrorists threaten more attacks on Britain 2006. News from Russia, September 2, 2005 [cited August 4 2006]. Available from http://newsfromrussia.com/.

Paz, Reuven. 2006. Reading their lips: The credibility of jihadi Web sites in Arabic as a source for information. PRISM 2006 [cited August 4 2006]. Available from http://www.e-prism.org/images/Read_Their_Lips.doc.

Reid, Edna, et al. 2005. Collecting and analyzing the presence of terrorists on the Web: A case study of jihad Websites. In Proceedings of the IEEE International Conference on Intelligence and Security Informatics (ISI 2005). Atlanta GA: Springer.

Robertson, N. 2006. Tapes shed new light on bin laden's network. CNN.com August 19, 2002. [cited August 4 2006]. Available from http://archives.cnn.com/2002/US/08/18/terror.tape.main/.

Rocket propelled grenade. 2006. [cited August 4 2006]. Available from http://www.militaryfactory.com/smallarms/detail.asp?smallarms_id=10.

Salem, Arab, Reid, Edna, and Chen, Hsinchun. 2006. Content analysis of jihadi extremist groups' videos. In Proceedings of the IEEE International Conference on Intelligence and Security Informatics (ISI 2006). San Diego, CA: Springer. [cited June 12 2006]. Available from http://ai.arizona.edu/research/terror/publications/isi_content_analysis_jihadi.pd (This paper - Multimedia Content Coding and Analysis is an expanded version of the ISI 2006 conference paper).

SITE Institute. 2004. Manufacture of the Explosive Belt for Suicide Operations, December 22 2004 [cited August 4 2006]. Available from http://siteinstitute.org/.

Tracking Al Qaeda. 2006. [cited January 23, 2006]. Available from http://trackingalqueda.blogspot.com/.

Weimann, Gabriel. 2006. Terror on the Internet: The new arena, the new challenges. Washington, D.C.: U.S. Institute of Peace.

Zhou, Yilu et al. 2006. Exploring the dark side of the Web: Collection and analysis of U.S. extremist online forums. In Proceedings of the IEEE International Conference on Intelligence and Security Informatics (ISI 2006). San Diego, CA: Springer.

Chapter 15
Extremist YouTube Videos

1 Introduction

With the emergence of Web 2.0, Web 2.0 sites such as forums, blogs, video-sharing web sites, and wikis have become more and more popular in the past few years. User behavior in Web 2.0 communities has changed from just browsing web pages to generating and spreading their own content and ideas. As noted by O'Reilly (2005), one of the main features of Web 2.0 is the "architecture of participation," which refers to online content that is generated by those who are motivated by their own personal interests. Numerous user-generated data provide valuable personal and up-to-date information, such as user preferences, sentiments, and opinions, which previously could be obtained only through surveys and interviews. Collecting and analyzing this considerable quantity of user-generated information (for research) is a challenge. Classification technologies provide promising methods to organize data according to different perspectives. Many studies have used classification technologies to analyze text-based data collected from blogs and forums and obtain insights. For example, Abbasi et al. (2008b) applied sentiment analysis to improve opinion classification of web forums in multiple languages. Zheng et al. (2006) adopted writing style features to identify online authorship.

Like blogs and forums, video-sharing web sites are an important part of Web 2.0. For example, YouTube, the world's largest video-sharing web site, receives more than 65,000 videos and 100 million video views every day. Video classification techniques can be used to improve user experiences with video-sharing web sites by identifying videos more closely related to users' personal interests and distinguishing them from the many irrelevant videos that are obtained by using keyword searches alone.

Another challenging issue for Web 2.0 sites is the issue of illegal content such as child pornography or threatening content such as sites exhorting violence and extremism. Among these, violent extremism content is considered to be among the most dangerous, especially after the attack of September 11. The US government invests many resources in detecting potential terrorism and protecting the USA from extremist violence. Chen et al. (2008) found that extremists use Web 2.0 as an effective

H. Chen, *Dark Web: Exploring and Data Mining the Dark Side of the Web*,
Integrated Series in Information Systems 30, DOI 10.1007/978-1-4614-1557-2_15,

platform to share resources, promote their ideas, and communicate among each other. For now, YouTube provides only the "flag" mechanism for users to mark inappropriate videos (Chen et al. 2008). Video classification can help video-sharing web sites manage videos automatically by classifying illegal or offensive videos and distinguishing them from acceptable ones.

Moreover, accurate video classification results are very useful for identifying implicit cyber communities on video-sharing web sites (Kumar et al. 1999). Implicit cyber communities can only be defined by the interactions among users, such as subscription, linking, or commenting. Chau and Xu (2007) studied implicit cyber communities for blogs, while Fu et al. (2008) used interaction coherence information to identify user communities for web forums. However, few studies have addressed the cyber communities on video-sharing web sites.

Different from the studies of forums and blogs which used text features to represent collected data, most studies in video analysis have used nontext features extracted from video clips and audio tracks (e.g., Messina et al. 2006). However, video-sharing communities not only allow users to upload and share videos but also provide functions to enable users to interact with other users, which generate additional text information. For instance, YouTube allows its users to comment on and rate videos, create personal video collections, and categorize and tag videos they upload. Such user-generated text information may contain explicit information related to video content and hence can be used to classify videos. In addition, this information can be easily obtained by parsing web pages or using various web APIs (Chen et al. 2008). But for now, few studies have explored user-generated text features in video classification.

In order to make use of the information provided by user-generated data and evaluate their effectiveness in online video classification, we propose a framework of video classification for video-sharing web sites by using user-generated text data such as comments, descriptions, video titles, etc. We evaluated the performance of different classification techniques and text feature sets. In addition, we conducted key feature analysis to identify the most useful user-generated data for online video classification and showed how our framework can help identify implicit cyber communities on video-sharing web sites.

The remainder of this chapter is organized as follows: Sect. 2 presents a review of current video classification research. Section 3 describes research gaps and questions, while Sect. 4 shows our research design. The test bed created and used in our experiment is discussed in Sect. 5. Experiments used to evaluate the effectiveness of the proposed approach and discussions of the results are illustrated in Sect. 6. A case study showing how the proposed framework can help identify implicit cyber extremist communities on video-sharing web sites is presented in Sect. 7. Section 8 concludes with closing remarks and future directions.

2 Literature Review

Among all data types, such as text, audio, and image, video has the highest capacity in terms of the volume and richness of the content. Videos not only contain diverse data types, i.e., image, audio, and text data, but also combine these data types

Table 15.1 Taxonomy of video classification studies

Category	Description	Label	
Domains			
General TV program	Sports games, news, weather reports, commercials	D1	
Movie and movie preview	Movies, movie preview videos	D2	
Specific scenario video	Staff meeting videos	D3	
Archival video	Videos of TRECVID, Internet Archive, or open video	D4	
Video-sharing Web site video	YouTube, MySpace, and Flicker videos	D5	
Feature types			
Nontext features	Low-level video features	Nontext features extracted from row clips, such as color, motion, and texture features	NT-L
	Mid-level/high-level video features	Semantic features, such as face, object, and anchor detection	NT-MH
Text features	Video-embedded text features	Text features from video-embedded information, such as subtitles and closed-caption	T-E
	User-generated text features	Text features from user-generated information, such as video titles, descriptions, tags, and category names	T-U
Techniques			
Machine learning techniques for classification	Hidden Markov models (HMM), Gaussian mixture model (GMM), and support vector machine (SVM)	T1	

together and further create deeper semantic meanings. These semantic meanings and information can be easily recognized by human beings, but how to leverage information technologies to process videos and extract these semantic meanings automatically is a challenging issue.

The semantic gap, referring to the gap between video features (e.g., color, texture, and volume of audio) and semantic concepts (Lew et al. 2006) (i.e., concepts meaningful to human beings, such as cars, faces, buildings, etc.), is one of the most challenging issues of video classification studies. To bridge the semantic gap and obtain a better understanding of video content, many different techniques have been developed to enhance classification accuracy (which refers to the percentage of correctly classified instances), and different video features have been identified to represent videos better (Dimitrova et al. 2000; Hsu and Chang 2005; Hung et al. 2007). Common video classification research characteristics include domains, feature types, and classification techniques. Table 15.1 shows the taxonomy of these important video classification analysis characteristics. The taxonomy and related studies are discussed in detail below.

2.1 Video Domains

There are five main categories of video domains: general TV programs, movies and movie previews, specific scenario videos, archival videos, and video-sharing web site videos. In addition to the basic components of videos, i.e., image and audio, videos within some domains provide extra information which can be utilized to classify videos more accurately. For example, some general TV programs contain subtitles and closed-captioning which can be extracted to help understand the video content.

TV programs are the most traditional video resources, and therefore, most studies have used general TV programs as their experiment data. Some studies in this domain have classified TV programs according to program types, such as news, sports games, weather reports, and commercial advertisements. Montagnuolo and Messina (2007) classified 700 broadcasted programs into seven TV program types and reported 86.2% average precision (which refers to the average percentage of correctly classified instances, which are programs in this case, across all predicted classes). Other studies focused on classifying a single type of TV programs into different specific events. For example, Hung et al. (2007) classified baseball videos into several important events, such as home run, hit, strikeout, etc., and achieved 95% average precision.

The second domain is movies and movie previews. Movies play an important role in the entertainment industry. Approximately 4,500 films (about 9,000 h of video) are produced every year (Rasheed et al. 2003). Hence, some studies have focused on classifying movies according to their genres. Vasconcelos and Lippman (2000) classified movies into three categories: romance/comedy, action, and others, including horror, drama, and adventure. In addition to movies, movie previews, previews of upcoming movies, or previews provided by DVD rental companies have also been used as test beds in previous studies. For instance, Rasheed et al. (2003) classified movie previews into different categories such as comedies, action films, dramas, and horror films.

Specific scenario videos are videos generated by individuals for specific events, such as meeting or lecture videos. For example, business meetings captured in video were used by Girgensohn and Foote (1999) to classify them into presenter, slides, and audience scenes.

Archival videos are generally collected and provided by organizations (e.g., Internet Archive). These videos are collected from various media sources (e.g., movies, TV programs, and personally made videos) and well organized into different categories (e.g., cartoons, movies, news, etc.) before being provided to the public. Some organizations will provide preprocessed videos for researchers to use for experiments and studies. For example, TRECVID was founded in 2003 and is now a well-known workshop that provides large testing datasets scored by uniform scoring procedures for video information retrieval studies.

As Web 2.0 gains in popularity, the study of video-sharing web sites has become an emerging domain of interest. Videos are uploaded by online users and reviewed by the public. Video-sharing web sites, such as YouTube and Yahoo! Video, usually provide a convenient environment for users to discuss and comment on videos. Some sites even provide APIs that allow people to easily extract relevant information from videos of interest.

Consequently, videos from video-sharing web sites contain several unique characteristics. First, most online videos are short, and their contents are highly diverse. Second, much user-generated data, such as descriptions and comments, can be collected easily for each online video. Information about video authors and reviewers is sometimes available, including other videos uploaded by the same person, etc. Third, due to copyright issues, online videos on video-sharing sites may not be always available for people to download and analyze. Hence, difficulties in collecting training datasets may be encountered when applying features extracted from videos to online video classification. User-generated data can be used as an alternative because the data can be easily obtained and may contain more explicit information about the content of associated videos. Currently, few studies have emphasized this kind of approach.

2.2 *Feature Types*

Features used in video classification studies can be divided into two main categories, nontext features and text features. While nontext features can be further split into low-level video features and semantic video features, text features contain text features extracted from videos and user-generated text features.

2.2.1 Nontext Features

Nontext features are features extracted from the two basic components of videos: audio and image. Djeraba (2002) stated that low-level video features are features extracted from the video clips and audio tracks without reference to any external knowledge. For example, color, texture, and motion are major low-level features extracted from video clips (Gibert et al. 2003; Huang et al. 1999; Ma and Zhang 2003). Fischer et al. (1995) utilized audio features such as volume of audio, audio wave forms, and audio frequency spectrum. Other features such as edge, lighting, and shot length were also adopted in some studies (e.g., Rasheed et al. 2003). Moreover, the text trajectory feature, which refers to the motion of texts in continuous video clips, is considered to be a low-level nontext feature as opposed to a text feature (e.g., Dimitrova et al. 2000).

Zhou et al. (2000) and Luo and Boutell (2005) claimed that low-level video features lack the capacity to identify semantic concepts, which make them inefficient to use for video classification alone. To solve this problem, mid-level and high-level video features are proposed to bridge the "semantic gap" (Lew et al. 2006), the gap between low-level video features and semantic concepts (Hsu and Chang 2005). These two feature types are generated from low-level features and are also known as semantic features.

Mid-level video features are extracted by mid-level feature detectors or sensors, which are pretrained classifiers used to capture mid-level features from input data, and each of them represents an atomic semantic concept, which cannot be represented by combinations of other semantic concepts. Some examples used in previous studies include cityscape, landscape, face, object, indoor, and outdoor (Chellappa et al. 1995; Lin and Hauptmann 2002; Samal and Iyengar 1992). Xu and Chang (2008) developed 374 mid-level feature detectors to detect video events. The average precision for event detection was between 24.4% and 38.2%. Mid-level features have been adopted in many studies. Dimitrova et al. (2000) used text and face trajectories to classify videos into four categories (i.e., news, commercials, sitcoms, and soaps) and reported 80% accuracy.

High-level video features are features containing multiple semantic concepts, which generally require a human to define (Borgne et al. 2007). Some studies relied on domain knowledge to achieve high-level analysis. Duan et al. (2003) combined sport domain knowledge with mid-level features to conduct high-level video analysis to categorize segments of videos of fieldball sports into different events. They constructed several mid-level feature detectors to capture semantic shots (e.g., field view, audience, goal view, player following, and replay) from videos of soccer games. With the help of sport domain knowledge, they first defined "in play segments," video segments consisting of shots taken when a game is playing (e.g., field views and player following), and "out of play segments," video segments containing shots taken when a game has been stopped by a referee (e.g., audience and replay). Further, specific events of each segment were identified. For example, kickoffs, passing, and shots were captured from "in play segments," while penalty kicks, throw-ins, and corner kicks were identified from "out of play segments."

2.2.2 Text Features

In addition to nontext features, some studies adopted text features to enhance the classification performances. Subtitles and closed-captions are the typical text information that can be extracted from videos of various types, such as TV programs and movies. Lin and Hauptmann (2002) extracted closed-captions from CNN broadcasts and treated each word of the closed-captions as a feature. Bag-of-words was used to represent the broadcasts, and their experiment results demonstrated that text features can improve the precision of classification results.

In addition, text information can also be obtained from audio tracks using speech recognition techniques (Smoliar and HongJiang 1994). For example, Amir et al. (2004) transcribed audio recordings, generated a continuous stream of timed words, and included the text information for video event detection.

User-generated text information is a new text data source emerging only recently with video-sharing web site videos. Different from the other four video classification domains, online videos are generally shorter but contain more user-generated information. In the Web 2.0 architecture of participation, online users not only review videos but also comment on videos and exchange opinions with other reviewers.

Through the user-participation process, much video-related text data are created. These data often contain explicit information about the video content and can be utilized to classify videos. In addition, more and more user-generated text information can be easily collected from video-sharing web sites. For example, the YouTube API allows users to extract information such as titles, user comments, descriptions, tags, etc. (Chen et al. 2008).

Sharma and Elidrisi (2008) recently used video tag information to classify YouTube videos into predefined YouTube categories such as education and comedy. They claimed that video tags are given by users and therefore contain highly user-centric information and can be used as the metadata of videos. Their results achieved approximately 65% accuracy. To the best of our knowledge, user-generated text information has not been used in other video-sharing web site video classification studies.

Four types of text features, lexical, syntactic, structural, and content-specific features, have often been used in previous text classification tasks. These four types of text features can be categorized into two broad categories: content-free features and content-specific features. Content-free features are features independent of the topics or domains of the text data and hence can be regarded as generic features. They include lexical features, syntactic features, and structural features (Zheng et al. 2006). Lexical features are used to capture lexical variations of an article in both character-level and word-level (e.g., the average word length and the total number of characters) (Argamon et al. 2003; Zheng et al. 2006). Syntactic features show syntactical patterns of sentences (Hirst and Feiguina 2007; Koppel et al. 2009). These patterns can be captured by identifying function words or punctuation within sentences. Structural features represent user habits of organizing an article (e.g., paragraph length and use of signature), which have been shown especially useful for online text (Abbasi and Chen 2005). These features can be used to identify writing styles of different authors. Content-specific features, on the other hand, are features that can be used to represent specific topics. For example, baseball videos can be easily classified into different baseball events by identifying informative content-specific keywords, such as "home run," "double play," "strikeout," and "hits." Content-specific features can be either manually selected (Zheng et al. 2006) or n-gram features extracted automatically from the collection (Peng et al. 2003; Abbasi and Chen 2008). Most of these text features can be considered for video classification on video-sharing sites based on user-generated content.

2.3 *Classification Techniques*

Based on our literature review, machine learning dominated the classification techniques of previous video classification studies. Among these techniques, hidden Markov model (HMM), Gaussian mixture model (GMM), and support vector machine (SVM) were the most adopted (Guironnet et al. 2005; Lu et al. 2001; Montagnuolo and Messina 2007; Zhou et al. 2005).

HMM is a popular technique widely used in pattern recognition (Rabiner and Juang 1986). The basic idea of HMM is to construct a model with hidden state variables which can be used to explain the observable variables in a sequential data. The model then can be applied to other sequential data for prediction tasks. Some researchers have applied HMM for video analysis and classification. Dimitrova et al. (2000) proposed to use HMM along with text and face features for video classification. Huang et al. (1999) presented four different methods for integrating audio and visual information for video classification based on HMMs. Gibert et al. (2003) used an HMM-based approach to classify sport videos into predefined genres using motion and color features. Eickeler and Muller (1999) classified TV broadcast news by using HMMs.

GMM can be used to model a large number of statistical distributions, including nonsymmetrical distributions. Given feature data, a class can be modeled with a multidimensional Gaussian distribution. In image processing applications, researchers used both unsupervised (Caillol et al. 1997; Pieczynski et al. 2000) and supervised versions (Oliveira de Melo et al. 2003) of mixture models. For example, Xu and Chang (2008) adopted GMM to classify TV broadcast programs. Girgensohn and Foote (1999) used a GMM classifier to classify staff meeting videos into different shot categories (slides, audiences, and presenters).

SVM has been shown to be a powerful statistical machine learning technique (Vapnik 1998). The basic idea of SVM is to find a linear decision boundary to separate instances of two classes within a space. While there are multiple linear decision boundaries existing in the space, SVM will select the one with the largest margin, which is the total distance between a decision boundary and the closest instances on each side. Ideally, while new instances are added into the space, a lower classification error is suggested by a larger margin. SVM has two characteristics which make it efficient for classification tasks. First, prior knowledge is not required for it to obtain a high generalization performance, and it can perform consistently with very high input dimensions. Second, SVM is not only efficient for solving classification problems with small samples but also can obtain a global optimal solution (Ma and Zhang 2003). In addition, SVM has shown excellent video classification performance (Jing et al. 2004; Lazebnik et al. 2006; Zhang et al. 2007). For example, Zhou et al. (2005) used SVM to classify soccer videos into different scenes (long shot, medium shot, or others) and reported over 92% average precision. Lin and Hauptmann (2002) applied SVM-based multimodal classifiers and probability-based strategies to continuous broadcast videos and classified them into news and weather report categories. The results showed that the precision of SVM-based multimodal classifiers was up to 1 and significantly better than probability-based strategies.

3 Research Gaps and Research Questions

Table 15.2 shows selected major studies in video classification, and some general conclusions can be drawn from it. For video domains, most video classification studies focused on TV program videos (D1), while few studies paid attention to the

Table 15.2 Selected major studies of video classification

	Domains					Feature types				Techniques
Previous studies	D1	D2	D3	D4	D5	NT-L	NT-MH	T-E	T-U	T1
Huang et al. (1999)	√					√				HMM
Girgensohn and Foote (1999)			√			√				GMM
Zhou et al. (2000)	√					√				Rule-based classifier
Dimitrova et al. (2000)	√						√			HMM
Lu et al. (2001)	√					√				HMM
Pan and Faloutsos (2002)	√					√				Vcube
Lin and Hauptmann (2002	√					√		√		SVM
Ma and Zhang (2003)	√					√				SVM, KNN
Rasheed et al. (2003)		√				√				Mean-shift classification
Gibert et al. (2003)	√					√				HMM
Duan et al. (2003)	√					√	√			C-support vector
Xu and Li (2003)	√					√				GMM
Hsu and Chang (2005)				√		√	√			SVM
Luo and Boutell (2005)				√		√	√			SVM and Bayesian network
Messina et al. (2006)	√					√				Fuzzy C-means
Hung et al. (2007)	√					√	√			Bayesian belief network
Xu and Chang (2008)	√					√	√			SVM
Sharma and Elidrisi (2008)					√				√	M5P Trees

D1 general TV program, *D2* movie and movie preview, *D3* specific scenario video, *D4* archival video, *D5* video-sharing web site video, *NT-L* low-level video features (a subcategory of nontext features), *NT-NH* mid-level/high-level video features (a subcategory of nontext features), *T-E* video-embedded text features (a subcategory of text features), *T-U* user-generated text features (a subcategory of text features), *T1* machine learning techniques for classification

other four domains, which are movies and movie previews (D2), specific scenario videos (D3), archival videos (D4), and video-sharing web site videos (D5). In terms of feature types, nontext features (i.e., low-level video features (NT-L) and mid-level/high-level video features (NT-MH)) were adopted by the majority of previous studies. Text features, video-embedded text features (T-E) and user-generated text features (T-U), were rarely explored. As for classification techniques, machine learning classification techniques (T1) dominate the area. Among various machine learning classification techniques, SVM, GMM, and HMM were the most used.

Based on our review of previous literature and conclusions, we have identified several important research gaps. First, with the emergence of Web 2.0, online videos from video-sharing web sites (D5) surprisingly have seldom been addressed.

Second, to the best of our knowledge, the work of Sharma and Elidrisi (2008) is the only research that used user-generated information for online video classification. However, their classifier was designed for YouTube predefined categories only, and the performance was not high. Third, among various user-generated text information, only video tags have been used (Sharma and Elidrisi 2008). Geisler and S. Burns (2007) showed that the majority of YouTube tag terms can provide additional information about videos. Ding et al. (2009) also showed that YouTube taggers like to identify specific information such as date, geographical locations, scientific domains, religions, and opinion terms for videos. We believe other user-generated text information, such as video descriptions and comments, are also useful in video classification and can help address the video semantic gap problem.

To address the research gaps mentioned above, this chapter proposes a text-based video content classification framework for online video-sharing sites. The proposed framework can be used to identify videos for any topic or user interest. It aims to answer the following research questions:

- Are user-generated text features useful for online video classification?
- What user-generated text data and feature sets are most effective for online video classification?
- Can feature selection improve the video classification results?
- Which text classification technique is best for online video classification?
- Can accurate video classification results help identify cyber communities on video-sharing sites?

4 System Design

Figure 15.1 illustrates our proposed system design. Our design consists of three major steps: data collection, feature generation, and classification and evaluation.

4.1 Data Collection

The data collection process is designed to identify candidate videos for the classification task and collect associated user-generated text data. The input of our system is a set of selected keywords that represent users' preferences and interests. The keywords are used to identify candidate videos. Various types of user-generated text information, including video titles, comments, video descriptions, etc., are then collected for those videos and stored in a database. Finally, users who generated the keywords are asked to create video categories based on their preferences, and a subset of the collection is randomly selected and manually classified into those categories by the users. The classification results are split into training and testing datasets which will be used later for building and evaluating classifiers, respectively.

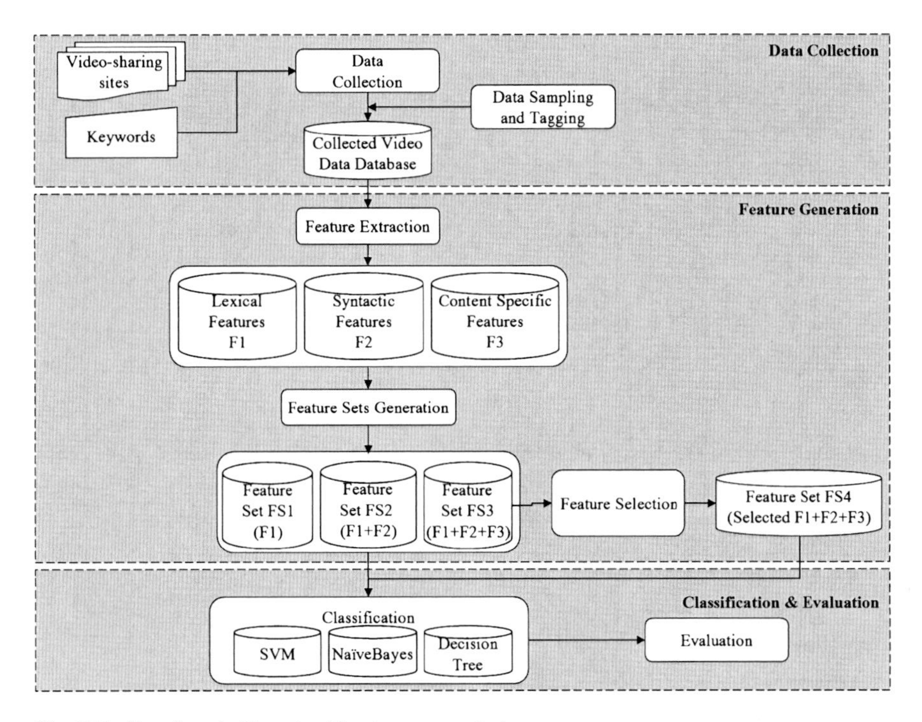

Fig. 15.1 Text-based video classification system design

4.2 *Feature Generation*

The feature generation process aims to generate text features from the collected text data that can best represent candidate videos. Three types of text features, i.e., lexical features, syntactic features, and content-specific features, are adopted in our system and denoted as F1, F2, and F3, respectively. These features have been considered in various text classification research (Zheng et al. 2006; Abbasi and Chen 2008; Abbasi et al. 2008), but rarely in video classification studies. Several feature sets are constructed by combining different feature types and applying feature selection techniques.

4.2.1 Feature Extraction

In this chapter, we examined three types of features: lexical features, syntactic features, and content-specific features. Structure features are not considered because such features (e.g., font size, font color, greetings, etc.) are not present in video text.

Lexical features consist of character-based and word-based features (Zheng et al. 2006) and have been widely used in previous authorship research. For example, different character-based lexical features were adopted in the studies of de Vel (2000), Forsyth and Holmes (1996), and Ledger and Merriam (1994). Mendenhall (1887) and de Vel (2000) utilized some word-length frequency features in their study. In this chapter, we adopted 82 lexical features, including both character-based and word-based lexical features used in previous studies.

As suggested by Zheng et al. (2006), syntactic features, which include function words and punctuation words, are often used to identify styles of articles in the sentence level. Several sets of function words have been proposed in previous research (Baayen et al. 1996; Tweedie and Baayen 1998). We adopted the set of 149 function words used in Zheng et al. (2006) because of its coverage. In addition, eight punctuation words suggested by Baayen et al. (1996) were used in our syntactic feature set.

Content-specific features are relevant to specific application domains and are important for online video classification. In this study, we adopted word-level, character-level, and POS tag unigrams, bigrams, and trigrams. In addition to n-gram features, specific user-provided video tags and video categories were included as binary features.

The complete feature list used in our study is shown in Table 15.3. We believe our study is one of the few to examine comprehensive text features for video classification on video-sharing sites.

4.2.2 Feature Set Generation

This step aims to generate feature sets by combining different types of features. These feature sets are then evaluated in the classification and evaluation process. In this chapter, we adopted an incremental strategy to generate feature sets. Three feature sets were first created. The first feature set contains lexical features only (FS1). Lexical features and syntactic features are combined to generate the second feature set (FS2). The third set is constructed by combining lexical, syntactic, and content-specific features (FS3). This incremental approach has been frequently adopted in previous authorship studies (Abbasi and Chen 2005; Zheng et al. 2006) as it includes increasingly more complex and topic-relevant feature groups. Through this approach, we can obtain better insights into the effects of adding new feature sets to the previous ones.

For content-free features (F1 and F2), the total number of features is predefined as is shown in Table 15.3. However, for n-gram-based content-specific features (F3), the feature size varies and is usually much larger than the number of content-free features. An effective approach to reduce the number of such features is to adopt a minimum frequency threshold (Mitra et al. 1997; Jiang et al. 2004). We set the minimum frequency as 10 for n-gram-based parameters by following the setting adopted in Abbasi et al. (2008).

Table 15.3 Text features adopted in this research

Features	Descriptions	Feature counts
Lexical features (F1)		
Character-based features		
1. Total number of characters		1
2. Total number of alphabetic characters		1
3. Total number of uppercase characters		1
4. Total number of digit characters		1
5. Total number of white-space characters		1
6. Total number of tab spaces		1
7–32. Frequency of letters		26
33–53. Frequency of special characters		21
Word-based features		
54. Total number of words		1
55. Total number of short words	Words less than four characters	1
56. Total number of characters in words		1
57. Average word length		1
58. Average sentence length in terms of word		1
59. Average sentence length in terms of character		1
60. Total number of different words		1
61. Hapax legomena	Frequency of once occurring words	1
62. Hapax dislegomena	Frequency of twice occurring words	1
63–82. Word length frequency distribution	Frequency of words in length of 1–20	20
Syntactic features (F2)		
83–90. Frequency of punctuation		8
91-239. Frequency of function words		149
Content-specific features (F3)		
POS tag n-grams	Unigram, bigrams, and trigrams	Various
Character-level n-grams	Unigram, bigrams, and trigrams	Various
Word-level n-grams	Unigram, bigrams, and trigrams	Various
Video tags	Binary features	Various
Video categories	Binary features	Various

4.2.3 Feature Selection

Feature selection techniques have been shown to be effective in improving classification performances by removing irrelevant or redundant features in a large feature set. Duan et al. (2003) used feature selection to identify discriminating audio signals, while Borgne et al. (2007) adopted feature selection to reduce the number of image features. Hundreds of thousands, or even more, of online videos are generated every day, and the efficiency of classifiers is therefore an extremely important consideration. By taking advantage of feature selection, we can expect to identify a small set of features which can not only perform as good as or even better than the whole feature set but also minimize the time needed to perform classification. In order to evaluate how feature selection can improve the performance of online video classification, the fourth feature set (FS4) was built by applying feature selection to FS3.

The information gain (IG) heuristic was adopted to perform feature selection. It has been shown to be an efficient feature selection method used in many text categorization studies (e.g., Abbasi and Chen 2005; Koppel and Schler 2003; Yang and Pedersen 1997). In this chapter, we used the Shannon entropy measure (Shannon 1948), in which

$$IG(C,F) = H(C) - H(C \mid F),$$

where

$IG(C,F)$ information gain for feature F,

$H(C) = -\sum_{i=1}^{n} p(C=i)\log_2 p(C=i)$ entropy across video category C,

$H(C \mid F) = -\sum_{i=1}^{n} p(C=i \mid F)\log_2 p(C=i \mid F)$ specific feature conditional entropy,

n total number of video categories.

If videos are classified into two categories in the data collection process and the numbers of videos in the two categories are the same, $H(C)$ is 1. Then a specific feature conditional entropy $H(C|F)$ is calculated for each feature F. If videos containing feature F are all in the same category, $H(C|F)$ is 0, and $IG(C, F)$ is 1. However, if the numbers of videos containing feature F are the same, $H(C|F)$ is 1, and $IG(C,F)$ is 0. Therefore, $IG(C,F)$ is between 0 and 1, and features with higher IG scores have better abilities to distinguish videos in different categories. All features with IG greater than 0 are selected.

4.3 *Classification and Evaluation*

As mentioned in the literature review, HMM, GMM, and SVM were the most adopted classification techniques in the previous video classification studies using nontext features. However, in our framework, user-generated text data were used as proxies of videos. Therefore, even though these three techniques can also be applied to text features, we prefer to choose classification techniques which have been frequently used in text analysis. Three state-of-the-art classification techniques in text analysis studies (e.g., Das and Chen 2007; Zheng et al. 2006) were used to construct video classifiers (i.e., SVM, C4.5, and Naïve Bayes), and their performances were also compared. SVM, widely adopted in text classification studies, is a powerful statistical machine learning technique first introduced by Vapnik (1995). Due to its ability to handle millions of inputs and its good performance, SVM was used in previous authorship analysis studies (e.g., de Vel 2000; Diederich et al. 2000). In addition, some studies have shown the excellent performance of SVM in video classification (Jing et al. 2004; Lazebnik et al. 2006). ID3 is a symbolic learning algorithm which has been extensively tested and shown its ability to compete with other machine learning techniques in predictive power (Chen et al. 1998; Dietterich et al. 1990). C4.5, an extension of ID3, is a decision-tree-building algorithm developed by Quinlan (1986). Based on a divide-and-conquer strategy and the entropy measure, C4.5 focuses on classifying mixed objects into categories according to attribute values of objects. Based on Bayes' theorem with strong independence assumptions, the Naïve Bayes classifier is a probabilistic classifier and uses the feature values of a new instance to estimate the probability of each category. It has also been used to perform text classification tasks in previous studies (Lewis 1998; McCallum and Nigam 1998; Sahami 1996). Tenfold cross-validation was used to evaluate all classifiers.

To evaluate the prediction performance, accuracy is adopted to evaluate the overall classification correctness of each classification technique (Abbasi et al. 2008). We use the average classification accuracy across all tenfolds as shown below:

$$\text{Accuracy} = \frac{\text{Number of correctly classified videos}}{\text{Total number of videos}} \tag{15.1}$$

5 Test Bed and Hypotheses

5.1 Test Bed

To evaluate our video classification framework, we chose YouTube as our data source. YouTube is the world's largest video-sharing web site. It provides robust APIs for searching videos and downloading user-generated text information about these videos. In this chapter, we collected the following seven types of user-generated data for each video: descriptions, titles, author names, names of other videos uploaded by the video author (AuthorVideoName), comments, categories, and tags. The difference between tags and categories is that tags are given by authors of videos and could be any term, while categories are predefined by YouTube and selected by authors when uploading a video. We found these seven data types to be the most content rich and carefully populated by the video authors.

The proposed framework can be used to identify videos for any topic or user interest. In this chapter, we aimed to identify extremist videos on YouTube. Many previous studies have demonstrated the need to identify illegal, extreme, or violent content on the Internet (Burris et al. 2000; Schafer 2002). Chen et al. (2008) showed that Web 2.0 has become an effective grassroots communication platform for extremists to promote their ideas, share resources, and communicate with each other. Extremist videos, such as suicide bombing, attacks, and other violent acts, can often be found on YouTube. Therefore, automatically identifying online extremist videos has become a major research challenge for Web 2.0 (Chen et al. 2008; Salem et al. 2008).

Our test bed was created by using 78 extremism-related English keywords selected by extremism study experts, who have studied extremist groups for many years and who have previously participated in government-funded and other research projects, to search for videos on YouTube. These keywords represent major topics, ideas, and issues of interest to many domestic and international extremist groups. In total, user-generated metadata for 31,265 potentially relevant videos were collected. These videos included query-related videos (videos directly retrieved from YouTube using given keywords), related videos (videos related to the query videos), and author-uploaded videos (videos uploaded by the authors of query-related videos).

To evaluate our video classification framework, 900 videos were randomly selected and tagged by extremism study experts as extremist or nonextremist videos. Among these 900 videos, 224 videos were tagged as extremist videos and 676 as nonextremist videos. In this chapter, we included 224 extremist videos and 224 randomly selected nonextremist videos as our test bed.

5.2 Hypotheses

We developed the following hypotheses to examine the performances of different feature sets and classification techniques for video classification:

H1: By progressively adding more advanced and content-rich feature sets and applying feature selection, video classification performances can be improved:

- H1.1: A combination of lexical and syntactic features outperforms lexical features alone in video classification, i.e., FS2 (F1 + F2) > FS1 (F1).
- H1.2: A combination of content-free and content-specific (lexical and syntactic) features outperforms the combination of content-free features alone in video classification, i.e., FS3 (F1 + F2 + F3) > FS2 (F1 + F2).
- H1.3: Applying feature selection on all feature sets improves online video classification, i.e., FS4 (Selected F1 + F2 + F3) > FS3 (F1 + F2 + F3).

H2: By using user-generated text data, SVM outperforms other classification techniques in video classification:

- H2.1: SVM outperforms C4.5 in video classification by using user-generated text data, i.e., SVM > C4.5.
- H2.2: SVM outperforms Naïve Bayes in video classification by using user-generated text data, i.e., SVM > Naïve Bayes.

6 Experiment Results and Discussion

For the 448 videos in our test bed, feature counts of four feature sets (FS1, FS2, FS3, and FS4) are shown in Table 15.4. The feature size was reduced from 34,229 (FS3) to 3,187 (FS4) after feature selection.

The experiment results of different feature types and techniques are summarized in Table 15.5. We observed the increase in accuracy for all three classification techniques as more advanced, and content-rich feature types were used, except when using C4.5 with FS2. In addition, after applying feature selection, the accuracy increased about 5.7% (C4.5) to 13.8% (SVM). In terms of classification techniques, SVM consistently outperformed C4.5 and Naïve Bayes with all feature sets. The best performance was achieved by using SVM with selected features of all feature types (FS4). Also, by comparing the best performances of these three techniques, we found that C4.5 had the worst performance. The best performance of C4.5 was only 66.09%, while Naïve Bayes and SVM had accuracy rates of 83.22% and 87.2%, respectively. The results indicated that C4.5 was not as efficient as the other two techniques in solving this problem. We discuss the results based on three aspects: feature types, classification techniques, and key features.

7 A Case Study: Domestic Extremist Videos

Similar to blogs and forums, implicit cyber communities in online video-sharing web sites can be defined by the interactions among users who have similar interests, including commenting, linking, or subscriptions (Chau and Xu 2007; Fu et al. 2008). Video classification is very important for community detection and social network analysis in video-sharing web sites because its results can be used to identify users

Table 15.4 Feature counts of experiment feature sets

Feature sets	Feature counts
FS1 (F1)	574
FS2 (F1 + F2)	1,673
FS3 (F1 + F2 + F3)	34,229
FS4 (selected F1 + F2 + F3)	3,187

Table 15.5 Accuracy for different feature sets and different techniques

	C4.5 (%)	Naïve Bayes (%)	SVM (%)
FS1	59.70	59.21	61.39
FS2	58.05	61.61	62.51
FS3	61.33	68.80	73.42
FS4	66.09	83.22	87.2

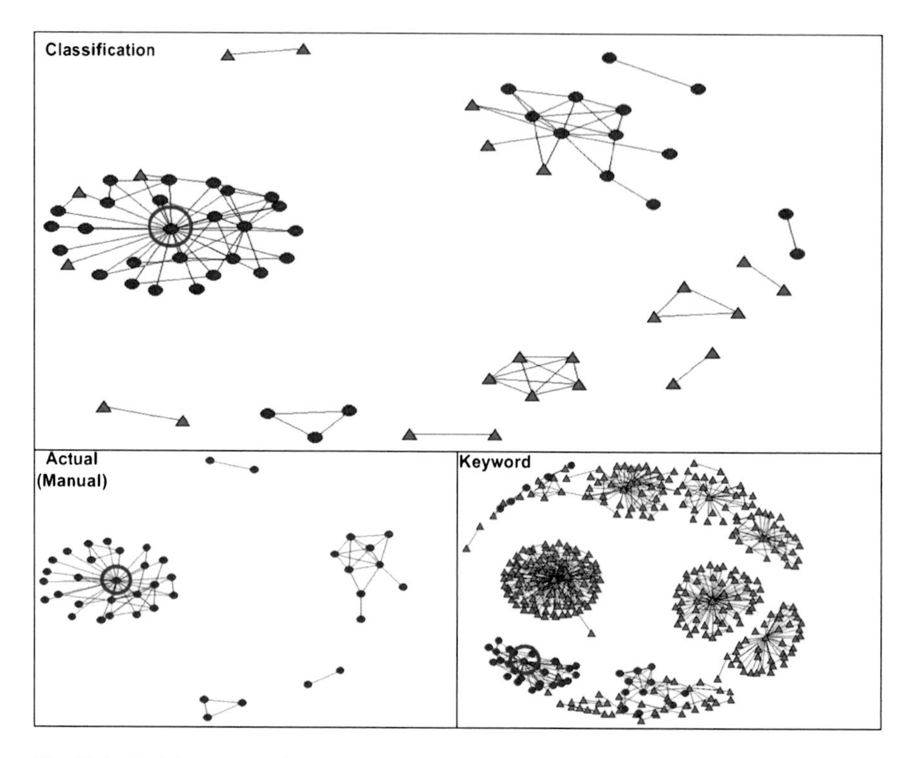

Fig. 15.2 Social networks of white supremacy groups on YouTube

of similar interests. Inaccurate video classification results affect not only the overall network topology of implicit cyber communities in video-sharing web sites but also individual node analyses such as centrality measures and participant roles, which are important units of cyber content analysis (Henri 1992; Rourke et al. 2001).

In order to illustrate how the proposed video classification framework can improve social network analysis as compared to the keyword-based query approach, we present an example from YouTube. User-generated text data for a total of 543 videos were collected by searching the phrase "white power" that refers to white supremacy groups on YouTube. Again, these videos included query-related videos, related videos, and author-uploaded videos. Relevant videos identified by our domain experts (through manual tagging), our video classification framework, and the keyword-based query approach which assumed all these videos were relevant, were used to construct the social networks, respectively. Authors and reviewers of each identified relevant video were considered to have interactions and thus linked with each other. Considering the size of the generated social networks, we excluded links between pairs of YouTube users who had only one interaction. Figure 15.2 shows the social networks generated by using a spring layout algorithm, which places more central nodes near the middle.

Our video classification framework performed well in this example, with the classification accuracy as high as 76.43%. Consequently, the generated social network was very similar to the actual network and revealed the overall network topology of the white supremacy group community on YouTube. For example, the actual network had 42 users and five connected components, while ours had 66 users and 12 components. In contrast, the keyword-based query approach generated a social network with a very different network topology due to many irrelevant videos. Its network contained as many as 379 users and 28 connected components.

Our video classification framework was also more reflective of users' actual involvement in the community, with a more approximate measurement of betweenness, closeness, and degree ranks, compared with the keyword-based query approach. For example, both the network of our video classification framework and the actual network ranked user "barbituraSS" first for all three measures mentioned above. This meant that user "barbituraSS" was the most important person in the white supremacy group community on YouTube. However, the keyword-based approach network underestimated his/her importance by ranking this user sixth for betweenness and degree measures and 237th for closeness measures. This disparity is attributable to the keyword approach exaggerating the size of the community by incorporating many irrelevant videos. In sum, the results suggest that the proposed video classification framework will result in a more accurate representation of the social network structure of implicit cyber communities for online video-sharing web sites and be helpful for individual node analyses, which is important for cyber content analysis.

8 Conclusions and Future Directions

In this chapter, we proposed a framework for text-based video content classification for online video-sharing sites. Different types of user-generated data (e.g., titles, descriptions, and comments) were used as proxies for online videos, and three types of text features (lexical, syntactic, and content-specific features) were extracted. We also adopted feature selection to improve accuracy and identify key features for online video classification. In addition, three feature-based classification techniques (C4.5, Naïve Bayes, and SVM) were compared. Experiments conducted on extremist videos on YouTube demonstrated the good performance of our proposed framework.

Several conclusions can be drawn from our findings. First, our results show that user-generated text data are an effective resource for classification of videos on video-sharing sites. The proposed framework was able to classify online videos based on users' interests with accuracy rates up to 87.2%, achieved by using SVM with selected features of all feature types (FS4). Second, adding more advanced and content-rich feature sets can improve the classification performance, no matter which classification technique was adopted. In comparison with using only lexical features, using all text features increased accuracies up to 12%. Third, the feature selection process can significantly improve the classification performance.

After applying feature selection, the accuracies sharply increased about 5.7% (C4.5)–13.8% (SVM). Fourth, among SVM, Naïve Bayes, and C4.5, SVM was the best classification technique for most cases, which was consistent with the findings of previous studies (Diederich et al. 2000; Zheng et al. 2006). Finally, our case study showed that an accurate video classification method can help identify and understand cyber communities on video-sharing sites.

In the future, we would like to consider both text features and nontext features for online video classification. We also intend to explore additional available user-generated data types, such as the user information of video authors and video reviewers. Moreover, we plan to investigate other classification techniques and feature selection methods which may be appropriate for text-based classification tasks.

Acknowledgments This work was supported by the NSF Computer and Network Systems (CNS) Program, "(CRI: CRD) Developing a Dark Web Collection and Infrastructure for Computational and Social Sciences" (CNS-0709338), September 2007–August 2010.

References

Abbasi, A., and Chen, H. (2005). Applying authorship analysis to extremist-group web forum messages. IEEE Intelligent Systems, 20(5), 67–75.

Abbasi, A., and Chen, H. (2008). Writeprints: A stylometric approach to identity-level identification and similarity detection in cyberspace. ACM Transactions on Information Systems, 26(2), 1–29.

Abbasi, A., Chen, H.-M., and Nunamaker, J. (2008a). Stylometric identification in electronic markets: Scalability and robustness. Journal of Management Information Systems, 25(1), 49–78.

Abbasi, A., Chen, H., and Salem, A. (2008b). Sentiment analysis in multiple languages: Feature selection for opinion classification in Web forums. ACM Transactions on Information Systems, 26(3), 1–34.

Amir, A., Basu, S., Iyengar, G., Lin, C.-Y., Naphade, M., Smith, J.R., et al. (2004). A multi-modal system for the retrieval of semantic video events. Computer Vision and Image Understanding, 96(2), 216–236.

Argamon, S., Šarić, M., and Stein, S. S. (2003). Style mining of electronic messages for multiple authorship discrimination: First results. Proceedings of the 9th ACM SIGKDD International Conference on Knowledge Discovery and Data Mining, Washington, DC, 475–480.

Baayen, H., van Halteren, H., and Tweedie, F. (1996). Outside the cave of shadows: Using syntactic annotation to enhance authorship attribution. Literary and Linguistic Computing, 11(3), 121–132.

Borgne, H.L., Guérin-Dugué, A., and O'Connor, N.E. (2007). Learning midlevel image features for natural scene and texture classification. IEEE Transactions on Circuits and Systems forVideo Technology, 17(3), 286–297.

Burris, V., Smith, E., and Strahm, A. (2000). White supremacist networks on the Internet. Sociological Focus, 33(2), 215–235.

Caillol, H., Pieczynski, W., and Hillion, A. (1997). Estimation of fuzzy Gaussian mixture and unsupervised statistical image segmentation. IEEE Transactions on Image Processing, 6(3), 425–440.

Chau, M., and Xu, J. (2007). Mining communities and their relationships in blogs: A study of online hate groups. International Journal of Human–Computer Studies 65, 57–70.

Chellappa, R., Wilson, C.L., and Sirohey, S. (1995). Human and machine recognition of faces: A survey. Proceedings of the IEEE, 83(5), 705–741.

Chen, H., Shankaranarayanan, G., She, L., and Iyer, A. (1998). A machine learning approach to inductive query by examples: An experiment using relevance feedback, ID3, genetic algorithms, and simulated annealing. Journal of the American Society for Information Science and Technology, 49(8), 639–705.

Chen, H., Thoms, S., and Fu, T. (2008). Cyber extremism in Web 2.0: An exploratory study of international Jihadist groups. IEEE International Conference on Intelligence and Security Informatics, 98–103.

Das, S.R., and Chen, M.Y. (2007).Yahoo! for Amazon: Sentiment extraction from small talk on the web. Management Science, 53(9), 1375–1388.

De Vel, O. (2000). Mining e-mail authorship. Proceedings of Workshop on Text Mining, ACM International Conference on Knowledge Discovery and Data Mining (KDD'2000), Boston, MA.

Diederich, J., Kindermann, J., Leopold, E., and Paass, G. (2000). Authorship attribution with support vector machines. Applied Intelligence, 19(1), 109–123.

Dietterich, T.G., Hild, H., and Bakiri, G. (1990). A comparative study of ID3 and backpropagation for English text-to-speech mapping. Proceedings of the 7th International Conference on Machine Learning, 24–31.

Dimitrova, N., Agnihotri, L., and Wei, G. (2000). Video classification based on HMM using text and faces. European Signal Processing Conference, Tampere, Finland.

Ding, Y., Jacob, E.K., Zhang, Z., Foo, S., Yan, E., George, N.L., et al. (2009). Perspectives on social tagging. Journal of the American Society for Information Science and Technology, 60(12), 2388–2401.

Djeraba, C. (2002). Content-based multimedia indexing and retrieval. Multimedia, IEEE, 9(2), 18–22.

Duan, L.-Y., Xu, M., Chua, T.-S., Tian, Q., and Xu, C.-S. (2003). A mid-level representation framework for semantic sports video analysis. Proceedings of the 11th ACM international Conference on Multimedia, 33–44.

Eickeler, S.,andMuller, S. (1999). Content-based video indexing of TV broadcast news using hidden Markov models. IEEE International Conference on Acoustics, Speech, and Signal Processing, 6, 2997–3000.

Fischer, S., Lienhart, R., and Effelsberg, W. (1995). Automatic recognition of film genres. Proceedings of the 3rd ACM International Conference on Multimedia, 295–304.

Forsyth, R.S., and Holmes, D.I. (1996). Feature finding for text classification. Literary and Linguistic Computing, 11(4), 163–174.

Fu, T., Abbasi, A., and Chen, H. (2008). A hybrid approach to web forum interactional coherence analysis. Journal of the American Society for Information Science and Technology, 59(8), 1195–1209.

Geisler, G., and Burns, S. (2007). Tagging video: Conventions and strategies of the YouTube community. Proceedings of the 7th ACM/IEEE-CS Joint Conference on Digital Libraries, 480–480.

Gibert, X., Li, H., and Doermann, D. (2003). Sports video classification using HMMS. Proceedings of the 2003 International Conference on Multimedia and Expo, Baltimore, MD, 345–348.

Girgensohn, A., and Foote, J. (1999). Video classification using transform coefficients. Proceedings of the Acoustics, Speech, and Signal Processing, 3045–3048.

Guironnet, M., Pellerin, D., and Rombaut, M. (2005). Video classification based on low-level feature fusion model. Proceedings of the 13th European Signal Processing Conference, Antalya, Turkey.

Henri, F. (1992). Computer conferencing and content analysis. In A. Kaye (Ed.), Collaborative Learning Through Computer Conferencing: The Najaden Papers, NewYork: Springer-Verlag, 117–136.

Hirst, G. and Feiguina, O. (2007). Bigrams of syntactic labels for authorship discrimination of short texts. Literary and Linguistic Computing, 22, 405–417.

Hsu, W., and Chang, S.-F. (2005). Visual cue cluster construction via information bottleneck principle and kernel density estimation. International Conference on Content-Based Image and Video Retrieval, 3568, 82–91.

Huang, J., Liu, Z., Wang, Y., Chen, Y., and Wong, E. (1999). Integration of multimodal features for video scene classification based on HMM. In IEEE Workshop Multimedia Signal Processing (MMSP-99) Copenhagen, Denmark, 53–58.

Hung, M.-H., Hsieh, C.-H., and Kuo, C.-M. (2007). Rule-based event detection of broadcast baseball videos using mid-level cues. Proceedings of the 2nd International Conference on Innovative Computing, Information and Control, 240–240.

Jiang, M., Jensen, E., Beitzel, S., and Argamon, S. (2004). Choosing the right bigrams for information retrieval. In Proceedings of the Meeting of the International Federation of Classification Societies.

Jing, F., Li, M., Zhang, H.-J., and Zhang, B. (2004). An efficient and effective region-based image retrieval framework. IEEE Transactions on Image Processing, 13(5), 699–709.

Koppel, M., and Schler, J. (2003). Exploiting stylistic idiosyncrasies for authorship attribution. Proceedings of the IJCAI Workshop on Computational Approaches to Style Analysis and Synthesis, 69–72.

Koppel, M., Schler, J., and Argamon, S. (2009). Computational methods in authorship attribution. Journal of the American Society for Information Science and Technology, 60, 9–26.

Kumar, R., Raghavan, P., Rajagopalan, S., and Tomkins, A. (1999). Trawling the web for emerging cyber-communities. Computer Network, 31, (11–16), 1481–1493.

Lazebnik, S., Schmid, C., and Ponce, J. (2006). Beyond bags of features: Spatial pyramid matching for recognizing natural scene categories. IEEE Computer Society Conference on Computer Vision and Pattern Recognition, 2, 2169–2178.

Ledger, G.R., and Merriam, T.V.N. (1994). Shakespeare, Fletcher, and the two Noble Kinsmen. Literary and Linguistic Computing, 9, 235–248.

Lew, M.S., Sebe, N., Djeraba, C., and Jain, R. (2006). Content-based multimedia information retrieval: State of the art and challenges. ACM Transactions on Multimedia Computing, Communications and Applications, 2(1), 1–19.

Lewis, D. (1998). Naive (Bayes) at forty: The independence assumption in information retrieval. Machine Learning, 4–15.

Lin,W.-H.,and Hauptmann, A. (2002). News video classification using SVM-based multimodal classifiers and combination strategies. Proceedings of the 10th ACM international Conference on Multimedia, 323–326.

Lu, C., Drew, M.S., and Au, J. (2001). Classification of summarized videos using hidden Markov models on compressed chromaticity signatures. Proceedings of the 9th ACM International Conference on Multimedia, 479–482.

Luo, J., and Boutell, M. (2005). Automatic image orientation detection via confidence-based integration of low-level and semantic cues. IEEE Transactions on Patent Analysis and Machine Intelligence, 27(5), 715–726.

Ma,Y.-F., and Zhang, H.-J. (2003). Motion pattern-based video classification and retrieval. EURASIP Journal on Applied Signal Processing, 2003(1), 199–208.

McCallum, A., and Nigam, K. (1998). A comparison of event models for Naïve Bayes text classification. Proceedings of the AAAI Workshop on Learning for Text Categorization, 41–48.

Mendenhall, T.C. (1887). The characteristic curves of composition. Science, 11(11), 237–249.

Messina, A., Montagnuolo, M., and Sapino, M.L. (2006). Characterizing multimedia objects through multimodal content analysis and fuzzy fingerprints. In IEEE International Conference on Signal-Image Technology and Internet-Based Systems, IEEE Computer Society Press, Los Alamitos.

Mitra, M., Buckley, C., Singhal, A., and Cardie, C. (1997). An analysis of statistical and syntactic phrases. Proceedings of the 5th RIAO Conference, Computer-Assisted Information Searching on the Internet, 200–214.

Montagnuolo, M., and Messina, A. (2007). Automatic genre classification of TV programmes using Gaussian mixture models and neural networks. Proceedings of the 18th International Conference on Database and Expert Systems Applications, 99–103.

Oliveira de Melo, A.C., Marcos de Moraes, R., and dos Santos Machado, L. (2003). Gaussian mixture models for supervised classification of remote sensing multispectral images. Lecture Notes in Computer Science, 2905, 440–447.
O'Reilly, T. (2005). What is Web 2.0? Design patterns and business models for the next generation of software. Available at: http://www.oreillynet.com/lpt/a/6228
Pan, J.-Y., and Faloutsos, C. (2002). VideoCube: A novel tool for video mining and classification. Proceedings of the 5th International Conference on Asian Digital Libraries: Digital Libraries: People, Knowledge, and Technology, 194–205.
Peng, F., Schuurmans, D., Keselj, V., and Wang, S. (2003). Automated authorship attribution with character level language models. Proceedings of the 10th Conference of the European Chapter of the Association for Computational Linguistics, 267–274.
Pieczynski,W., Bouvrais, J.,andMichel, C. (2000). Estimation of generalized mixture in the case of correlated sensors. IEEE Transactions on Image Processing, 9(2), 308–312.
Quinlan, J.R. (1986). Induction of decision trees. In J.W. Shavlik and T.G. Dietterich (Ed.), Readings in Machine Learning, Morgan Kaufmann, San Mateo, CA, 81–106.
Rabiner, L.R., and Juang, B.H. (1986). A tutorial on hidden Markov models. IEEE ASSP Magazine, 4–15.
Rasheed, Z., Sheikh, Y., and Shah, M. (2003). Semantic film preview classification using low-level computable features. Proceedings of the 3rd International Workshop on Multimedia Data and Document Engineering.
Rourke, L., Anderson, T., Garrison, D.R., and Archer, W. (2001). Methodological issues in the content analysis of computer conference transcripts. International Journal of Artificial Intelligence in Education, 12, 8–22.
Sahami, M. (1996). Learning limited dependence Bayesian classifiers. Proceedings of the 2nd International Conference on Knowledge Discovery and Data Mining, 335–338.
Salem, A., Reid, E., and Chen, H. (2008). Multimedia content coding and analysis: Unraveling the content of Jihadi extremist groups' videos. Studies in Conflict and Terrorism, 31(7), 605–626.
Samal, A., and Iyengar, P.A. (1992). Automatic recognition and analysis of human faces and facial expressions: A survey. Pattern Recognition, 25(1), 65–77.
Schafer, J. (2002). Spinning the web of hate: Web-based hate propagation by extremist organizations. Journal of Criminal Justice and Popular Culture, 9(2), 69–88.
Shannon, C.E. (1948). A mathematical theory of communication. Bell System Technical Journal, 27(4), 379–423.
Sharma, A.S., and Elidrisi, M. (2008). Classification of multi-media content (videos onYouTube) using tags and focal points. Unpublished manuscript.
Smoliar, S.W., and HongJiang, Z. (1994). Content based video indexing and retrieval. Multimedia, IEEE, 1(2), 62–72.
Tweedie, F., and Baayen, R. (1998). How variable may a constant be? Measures of lexical richness in perspective. Computers and the Humanities, 32(5), 323–352.
Vapnik, V.N. (1995). The nature of statistical learning theory. New York: Springer-Verlag.
Vapnik, V.N. (1998). Statistical learning theory. New York, NY, Wiley-Interscience.
Vasconcelos, N., and Lippman, A. (2000). Statistical models of video structure for content analysis and characterization. IEEE Transactions on Image Processing, 9(1), 3–19.
Xu, D., and Chang, S.-F. (2008).Video event recognition using kernel methods with multilevel temporal alignment. IEEE Transactions on Patent Analysis and Machine Intelligence, 30(11), 1985–1997.
Xu, L.-Q., and Li, Y. (2003). Video classification using spatial-temporal features and PCA. International Conference on Multimedia and Expo, 3, 485–488.
Yang,Y., and Pedersen, J. O. (1997). A comparative study on feature selection in text categorization. Proceedings of the 14th International Conference on Machine Learning, 412–420.
Zhang, J., Marszalek, M., Lazebnik, S., and Schmid, C. (2007). Local features and kernels for classification of texture and object categories: A comprehensive study. International Journal of Computer Vision, 73(2), 213–238.

Zheng, R., Li, J., Chen, H., and Huang, Z. (2006). A framework for authorship identification of online messages: Writing-style features and classification techniques. Journal of the American Society for Information Science and Technology, 57(3), 378–393.

Zhou, W., Vellaikal, A., and Kuo, C.C.J. (2000). Rule-based video classification system for basketball video indexing. Proceedings of ACM Workshops on Multimedia, 213–216.

Zhou,Y.-H., Cao,Y.-D., Zhang, L.-F., and Zhang, H.-X. (2005). An SVM-based soccer video shot classification. Proceedings of 2005 International Conference on Machine Learning and Cybernetics, 9, 5398–5403.

Chapter 16
Improvised Explosive Devices (IED) on Dark Web

1 Introduction

Since the 1990s, the Internet has been instrumental in the advancement of many social movements. An early and well-cited example of the effective use of computer-mediated communication (CMC) for activism and protest is the case of the Zapatista movement and the Zapatista National Liberation Army (EZLN) in their rebellion against the ongoing colonial repression of the indigenous people in Mexico (Collier 1994; Cleaver 1995). The Zapatista movement leveraged the Internet and CMC "for the dissemination of information, the sharing of experience, and the facilitation of discussion and organizing … (and) for the amplification and archiving of the developing history of the struggle" (Cleaver 1999). Through their use of the Internet and CMC, the Zapatista movement demonstrated "a new capacity for this and other social movements to communicate across borders and to operate at a transnational level" and was considered a "prototype" for other movements (Cleaver 1999). Related antiglobalization demonstrations inspired by the Zapatistas, such as the J18 and the protests against the World Trade Organization in Seattle during 1999, similarly utilized the Internet and CMC to promote and coordinate their activism, and were also regarded as models of demonstration to be followed in subsequent years (Starr 2000).

Recognizing the importance of CMC to social movements, researchers have focused studies specifically on the use of the Internet by various movements, from grassroots activism (Tesh 2002) to transnational mass protest movements (Clark and Themudo 2006). Social movement activism mediated by the Internet is often referred to as cyber protest and represents an emerging field of social movement research (Meikle 2002; Jordan and Taylor 2004; van de Donk et al. 2004). Cyber protest is considered an extension of a social movement into a new media space, an articulation of the movement on the Internet (Castells 1998). Cyber protest is diverse, and many forms have been identified in the literature, from hacktivism – electronic civil disobedience, as in the case of the 1998 blockade of the Pentagon Web site by the Electronic Disturbance Theater (EDT) – to culture jamming – tactical media using techniques of appropriation, collage, ironic inversion, and juxtaposition,

H. Chen, *Dark Web: Exploring and Data Mining the Dark Side of the Web*, Integrated Series in Information Systems 30, DOI 10.1007/978-1-4614-1557-2_16,

as in the case of the ®™ark Web sites protesting the Bush presidential campaign of 2000 (Meikle 2002). Although the Internet has empowered social movements to produce novel expressions of protest, they continue to rely upon the traditional tactics proven effective in the pre-Internet world adapted to a virtual environment (Meikle 2002). For example, the EDT blockade was regarded by researchers as a virtual sit-in, a traditional and successful method of protest. Following the Zapatista model, the most abundant forms of cyber protest leverage the Internet as a platform for communication, to disseminate information, share experiences, develop solidarity and identity, and promote, facilitate, and organize activism. The extensive usage of CMC by a social movement creates a widespread cyber culture on the Internet, establishing many linked virtual communities for members to congregate and socialize (Meikle 2002), forming rich and complex social systems (Jones 1994), and representing fertile areas for conducting social research.

While researchers have recognized the importance of the Internet to social movements and the value of studying virtual communities and cyber protest to further their understanding, they have also identified challenges in conducting social research online. Manifestations of social movements on the Internet and their expressions of cyber protest are dynamic, appearing and disappearing without warning, and distributed across multiple related Web sites, forums, or virtual communities, some of which may not be readily apparent to the researcher.

> "Although they may include formal organizations as components, on the whole they are not an organization. A social movement typically lacks membership forms, statutes, chairpersons, and the like. It may expand or shrink considerably over periods of time, and exhibit phases of visibility and latency. Also, unlike political parties, social movements may have significant overlaps with other movements. Moreover, a social movement may quickly change its forms, strategy, tactics, and even some of its goals" (van de Donk et al. 2004). The cyber protest of social movements is "fuzzy and fluid phenomena often without clear boundaries," and a "moving target, difficult to observe" (van de Donk et al. 2004)

The dynamic and open nature of social movements on the Internet, as described by cyber protest researchers, presents unique challenges to conducting social research online.

Researchers have proposed approaches to social research on the Internet, specifically attempting to address the challenges presented by the scale and complexity of the online environment. Paccagnella (1997) suggests that scholars conducting social research on virtual communities "exploit the possibilities offered by new, powerful, and flexible tools for inexpensively collecting, organizing, and exploring digital data." Such tools would facilitate the analysis of "cultural cyber-artifacts," the member communications and creations representing the archived history of the virtual community. Jones (1997) presents a theoretical outline for conducting social research on virtual communities, adapting the perspectives of archaeology to CMC. Similar to Paccagnella, the approach focuses on the cultural cyber-artifacts of a virtual community, which provide an integrated framework for analysis to understand the life of the community and its members (Jones 1997). Due to its focus on the study of cultural cyber-artifacts, the approach was called cyber-archaeology. By focusing on the cyber-artifacts of virtual communities and developing sophisticated tools for their collection and analysis, issues of scale and complexity surrounding cyber protest may be resolved, enabling broad and longitudinal research on social movements. However, few approaches have been proposed in the literature with the technical sophistication and capabilities to assist in social research on modern virtual communities and forms of cyber protest, and social researchers continue to be slow to react to social movements as they emerge (Edelman 2001).

In this chapter, we present a cyber-archaeology framework for social movement research. We refer to the approach as cyber-archaeology, following Jones (1997), because of our focus on the cultural cyber-artifacts of cyber protest. Our framework overcomes many of the issues of scale and complexity facing social research on the Internet and specifically addresses many of the suggestions of Paccagnella (1997), enabling broad and longitudinal social research on virtual communities. The framework is not intended to replace the social researcher or directly produce theoretical conclusions regarding social movements, but instead, we offer a collection of powerful tools and techniques to assist in conducting social movement research on the Internet. The approach is automated and scalable, allowing a researcher to easily collect and analyze social movement cyber-artifacts across multiple linked virtual communities. The techniques for cyber-artifact analysis adopt perspectives of social movement theory, specifically the network (Diani and McAdam 2003) and self-organization (Fuchs 2006a) models of social movements.

To demonstrate the cyber-archaeology framework for social movement research and provide a detailed instantiation of the proposed approach for evaluation, a Dark Web case study on a broad group of related IED virtual communities is also presented.

2 Cyber-Archaeology Framework for Social Movement Research

We present in this section our cyber-archaeology framework for social movement research. The proposed framework has three phases, as shown in Fig. 16.1. In phase 1, social movement research design, social researchers identify the social movement of interest, associated virtual communities, and target cyber-artifacts for collection and analysis. Phase 2 consists of the automated collection and classification of cyber-artifacts across one or many identified virtual communities. Phase 3 supports the analysis of cyber-artifacts by social researchers from the perspectives of social movement theory, including various network analysis and visualization techniques, to perform multilevel, link, time, and homophily analysis, as suggested by Diani (2003). Each of these phases is discussed in detail below.

The first phase of the cyber-archaeology framework consists of the social research design on the movements of interest. Researchers identify a social movement of interest and its related virtual communities, in consultation with domain experts when necessary, to develop a seed set of Web sites for analysis. The list of targeted virtual communities can then be expanded through link analysis of the seed set of sites, to include Web sites linked to and from the virtual communities already identified.

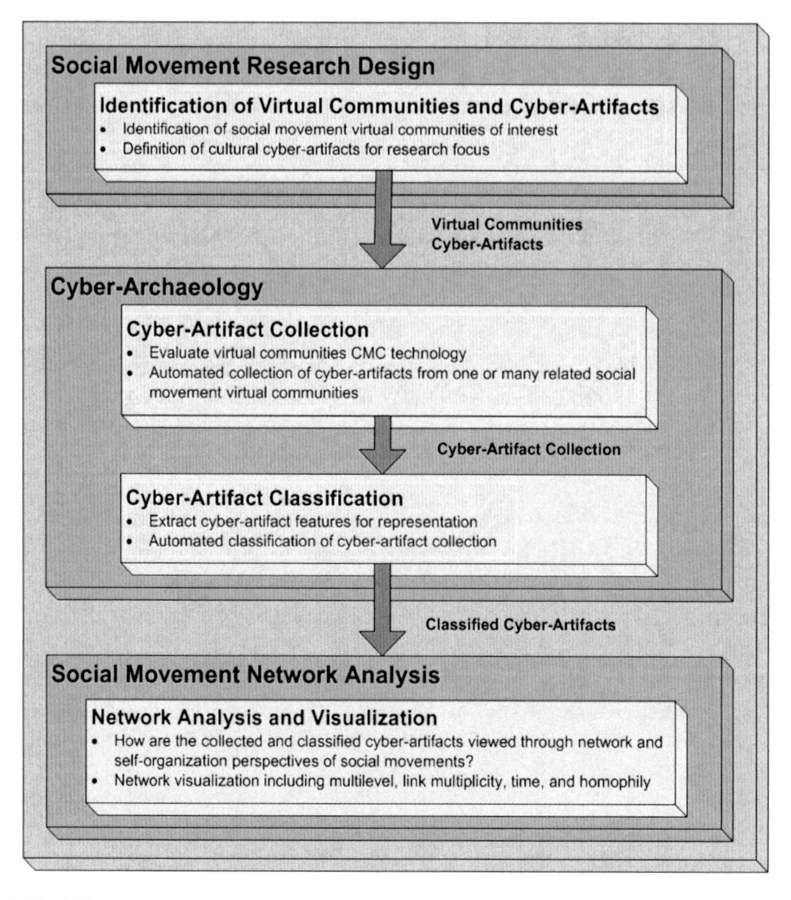

Fig. 16.1 Cyber-archaeology framework

Once a satisfactory list of virtual communities has been defined, the targeted cultural cyber-artifacts for research focus must be determined. A lexicon of terms pertaining to the cyber-artifacts of interest must also be developed to assist in the collection process by enabling the identification of relevant content. By limiting the search to specific cyber-artifacts of interest, the social researcher can focus their investigation, thereby alleviating issues of scale impacting social research on the Internet. The cyber-artifacts selected could be based upon, for example, influential figures in the movement, meaningful events, or core ideologies and symbolism. Focusing on specific cyber-artifacts that develop the identity of the movement and create solidarity among members deepens the social research. However, researchers need not restrict their focus to specific cyber-artifacts; the framework supports a more comprehensive approach. Without restriction, all cyber-artifacts of the virtual communities may be collected for analysis, potentially overwhelming the researcher but providing completeness in the research.

The second phase consists of the automated collection and categorization of the targeted cyber-artifacts of interest from the virtual communities identified in the first phase. We consider the collection and classification tasks to be core to any study on social movements on the Internet. Researchers have highlighted the promise for enriching the virtual community research field by developing automated techniques to assist with these tasks (Paccagnella 1997). Researchers have also identified opportunities to improve the understanding of social movements by broadening the research in scope, beyond a single-organization or single-issue focus (Edelman 2001). To aid in the analysis of the collected cyber-artifacts, automated models for classification can be developed to categorize cyber-artifacts across numerous dimensions depending upon their nature and the goal of the research. Developing automated techniques for cyber-artifact collection and categorization allows social researchers to ask questions that would be infeasible to answer by traditional, manual methods, expanding the scope of research and the understanding of social movements. The outputs of the second phase, the collected and classified cyber-artifacts of the virtual communities, are then analyzed manually by the social researcher to learn about the movement. The third phase of the framework is considered optional, although these analyses are guided by social movement theory. Therefore, phases 1 and 2 of the framework could also support other forms of analysis, if desired by the social researcher.

The third phase of the framework consists of analysis of the collected and classified cyber-artifacts by social researchers from the perspectives of social movement theory. In particular, we adopt the network and self-organization models of social movements. Three forms of network analysis and visualization techniques are presented, providing insight into the patterns of self-organization exhibited by the virtual community with respect to the collected and classified cyber-artifacts. These network analyses and visualizations follow the research program proposed by Diani (2003) and cover multilevel analysis, link multiplicity, the evolution of networks over time, and structural homophily. The methods developed to support the third phase of the framework also utilize structured approaches and automation, similar to the techniques developed in the second phase, to support scalability in social research.

3 Case Study

To demonstrate the efficacy of the cyber-archaeology framework and provide a detailed instantiation of the proposed approach for evaluation, we present a Dark Web case study on a broad group of related IED virtual communities.

3.1 *Introduction*

Some of the most sophisticated virtual communities with the strongest membership and social ties between community members belong to socially extreme movements. There are many social movements throughout the world that could be considered socially extreme or extremist, struggling for various causes based on race, gender, sexuality, and religion. Extremist movements actively use the Internet as a quick, inexpensive, and anonymous means of communication, overcoming the limitations of traditional media and allowing their ideological information to reach new audiences around the world. Studies have shown that these groups promote their positions through their Internet communications and publications (Glaser et al. 2002) and found evidence of resource sharing, fundraising, propaganda, recruitment, and training materials (Burris et al. 2000; Schafer 2002; Zhou et al. 2005). Social scientists have long been interested in Middle Eastern movements supporting anti-Western positions, some of which contain participants or subgroups with socially extreme perspectives whose activism includes high-risk behavior, applying social movement theory to understand these societies and individuals (Snow and Marshall 1984; Wiktorowicz 2002). To be clear, we do not consider all anti-Western movements of the Middle East to be extreme, nor do we believe that all members of such movements engage in extremist behavior in their activism. However, some of the participants in these social movements hold extreme positions and engage in high-risk behavior in their activism, with grave cost. Social researchers have specifically studied the motivations and dynamics of high-risk behavior in social movements (McAdam 1986). Communities supporting high-risk activism stress the importance of social networks, as strong social ties are required to provide the trust and solidarity among activists to encourage the high-risk behavior (Wiktorowicz 2002). To study these social movements with some participants whose activism includes high-risk behavior, we perform a cyber-archaeological analysis on a specific theme of cultural cyber-artifacts significant to some anti-Western Middle Eastern movements: content related to improvised explosive devices (IEDs). IEDs have social significance among some movement participants as an enabler of violent uprising and resistance, equalizer of power, and as the focus of high-risk activism and related symbolism. Examining IED-related cyber-artifacts in a global social context can provide a deeper understanding of these social movements and the high-risk behavior of some of the participants.

3.2 Cyber-Archaeology Framework Phase 1

In the first phase of the cyber-archaeology framework, a list of social movements of interest and related virtual communities is defined. For the case, the social movements and virtual communities were identified using the expertise of a number of government and research organizations, including the United States Committee for a Free Lebanon, Counter-Terrorism Committee of the UN Security Council, US State Department reports, Official Journal of the European Union, and other government reports. The seed set of social movement virtual communities was expanded through link analysis. In-links to and out-links from the virtual communities were collected and examined manually to identify additional related virtual communities to be included in the social movement research. Utilizing an automated approach to cyber-artifact collection and analysis, social movement research can extend across multiple related virtual communities.

Table 16.1 Sample from Arabic IED lexicon

Keyword	Variations	Translation
المدى الفعّال	المدى الفعال ألمدى ألفعال ألمدى ألفعَال	Effective range
ألعبوات	العبوات	Explosives
الأسلحة	الاسلحة اسلحة أسلحة	Weapons, "the weapons"
التصنيع الكيماوي	ألتصنيع ألكيماوي تصنيع كيماوي	Chemical manufacturing

Since the social movements identified for this research communicate primarily in Arabic, a lexicon of terms relevant to IEDs was developed with assistance from domain and language experts to direct the cyber-artifact collection effort. The lexicon for IED-related cyber-artifacts consisted of more than 100 terms and their linguistic variations; a subset of the terms is shown in Table 16.1.

3.3 *Cyber-Archaeology Framework Phase 2*

The second phase of the cyber-archaeology framework consists of the automated collection and categorization of IED-related cyber-artifacts from the identified social movement virtual communities. The automated approach provides the ability to expand the research to multiple virtual communities and acquire a more complete collection of cyber-artifacts of interest.

3.3.1 Cyber-Artifact Collection

The collection of IED-related cyber-artifacts from social movement virtual communities entails the use of focused crawling techniques. Focused crawlers are intelligent Web spiders designed to retrieve content relating to a specific subject. They "seek, acquire, index, and maintain pages on a specific set of topics that represent a narrow segment of the web" (Chakrabarti et al. 1999).

A cyber-archaeology focused crawler for collection of cyber-artifacts from social movements faces several design challenges including accessibility, Web crawling features, and techniques. In terms of accessibility, the collection of CMC cyber-artifacts requires acquiring and maintaining access to social movement virtual communities. Many virtual communities do not allow anonymous access. To access these virtual communities using a focused crawler, user IDs and passwords are solicited from community administrators. These credentials are embedded within

Table 16.2 Cyber-artifact collection

Number of Web sites: 30	Number of Web pages: 2,541
Core Web sites	
Web site	No. of Web pages
www.qudsway.com	1,209
www.albasrah.net	332
www.khayma.com	162
www.jamaat.org	141
www.hilafet.com	66
www.geocities.com	51

the focused crawler and provided when access to a virtual community is required. Crawling parameters, such as the number of concurrent connections, download intervals, time-out, and speed, are set experimentally based upon the tolerances of the virtual community to the collection activity of the focused crawler.

Web crawling features identified through manual analysis of the virtual communities are utilized by the focused crawler to characterize and control the crawl space. Aggarwal et al. (2001) suggested the use of URL tokens for link ordering in the crawl space. For example, in the case of Web forums, URLs containing words such as "board," "thread," or "message" are tokens of interest in identifying pages with forum communications to be crawled. Additionally, the domain names of third-party file hosting Web sites and file extension tokens such as .wmv and .mpg are significant in locating multimedia cyber-artifacts. Ester et al. (2001) suggested the use of page levels in the collection of cyber-artifacts, which are often stored on third-party hosts. In such cases, a rule-based approach can be used to allow the focused crawler to search a few additional domain levels and "tunnel" to the desired cyber-artifacts for collection.

These Web crawling features are utilized in concert with link ordering techniques to direct the crawling effort. Link ordering techniques provide the ability to adapt the crawling process and accommodate the structure of various social movement virtual communities. For example, breadth-first search can be used for collection of CMC cyber-artifacts from board page Web forums. Depth-first search is typically more appropriate for Internet service provider CMC forums, due to the prevalence of ad pages that need to be traversed to reach the page containing the CMC cyber-artifacts of interest.

Addressing the issues of accessibility, Web crawling features, and techniques employed in the analysis of the Web technologies used in virtual communities allows for Web site wrappers to be developed and used in tuning the focused crawler for cyber-artifact collection. Once the focused crawler has been properly configured, cyber-artifacts of interest can be collected from multiple virtual communities in an automated fashion with little supervision. For the purposes of the case study, these focused crawling techniques were applied in the collection of IED-related cyber-artifacts from 30 virtual communities identified in phase 1. In total, 2,541 IED-related Web page cyber-artifacts were collected and qualified through manual inspection by a domain expert. Over 90% of the cyber-artifacts were gathered from a core set of six Web sites, as listed in Table 16.2 and by the frequency distribution in Fig. 16.2.

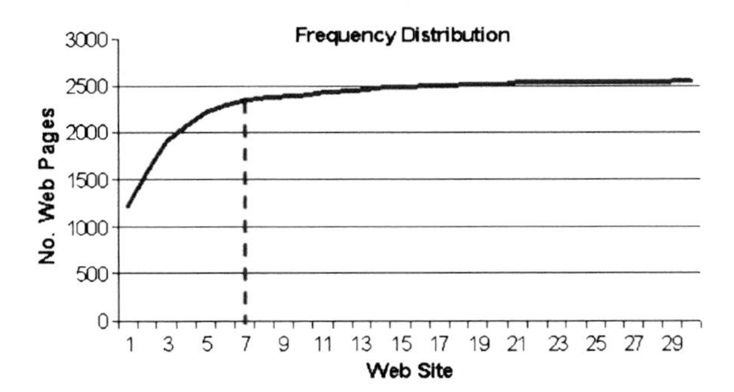

Fig. 16.2 Cyber-artifact frequency distribution

Table 16.3 Description of core Web sites with IED-related content

Web site	Description
www.qudsway.com	Web site contains articles and discussions on Middle East politics
www.albasrah.net	Web site with links to the former Iraqi regime
www.khayma.com	Arabic directory. Much of the content pertains to diverse political topics
www.jamaat.org	Web site of the Jamaat group of Pakistan
www.hilafet.com	Web site of the Hizb-ut-Tahrir party
www.geocities.com	/m_ale3dad4: Is a collection of military training materials
	/maoso3ah: Is an "encyclopedia" for military training

The core Web sites containing IED-related cyber-artifacts include major Iraqi resistance sites such as albasrah.net, Web portals such as said.net and khayam.com, and Web sites for political parties such as qudsway.com, jamaat.org, and hilafet.com. A description of each of these core Web sites is presented in Table 16.3.

3.3.2 Cyber-Artifact Classification

To assist in the analysis of the collected cyber-artifacts, automated models for classification can be developed to categorize cyber-artifacts across numerous dimensions depending upon their nature. For CMC communications, a useful categorization for analysis of cyber-artifacts is by genre of communication. Genre is a literary and rhetorical concept that describes widely recognized types of discourse (Yates and Orlikowski 1992). Genres of communication are traditionally defined in terms of shared purpose and common conventions of substance and form (Herring and Paolillo 2006). Shared purpose refers to the social motives and intentions behind the communication. Substance refers to the themes and topics expressed in the communication. Form refers to observable physical and linguistic features of the communication, including structural attributes, communication medium, and language, such as degree of formality and use of a specialized vocabulary. Genres of communication are used within a community to establish expectations about the purpose, content, participants, form, time, and place of the interaction, describing the who, what, where, when, why, and how of the communication (Yates and Orlikowski 2002).

Table 16.4 Description of the genres of IED Web pages

Genre	Type	Description
Material pages	Weapons	Detailed descriptions of various *weapons and IEDs*. Include training materials on how to build, use, and maintain devices
	Attack reports	Reports on the *occurrence of an attack* with some elaboration on the specific details of the operation
	Military and recon reports	Reports on the latest *military intelligence* (provides statistics of operations, troop, and vehicle movements, etc.)
	Tactics	Discusses the pros and cons of various *military tactics* and provides examples from previous operations
Discussion pages	General discussion	General discussion of *IEDs* and *IED-related events*

Traditionally, researchers have analyzed genres to understand the communication dynamics of a community or organization. Yates and Orlikowski (1992) observed that the emergence of genres within an organization is due to particular sociohistorical contexts, which are reinforced over time as the situation recurs. Researchers have also developed techniques for the automated classification of documents by genre. While genre classification is similar in nature to topic or author classification, the representative features used to characterize the document differ according to the type of classification being performed. In topical classification, features are dependent upon the topic, whereas in author and genre classification, features are often considered orthogonal to topic, and greater emphasis is placed on the stylistic and structural elements of the document. Researchers have improved methods for automated genre classification through examination of feature representations. Santini (2004) studied the contribution of a class of features representing syntactic structure in discriminating among different genres of the benchmark British National Corpus dataset. Shepherd et al. (2001) adapted genre classification to Web pages by incorporating functionality into the traditional genre representation tuple of form and content, including aspects of user interaction and processing capability. Automated genre classification has also been applied by researchers to other forms of computer-mediated communication. Finn and Kushmerick (2006) applied genre classification to news articles and reviews collected from the Web. Herring and Paolillo (2006) investigated the relationship between genre, language, and gender through analysis of Weblogs.

Classification of collected CMC cyber-artifacts by communication genre assists in the analysis of the social movement virtual community. Through manual analysis of the collected IED-related CMC cyber-artifacts by a domain expert, two distinctive genres of communication were identified: Web pages containing descriptions on IED-related weapons, attacks, and military tactics (materials pages) and Web pages with general discussion of IEDs or IED-related events (discussion pages). The two genres of Web pages identified are described briefly in Table 16.4. Sample Web pages from each type are presented in Figs. 16.3, 16.4, 16.5, 16.6 and 16.7.

* يوم 18 / 1 / 2004م ؛
1 هجوم على دورية عسكرية أمريكية خارج مدينة "الديوانية" جنوب بغداد ، اسفر عن مقتل جندي امريكي واحد واصابة ثلاثة آخرين واعطاب عربة نوع "همفي"
2 هجوم على دورية عسكرية امريكية استدرجت الى كمين مزروع بالمتفجرات شمال مدينة تكريت ، واسفرت العملية عن مقتل جنديين امريكيين وجرح اربعة آخرين وتدمير ثلاث عربات مختلفه
3 فجرت سيارة نوع "مرسيدس" عن بعد اثناء مرور دورية عسكرية امريكية بمحاذاتها وسط مدينة تكريت واسفر الانفجار عن مقتل جنديين امريكيين وجرح ثلاثة آخرين وتدمير عربة نوع "همفي"

Fig. 16.3 Sample general discussion page – discussion on an attack on a military patrol north of the city of Tikrit

معركة قهر الصليب

قال الله تعالى: (قاتلوهم يعذبهم الله بأيديكم ويخزهم وينصركم عليهم ويشف صدور قوم مؤمنين) . تأتي هذه العملية المباركة بين رأس الصليبية العالمية (أمريكا) ورأس حربة الأمة الإسلامية ألا وهم المجاهدون ضمن الهجمات المستمرة على قوات الصليب المحتلة التي اقضت مضاجعهم. وتجدر الإشارة هنا إلى كون هذه المعركة التي سنتكلم عن أحداثها, إنما هي مثال من بين المعارك الشبه يومية بين المجاهدين وقوات التحالف الصليبي في أقاليم شتى أفغانية. أما بالنسبة لهذه العملية التي سميت بمعركة قهر الصليب لما سيأتي سببه لاحقا إن شاء الله, فهي تعتبر أكبر العمليات حتى الآن ضد العدو الصليبي في الإقليم الجنوبي من أفغانستان حيث أنه في السابق كانت العمليات تقتصر على رماية الصواريخ المتوسطة والبعيدة المدى على معاقل الأمريكان ونصب الكمائن لقوافلهم العسكرية. وبالرغم من البركة العظيمة التي وضعها الله لتلك العمليات على صغر حجمها النسبي حيث تكبدت القوات الصليبية وما تزال تتكبد خسائر ليست هينة في الأفراد والطائرات والآليات, إلا إن المجاهدين حرصوا على الإنتقال بالمعركة إلى مستوى آخر لزيادة لهيب النار الذي باتت تعيش فيه جيوش الصليب بقيادة أمريكا ببركة هذا الجهاد المبارك والإستمرار بإذن الله في تس***ها حتى يقضي الله بانهيار هذا الصنم الذي ملأ الدنيا بالإنحلال والدمار.

Fig. 16.4 Sample attack report page – report on the events of a major attack against forces in southern Afghani provinces

1 - ثلاثي نترطولوين *TNT* :

النسب :

طولوين C_7H_8	المحول الأول كبرتيك = 41.4ملم	المحلول الثاني كبريتيك =23.4ملم
114ملم	نيتريك =15.1 ملم	نيتريك = 37.5 ملم
	يضاعف 2 مرة	يضاعف 3 مرات

خطوات العمل :

1- نضع 57 ملم من المحلول الأول في حماك ثلجي ثم نصب عليه 114 من الطولوين قطرة قطرة ، تدريجيا وبهدوء وببطء حتى لا ترتفع الحرارة عن 20° م ثم نقلب ذلك الخليط جيداً لمدة 15 دقيقة .

2- نسخن الخليط لرفع درجة الحرارة إلى 50° م مع التقليب .

3- نضيف 280ملم من المحلول الأول إلى ذلك الخليط ونرفع درجة الحرارة إلى 55° م لمدة 15 دقيقة .

Fig. 16.5 Sample weapons page – composition of trinitrotoluene (TNT) and procedure for preparation of the explosive

إحصائية جديد ودقيقة
لعدد الدبابات والمدرعات المجنزرة المدمرة للعدو خلال أربع أيام
شبكة البصرة
عبد الله محب المجاهدين
تم حشد في الأونة الأخيرة حول الفلوجة 730 دبابة أبرامز و 570 مدرعة مجنزرة من طراز برادلي أي مجموع عدد الدروع 1300 ...

Fig. 16.6 Sample recon report page – new statistics on the number of enemy tanks and armored carriers spotted during the last 4 days

المقال بعنوان : درس قاس في المعركة : 150من المارينز يواجهون قناص واحد
شبكة البصرة
نشرة بدر الرافدين
هذا القناص أبقى 150 من المارينز في رعب و حذر لاغلب اليوم و هو يعطي صورة عن طبيعة العدو (يقصد المجاهدين)
حيث أن بعضهم قتلة محترفون و يعرفون جيدا كيف يضربون و يهربون
قال الليوتينانت أندي ايكرت : "انه يجلس في مكان ما بالأعلى يتناول شطيرة بينما نحن في سعي محموم للعثور عليه".

Fig. 16.7 Sample tactics page – 150 marines face one sniper in Iraq

Developing automated methods for the classification of these cyber-artifacts provides the ability to enhance and expand the scope of social movement research on virtual communities. Our genre classification approach for IED-related Web pages is comprised of an extended feature representation coupled with machine learning algorithms. The extended feature representation incorporates rich structural information with stylistic and topical linguistic features, intended specifically to represent the purpose, substance, and form of the Web page, indicative of genres of communication (Yates and Orlikowski 2002). Structural features include HTML tags and information about the technical structure of the Web page, which can specifically express genre information to differentiate materials and discussion pages. Stylistic features include word- and character-level lexical measures, word-length distributions, vocabulary richness, function words, and punctuation usage, and effectively capture the stylistic tendencies of the Web page (Abbasi and Chen 2005; Zheng et al. 2006). Topical features such as word n-grams are used to identify themes and topics in Web pages. Only features exceeding a threshold number of occurrences (three) were included in the representation. The extended feature set is summarized in Table 16.5.

The support vector machine (SVM) learning algorithm was used with the extended feature representation to automatically classify the collected IED-related Web pages. The SVM is a statistical-based algorithm that has been used successfully in text categorization (Joachims 1998) and is particularly suited to learning with extended feature representations. Classification models that utilize extended feature representations often leverage methods for feature selection to identify the most relevant features for inclusion in the model, thereby eliminating unnecessary information and reducing model complexity. For the classification of the IED-related Web page cyber-artifacts, we evaluate two SVM models: The first included the entire extended feature set (labeled SVM), while the second used feature selection on the extended feature set based on the information gain heuristic, a measure of entropy (labeled SVM-IG). Both SVM classification models utilized linear kernel functions.

The test bed of 2,541 collected IED-related cyber-artifact Web pages was used for the development and evaluation of the automated genre classifier. As previously stated, 2,501 discussion and 40 materials Web pages were identified through manual categorization by a domain expert. Due to the large number of discussion pages relative to materials pages, adjustments were made to provide equal representation

Table 16.5 Extended feature set for genre classification

Group	Category	Quantity	Description/examples
Stylistic	Word-level lexical	5	Total words, % char. per word
	Character-level lexical	5	Total char., % char. per message
	Character n-grams	<18,278	Count of letters, char. bigrams, trigrams (e.g., كك,اب)
	Digit n-grams	<1,110	Count of digits, digit bigrams, digit trigrams (e.g., 1, 12, 123)
	Word-length distribution	20	Frequency distribution of 1–20- letter words
	Vocabulary richness	8	Richness (e.g., hapax legomena, Yule's K, Honore's H)
	Special characters	21	Occurrences of special characters (e.g., @, #, $,%, ^, &, *, +, and =)
	Function words	300	Frequency distribution of function words (e.g., of, for, to)
	Punctuation	12	Occurrence of punctuation marks (e.g., !;:,.?)
	Word root n-grams	Varies	Roots, bigrams, trigrams (e.g., كتب, كسب)
Structural	Message-level	6	For example, has greeting, has URL, requoted content
	Paragraph-level	8	For example, number of paragraphs, sentences per paragraph
	Technical structure	50	For example, file extensions, fonts, use of images, HTML tags
	HTML tag n-grams	<46,656	For example, <head>, , <td>, <message>
Topical	Word n-grams	Varies	Bag-of-words n-grams (e.g., "explosive," "explosive device")

of materials and discussion pages in the classification experiments. One hundred iterations of the experiment were performed, and in each iteration, 40 discussion pages were randomly selected from the 2,501 to complement the 40 materials Web pages; a different set of 40 discussion pages was used in each iteration of the experiment. Tenfold cross-validation was performed to test classification and generate performance results, measured by computing the percentage of correctly classified Web pages. Overall performance values were calculated as simple averages of the 100 iterations.

Overall, both the SVM and SVM-IG models performed well in genre classification of IED-related materials and discussion Web pages, with better than 81% accuracy. The SVM-IG model using the extended feature set and the information gain heuristic for feature selection performs particularly well, with over 88% accuracy, and outperforms the standard SVM model by 7%. A t test was performed to determine whether the performance increase gained by utilizing feature selection was significant. The performances of the SVM-IG model were a statistically significant improvement ($p<0.001$) over the model that used the entire feature representation without feature selection. Furthermore, the SVM-IG model was more reliable, performing with better than 80% classification accuracy in every iteration of the

Table 16.6 Results of genre classification of IED Web pages

Model	Features	Mean accuracy (%)	Standard deviation (%)	Range (%)
SVM	21,333	81.938	5.313	65.0–92.5
SVM-IG	9,268	88.838	3.238	80.0–96.3

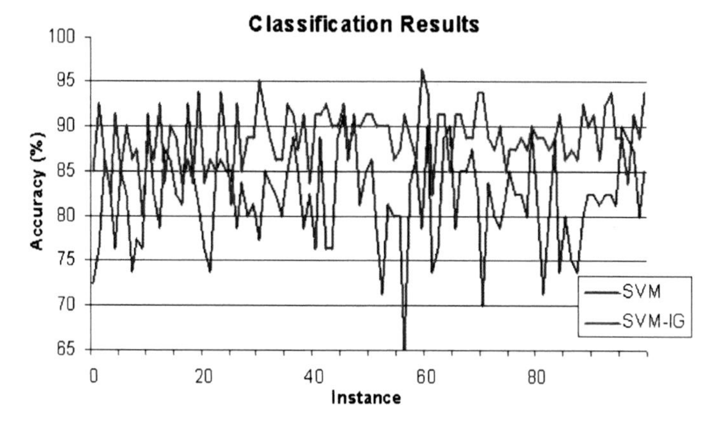

Fig. 16.8 Results of genre classification experiments

experiment. The information gain heuristic identified 9,268 of the 21,333 features as key and relevant to the genre classification. The results of the experiments are presented in Table 16.6 and Fig. 16.8.

3.4 *Cyber-Archaeology Framework Phase 3*

The third phase of the cyber-archaeology framework consists of the analysis of collected IED-related cyber-artifacts through the perspectives of social movement theory. In particular, we adopt the network and self-organization models of social movements. Three forms of network analysis and visualization techniques are presented, providing insight into the patterns of self-organization exhibited by the virtual community with respect to the collected and classified cyber-artifacts. These network analyses and visualizations follow the research program proposed by Diani (2003) and cover multilevel analysis, link multiplicity, the evolution of networks over time, and structural homophily. The methods developed to support the third phase of the framework also utilize structured approaches and automation, similar to the techniques developed in the second phase, to support scalability in the social research.

In the case of IED-related cyber-artifacts, learning how the highly specialized knowledge pertaining to these technologies is organized, distributed, and leveraged throughout the related social movement virtual communities is of particular research interest. Furthermore, an understanding of the communication patterns related to

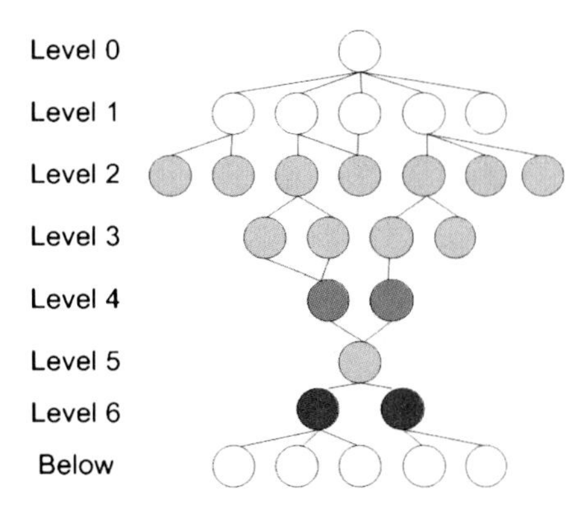

Fig. 16.9 Site map featuring intensity of discussion pages in page levels

these cultural artifacts is significant to the understanding of the social movements and the relations the communities have with IEDs and related high-risk activism.

The third phase of the framework consists of three forms of network analysis and visualization, addressing multilevel analysis, link multiplicity, time, and homophily (Diani 2003). The first type of network analysis, site map analysis, visualizes the self-organization of discussions and resources pertaining to the cyber-artifact of interest within virtual communities. Site map analysis addresses the multilevel analysis suggested by Diani (2003). The second type of network analysis, link analysis, visualizes the ways virtual communities interact, self-organize, and share resources. This form of network visualization addresses the link multiplicity analysis proposed by Diani (2003). The third type of network analysis visualizes the growth of virtual communities over time, addressing the temporal analysis suggested by Diani (2003). By leveraging the classification of the cyber-artifacts performed in the second phase, the three forms of network analysis and visualization can also provide insight into the structural homophily of the various classes of cyber-artifacts in the virtual communities. Comparing the classes of collected cyber-artifacts, patterns and differences in structural homophily may be revealed in the three forms of network analysis.

The first type of network analysis, site map analysis, determines how IED-related cyber-artifacts are organized within a single virtual community domain, providing a form of multilevel analysis (Diani 2003). The average site maps for Web sites containing IED-related cyber-artifacts were constructed to analyze the intensities of discussion and materials Web pages at various levels within the domain, shown in Figs. 16.9 and 16.10, respectively. The site map was generated by counting the frequency of pages for each level and dividing by the total number of Web sites. These site maps are indicative of the communication patterns of the participants, showing where within their virtual community these issues are being discussed and specific IED-related resources mobilized. The site maps were drawn to a 20:1 scale; thus, five level-1 pages on the map represent 100 actual pages. The intensities of discussion and materials Web pages at various levels of the site map are represented using different color shades, with dark colors being more intense concentrations. The distribution of discussion and materials pages among the page levels in the average site maps is also presented in Table 16.7.

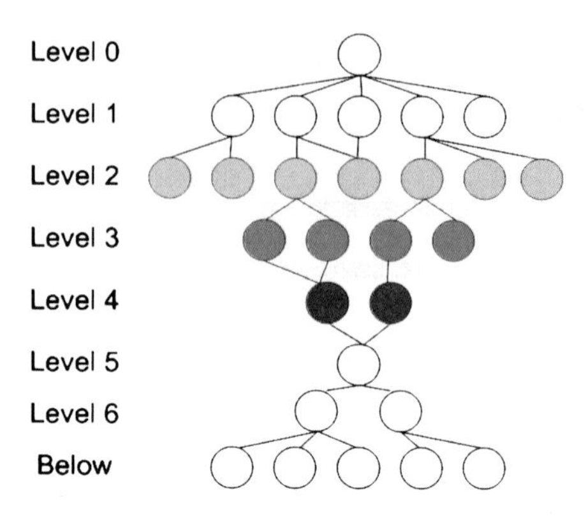

Fig. 16.10 Site map featuring intensity of materials pages in page levels

Table 16.7 Distribution of Web pages in site maps

	Discussion pages		Materials pages	
Page level	% Pages	No. of pages	% Pages	No. of pages
0	0.000	0	0.000	0
1	1.465	37	0.000	0
2	14.178	358	17.500	7
3	11.485	290	25.000	10
4	22.574	570	57.500	23
5	15.168	383	0.000	0
6	33.227	839	0.000	0
Below	1.782	45	0.000	0

The majority of materials Web pages occur in domain levels 2–4, while discussion content is often located in levels 2–6. The occurrence of IED-related cyber-artifacts at the middle levels of the domain may indicate that authors do not want their content to be too readily visible but also do not want to bury their content too deeply to be seen by community members. The concentration of discussion pages at lower levels in the domain than materials pages may be attributed to the strong identity traces in discussion content left by the message author, which the author wishes to hide. Materials pages are usually formally written manuals, often in document format, while discussion pages are casual free text with the author's stylistic tendencies evident. These observed differences between the organization of materials and discussion pages may be indicative of aspects of structural homophily (Diani 2003). The site map analysis also identified key terms in the IED lexicon and their occurrence distribution across site levels. Results revealed that certain terms occur more frequently at specific site levels. A selection of terms is presented in Table 16.8. Also shown are the common levels where the terms are likely to occur, with their average occurrence per page.

The second type of network analysis, link analysis, determines how multiple related virtual communities interact, share cyber-artifacts, and develop their collective identity. This visualization addresses Diani's (2003) suggestion to analyze link

Table 16.8 Key lexicon terms and levels of occurrence

Term	Translation	Levels	Average occurrence/page
الإشارة	Signaling	6–8	0.951
الدرس	Lesson	7–8	0.503
طلقة	Shot	Below 8	0.333
التغذية	Loading	1–3	0.268
الذخيرة	Ammunition	3–7	0.189
العيار	Caliber	5	0.113

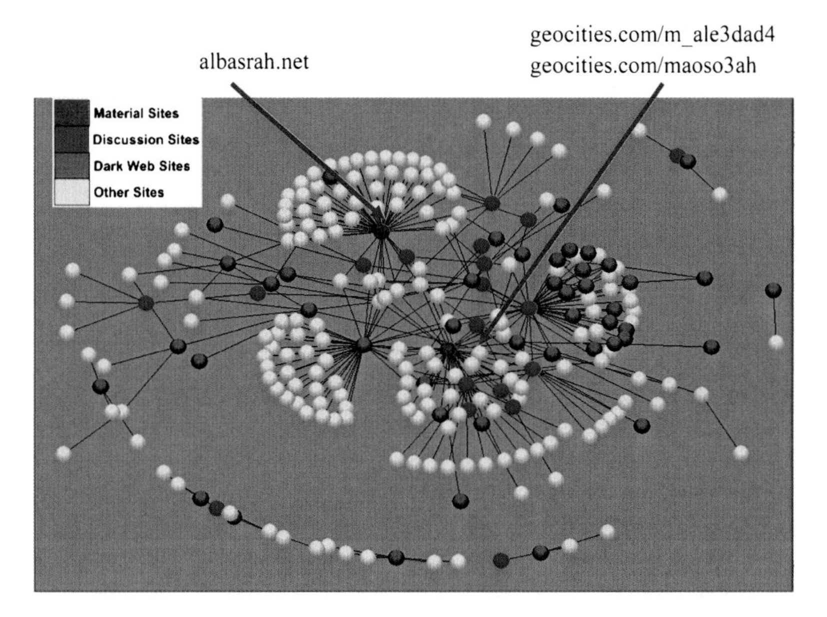

Fig. 16.11 Link network structure of social movement virtual communities

multiplicity. The link network structure, as displayed in Fig. 16.11, was constructed to show how virtual communities containing IED-related cyber-artifacts were connected, organizing their resources across multiple related sites of the social movement.

Each node of the link network represents a Web site and is color-coded depending upon the IED-related cyber-artifacts collected from the virtual community. The link network also includes Web sites not previously analyzed in the social movement study but with linkages to the investigated virtual communities. These Web sites, particularly those with extensive linkages to multiple virtual communities from which IED-related cyber-artifacts were already collected, are strong candidates for future related studies as their linkages may represent the presence of IED-related cyber-artifacts of interest.

Three major hubs for materials-related IED cyber-artifacts are identified in the network map. Albasrah.net is a Web site with links to the former Iraqi regime that

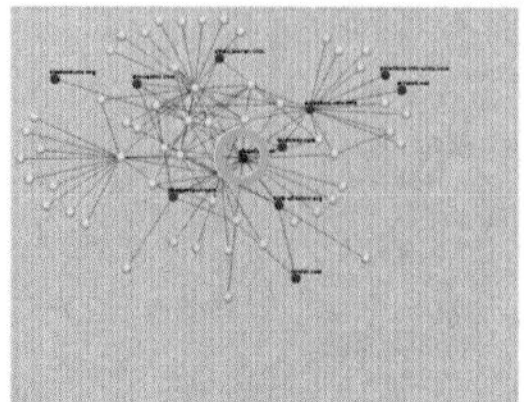

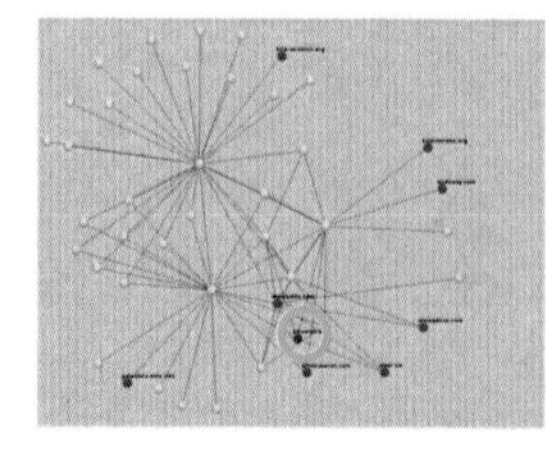

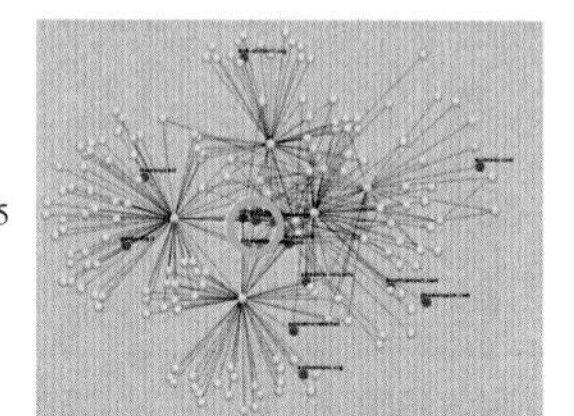

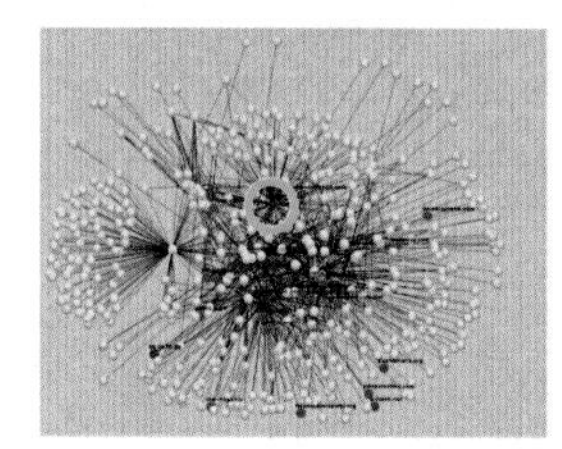

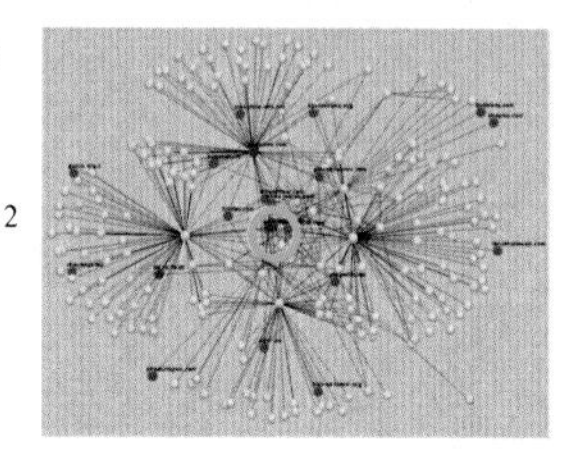

Fig. 16.12 Snapshots of link networks over a 1-year period

contains a large collection of war images and reports of military operations. Geocities.com/m_ale3dad4 is a collection of training materials, with topics including weapons, their usage, and the manufacturing of IEDs. The Web site also contains video demonstrations, manuals, and books in English and Arabic. Geocities.com/maoso3ah is an "encyclopedia" of military training. As shown in the link network, these Web sites act as major hubs for the mobilization of IED-related resources. The link network map provides insight into the relationship between virtual communities and how the production and reproduction of cyber-artifacts occurs between related constituents (Fuchs 2006b).

To understand the evolution of the virtual communities and their self-organization over time (Fuchs 2006b) and address the temporal dimension of network analysis proposed by Diani (2003), link network snapshots were taken over a one-year period for two major virtual communities, presented in Fig. 16.12.

In the link network diagrams, the virtual community Web sites containing IED-related discussions and materials, almuqatila.com and albasrah.net, respectively, are circled in green. Three snapshots were taken during the 1-year period, in January, July, and December of 2005. Both virtual communities experienced significant growth in their link networks over that time. In the case of almuqatila.com, the discussion Web site, total links to and from the Web site increased from 16 to 35. Diani suggested one of the significant aspects of social movement networks was centralization (2003). Betweenness is a measure of centrality in a link network, representing the degree to which a node provides a path to information resources to the rest of the network. The betweenness of almuqatila.com increased from 86.91 to 947.16, signifying an expanded role as information broker to the network. The link network of albasrah.net, the virtual community Web site containing IED-related materials, expanded dramatically in comparison. Total links to and from albasrah.net increased from 39 to 196, and the Web site's betweenness measure increased from 595.14 to 31,857.12.

The explosion in the albasrah.net link network may be due to the specialized knowledge contained in IED-related materials cyber-artifacts. These resources are considered particularly valuable to the virtual communities, and they are mobilized quickly and diffused throughout the movement. The differences observed in the growth of the link network supporting the virtual communities containing discussion and materials pages are revealing of structural homophily (Diani 2003). Temporal analysis of the virtual community link networks provides specific insight into the growth and integration of the community in the larger movement, the production and reproduction of movement participation, and the emergence and distribution of its specialized resources (Fuchs 2006b).

4 Conclusion

The Internet, virtual communities, and expressions of cyber protest are significant to social movements and offer ideal environments for conducting social research. In this chapter, we present a cyber-archaeology framework for social movement research. Our framework overcomes many of the issues of scale and complexity facing social research on the Internet and specifically addresses many of the suggestions of Paccagnella (1997), enabling broad and longitudinal social research on virtual communities. We offer a collection of powerful tools and techniques to assist in conducting social movement research on the Internet. The approach is automated and scalable, allowing a researcher to easily collect and analyze social movement cyber-artifacts across multiple linked virtual communities. The techniques for cyber-artifact analysis adopt perspectives of social movement theory, specifically the network (Diani and McAdam 2003) and self-organization (Fuchs 2006a) models of social movements.

To demonstrate the cyber-archaeology framework for social movement research and provide a detailed instantiation of the proposed approach for evaluation, a case study on a broad group of related social movement virtual communities is presented. Utilizing focused crawling, cyber-artifacts of interest were collected from multiple linked virtual communities. To categorize the collected cyber-artifacts by genre of communication, an extended feature representation and machine learning models were developed and tested, with nearly 90% accuracy in genre classification. Finally, the cyber-artifacts were analyzed through perspectives of self-organization and the network model of social movements, using network analysis and visualization techniques.

We hope the case study has demonstrated the efficacy of the proposed framework for social movement research and promotes its future use by researchers. Utilizing the proposed automated approaches to cyber-artifact collection and classification allows for large-scale social research across a greater number of virtual communities compared to traditional methods. A cyber-archaeology approach guided by social movement theory can produce high-quality social research and a deeper understanding of widespread movements.

Acknowledgments Funding for this research was provided by (1) NSF, "CRI: Developing a Dark Web Collection and Infrastructure for Computational and Social Sciences," 2007–2010 and (2) NSF, "EXP-LA: Explosives and IEDs in the Dark Web: Discovery, Categorization, and Analysis," 2007–2010.

References

Abbasi, A. and Chen, H. (2005). Identification and comparison of extremist-group web forum messages using authorship analysis. *IEEE Intelligent Systems*, 20(5), 67–75.

Aggarwal, C. C., Al-Garawi, F., and Yu, P. S. (2001). Intelligent crawling on the world wide web with arbitrary predicates. In *Proceedings of the 10th WWW Conference*. Hong Kong.

Burris, V., Smith, E., and Strahm, A. (2000). White supremacist networks on the Internet. *Sociological Focus*, 33(2), 215–235.

Chakrabarti, S., Van Den Berg, M., and Dom, B. (1999). Focused crawling: a new approach to topic-specific resource discovery. In *Proceedings of the Eighth WWW Conference*. Toronto, Canada.

Castells, M. (1998). *End of millennium – The information age: Economy, society, and culture*. Blackwell Publishers.

Collier, G. (1994). *Basta! Land and the Zapatista rebellion in Chaipas*. Food First Books.

Clark, D. and Themudo, N. (2006). Linking the web and the street: Internet-based 'dotcauses' and the 'anti-globalization' movement. *World Development*, 34(1), 50–74.

Cleaver, H. (1995). The Zapatistas and the electronic fabric of struggle. Online, http://www.eco.utexas.edu/faculty/Cleaver/zaps.html.

Cleaver, H. (1999). Computer-linked social movements and the global threat to capitalism. Online, http://www.eco.utexas.edu/faculty/Cleaver/polnet.html.

Diani, M. (2003). Networks and social movements: A research programme. In Diani, M. and McAdam D. (eds.), *Social Movements and Networks. Relational approaches to collective action*. Oxford University Press.

Diani, M. and McAdam D. (2003). *Social movements and networks. Relational approaches to collective action*. Oxford University Press.

Edelman, M. (2001). Social movements: changing paradigms and forms of politics. *Annual Review of Anthropology*, 30, 285–317.

Ester, M., Grob, M., and Kriegel, H. (2001). Focused web crawling: a generic framework for specifying the user interest and for adaptive crawling strategies. In *Proceedings of the International Conference on Very Large Databases*.

Finn A. and Kushmerick, N. (2006). Learning to classify documents according to genre. *Journal of the American Society for Information Science and Technology*, 57(11), 1506–1518.

Fuchs, C. (2006a). The self-organization of social movements. *Systemic Practice and Action Research*, 19(1), 101–137.

Fuchs, C. (2006b). The self-organization of cyberprotest. In Morgan, K., Brebbia, C., and Spector, J. (eds.), *The Internet Society II*. Southampton/Boston. WIT Press. pp. 275-295.

Glaser, J., Dixit, J., and Green, D. (2002). Studying hate crime with the Internet: what makes racists advocate racial violence? *Journal of Social Issues*, 58(1), 177–193.

Herring, S. and Paolillo, J. (2006). Gender and genre variation in weblogs. *Journal of Sociolinguistics*, 10(4), 439–459.

Joachims, T. (1998). Text categorization with support vector machines: learning with many relevant features. In *Proceedings of the ECML*, (pp. 137–142).

Jones, Q. (1997). Virtual-communities, virtual settlements, and cyber-archaeology: a theoretical outline. *Journal of Computer-Mediated Communication*, 3(3).

Jones, S. (1994). *Cyber-society: Computer-mediated communication and community*. Sage Publications.

Jordan, T. and Taylor, P. (2004). *Hacktivism and cyberwars: Rebels with a cause?* Routledge.

McAdam, D. (1986). Recruitment to high-risk activism: the case of freedom summer. *The American Journal of Sociology*, 92(1), 64–90.

Meikle, G. (2002). *Future active: Media activism and the Internet*. Routledge.

Paccagnella, L. (1997). Getting the seats of your pants dirty: strategies for ethnographic research on virtual communities. *Journal of Computer-Mediated Communication*, 3(1).

Santini, M. (2004). A shallow approach to syntactic feature extraction for genre classification. In *Proceedings of the 7th Annual Colloquium for the UK Special Interest Group for Computational Linguistics (CLUK)*.

Schafer, J. (2002). Spinning the web of hate: web-based hate propagation by extremist organizations. *Journal of Criminal Justice and Popular Culture*, 9(2), 69–88.

Shepherd, M., Watters, C., and Kaushik, R. (2001). Lessons from reading E-News for browsing the web: the role of genre and task. In *Proceedings of the American Society for Information Science*, 38, 256–267.

Snow, D. and Marshall, S. (1984). Cultural imperialism, social movement, and the Islamic revival. In L. Kriesberg (ed.), *Research in Social Movements, Conflicts, and Change*, Vol. 7. London: Jai Press.

Starr, A. (2000). *Naming the enemy: Anti-corporate movements confront globalization*. Zed Books.

Tesh, S. (2002). The Internet and the grass roots. *Organization and Environment*, 15(3), 336–339.

Van de Donk, W., Loader, D., Nixon, P., and Rucht, D. (2004). *Cyberprotest: New media, citizens and social movements*. Routledge.

Wiktorowicz, Q. (2002). Islamic activism and social movement theory: a new direction for research. *Mediterranean Politics*, 7(3), 187–211.

Yates, J. and Orlikowski, W. (1992). Genres of organizational communication: a structurational approach to studying communication and media. *Academy of Management Review*, 17(2), 299–326.

Yates, J. and Orlikowski, W. (2002). Genre systems: structuring interaction through communicative norms. *Journal of Business Communication*, 39(1), 13–35.

Zheng, R., Qin, Y., Huang, Z., and Chen, H. (2006). A framework for authorship analysis of online messages: writing-style features and techniques. *Journal of the American Society for Information Science and Technology*, 57(3), 378–393.

Zhou, Y., Reid, E., Qin, J., Chen, H., and Lai, G. (2005). U.S. extremist groups on the web: link and content analysis. *IEEE Intelligent Systems*, 20(5), 44–51.

Chapter 17
Weapons of Mass Destruction (WMD) on Dark Web

1 Introduction

The tragic events of September 11 have caused drastic effects on many aspects of society. Academics in the fields of natural sciences, computational science, information science, social sciences, engineering, medicine, and many others have been called upon to help enhance the government's ability to fight terrorism and other crimes. Six critical mission areas have been identified where information technology can contribute, as suggested in the National Strategy for Homeland Security report (Office of Homeland Security 2002), including intelligence and warning, border and transportation security, domestic counterterrorism, *protecting critical infrastructure, defending against catastrophic terrorism,* and *emergency preparedness and response.* Facing the critical missions of national security and various data and technical challenges, we believe there is a pressing need to develop the science of "Intelligence and Security Informatics" (ISI) (Chen 2006), with its main objective being the "development of advanced information technologies, systems, algorithms, and databases for national security–related applications, through an integrated technological, organizational, and policy-based approach."

In the area under *defending against catastrophic terrorism*, weapons of mass destruction (WMD), especially nuclear weapons, have been considered one of the most dangerous threats to US homeland security and international peace and prosperity. There is a critical need to advance fundamental knowledge in new technologies for the detection of nuclear threats and to develop intellectual capability in fields relevant to long-term advances in nuclear detection capability.

In this research, we propose to develop a Capability-Accessibility-Intent Model to identify and analyze: (1) the unique capabilities of countries, institutions, and researchers in developing nuclear WMD, (2) the accessibility of nuclear facilities and materials in high-risk countries (e.g., Iran, North Korea, and other Middle Eastern countries) and by potential international and domestic terrorist groups, and (3) the stated intent (and threat) of selected rogue countries or terrorist groups in obtaining and using nuclear materials. Based on open source publications, reports, and web

H. Chen, *Dark Web: Exploring and Data Mining the Dark Side of the Web*,
Integrated Series in Information Systems 30, DOI 10.1007/978-1-4614-1557-2_17,

sites, we aim to develop a knowledge base of the "Nuclear Web" to represent the major high-risk countries, organizations, institutions, researchers, and their nuclear capabilities. In addition, we plan to leverage our highly successfully and internationally acclaimed "Dark Web" project, which collects international jihadist-generated contents (web sites, forums, blogs, etc.) on the Internet, to identify terrorist and extremist groups and members who may have expressed their illicit intent to develop or use such nuclear WMD capabilities.

2 Literature Review: Knowledge Mapping and Focused Web Crawling

In this section, we review research that is relevant to open source content collection and analysis. The research is grouped here into one of two streams of academic research: *knowledge mapping* and *focused web crawling*.

2.1 Knowledge Mapping

In Diane Crane's seminal book on "Invisible Colleges: Diffusion of Knowledge in Scientific Communities" (Crane 1972), she suggests that it is the "invisible college," which consists of a small group of highly productive scientists and scholars, that is responsible for growth of scientific knowledge. The productive scientists and scholars form a network of collaborators in promoting and developing their fields of study. The presence of an invisible college or network of productive scientists linking separate groups of collaborators within a research area has been evident in many studies (Chen 2003; Shiffrin and Börner 2004). In nuclear physics research, we believe this is equally true. Productive researchers and scholars in developed countries often form the nucleus of the field; however, nuclear scholars and researchers in many developing and volatile regions (e.g., India, Pakistan, Iran, North Korea, etc.) also follow such developments closely and have often developed their own nuclear capabilities. *Knowledge mapping*, based on text mining, network analysis, and information visualization, has become an active area of research that helps reveal such an interconnected, invisible college or network of scholars and their seminal publications, important ideas, and critical capabilities.

2.1.1 Text Mining

For knowledge mapping research, text mining can be used to identify critical subject and topic areas that are embedded in the title, abstract, and text body of published articles. Based on automatic indexing or information extraction techniques, documents are often represented as a vector of features (i.e., keywords, noun phases, or entities). Articles that are collected and grouped based on authors, institutions, topic areas, countries, or regions can be analyzed to identify the underlying themes, patterns, or trends. Popular content analysis techniques include clustering algorithms, self-organizing map (SOM), multidimensional scaling (MDS), principal component analysis (PCA), co-word analysis, and pathfinder network (Chen 2003; Chen 2001).

2.1.2 Network Analysis

Recent advances in *social network analysis* and *complex networks* have provided another means for studying the network of productive scholars in the invisible college. A collection of methods that are recommended in literature for studying networks is the social network analysis (SNA) techniques (Sparrow 1991; Xu and Chen 2005). Because SNA is designed to discover patterns of interactions between social actors in social networks, it is especially apt for coauthorship network analysis. Specifically, SNA is capable of detecting subgroups (of scholars), discovering their pattern of interactions, identifying central individuals, and uncovering network organization and structure. It has also been used to study criminal networks (Xu and Chen 2005).

2.1.3 Information Visualization

The last step in the knowledge mapping process is to make knowledge transparent through the use of various information visualization (or mapping) techniques. Shneiderman (1996) proposed seven types of information representation methods including the *1D (one-dimensional)*, *2D (two-dimensional)*, *3D (three-dimensional)*, *multidimension*, *tree*, *network*, and *temporal* approaches. The two commonly used interaction approaches are: *overview + detail* and *focus + context* (Card et al. 1999).

We believe that knowledge mapping research can help us identify open source intelligence (OSINT) content of relevance to nuclear physics and WMD, especially for assessing the capabilities of those high-risk regions, countries, institutions, groups, and researchers.

2.2 *Focused Web Crawling*

Focused crawlers "seek, acquire, index, and maintain web contents on a specific set of topics that represent a narrow segment of the web" (Chakrabarti et al. 1999). Unlike knowledge mapping research that often relies on existing information sources, *focused web crawling* aims to collect from the web previously disorganized and disparate information of relevance to a particular domain. For nuclear threat detection, it is critical to develop a knowledge base of the "Nuclear Web" (of people, organizations, capabilities, threat levels, etc.) based on open source content from the web. We briefly review previous research pertaining to these important considerations, which include *accessibility, content richness,* and *URL ordering techniques.*

2.2.1 Accessibility

As noted by Lawrence and Giles (1998), a large portion of the Internet is dynamically generated. Such content typically requires users to have prior authorization, fill out forms, or register (Raghavan and García-Molina 2001). This covert side of the Internet is commonly referred to as the hidden/invisible web. Two general strategies have been introduced to access the hidden web via automated web crawlers. The first approach entails use of automated form-filling techniques. A second alternative for accessing the hidden web is a task-specific human-assisted approach. This approach provides a semiautomated framework that allows human experts to assist the crawler in gaining access to hidden content.

2.2.2 Content Richness

The web is rich in indexable and multimedia files. Difficulties in indexing make multimedia content difficult to accurately collect (Baeza-Yates 2003). Many previous studies have ignored multimedia content altogether. However, we observe that multimedia files have been heavily used by terrorist groups for their propaganda and recruiting purposes. For nuclear-related contents, we anticipate a need for multimedia content collection and processing.

2.2.3 URL Ordering Techniques

URL ordering helps guide the crawlers toward the targeted documents and contents. Numerous link analysis techniques have been used for URL ordering. For example, Cho et al. (1998) evaluated the effectiveness of PageRank and backlink counts. Chau and Chen (2003) used a Hopfield net crawler that collected pages related to the medical domain based on link weights.

3 The Capability-Accessibility-Intent Model for Nuclear Threat Detection: Nuclear Web and Dark Web

In this research, we propose a framework that aims to investigate the capability, accessibility, and intent of critical high-risk countries, institutions, researchers, and extremist or terrorist groups (Fig. 17.1).

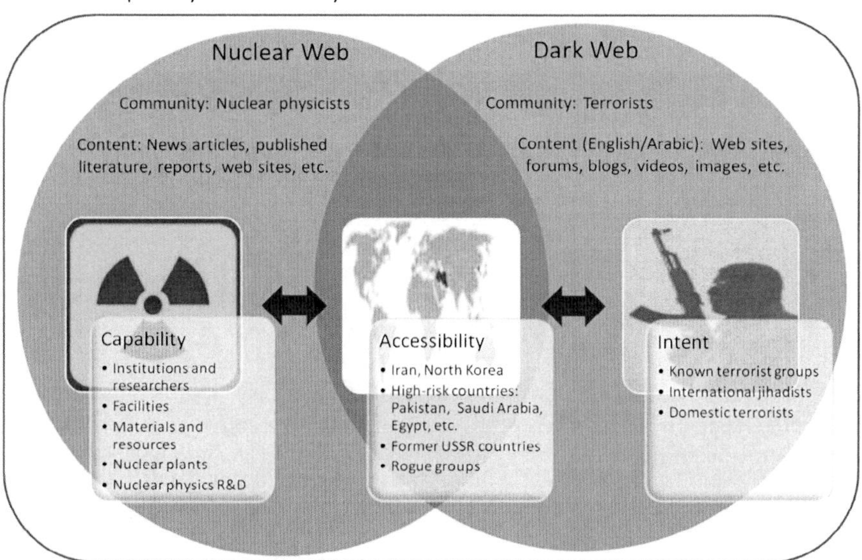

Fig. 17.1 Capability-accessibility-intent model for nuclear threat detection

3.1 Capability

Nuclear physics and bomb making demand significant scientific knowledge, engineering expertise, and material and resource availability. Such nuclear weapon capabilities are not easily obtained and can often take many years of heavy resource investment. Only selected countries, institutions, and researchers have access to such know-how, materials, resources, and facilities. The USA and European countries clearly have the best capabilities. However, they are also typically under better management and oversight. The threat potentials are significantly lower. On the other hand, selected rogue countries and volatile regions are also developing such capabilities, including Iran, North Korea, Pakistan, India, and other Middle Eastern countries. Although their capabilities are lower, their threat levels are much higher. By collecting and analyzing (using *knowledge mapping* techniques) nuclear-related publications (journal articles, conference proceedings, reports, press articles, etc.) that are generated by scholars and researchers in these high-risk regions, we will be able to identify the "invisible college of nuclear scholars" and their capabilities.

3.2 Accessibility

Although many countries and institutions have exhibited strong capabilities in nuclear research, their know-how, processes, materials, and facilities are inaccessible to most outsiders. In selected Middle Eastern and Muslim countries, local nuclear personnel may have a higher chance of being coerced or influenced by local radical groups. Their facilities may have higher accessibility to hostile agents, thus posing significant threats. Similarly, in the former East-bloc USSR countries, nuclear materials and know-how may be more accessible to local gangs and mafias for illicit purposes. A systematic accessibility analysis of institutions and facilities in various high-risk countries and regions is needed to gauge their risk level.

3.3 Intent

In addition to capability and accessibility, the potential vicious intent of selected radical, extremist, and terrorist groups needs to be better studied. It is well known that some terrorist organizations have more highly educated recruits and can perform more sophisticated and coordinated operations (e.g., al-Qaeda). In many terrorist web sites, forums, and blogs, training manuals and instructions for creating explosives (Chen 2006), bio/chemical agents, and even nuclear bombs can be easily found. Opinion leaders, followers, sympathizers, and wannabes in these Dark Web sites and forums often discuss and exchange radical and violent ideas of relevance to global jihad and other destructive acts. We believe *focused web crawling* techniques can be extremely useful in identifying nuclear threat intent of actors in the Dark Web cyberspace.

4 Case Study: Nuclear Web and Dark Web

4.1 Knowledge Mapping for Nuclear Web

Knowledge mapping data sources: Based on our initial analysis of the nuclear physics–related information content, we have identified the following relevant knowledge mapping sources:

- *Inspec*: Inspec data is available from 1896 to 2008. Using "nuclear physics" as the search keyword, we were able to identify 552,885 bibliographic records, with country, author, and institution information.
- *Physics Preprint*: Nuclear experiment and nuclear theory are two subject areas of relevance in Physics Preprint. Dated from 1994 to 2008, we were able to identify about 11,000 full-text articles and reports in the full-text format (with substantial content details).
- *Thomson SCI Database*: 69,936 nuclear-related records were found in the SCI database, from 1952 to 2008.
- *Energy Citations Database*: In the Energy Citations Database, we found more than 8,000 nuclear physics–related records. Again, 2,107 records were found at the peak of 2005.

In addition to these publication sources, we have also identified other Nuclear Web contents generated by various agencies, e.g., International Atomic Energy Agency, National Nuclear Security Administration, Defense Nuclear Facilities Safety Board, etc. Although most of these contents are general and benign, we have also identified foreign nuclear contents of selected at-risk countries that may demand systematic monitoring and analysis, e.g., Atomic Energy Organization of Iran (http://www.aeoi.org.ir; Arabic content) and Korean Peninsula Energy Development Organization (http://www.kedo.org; North Korea). Selected reports, photos, and news posted in these web sites may bring valuable contextual intelligence.

Results: We conducted a preliminary study to analyze the nuclear-related research literature in the Thompson SCI Database, which provides approximately 5,900 of the world's leading scholarly science and technical journals covering more than 150 disciplines. In the SCI Database, there are four nuclear-related areas: (1) radiology, nuclear medicine, and medical imaging; (2) chemistry, inorganic, and nuclear; (3) physics, nuclear; and (4) nuclear science and technology. The total number of nuclear-related articles in SCI is 69,936. We analyzed the research literature published by authors from selected high-risk countries. For example, we found 184 nuclear publications from Iran and 196 from Pakistan. We also analyzed the top researchers in these countries. Table 17.1 shows the top five first authors and the top five general authors (regardless of author order) and the number of articles they published. In Iran, Modarres, M had the largest number of publications as both first author and general author. In Pakistan, Khan, HA had the largest number of publications as both first author and general author.

Besides individual author's information, we analyzed their organizations as well. Table 17.2 lists the top five organizations with the most publications in Iran and Pakistan. Most of them were university departments or government research centers.

To study the collaboration status, we analyzed the coauthorship relationship among researchers. Figure 17.2 is an example of a coauthorship network among prominent Iranian nuclear researchers. The node in the network represents an individual researcher. The bigger the node, the more publications the researcher has. The link between two researchers indicates that these two researchers have published scientific article(s) together. The thicker the link, the more articles these two authors have published together. There are two large subgroups in the center of the graph with 22 and 16 researchers, respectively. These are clearly the key people to watch in this volatile country.

4.2 *Focused Web Crawling for Dark Web*

Analysis of web content is becoming increasingly important due to augmented communication via Internet computer-mediated communication (CMC) sources such as e-mail, web sites, forums, and chat rooms. The numerous benefits of the Internet and CMC have been coupled with the realization of some vices. In addition

Table 17.1 The top five first authors and the top five general authors in Iran and Pakistan

Country	Rank	First author	Number of publications	Author	Number of publications
Iran	1	Modarres, M	13	Modarres, M	20
	2	Jalilian, AR	7	Shamsipur, M	11
	3	Sohrabi, M	7	Moshfegh, HR	9
	4	Bordbar, GH	6	Jalilian, AR	9
	5	Boroushaki, M	6	Sabet, M	8
Pakistan	1	Khan, HA	27	Khan, HA	67
	2	Qureshi, IE	8	Qureshi, IE	39
	3	Gul, K	6	Manzoor, S	21
	4	Khan, MJ	5	Shahzad, MI	18
	5	Ansari, SA	5	Qureshi, AA	13

Table 17.2 Top five nuclear research organizations in Iran and Pakistan

Country	Rank	Organization	Number of publications
Iran	1	Shiraz Univ, Dept Phys, Shiraz, Iran	11
	2	Atom Energy Org Iran, Ctr Nucl Res, Tehran, Iran	10
	3	Sharif Univ Technol, Dept Mech Engn, Tehran, Iran	7
	4	Razi Univ, Dept Chem, Kermanshah, Iran	6
	5	Univ Teheran, Dept Elect and Comp Engn, Tehran, Iran	5
Pakistan	1	Pinstech, Radiat Phys Div, Islamabad, Pakistan	17
	2	Pakistan Inst Nucl Sci and Technol, Div Nucl Chem, Islamabad, Pakistan	12
	3	Quaid I Azam Univ, Dept Chem, Islamabad, Pakistan	8
	4	Punjab Univ, Dept Phys, Lahore, Pakistan	6
	5	Pakistan Inst Nucl Sci and Technol, Islamabad, Pakistan	6

to misuse in the form of cybercrime, identity theft, and the sales and distribution of pirated software, the Internet has also become a popular communication medium and haven for extremist and hate groups. This problematic facet of the Internet is often referred to as the Dark Web (Chen 2006).

Extremist and terrorist groups often use the Internet to promote hatred and violence (Glaser et al. 2002). The Internet offers a ubiquitous, quick, inexpensive, and anonymous means of communication for extremist groups. Many studies have conducted content analysis on the Dark Web (e.g., Schafer 2002; Zhou et al. 2005) and found evidence of ideological resource sharing, fundraising, propaganda, training,

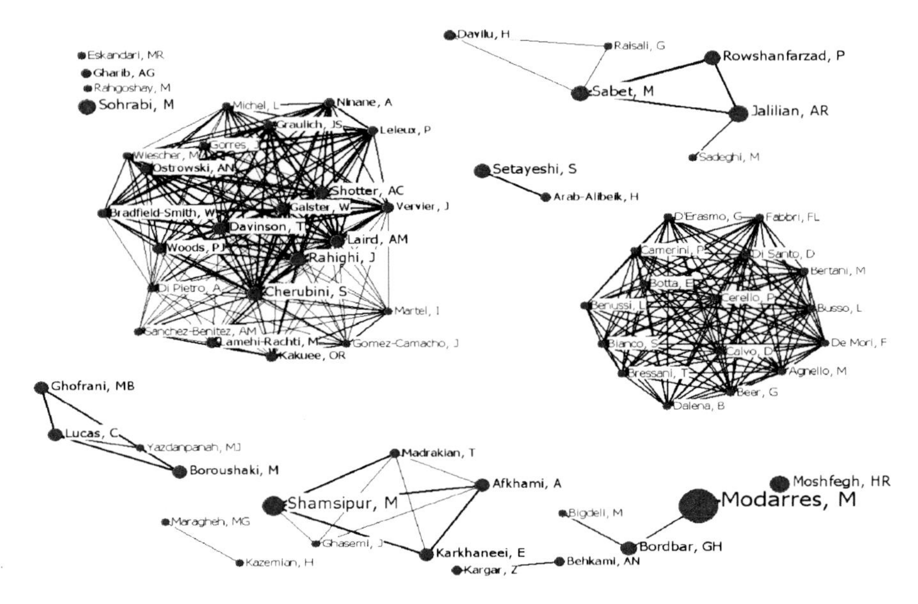

Fig. 17.2 A coauthorship network of prominent Iranian nuclear researchers

and recruitment-related material. Other studies also measured the violent and hateful affect intensities and found considerable presence of both in US supremacist and Middle Eastern extremist group forums (Abbasi and Chen 2005). The aforementioned studies present important content analysis findings that provide insight into the communication and propaganda dissemination dynamics of Dark Web forums and web sites. However, there has been limited work on identifying and analyzing content pertaining to nuclear and WMD threats. In this section, we summarize our proposed system design for the collection of nuclear-related content in the Dark Web.

A focused crawling system for dark web nuclear content: We propose to develop a focused crawling system for the Dark Web nuclear content as shown in Fig. 17.3. The site identification phase is intended to identify extremist groups and their web sites. Sources for the US domestic extremist groups include the Anti-Defamation League, FBI, Southern Poverty Law Center, Militia Watchdog, etc. Sources for the international extremist groups include the United States Committee for a Free Lebanon, Counter-Terrorism Committee of the UN Security Council, and US State Department reports, as well as government reports from other countries. Once groups have been identified, we create an initial set of URLs and their related in-link and out-link web sites. We also plan to search major search engines to identify other web sites by using a carefully developed lexicon of Arabic and English nuclear keywords.

The site preprocessing phase has three components: accessibility, structure, and wrapper generation. The accessibility component deals with acquiring and maintaining access to Dark Web sites and forums. The structure component is designed

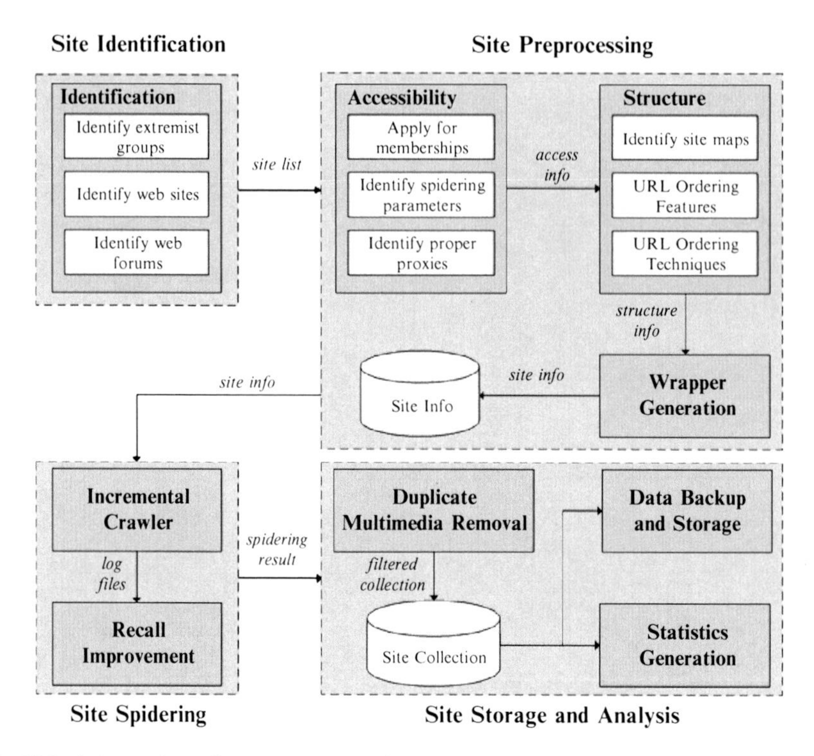

Fig. 17.3 A focused crawling system design for Dark Web nuclear content

to identify the URL mapping and devise the crawl space URL ordering using the relevant features and techniques. Many Dark Web forums do not allow anonymous access (Zhou et al. 2005). In order to access and collect information from those forums, one must create a user ID and password, send an application request to the web master, and wait to get permission/registration to access the forum. Once access has been attained, spidering parameters such as number of connections, download intervals, timeout, speed, etc., need to be set appropriately according to server and network limitations and the various site and forum blocking mechanisms. However, we may still be blocked based on our IP address. Therefore, we will use proxies to increase not only our recall but also our anonymity.

The site structure component involves identifying the site map and URL ordering features and techniques. In addition to our nuclear keyword lexicon, we intend to use URL tokens. For example, for web forums, we are interested in URLs containing words such as "board," "thread," "message," etc. (Glance et al. 2005). Additional relevant URL tokens include domain names of third-party file hosting web sites. These third parties often contain multimedia files. File extension tokens (e.g. ".jpg" and ".wmv") are also important.

"الاعتراف الأخير" للعقل العلمي العراقي	The Last Disclosure of the Iraqi Scientific Mind
محمد عارف	Muhammad Aref
مستشار في العلوم والتكنولوجيا	Science and Technology Advisor
"أيها الوطن العراقي العظيم، تكوّر على جراحك ولا تنحن لأي كان. كن شامخاً مثل نخيلك، أول من يستقبل الشمس والمطر". بهذه الكلمات اختتم كُتاب "الاعتراف الأخير" رواية القصة الحقيقية للبرنامج النووي العراقي. السخرية المريرة واضحة في عنوان الكتاب، الذي شارك في تأليفه عالم الفيزياء النووية جعفر ضياء جعفر، وزميله عالم الكيمياء النووية نعمان النعيمي. وهل هناك أجدر بالسخرية من كتب "صانع قنبلة صدّام" و"القنبلة في حديقتي المنزلية"، وعشرات الاعترافات والشهادات المزورة عن "أسلحة الدمار الشامل العراقية"؟ و"من غيرنا يعرف الحقيقة على حقيقتها"، يتساءل جعفر والنعيمي، اللذان ساهما في تأسيس وإدارة "البرنامج النووي العراقي" منذ البداية حتى النهاية.	"Dear great nation of Iraq ..." with those words the "last disclosure" book was concluded, the real story of the Iraqi nuclear program ...

Fig. 17.4 Excerpt from the review of a book authored by Iraqi nuclear scientists Nu'man Al Nu'aymi and Jaafar Dia Jaafar

The URL ordering techniques are more important for forums as compared to web sites. We use rules based on URL tokens and levels to control the crawl space. Moreover, to adapt to different forum structures, we need to use different crawl space traversal strategies. Breadth first (BFS) is used for board page forums, while depth first (DFS) is used for Internet service provider (ISP) forums. DFS is necessary for many ISP forums due to the presence of ad pages that periodically appear within these forums. When such an ad page appears, it must be traversed in order to get to the message pages (typically, the ad pages have a link to the actual message page).

The incremental crawler fetches only new and updated pages. A log file is sent to the recall improvement component. The log shows the spidering status of each URL. A parser is used to determine the overall status for each URL (e.g., "download complete," "connection timed out"). Uncollected pages are respidered. Multimedia files are occasionally manually downloaded, particularly larger video files that may otherwise timeout. The forum storage and analysis phase will consist of a statistics generation and duplicate multimedia removal components. Once files have been collected, they must be stored and analyzed. Statistics are generated for the number of static and dynamic indexable files, multimedia files (e.g., image, audio, and video), archive files (e.g., RAR, ZIP), and nonstandard files (unrecognized file formats).

Preliminary results: In our recent preliminary study of the Dark Web, we developed a small lexicon of nuclear-related English and Arabic keywords, e.g., "نووي (nuclear)," "انشطار (fission)," "كتلة حرجة (critical mass)," etc. Using the spidering process described above, we identified 128 Arabic web sites and 95 English web sites with potentially relevant nuclear content. The majority of the relevant web pages discussed international nuclear policies, in particular, the nuclear standoffs between the West and North Korea and Iran. Other web pages discussed the former Iraqi nuclear program. For instance, one of the web sites posted an interview with Iraqi nuclear scientists who participated in the former Iraqi regime's nuclear

Fig. 17.5 NTM front page: "The jihadi nuclear bomb and the method of nuclear enrichment; Volume 1: The jihadi nuclear bomb"

weapons program (Fig. 17.4). Moreover, some jihadists consider nuclear weapons to be an important component in their future operations. Although it is considerably more difficult to uncover terrorist-generated data on nuclear technology, we were able to find a handful of primers written specially for jihadists, e.g., The "Nuclear Tutorial for the Mujahedeen" (NTM) ("دورة الاعداد النووي للمجاهدين"). This set of lessons was found on the "Encyclopedia of Training and Preparation," a web site dedicated to provide future jihadists with basic military training and useful manuals. (Source: http://www.geocities.com/m_alu3dad4/).

The NTM is a 19-lesson workshop on nuclear technology (see Fig. 17.5 for the front page). The lessons are collected in 14 pdf files with a total of 477 pages. The author declares that the purpose of this tutorial is to teach the Mujahedeen (Holy Warriors) the basics of nuclear and missile technology. He claims that he relied on various Western web sources and references (which he did not acknowledge specifically). The topics discussed in this jihad nuclear primer are: introduction to nuclear physics, Fermi physics, natural radiation, nuclear characteristics of some elements, the nuclear bomb, nuclear material used in the bomb, preparation of the radium nuclear bomb, nuclear and EM bombs, and basic missile technology.

5 Conclusions

In this research, we propose a research framework that aims to investigate the capability, accessibility, and intent of critical high-risk countries, institutions, researchers, and extremist or terrorist groups. Selected knowledge mapping and focused web crawling techniques and preliminary findings are presented in this chapter. We believe our proposed framework and techniques can help shed light on the proliferation and threats of global WMD terrorism.

Acknowledgments Funding for this research was provided by (1) NSF, "CRI: Developing a Dark Web Collection and Infrastructure for Computational and Social Sciences," NSF CNS-0709338, 2007–2010; and (2) NSF, "EXP-LA: Explosives and IEDs in the Dark Web: Discovery, Categorization, and Analysis," NSF CBET-0730908, 2007–2010.

References

Abbasi, A. and Chen, H. (2005). Identification and Comparison of Extremist-Group Web Forum Messages using Authorship Analysis. IEEE Intelligent Systems, 20(5), 67–75.

Baeza-Yates, R. (2003). Information Retrieval in the Web: Beyond Current Search Engines. International Journal of Approximate Reasoning, 34, 97–104.

Card, S. K., Mackinlay, J. D. and Shneiderman, B. (1999). *Readings in Information Visualization: Using Vision to Think*, Morgan Kaufmann Publishers, Inc., San Francisco, CA.

Chakrabarti, S., Van Den Berg, M., and Dom, B. (1999). Focused Crawling: A New Approach to Topic-Specific Resource Discovery. In Proceedings of the Eight World Wide Web Conference, Toronto, Canada.

Chau, M. and Chen, H. (2003). Comparison of Three Vertical Search Spiders. IEEE Computer, 36(5), 56–62.

Chen, C. (2003). *Mapping Scientific Frontiers*. London, UK, Springer-Verlag.

Chen, H. (2001). *Knowledge Management Systems: A Text Mining Perspective*, University of Arizona, Tucson, Arizona.

Chen, H. (2006). *Intelligence and Security Informatics*, Springer-Verlag.

Cho, J., Garcia-Molina, H., and Page, L. (1998). Efficient Crawling Through URL Ordering. In Proceedings of the 7th World Wide Web Conference, Brisbane, Australia.

Crane, D. (1972). *Invisible Colleges: Diffusion of Knowledge in Scientific Communities*. Chicago, University of Chicago Press.

Glance, N., Hurst, M., Nigam, K. Siegler, M., Stockton, R. and Tomokiyo, T. (2005). Analyzing Online Discussion for Marketing Intelligence, In Proceedings of the 14th International World Wide Web Conference, Chicago, Illinois.

Glaser, J., Dixit, J., and Green, D. P. (2002). Studying Hate Crime with the Internet: What Makes Racists Advocate Racial Violence? Journal of Social Issues, 58(1), 177–193.

Lawrence, S. and Giles, C. L. (1998). Searching the World Wide Web. *Science*, 280(5360), 98–100.

Office of Homeland Security. (2002). National Strategy for Homeland Security. Washington D.C.: Office of Homeland Security.

Raghavan, S. and Garcia-Molina, H. (2001). Crawling the Hidden Web. In Proceedings of the 27th International Conference on Very Large Databases.

Schafer, J. (2002). Spinning the Web of Hate: Web-Based Hate Propagation by Extremist Organizations. Journal of Criminal Justice and Popular Culture, 9(2), 69–88.

Shneiderman, B. (1996). The Eyes Have It: A Task by Data Type Taxonomy for Information Visualization, Proceedings of IEEE Workshop on Visual Languages'96, 336–343.

Shiffrin, R. M. and Börner, K., (2004). *Mapping Knowledge Domains*. Arthur M. Sackler Colloguia of the National Academy of Sciences, National Academies of Sciences.

Sparrow, M. K., (1991). The Application of Network Analysis to Criminal Intelligence: An Assessment of the Prospects, Social Networks, 13, 251–274.

Xu, J. and Chen, H. (2005) CrimeNet Explorer: A Framework for Criminal Network Knowledge Discovery. ACM Transactions on Information Systems, 23(2), pp. 201–226.

Zhou, Y., Reid, E., Qin, J., Chen, H., and Lai, G. (2005). U.S. Extremist Groups on the Web: Link and Content Analysis. IEEE Intelligent Systems, 20(5), 44–51.

Chapter 18
Bioterrorism Knowledge Mapping

1 Introduction

Since the anthrax attacks following 9/11, bioterrorism has been given a high priority in national security. Bioterrorism involves bioweapon attacks against a civilian population. Such attacks against civilians are usually intended to cause widespread panic and terror (Lane et al. 2001). Given the potential significance of bioterrorism events, biodefense initiatives have received significant attention from both government agencies and research communities. One critical problem in biodefense research and practice is that biomedical research used for defense purposes can also be applied to biological weapons development. To mitigate risk, the US Government has attempted to monitor and regulate biomedical research labs, especially those that study bioterrorism agents/diseases (United States Centers for Disease Control and Prevention 2005). However, monitoring worldwide biomedical researchers and their work is still an issue.

With explosive growth of scientific information, there is an overwhelming amount of journal articles in relevant research areas (Bruijn and Martin 2002; Cohen and Hersh 2005). Literature resources are very important to scientific research. Scientific literature always contains the newest information and intelligence essential to researchers. It may be used as a resource to monitor bioterrorism research.

In this chapter, we develop an integrated approach to monitor and analyze worldwide bioterrorism research literature by using knowledge mapping techniques. Our objectives are to identify:

- Researchers who have expertise in the bioterrorism agents/diseases research domain
- Major institutions and countries where these researchers reside
- Emerging topics and trends in bioterrorism agents/diseases research

The remainder of this chapter is organized as follows: Sect. 2 reviews previous research on bioterrorism literature analysis and knowledge mapping techniques. Section 3 describes the test bed for this chapter. The research design is introduced

H. Chen, *Dark Web: Exploring and Data Mining the Dark Side of the Web*,
Integrated Series in Information Systems 30, DOI 10.1007/978-1-4614-1557-2_18,

in Sect. 4. Analysis results are presented and discussed in Sect. 5. In Sect. 6, we present conclusions and directions for future work.

2 Literature Review

2.1 *Bioterrorism Literature Analysis*

Terrorism is not an easy research topic because of the clandestine nature of terrorist groups (Merari 1991; Silke 2001). It is also difficult to identify the intellectual structure and characteristics of contemporary terrorism literature (Reid and Chen 2007). Some previous studies used bibliometrics to examine terrorism research publications and offer an evolutional view of the development of the field (Kennedy and Lum 2003; Reid 1983, 1997). Reid (1983) used both content analysis and citation analysis to identify the most frequently cited (MFC) terrorism publications.

In the bioterrorism domain, biodefense research and practice may also be applied toward biological weapon development. Mining implicit knowledge/information from biomedical literature was started by Swanson's pioneering work on Raynaud's disease/fish-oil discovery in 1986 (Swanson 1986). Since then, biomedical informatics tools have been developed to protect our populations from bioterrorism attacks (Kohane 2002). Hu and his team (2005a, b, 2006) used a text mining approach on PubMed literature to identify candidate viruses and bacteria as potential bioterrorism weapons.

However, few efforts have been made to monitor and analyze worldwide bioterrorism research and researchers. Few studies have used knowledge mapping techniques to analyze the research status of the bioterrorism area. In this chapter, we focus on 58 bioterrorism agents/diseases from the Centers for Disease Control and Prevention (CDC) list for humans and 58 diseases from the World Organization for Animal Health (OIE) list for animals. The related literature were collected and analyzed. We also highlight the research status of state sponsors of terrorism.

2.2 *Knowledge Mapping*

Knowledge mapping techniques can reveal the interconnected network of scholars and their publications and ideas. Three types of analysis are often adopted in knowledge mapping research: text mining, network analysis, and information visualization (Chen and Roco 2009).

2.2.1 Text Mining

For knowledge mapping research, text mining can be used to identify critical subjects and topic areas that are embedded in the title, abstract, and text body of published articles (Chen and Roco 2009). Text mining consists of two significant classes of techniques: natural language processing (NLP) and content analysis.

In NLP, information extraction is a computationally effective method to identify important concepts from text documents. Most existing information extraction approaches combine machine learning algorithms such as neural networks, decision tree, hidden Markov model, and entropy maximization with a rule-based or a statistical approach. The best systems have been shown to achieve more than 90% accuracy in both precision and recall rates when extracting persons, locations, organizations, dates, times, currencies, and percentages from newspaper articles (Chinchor 1998).

In content analysis, articles that are collected and grouped based on authors, institutions, topic areas, countries, or regions can be analyzed to identify the underlying themes, patterns, or trends. Popular techniques include clustering algorithms, self-organizing map (SOM), multidimensional scaling (MDS), principal component analysis (PCA), co-word analysis, and PathFinder network (PFNET). For example, Chen and his team (1996) developed a multilayered SOM (ET-Map) to categorize 110,000 Internet web pages. Kohonen and his colleagues (2000) adopted SOM to map 6.8 million patent abstracts onto a one-million-node SOM.

2.2.2 Network Analysis

Social network analysis (SNA) has provided the means for studying the network of productive scholars. It is capable of detecting subgroups (of scholars), discovering their pattern of interactions, identifying central individuals, and uncovering network organization and structure (Chen and Roco 2009).

Given a collaboration or coauthorship network, traditional data mining techniques such as cluster analysis may be used to detect underlying groups that are not otherwise apparent in the data. Burt (1976) applied hierarchical clustering methods based on structural equivalence measures (Lorrain and White 1971) to detect subgroups in a social network.

Several measures, such as degree, betweenness, and closeness, are related to the roles of individuals in a network (Wasserman and Faust 1994). The degree of a particular node is the number of direct links it has, its betweenness is the number of geodesics (the shortest paths between any two nodes) passing through it, and the closeness is the total number of all the geodesics between the node and every other node in the network. These measures imply the importance of individuals in the network.

2.2.3 Information Visualization

The last step in the knowledge mapping process is to make knowledge transparent through the use of various information visualization techniques (Zhu and Chen 2005).

Shneiderman (1996) proposed seven types of information representation methods, including the 1D (one-dimensional), 2D, 3D, multidimensional, tree, network, and temporal representation. The 1D representation has been applied to display either the content of a single document (Hearst 1995) or an overview of a collection of documents (Eick et al. 1992). The output of a SOM has been used as 2D representation in some visualization systems (Chen et al. 1996; Kohonen 1995; Huang et al. 2004). In 3D representation, realistic metaphors such as rooms or bookshelves (Card et al. 1996) or buildings (Andrews 1995) are employed to depict abstract information. The multidimensional approach represents information into a three-dimensional or a two-dimensional space. The tree approach is often used to represent hierarchical relationships. The network representation method is often applied when a simple tree structure is insufficient for representing complex relationships. The temporal approach visualizes information based on temporal order. Location and animation are commonly used as visual variables to reveal the temporal aspect of information.

The seven types of representation methods turn abstract textual documents into objects that can be displayed. To achieve effectiveness, an effective information representation method needs to be integrated with user-interface interaction. Recent advances in hardware and software allow quick user-interface interaction, and various combinations of representation methods and user-interface interactions have been developed.

3 Research Test Bed

We built two sets of test data based on human- and animal-related bioterrorism agents/diseases, respectively. Related research articles were retrieved from the MEDLINE database. For the human bioterrorism agents/diseases dataset, we retrieved 178,599 publication records from MEDLINE (1964–2005) by searching article abstracts and titles using 58 keywords from the CDC's list of agents by category (http://www.bt.cdc.gov/ Agent/agentlist.asp). For the animal bioterrorism agents/diseases dataset, we retrieved 135,774 publication records from MEDLINE (1965–2005) by searching article abstracts and titles using 58 keywords from OIE's list of diseases by species (http://www.oie.int/eng/maladies/en_classification.htm). Tables 18.1 and 18.2 show the characteristics of the two datasets.

For human agents/diseases, *E. coli* and Q fever, both in CDC's agents category B, had the greatest number of publications among all the agents/diseases. Botulism had the highest number of publications in category A, followed by anthrax and plague. There were relatively fewer publications in category C. For animal diseases/agents, most publications were about Q fever. There were also many publications on vesicular stomatitis, foot-and-mouth disease, and rabies.

Table 18.1 Human agents/diseases dataset broken down by CDC's agents category

Agent/disease	Number of publications	Number of unique authors
Category A	8,635	23,891
Botulism	3,780	9,988
Anthrax	1,674	5,579
Plague	1,504	4,169
Smallpox	846	2,623
Viral hemorrhagic fever	678	1,945
Tularemia	494	1,454
Category B[a]	170,460	356,162
E. coli	106,479	212,338
Q fever	34,312	115,136
Category C (Only Nipah virus and hantavirus)	919	2,974
Overall[b]	178,599	381,684

[a] Only the two most researched diseases in category B are shown
[b] Some articles mention multiple diseases

Table 18.2 Animal agents/diseases dataset broken down by OIE's disease list

Agent/disease	Number of publications	Number of unique authors
Q fever	33,999	114,600
Vesicular stomatitis	2,374	7,281
Foot-and-mouth disease	2,338	7,159
Rabies	2,209	5,509
Brucellosis	1,955	5,585
Anthrax	1,240	4,236
Paratuberculosis	997	2,616
Japanese encephalitis	988	2,870
West Nile virus	944	2,086
Avian influenza	717	3,446
Overall[a]	135,774	320,630

[a] Only top ten diseases are shown

4 Research Design

Figure 18.1 shows the research design for mapping worldwide bioterrorism research literature. The design consists of three components. The first component, data acquisition, involves gathering the bioterrorism agents/diseases–related research literature from the MEDLINE database. The second component, data parsing and cleaning, contains methods to parse data into relational databases and consolidate the parsed facts. The last component, data analysis, involves identifying the productivity status, the collaboration status, and the emerging topics of bioterrorism research.

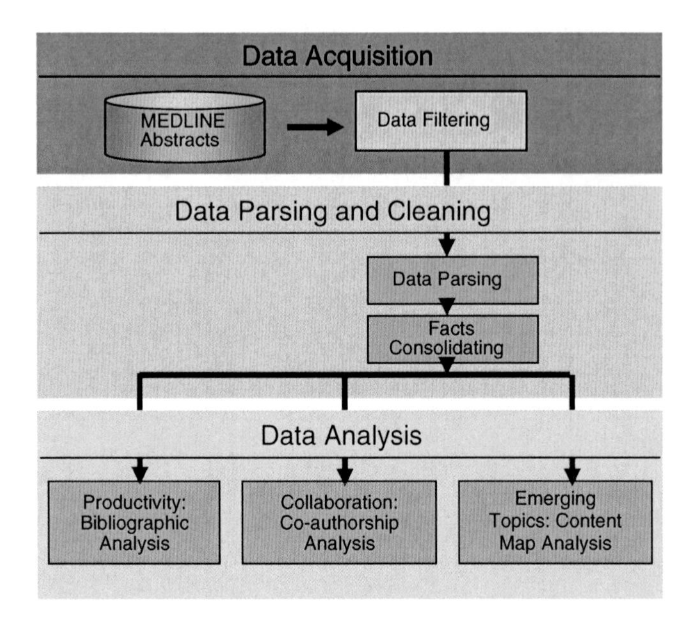

Fig. 18.1 Research design – mapping worldwide bioterrorism research literature

4.1 Data Acquisition

Research articles were retrieved from the MEDLINE database. Compiled by the US National Library of Medicine (NLM) and published on the web by Community of Science, MEDLINE is the world's most comprehensive source of life science and biomedical bibliographic information. It contains nearly 11 million records from over 7,300 different publications from 1965 to November 16, 2005 (http://medline.cos.com/). All the related articles were collected by using keyword filtering.

4.2 Data Parsing and Cleaning

In data parsing, the title, abstract, and authors' information for each article were parsed and stored in a relational database. The institutions and countries of the authors were parsed out by using dictionaries of countries, states, cities, and institutions. All the author names of an article were parsed out, but only the first author's institution was kept for later analysis

In facts consolidating, some variations of foreign institution names and city names were spot-checked and fixed manually.

4.3 *Data Analysis*

We used bibliographic analysis to study the productivity of authors, institutions, and countries. We also assessed the trends and evolution of bioterrorism agents/diseases research activities. We used coauthorship analysis to study collaboration between researchers. We also detected the independent or isolated research groups in the field. We used SOM to study the active research topics and discover the emerging research topics in different time spans

5 Analysis Results and Discussion

5.1 *Productivity Status of Bioterrorism Research*

Bibliographic analysis was used to identify the most productive bioterrorism research countries, institutions, and researchers for both human and animal agents/diseases.

For human agents/diseases research, the United States had the most publications, followed by Japan and the United Kingdom. At the institution level, Harvard University had the most publications, followed by the University of Wisconsin-Madison and Institute Pasteur, Paris. For animal agents/diseases research, the United States also had the most publications, followed by Japan and the United Kingdom. At the institution level, CDC-Atlanta had the most publications, followed by National Taiwan University and Institute Pasteur, Paris.

We also analyzed countries that were mentioned as state sponsors of terrorism by the government's country reports on terrorism, such as Iran, Cuba, Sudan, Libya, North Korea, and Syria, and countries in the Middle East and North Africa. For example, Table 18.3 shows the top five researchers in selected state sponsors of terrorism, Middle East and North Africa (excluding Israel), and Eurasia for human agents/diseases.

In Cuba, Guzman, M. G., whose expertise was in Q fever, published the most articles, followed by Campos, J. with expertise in *E. coli.*

In the Middle East and North Africa, Israeli researchers had the largest number of publications. Other than Israel, Memish, Z. A. and Al-Eissa, Y. A., both from Saudi Arabia, were among the most productive researchers.

In Eurasia, Ozen, S. from Turkey had the largest number of publications, followed by Avaeva, S. M. from Russia.

Overall, the top researchers in the state sponsors of terrorism focused more on *E. coli*; the top researchers in the Middle East and North Africa focused more on Q fever and brucellosis; and the top researchers in Eurasia focused more on *E. coli*, Q fever, and smallpox.

Table 18.3 Top five researchers in selected state sponsors of terrorism, Middle East and North Africa (excluding Israel), and Eurasia for human agents/diseases

Region	Rank	Researcher	Number of publications
State sponsors of terrorism:			
Cuba	1	Guzman, M. G. (Q fever)	18
	2	Campos, J. (*E. coli*)	14
	3	Fando, R. (cholera)	9
	4	Kouri, G. (Q fever)	9
	5	Silva, A. (*E. coli*)	5
Iran	1	Jafari, A. (*E. coli*)	8
	2	Katouli, M. (*E. coli*)	7
	3	Bouzari, S. (*E. coli*)	7
	4	Shokouhi, F. (*E. coli*)	5
	5	Farhoudi-Moghaddam, A. A. (*E. coli, Salmonella*)	5
Middle East and North Africa			
	1	Memish, Z. A., Saudi Arabia (brucellosis)	27
	2	Al-Eissa, Y. A., Saudi Arabia (Q fever)	21
	3	Majeed, H. A., Kuwait (Q fever)	15
	4	Botros, B. A., Egypt (Q fever)	13
	5	Araj, G. F., Kuwait (brucellosis)	13
Eurasia	1	Ozen, S., Turkey (Q fever)	30
	2	Avaeva, S. M., Russia (*E. coli*)	18
	3	Baykov, A. A., Russia (*E. coli*)	17
	4	Shchelkunov, S. N., Russia (smallpox)	16
	5	Skulachev, V. P., Russia (*E. coli*)	15

5.2 Collaboration Status of Bioterrorism Researchers

Coauthorship analysis was used to identify and visualize collaboration between researchers for both human and animal agents/diseases. We analyzed different collaboration groups based on different agents/diseases and different regions.

For example, Fig. 18.2 shows the collaboration status of researchers on anthrax. The node in the network represents an individual researcher. The bigger the node, the more publications the researcher has published. The link between two researchers means that these two researchers have published scientific article(s) together. The thicker the link, the more articles these two authors have published together. We only included researchers who published more than five articles. The largest group in the center consists of researchers from the United States. The second largest group is from France. The smaller groups are from India, Israel, Italy, and United Kingdom.

Figure 18.3 shows the collaboration status of researchers in state sponsors of terrorism on CDC's category A agents/diseases. There are six groups, all from Iran. The two largest groups with the most productive researchers are from Pasteur

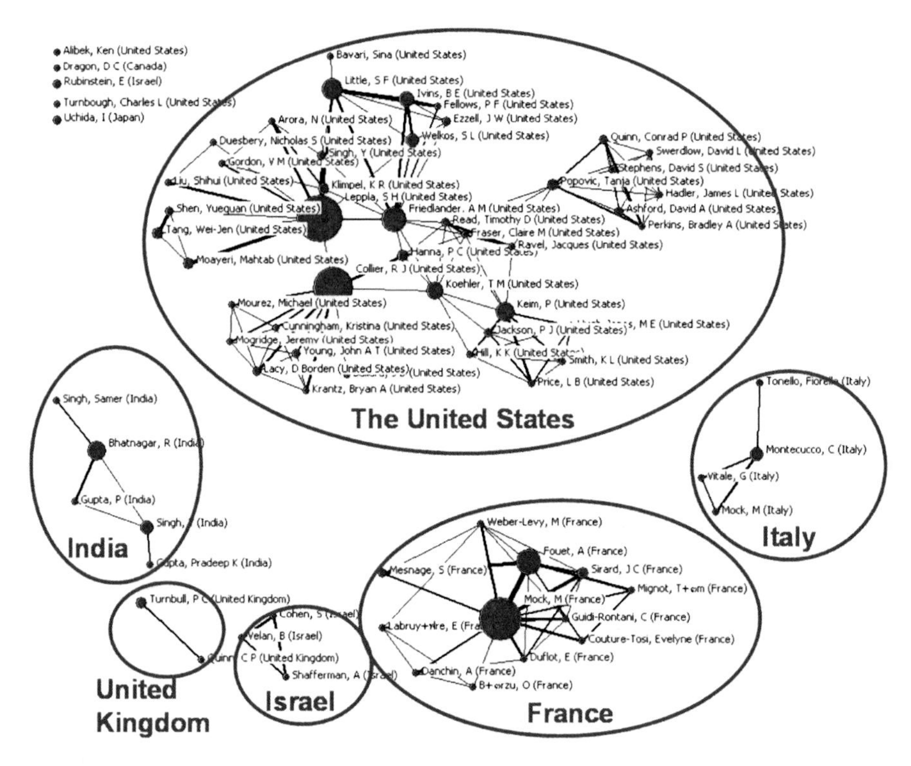

Fig. 18.2 Collaboration status of researchers on anthrax. Researchers with more than five articles are shown

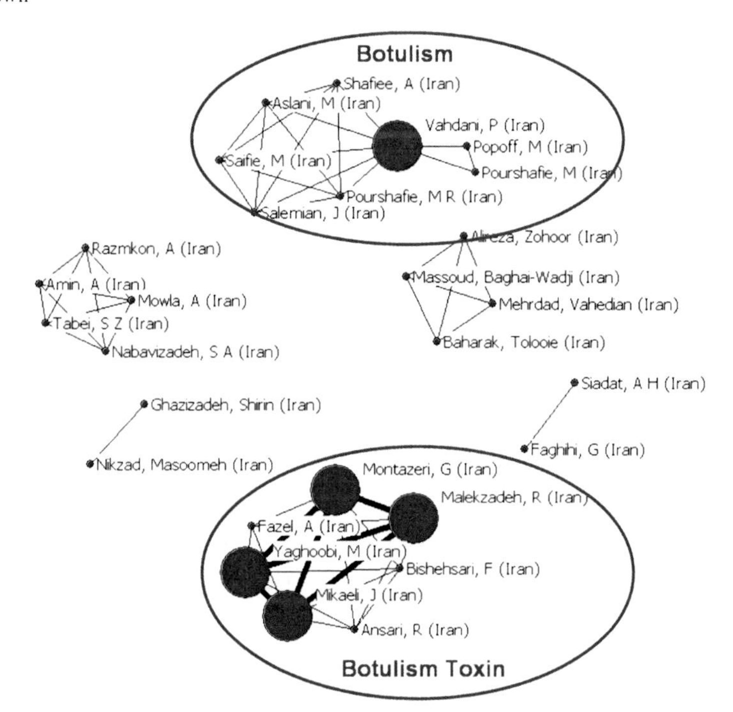

Fig. 18.3 Collaboration status of researchers in state sponsors of terrorism on CDC's category A agents/diseases

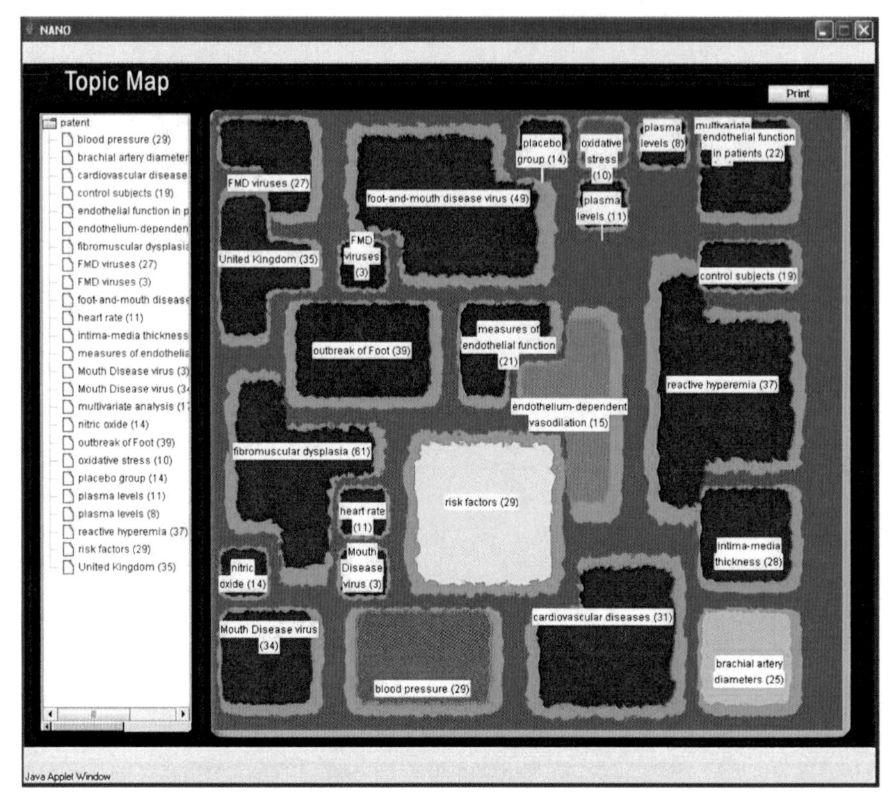

Fig. 18.4 Content map for human agents/diseases literature (1996–2000)

Institute of Iran (top) and Tehran University of Medical Sciences (bottom). Both groups focused on botulism.

5.3 Emerging Topics of Bioterrorism Research

Content map analysis was used to identify the emerging topics and trends for both human and animal agents/diseases research. Figures 18.4 and 18.5 show the evolution of major research topics in human agents/diseases literature for two time periods, 1996–2000 and 2001–2005, respectively.

The nodes in the folder tree and colored regions are topics extracted from research papers. The topics are organized by the multilevel self-organization map algorithm. The conceptually closer technology topics (according to co-occurrence patterns) are positioned closer geographically. Numbers of papers belonging to the topics are

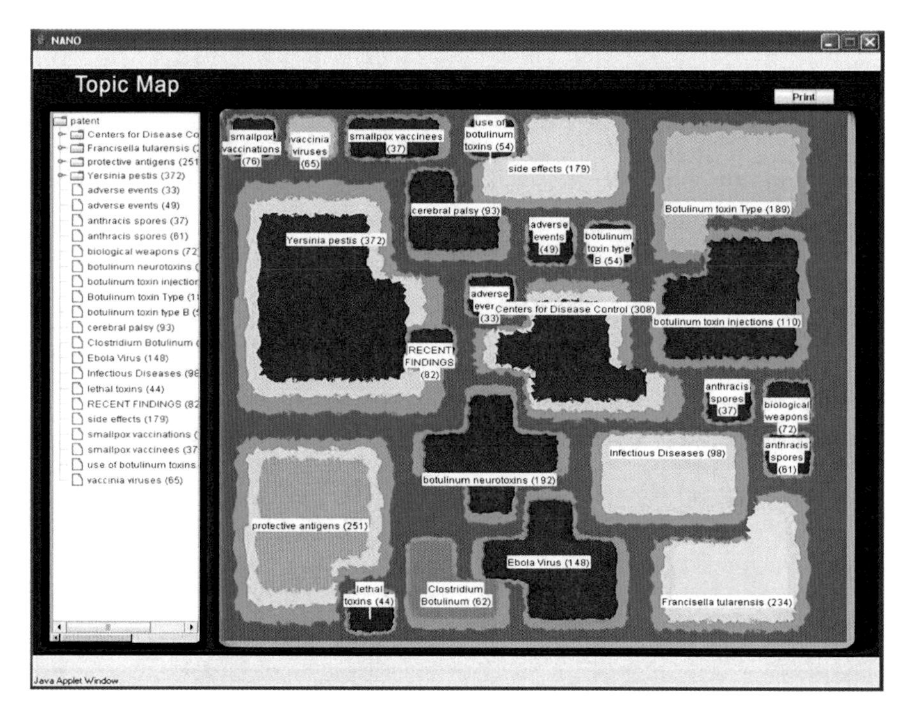

Fig. 18.5 Content map for human agents/diseases literature (2001–2005)

presented after the topic labels. The sizes of the topic regions also correspond to the number of documents assigned to the topics. Region color indicates the growth rate of the associated topic. The warmer the color, the higher the growth rate is. The growth rate is defined as the number of articles published in the previous time period/the number of articles published in the following time period for a particular topic (region).

It can be observed that dominating topic regions during 1996–2000 are "Botulinum toxin type," "*Francisella tularensis*," "*Clostridium botulinum*," "Effect of Botulinum toxin," "V antigens," and "Ebola viruses." The sizes of these topic regions suggest that they were the key technology topics during the 5 years before 2000. Among these dominating topics, "Botulinum toxin type," "Effect of Botulinum toxin," "Ebola viruses," and "V antigens" are emerging ones.

During 2001–2005, dominating topics are "*Yersinia pestis*," "Centers of Disease Control," "Protective antigens," "*Francisella tularensis*," and "Botulinum neurotoxin." The new important topics observed are "Biological weapons," "Anthracis spores," and "Smallpox vaccination." We can see a shift in research interest toward the use of anthrax spores and biological weapons after 2000.

6 Conclusions and Future Directions

Monitoring worldwide bioterrorism research is becoming increasingly important and urgent. In this chapter, we built an integrated approach to mapping worldwide bioterrorism literature and capabilities. We analyzed the productivity status, collaboration status, and emerging topics by using knowledge mapping techniques. Bibliographic analysis was used to analyze the productivity of researchers, institutions, and countries. Coauthorship analysis was used to analyze collaboration among researchers. Content map analysis was used to identify the emerging research topics and trends. The analysis results provide insights into the worldwide status of the bioterrorism research domain.

In the future, we plan to monitor and analyze more bioterrorism agents/diseases together with more literature sources. With the growth of focused bioterrorism agents/diseases and the number of literature sources, we also plan to develop and incorporate more advanced analysis and visualization techniques into our approach.

Acknowledgments We would like to thank all the members of the University of Arizona Artificial Intelligence Lab for their support and assistance of this research. This material is based upon work supported by the National Science Foundation's Information Technology Research (ITR) Program, grant number IIS-0428241, "BioPortal: A National Center of Excellence for Infectious Disease Informatics."

References

Andrews, K. "Visualizing cyberspace: Information visualization in the Harmony Internet browser," *Proceedings of IEEE Symposium on Information Visualization (InfoVis'95)*, pp. 97–104, 1995.

Biosafety in Microbiological and Biomedical Laboratories, Washington: U. S. Government Printing Office, Ed. 5th., 2007.

Bruijn, B. and Martin, J. "Literature mining in molecular biology," *Proceedings of the EFMI Workshop on Natural Language Processing in Biomedical Applications*, Nicosia, Cyprus, March 8–9, pp. 1–5, 2002.

Burt, R. S. "Positions in networks," *Social Forces*, vol. 55, no. 1, pp. 93–122, 1976.

Card, S.K., Robertson, G.G., and York, W. "The WebBook and the WebForager: An information workspace for the World Wide Web," *Proceedings of the ACM SIGCHI Conference on Human Factors in Computing Systems (CHI'96)*, pp. 111–117, 1996.

Chen, H. and Roco, M. *Mapping Nanotechnology Innovations and Knowledge: Global and Longitudinal Patent and Literature Analysis*. Springer, 2009.

Chen, H., Schuffels, C., and Orwig, R. "Internet categorization and search: A self-organizing approach," *Journal of Visual Communication and Image Representation*, vol. 7, no. 1, pp. 88–102, 1996.

Chinchor, N. "MUC-7 test scores introduction," In *Proceedings of the Seventh Message Understanding Conference*, 1998.

Cohen, A.M. and Hersh, W. R., "A survey of current work in biomedical text mining," *Briefings in Bioinformatics*, vol. 6, no. 1, pp. 57–71, 2005.

Eick, S.G., Steffen, J.L., and Sumner, E.E. "Seesoft: A tool for visualizing line-oriented software," *IEEE Transactions on Software Engineering*, vol. 18, no. 11, pp. 11–18, 1992.

Hearst, M. "TileBars: Visualization of term distribution information in full text information access," *Proceedings of the ACM SIGCHI Conference on Human Factors in Computing Systems*, pp. 59–66, 1995.

Hu, X., Yoo, I., Rumm, P., and Atwood, M. "Mining candidate viruses as potential bio-terrorism weapons from biomedical literature," in *2005 IEEE International Conference on Intelligence and Security Informatics (IEEE ISI-2005)*, Atlanta, Georgia, 2005.

Hu, X., Zhang, X., Yoo, I., Atwood, M. and Rumm, P. "A text mining approach for identifying candidate viruses as potential bio-terrorism weapons," *GESTS International Transaction on Compute Science and Engineering*, vol. 9, no. 1, pp. 109–120, 2005.

Hu, X., Zhang, X., Wu, D., Zhou, X., and Rumm, P. "Text mining the biomedical literature for identification of potential virus/bacterium as bioterrorism weapons," in *Terrorism Informatics: Knowledge Management and Data Mining for Homeland Security*, Ed. H. Chen, E. Reid, J. Sinai, A. Silke, and B. Ganor, Springer, 2006.

Huang, Z., Chung, W., and Chen, H. "Graph model for E-commerce recommender systems," *Journal of the Amercian Society for Information Science and Technology (JASIST)*, vol. 55, no. 3, pp. 259–274, 2004.

Kennedy, L.W. and Lum, C.M. *Developing a Foundation for Policy Relevant Terrorism Research in Criminology*. New Brunswick, Rutgers University, 2003.

Kohane, I. S. "The contributions of biomedical informatics to the fight against bioterrorism," *Biomedical Informatics and Bioterrorism*, 2002. 9: p. 116–119.

Kohonen, T., Kaski, S., Lagus, K., Salojärvi, J., Honkela, J., Paatero, V., and Saarela, A. "Self organization of a massive document collection," *IEEE Transactions On Neural Networks*, vol. 1, no. 3, pp. 574–585, 2000.

Kohonen, T. *Self-Organizing Maps*. Berlin: Springer-Verlag, 1995.

Lane, H.C., LaMontagne, J., and Fauci, A.S. "Bioterrorism: A clear and present danger," *Nature Medicine*, vol. 7, no. 12, pp. 1271–1273, 2001.

Lorrain, F. and White, H.S. "Structural equivalence of individuals in social networks," *Journal of Mathematical Sociology*, vol. 1, pp. 49–80, 1971.

Merari, A. "Academic research and government policy on terrorism," *Terrorism and Political Violence*, vol. 3, no. 1, pp. 88–102, 1991.

Reid, E. and Chen, H. "Mapping the contemporary terrorism research domain," *International Journal of Human-Computer Studies*, vol. 65, pp. 42–56, 2007.

Reid, E. *Analysis of Terrorism Literature: a Bibliometric and Content Analysis Study*. Dissertation, USC. School of Library and Information Management, University of Southern California, Los Angeles, 1983.

Reid, E. "Evolution of a body of knowledge: an analysis of terrorism research," *Information Processing and Management*, vol. 33, no. 1, pp. 91–106, 1997.

Shneiderman, B. "The eyes have it: A task by data type taxonomy for information visualizations," In *Proceedings of the IEEE Symposium on Visual Languages*, Washington: IEEE Computer Society Press, 1996.

Silke, A., "Devil you know: Continuing problems with terrorism research," *Terrorism and Political Violence*, vol. 13, no. 4, pp. 1–14, 2001.

Swanson, D.R. "Fish oil, Raynauds syndrome, and undiscovered public knowledge," *Perspectives in Biology and Medicine*, vol. 30, no. 1, pp. 7–18, 1986.

United States Centers for Disease Control. Possession, use, and transfer of select agents and toxins–Reconstructed replication competent forms of the 1918 Pandemic Influenza Virus containing any portion of the coding regions of all eight gene segments. Interim Final Rule. *Federal Register* 70(202): 61047–61049, Oct. 20, 2005.

Wasserman, S. and Faust, K. *Social Networks Analysis: Methods and Applications*. Cambridge: Cambridge University Press, 1994.

Zhu, B. and Chen, H. "Information visualization," *Annual Review of Information Science and Technology (ARIST)*, vol. 39, pp. 139–178, 2005.

Chapter 19
Women's Forums on the Dark Web

1 Introduction

The rapid development and evolution of the Internet have enabled people to access information whenever and wherever they want. Recently, with the advent of Web 2.0, the Internet has evolved toward multimedia-rich content delivery, end-user content generation, and community-based social interactions (O'Reilly 2005). More and more web forums, blogs, wikis, and other social media have been generated and become popular. Such Web 2.0 social media help enhance information sharing, opinion generation, and community-based discussions for various emerging social and political topics.

Although it has a male-dominated history, the Internet is becoming a new medium for women to share their concerns and express opinions about personal, social, and political issues (Harcourt 2000). With such a trend, the need for women to claim the Internet as an important space of their own has emerged. Women could gain equal presence or influence with men in the virtual community. In addition, their desire for gender equality continues to influence their Internet contributions and writing. Meanwhile, the increasing availability of the Internet offers marginalized groups and individuals a voice in the public sphere (Harp and Tremayne 2006; Mitra 2004). For example, Harcourt (2000) mentions the increasing voice of local Arab women on a global level through the Internet; Mitra (2004) argues that the Internet has allowed women in South Asia to be heard by the outside world.

In many disciplines, questions concerning gender differences in the context of online communication have been raised (Halbert 2004). Online gender differences (i.e., digital gender gap in some studies), which refers to the differences between women and men in Internet use, have been shown and studied in previous research (Fountain 2000; Fuller 2004; Harp and Tremayne 2006). Some studies point out that women are less likely to express political opinions and tend to have a less authoritative manner in their conversation style (Ogan et al. 2005). More research is critically needed to explain online gender differences in social, political, and even business (e.g., online shopping) activities.

H. Chen, *Dark Web: Exploring and Data Mining the Dark Side of the Web*,
Integrated Series in Information Systems 30, DOI 10.1007/978-1-4614-1557-2_19,

Text classification techniques can be used to identify and analyze online gender differences by examining the discrepancy between women's and men's writing styles. Previous studies on gender classification focus on using feature-based text classification methods to examine how women and men might use language(s) differently. Those studies have shown noticeable differences between women's and men's writing styles on different types of texts, including e-mails (Corney et al. 2002), web blogs (Nowson and Oberlander 2006; Schler et al. 2006), novels (Hota et al. 2006), and nonfiction articles (Koppel et al. 2002). In the web forum context, previous studies mainly used keyword-based analysis to examine the topic differences between males and females (Guiller and Durndell 2007; Seale et al. 2006). Few studies have investigated the gender differences in web forums using feature-based text classification techniques.

In this study, we propose a feature-based gender classification framework for web forums to examine the writing styles and contents (including different types of linguistic features) of female and male posters. Our analysis encompasses gender classification on an Islamic women's political forum. We compare and examine different feature sets consisting of content-free and content-specific features. We also study feature selection using the information gain (IG) heuristic. We further analyze the different topics preferred by women and men, respectively. The results of classification using support vector machine (SVM) indicate a high level of classification accuracy, demonstrating the efficacy of this framework of gender classification for web forums.

The remainder of this chapter is organized as follows: Sect. 2 provides a review of previous research in online gender differences and online text classification studies. Section 3 describes research gaps and questions, while Sect. 4 presents our research design. Section 5 describes the experiment used to evaluate the effectiveness of the proposed framework and discusses the results. Section 6 concludes the chapter with closing remarks and future directions.

2 Literature Review

In this section, we review previous research on online gender differences and online text classification studies.

2.1 *Online Gender Differences*

With the increasing availability and popularity of the Internet, as well as the advent of Web 2.0, more and more women participate in community-based social media (Consaluo and Paasonen 2002). The Internet, therefore, has become a medium for women to share their political pressure and knowledge (Harcourt 2000). They are also creating their own online networks to exchange information and opinions (Sherman 2001).

The Internet is not only useful as a fast communication medium; it is also a very crucial channel of information on women's rights issues. Women use the Internet to fight against violence by building a strong layer of support through which their personal struggles can be discussed and solutions shared (Harcourt 2000). As an example, in her study, Harcourt (2000) talks about a case regarding a Muslim woman's right of choice of marriage; she argues that "we could, within hours, receive case law on the issue from other Muslim countries as well as legal and scholarly opinions and references that prove critical in winning the case." As argued by Shade (2002; see also Seale et al. 2006), the Internet is to the third-wave feminists what independent feminist presses were to the second-wave, providing a means to form female networks and resist the relatively male-dominated networks in cyberspace. In sum, women want to be treated equally with men in the virtual society and claim the Internet as an important space of their own (Halbert 2004).

Along with such a trend, researchers have an increasing interest in studying online gender differences, which refers to the fact that there are differences between women and men in Internet use (Bimber 2000). The major online gender difference noted is that fewer women than men use the Internet. For example, the A. C. Nielsen CommerceNet consortium from 1999 showed that among US and Canadian Internet users, 53% were men and 47% were women; among online shoppers, 62% were men and 38% were women; and among people who reported having used the Internet in the last 24h for any purpose, 68% were men and 32% were women (CommerceNet 1999). In the realm of political activity, the National Election Study (NES) data shows that visitors to Internet campaign sites during the 1998 election season were 60% male and 40% female (National Election Study 1998). However, with the rapid development and increasing availability of the Internet, more and more women are accessing the Internet to acquire information, express their ideas, and share common concerns. The May 2008 survey by the Pew Internet and American Life Project found that 73% of men and 73% of women use the Internet (Pew Internet and American Life Project 2008). In contrast, its 2004 survey reported 66% and 61% Internet use for men and women, respectively.

Although access technology is not an issue today, women and men do have differences in Internet use depending on motivation and interest in the content being produced and consumed (Harp and Tremayne 2006). Jackson et al. (2001) found that women are more likely to use the Internet as a communication tool, and men are more likely to use it as a means of information seeking. According to Ogan et al. (2005), women are less likely to express political opinions and tend to have a less authoritative manner in their conversation style. Meanwhile, some studies (Fuller 2004; Youngs 2004) argue that women are responsible for and belong to the private sphere of life, i.e., the domestic sphere of home, family, private relations, and sexual reproduction; on the other hand, men are best suited and responsible for the public sphere and political realm including government and commercial establishments. Harp and Tremayne (2006) use network theory and feminist theory to study online gender differences in web blogs and offer suggestions for increasing the representation of female voices in the political blogosphere.

As to online communication on web forums, previous studies have used keyword analysis to show that women and men do have different topics that they are interested in and care about (Seale et al. 2006). Seale et al. (2006) analyzed cancer-related web forums and found that women's discussions are more likely to lean toward the exchange of emotional support, including concern with the impact of illness on a wide range of other people; however, men are more likely to participate in the threads on treatment information, medical personnel, and procedures. Guiller and Durndell (2007) analyzed an online course discussion board and found that women are more likely to explicitly agree and support others and make more personal and emotional contributions than men; on the other hand, men are more likely to use authoritative language and to respond negatively in interactions than women.

2.2 *Online Text Classification*

In this study, we adopt online text classification techniques to study the online gender differences in web forums by examining the writing style of the posted messages. Online text classification has several important characteristics, including various types of problems, features, and online texts. These are summarized in the taxonomy presented in Table 19.1. Based on the proposed taxonomy, Table 19.2 shows selected previous studies dealing with online text classification. We discuss the taxonomy and related studies in detail below.

2.2.1 Different Types of Online Text Classification Problems

With the advent of Web 2.0, more and more automatic classification studies using online text-based social media data have appeared. In those studies, the investigated classification problems mainly include authorship classification, sentiment classification, and gender classification. Unlike the classical topic-based classification problem in information retrieval, social media classification relies heavily on the information and fluid writing styles of authors in various online social media, such as e-mail, blog, and forum.

Authorship classification: Authorship classification aims at determining which author produced which piece of writing by examining the styles and contents of writings produced by different authors. Previous studies have applied authorship classification to various online social media texts. De Vel and his collaborators (2000) have conducted a series of experiments to identify the authors of e-mails. They apply the conventional text classification methods to authorship classification on online texts. Some more recent studies (Abbasi and Chen 2005; Li et al. 2006; Zheng et al. 2006) construct frameworks for authorship classification on online texts, with an emphasis on comparing different types of features and classification techniques. A recent comprehensive study conducted by Abbasi and Chen (2008a)

Table 19.1 A taxonomy of online social media text classification

Category	Description	Label
Problems		
Authorship classification	Determines which author produced which piece of writing by examining all the writings by all the authors in the collection	P1
Sentiment classification	Determines whether a text is objective or subjective, or whether a subjective text contains positive or negative sentiments	P2
Gender classification	Determines whether a piece of writing was produced by a female or male by examining the writing styles and contents	P3
Features		
Lexical features	Character- or word-based statistical measures of lexical variation	F1
Features		
Syntactic features	Function words; punctuation; part-of-speech (POS) tags	F2
Structural features	Features related to text organization and layout; technical features such as various file extensions, font sizes, and font colors	F3
Content-specific features	Important keywords and phrases (e.g., n-grams) on certain topics	F4
Online text types		
E-mail	Using e-mails as dataset	T1
Web blog	Using web blogs as dataset	T2
Web forum	Using web forum messages as dataset	T3
Online review	Online reviews of products, movies, music, etc.	T4
Online news	Online news articles and news web pages	T5

Table 19.2 Selected previous studies in online social media text classification

	Problems			Features				Online text types				
Study	P1	P2	P3	F1	F2	F3	F4	T1	T2	T3	T4	T5
de Vel 2000	✓			✓				✓				
del Vel et al. 2001	✓			✓		✓		✓				
Abbasi and Chen 2005	✓			✓	✓	✓	✓			✓		
Zheng et al. 2006	✓			✓	✓	✓	✓			✓		
Li et al. 2006	✓			✓	✓	✓	✓		✓			
Abbasi and Chen 2008	✓			✓	✓	✓	✓			✓		
Subasic and Huettner 2001		✓			✓							✓
Pang et al. 2002		✓					✓				✓	
Turney 2002		✓			✓						✓	
Dave et al. 2003		✓					✓				✓	
Hu and Liu 2004		✓			✓						✓	
Gamon 2004		✓		✓	✓	✓					✓	
Wiebe et al. 2004		✓		✓	✓	✓						✓
Grefenstette et al. 2004		✓			✓					✓		
Mishne 2005		✓		✓		✓	✓		✓			
Abbasi et al. 2008a		✓		✓	✓	✓	✓		✓			
Corney et al. 2002			✓	✓		✓	✓	✓				
Schler et al. 2006			✓	✓		✓	✓		✓			
Nowson and Oberlander 2006			✓		✓		✓		✓			

tested their newly developed Writeprints technique with a rich set of features on various online datasets, including e-mails, instant messages, feedback comments, and program codes. The accuracy of their algorithm reaches as high as 94% when differentiating between 100 authors.

Sentiment classification: Sentiment classification for online texts aims to analyze direction-based texts (i.e., texts containing opinions and emotions) to determine whether a text is objective or subjective, or whether a subjective text contains positive or negative sentiments. The common, two-class sentiment classification problem involves classifying sentiments as positive or negative (Pang et al. 2002; Turney 2002). However, additional variations include classifying sentiments as opinionated/subjective or factual/objective (Wiebe et al. 2001; Wiebe et al. 2004). Instead of sentiments, some other studies attempt to classify emotions, including happiness, sadness, anger, horror, etc. (Grefenstette et al. 2004; Mishne 2005; Subasic and Huettner 2001). As to web forums, Abbasi et al. (2008b) develop a system for sentiment classification with a new feature selection algorithm and test its performance on both English and Arabic web forums.

Gender classification: Gender classification aims to determine whether a piece of writing was produced by a female or male by examining the writing styles and contents of female and male authors. Previous gender classification studies using automatic text classification techniques have been done on both traditional articles (e.g., novels and nonfiction articles) and online social media texts (e.g., e-mails and web blogs).

As an example of gender classification on traditional articles, Koppel et al. (2002) use the exponential gradient (EG) algorithm to classify genders for both fiction and nonfiction documents. By using a feature set combining function words and part-of-speech (POS) tags, they achieve 79.5% accuracy for fiction documents and 82.6% accuracy for nonfiction documents. After feature selection, the accuracy increased to 98% for both fiction and nonfiction documents. Another study conducted by Hota et al. (2006) classifies the gender of Shakespeare's characters based on a collection of his plays. They achieve the highest accuracy of 74.28% using support vector machine (SVM) on the feature set consisting of both content-independent and content-based features. Argamon and his collaborators (2003a) analyze writing styles and identify a set of lexical and syntactic features that differ significantly according to author gender in both fiction and nonfiction documents. In particular, they find that although the total number of nominals used by female and male authors is virtually identical, females use many more pronouns, and males use many more noun specifiers.

For online social media text, most previous gender classification studies focus on e-mails (Corney et al. 2002) and web blogs (Nowson and Oberlander 2006; Schler et al. 2006). Corney et al. (2002) use SVM to classify genders for e-mails and achieve the highest F-measure of 71.1% using the combination of lexical features, structural features, and selected gender-specific features. Nowson and Oberlander (2006) use SVM to classify genders for web blogs and achieve the highest accuracy of 91.5% using the combination of part of speech (POS), bigrams, and trigrams as the features. Schler et al. (2006) also conduct gender classification on web blogs and emphasize the significant differences in writing styles and contents between female and male bloggers as well as among authors of different ages.

2.2.2 Features for Online Social Media Text Classification

Features are very important for online social media text classification. Good feature sets can improve the performance of the classifier. There are four types of features that are often used in previous online social media text classification studies: lexical features, syntactic features, structural features, and content-specific features. Among them, the first three types are content-free features; the fourth type contains features related to specific topics.

Lexical features: Lexical features are character- or word-based statistical measures of lexical variation. Lexical features mainly include character-based lexical features (Argamon et al. 2003b; Gamon 2004; Yule 1938), vocabulary richness measures (Yule 1944), and word-based lexical features (de Vel 2001; Mishne 2005; Zheng et al. 2006). Examples of character-based lexical features are the total number of characters, the number of characters per sentence, the number of characters per word, and the usage frequency of individual letters. Examples of vocabulary richness measures include the number of words that occur once (hapax legomena) and twice (hapax dislegomena) and some other statistical measures defined by Yule (1944). Examples of word-based lexical features are the total number of words, the number of words per sentence, and word-length distribution.

Syntactic features: Syntactic features indicate the patterns used to form sentences. Commonly used syntactic features include function words (Koppel et al. 2002; Koppel et al. 2006; Mosteller 1964), punctuation (Baayen et al. 2002), and part-of-speech (POS) tags (Argamon et al. 1988; Baayen et al. 2002; Gamon 2004; Nowson and Oberlander 2006). These studies also demonstrate that syntactic features may be more reliable compared with lexical features. To study the writing style differences between females and males, Argamon and his collaborators (2003a) use over 1,000 features including 467 function words and a set of POS tags. In their studies on authorship classification for online texts, Abbasi and Chen (2008) use up to 300 function words, 8 punctuations, and almost 2,300 POS tags as syntactic features.

Structural features: Structural features show the text organization and layout. They are especially useful for online social media texts (de Vel et al. 2001). Traditional structural features include greetings, signatures, the number of paragraphs, and the average paragraph length (de Vel et al. 2001; Zheng et al. 2006). Other structural features include technical features such as the use of various file extensions, font sizes, and font colors (Abbasi and Chen 2005). For example, Zheng et al. (2006) use 14 different structural features in their authorship classification study on both English and Chinese online messages. Abbasi and Chen (2008) adopt 64 structural features in their authorship classification on various types of online texts.

Content-specific features: Different from the content-free features (i.e., lexical features, syntactic features, and structural features), content-specific features are comprised of important keywords and phrases on certain topics (Martindale and McKenzie 1995; Zheng et al. 2006) such as word n-grams (Abbasi and Chen 2005; Abbasi and Chen 2008; Diederich et al. 2003; Nowson and Oberlander 2006). Usually, these features can express personal interest in a specific domain. For example, content-specific features on a discussion of computers may include "laptop" and "notebook." Previous studies have shown that content-specific features can improve the performance of online text classification (Abbasi and Chen 2005; Abbasi et al. 2008a; Schler et al. 2006; Zheng et al. 2006).

2.2.3 Different Types of Online Social Media Texts

In the taxonomy shown in Table 19.1, we summarize the major types of texts used in previous online text classification studies as e-mail, web blog, web forum, online review, and online news. Some previous studies use e-mails to form their datasets; others use web blogs, web forum messages, online reviews, or online news. In the taxonomy, the online review category includes the online reviews of products, movies, music, etc.; the online news category consists of online news articles and news web pages.

Some general conclusions can be drawn from Table 19.2 and the literature review. Most previous online text classification studies focus on authorship classification and sentiment classification; relatively less effort has been put on gender classification. For authorship classification, to improve the classification performance, the most recent studies (Abbasi and Chen 2005; Abbasi and Chen 2008; Zheng et al. 2006) have incorporated all four types of features, i.e., lexical features, syntactic features, structural features, and content-specific features. For sentiment classification, early studies (Dave et al. 2003; Hu and Liu 2004; Subasic and Huettner 2001; Turney 2002) often use one type of feature. Some later studies (Gamon 2004; Mishne 2005; Wiebe et al. 2004) add other types of features to improve the classification performance. Abbasi et al. (2008b) conduct sentiment classification using all four types of features. Previous gender classification studies also include different types of features; however, we have not seen a study using all four types of features.

According to different types of online texts, previous authorship classification studies mainly use e-mails (de Vel 2000), web blogs (Li et al. 2006), and web forums (Abbasi and Chen 2005; Abbasi and Chen 2008; Zheng et al. 2006); few have used online reviews or online news. In contrast, previous sentiment classification studies mainly use web blogs (Mishne 2005), web forums (Abbasi et al. 2008a; Grefenstette et al. 2004), online reviews (Gamon 2004; Hu and Liu 2004; Pang et al. 2002; Turney 2002), and online news (Subasic and Huettner 2001; Wiebe et al. 2004); few have used e-mails. The texts used in previous gender classification studies are relatively limited, mainly e-mails (Corney et al. 2002) and web blogs (Nowson and Oberlander 2006; Schler et al. 2006). Few studies have used web forums, online reviews, or online news.

3 Research Gaps and Questions

Our review of previous literature and our conclusions point to several notable research gaps. First, few studies have investigated online gender differences in the context of web forums using feature-based gender classification techniques. Previous studies have shown the existence and evolution of online gender differences and the importance of gender role in political movements on the web. However, in analyzing the gender differences in web forums, most studies used basic keyword-based analysis. Second, to the best of our knowledge, no previous study has used both content-free features (i.e., lexical, syntactic, and structural features) and content-specific features to conduct automatic gender classification for web forums.

Therefore, we raise the following research questions:

1. Can gender classification techniques be used to identify and analyze online gender differences in web forums?
2. Will the use of both content-free features (i.e., lexical, syntactic, and structural features) and content-specific features improve gender classification performance for web forums compared to using only the content-free features?
3. For relatively large feature sets, will feature selection that returns a smaller number of the most important features improve the gender classification performance for web forums?

4 Research Design

In order to address these questions, we develop a framework of feature-based gender classification on web forums. The framework includes several essential components which we will describe in detail: web forum message acquisition, feature generation, and classification and evaluation.

4.1 Web Forum Message Acquisition

This component consists of two steps: forum message collecting and forum message parsing. First, spidering programs are developed to collect all the messages in a given open source web forum as HTML pages. After that, we build parsers to parse out the message information from the raw HTML pages and store the parsed data in a relational database.

4.2 Feature Generation

In this component, we generate different feature sets containing different types of features. By doing this, we can compare and evaluate the performance of different feature sets in gender classification for web forums in order to answer our research questions 2 and 3.

There are several steps in this component: feature extraction, unigram/bigram preselection, and feature selection. Each of these steps leads to the generation of different feature sets.

Feature Extraction: Different types of features are extracted based on all messages collected from a given open source web forum. In this study, we extract the lexical features (denoted by F1), syntactic features (denoted by F2), and structural features (denoted by F3) as content-free features, and unigrams (denoted by F4(unigram)) and bigrams (denoted by F4(bigram)) as content-specific features.

As we described in Sect. 2.2.2, lexical features (i.e., F1) mainly include character-based lexical features, vocabulary richness measures, and word-based lexical features. In this study, we adopt the character-based lexical features used in de Vel (2000), Forsyth and Holmes (1996), and Ledger and Merriam (1994); the vocabulary richness features used in Tweedie and Baayen (1998); and the word-length frequency features used in Mendenhall (1887) and de Vel et al. (2001). In total, we use 87 lexical features.

Syntactic features (i.e., F2) are important because they can indicate people's different habits of organizing sentences (Zheng et al. 2006). Function words and punctuation are often used as syntactic features. Different sets of function words, ranging from 12 to 122, have been tested in various studies (Baayen et al. 1996; Burrows 1989; de Vel et al. 2001; Holmes and Forsyth 1995; Tweedie and Baayen 1998). However, there is no generally accepted good set of function words for different applications. In this study, we adopt a large set of 150 function words used in Zheng et al. (2006) since this study also focuses on web forum messages, although it is about authorship classification instead of gender classification. In addition to function words, we adopt eight punctuations suggested by Baayen et al. (1996). Therefore, we use 158 syntactic features in total.

Structural features (i.e., F3) represent the layout of writing. De Vel (2000) introduces several structural features specifically for e-mails. Zheng et al. (2006) use 14 structural features in their authorship classification study for web forums. In a series of online text classification studies, Abbasi and his collaborators (Abbasi and Chen 2005; Abbasi and Chen 2008; Abbasi et al. 2008a) use 62 structural features related to both word structures (e.g., the number of paragraphs, the number of sentences per paragraph, etc.) and technical structures (e.g., font colors, font sizes, use of images, etc.). Most of those 62 features are about technical structures, including 29 different font colors, 8 different font sizes, 4 types of image displays, and 7 types of hyperlinks. In this study, we adopt some of the structural features used in previous research. We choose five of the most common features that can be applied to a broad number of general web forums. All of them are related to word structures. Specifically, the five structural features are the total number of sentences in a message, the total number of paragraphs in a message, the number of sentences per paragraph in a message, the number of characters per paragraph in a message, and the number of words per paragraph in a message. We do not use many structural features related to technical structures (e.g., font colors and font sizes) since some web forums may not have the related characteristics. For example, some popular (but old) web forums do not have the functions for users to change the font colors and font sizes.

Unigram/bigram preselection: Although content-free features are important for online text classification, content-specific features that consist of important keywords and phrases on certain topics could be more meaningful, thus leading to relatively high representative ability. Content-specific features used in previous online text classification studies are either a relatively small number of manually selected, domain-specific keywords (Li et al. 2006; Zheng et al. 2006), or a relatively large number of n-grams automatically learned from the textual data collection (Abbasi and Chen 2005; Abbasi et al. 2008a; Abbasi et al. 2008b; Peng 2003; Schler et al. 2006). The large potential feature spaces of n-grams have been shown to be effective for online text classification (Abbasi and Chen 2008). Therefore, in this study, we use n-grams as content-specific features. Specifically, we use unigrams (i.e., F4(unigram)) and bigrams (i.e., F4(bigram)). The unigrams and bigrams are extracted from all messages in the web forum. After removing the stopwords, we keep the unigrams and bigrams that appear more than ten times in the whole forum as our content-specific features.

By conducting feature extraction and unigram/bigram preselection, we obtain five types of features. Based on those different types of features, we build three feature sets in an incremental way: (1) feature set F1 + F2 + F3 which includes lexical features,syntacticfeatures,andstructuralfeatures;(2)featuresetF1 + F2 + F3 + F4(unigram) which consists of lexical features, syntactic features, structural features, and unigrams; and (3) feature set F1 + F2 + F3 + F4(unigram) + F4(bigram) which is composed of lexical features, syntactic features, structural features, unigrams, and bigrams. This incremental order represents the evolutionary sequence of features used for online text classification (Abbasi and Chen 2008; Zheng et al. 2006). Studies (Abbasi and Chen 2008; Zheng et al. 2006) have shown that lexical and syntactic features are the foundation for structural and content-specific features. In this study, we use feature set F1 + F2 + F3 which contains only content-free features as the baseline feature set to assess the performance of the other two proposed feature sets: feature set F1 + F2 + F3 + F4(unigram) and feature set F1 + F2 + F3 + F4(unigram) + F 4(bigram), each of which also incorporates content-specific features.

The baseline feature set (i.e., feature set F1 + F2 + F3) contains 250 features with 87 lexical features, 158 syntactic features, and 5 structural features, as we described before. By adding unigrams and unigrams plus bigrams as content-specific features, respectively, feature sets F1 + F2 + F3 + F4(unigram) and F1 + F2 + F3 + F4(unigram) + F4(bigram) have much larger numbers of features than the baseline feature set in general. The exact numbers depend on the number of unigrams and bigrams collected from the text collection.

Feature selection: When the number of features is large, feature selection may improve the classification performance by selecting an optimal subset of features (Guo and Nixon 2009). Previous classification studies using n-gram features usually include some form of feature selection in order to extract the most important words or phrases (Koppel and Schler 2003). In this study, we use the information gain (IG) heuristic to conduct feature selection due to its reported effectiveness in previous online text classification research (Abbasi and Chen 2008; Koppel and Schler 2003), thus building two selected feature sets: selected feature set F1 + F2 + F3 + F4(unigram) and selected feature set F1 + F2 + F3 + F4(unigram) + F4(bigram).

As defined in the following formula, information gain *IG(C,A)* measures the amount of entropy decrease on a class *C* when providing a feature *A* (Quinlan 1986; Shannon 1948). The decreasing amount of entropy reflects the additional information gained by adding feature *A*. In the formula, *H(C)* and *H(C|A)* represent the entropies of class *C* before and after observing feature *A*, respectively. The information gain for each feature varies along the range 0–1 with higher values indicating more information gained by providing certain features.

All features with an information gain greater than 0.0025 (i.e., $IG(C,A)>0.0025$) are selected. The use of such a threshold is consistent with prior work using IG for text feature selection (Abbasi et al. 2008a; Yang and Pedersen 1997):

$$IG(C,A) = H(C) - H(C \mid A), \text{where}$$

$$H(C) = -\sum_{c \in C} p(c)\log_2 p(c),\ H(C \mid A) = -\sum_{a \in A} p(a) \sum_{c \in C} p(c \mid a)\log_2 p(c \mid a).$$

4.3 *Classification and Evaluation*

In this study, we build five different feature sets: feature set F1+F2+F3, feature set F1+F2+F3+F4(unigram), feature set F1+F2+F3+F4(unigram)+F4(bigram), selected feature set F1+F2+F3+F4(unigram), and selected feature set F1+F2+F 3+F4(unigram)+F4(bigram). We aim to study and compare the performance of these feature sets in order to identify the best one for web forum gender classification. Because of its often reported best performance in many previous online text classification studies (Abbasi and Chen 2008; Abbasi et al. 2008a; Li et al. 2006; Zheng et al. 2006), we choose SVM as the classifier.

To assess the performance of each feature set on gender classification for web forums, we adopt the standard classification performance metrics, i.e., accuracy, precision, recall, and F-measure. These metrics have been widely used in information retrieval and text classification studies (Abbasi and Chen 2008; Abbasi et al. 2008a; Li et al. 2008). In particular, accuracy measures the overall classification correctness, while precision, recall, and F-measure evaluate the correctness of each class:

$$\text{Accuracy} = \frac{\text{Number of all correctly classified Web forum messages}}{\text{Total number of Web forum messages}},$$

$$\text{Precision (i)} = \frac{\text{Number of correctly classified Web forum messages for class i}}{\text{Total number of Web forum messages classified as class i}},$$

$$\text{Recall (i)} = \frac{\text{Number of correctly classified Web forum messages for class i}}{\text{Total number of Web forum messages in class i}},$$

$$\text{F - Measure}\ (i) = \frac{2 \times \text{Precision}\ (i) \times \text{Recall}\ (i)}{\text{Precision}\ (i) + \text{Recall}\ (i)},\ \text{where}$$

$i = 1, 2$ with classes 1 and 2 being web forum messages written by female and male authors, respectively.

5 Experimental Study

To assess the effectiveness of the proposed research design, we conduct an experiment on a large and long-standing international Islamic women's political forum. In the following, we detail the test bed, hypotheses, experimental results, and discussion.

5.1 *Test Bed*

We conduct our experiment on a large, international Islamic women's political forum to evaluate our proposed framework of gender classification for web forums. We choose it for three reasons: First, it is a large, long-standing (about 4 years) international political forum and thus can be used to study the international cyber political movement; second, it has the self-reported gender information for each registered member, thus providing the gold standard to evaluate the performance of our automatic gender classifiers; third, since it is a women's forum, more females may participate, thus providing a larger number of messages written by female authors compared with other general, male-dominated web forums. We believe the international, political, and female-oriented nature of this large active forum makes it an ideal test bed for our research.

We collected and parsed all the messages in the forum posted up to March 2007. In total, we gathered 34,695 different messages in 4,352 unique threads. The numbers of messages written by females and males are quite balanced. There are 17,785 and 16,572 messages written by females and males, respectively. An additional 338 messages do not have gender information. The time span of the collected messages is from June 9, 2004 to March 13, 2007. Based on careful discussion with our political science collaborator, who has significant experience in such women's political forums, we believe that this test bed is of high quality and has credible participant-specified gender information.

To test the performance of our classifiers, we randomly selected 100 authors, 50 females and 50 males. In total, there are 12,690 messages posted by those 100 authors. On average, each female participant produced 142.26 messages, and each male participant wrote 111.54 messages.

5.2 Hypotheses

Drawing on the vast online social media classification literature we reviewed, we posit that adding content-specific features to the baseline content-free features can improve the performance of gender classification for web forums, and conducting feature selection on a relatively large number of features can improve the performance of gender classification for web forums.

5.3 Experimental Results

We built five different feature sets: F1+F2+F3, F1+F2+F3+F4(unigram), F1+F2+F3+F4(unigram)+F4(bigram), selected F1+F2+F3+F4(unigram), and selected F1+F2+F3+F4(unigram)+F4(bigram). As described before, feature set F1+F2+F3 contains 250 content-free features. For the content-specific features (i.e., unigrams and bigrams), we selected the ones appearing more than ten times among all the messages in the forum. In total, we get 6,012 unigrams and 4,022 bigrams. Therefore, there are 6,262 and 10,284 features in feature sets F1+F2+F3+F4(unigram) and F1+F2+F3+F4(unigram)+F4(bigram), respectively. After conducting feature selection using the information gain heuristic, the two selected feature sets F1+F2+F3+F4(unigram) and F1+F2+F3+F4(unigram)+F4(bigram) consist of 351 and 640 features, respectively. The feature selection is carried out by Weka's information gain attribute evaluator (Witten and Frank 2005). The two selected feature sets are much smaller than the corresponding ones without feature selection. Table 19.3 lists the number of features in each feature set.

The classification is carried out by using a linear kernel with the sequential minimal optimization (SMO) algorithm (Platt 1999) included in the Weka data mining package (Witten and Frank 2005). Testing is done via tenfold cross-validation. Table 19.4 shows the precision, recall, and F-measure of gender classification on each feature set.

All three types of measurement values keep increasing in the same way as the accuracy. The highest precision, recall, and F-measure for both classes (i.e., female and male) are achieved on the selected feature set F1+F2+F3+F4(unigram)+F4(bigram).

5.4 Different Topics of Interest: Females and Males

In our experimental study, we achieve the highest classification accuracy of 86% on the selected feature set F1+F2+F3+F4(unigram)+F4(bigram). This indicates that gender differences do exist in web forums, and the features used for classification, especially the content-specific features, have high discriminating capability to distinguish the online gender differences between female and male posters.

Table 19.3 The number of features in each feature set

Feature set	Number of features
F1 + F2 + F3	250
F1 + F2 + F3 + F4(unigram)	6,262
F1 + F2 + F3 + F4(unigram) + F4(bigram)	10,284
Selected F1 + F2 + F3 + F4(unigram)	351
Selected F1 + F2 + F3 + F4(unigram) + F4(bigram)	640

Table 19.4 Performance measures using different feature sets

Feature set	Class	Precision	Recall	F-measure
F1 + F2 + F3	Female	57.10%	72.00%	63.69%
	Male	62.20%	46.00%	52.89%
	Average	59.70%	59.00%	59.35%
F1 + F2 + F3 + F4(unigram)	Female	63.00%	58.00%	60.40%
	Male	61.10%	66.00%	63.46%
	Average	62.10%	62.00%	62.05%
F1 + F2 + F3 + F4(unigram) + F4(bigram)	Female	62.50%	70.00%	66.04%
	Male	65.90%	58.00%	61.70%
	Average	64.20%	64.00%	64.10%
Selected F1 + F2 + F3 + F4(unigram)	Female	90.20%	74.00%	81.30%
	Male	78.00%	92.00%	84.42%
	Average	84.10%	83.00%	83.55%
Selected F1 + F2 + F3 + F4(unigram) + F4(bigram)	Female	92.90%	78.00%	84.80%
	Male	81.00%	94.00%	87.02%
	Average	86.90%	86.00%	86.45%

By investigating the features in the selected feature set F1 + F2 + F3 + F4(unigra m) + F4(bigram), we observe that females talk more about family members, God, peace, marriage, and goodwill; on the other hand, males talk more about extremism, holy man, and belief.

Table 19.5 lists some examples of the unigrams and bigrams preferred by females and males, respectively, from the selected feature set F1 + F2 + F3 + F4(unigram) + F 4(bigram). They are among the features with the highest information gain values, therefore showing high discriminatory power. We conduct chi-square (χ^2) tests to examine the statistical significances of the differences between females and males using those unigrams and bigrams. A domain expert from an Islamic country provided the meanings of some of those unigrams and bigrams.

As summarized in Table 19.5, all the listed female-preferred unigrams and bigrams are statistically significant. Specifically, significant terms/words in female conversations include sis (i.e., sisters by Islamic), sister, mother, husband, flower, amen, alhamdulillah (i.e., thank God), inshaallaah (i.e., in God's will), ahhah kheir (i.e., God is good), and sexually defiled. Male-preferred unigrams and bigrams are statistically significant, except for "original Arabic." Specifically, significant terms/words in male discussions include Salafi (i.e., an extremist sect of Islam), Allah

Table 19.5 Examples of female- and male-preferred unigrams and bigrams from the selected feature set F1 + F2 + F3 + F4(unigram) + F4(bigram)

Keyword	χ^2 value	*P* value	Result	Meaning
Female-preferred unigrams and bigrams				
Sis	456.07	<0.0001**	Supported	Sisters by Islamic
Sister	165.08	<0.0001**	Supported	
Mother	123.88	<0.0001**	Supported	
Husband	51.87	<0.0001**	Supported	
Flower	9.00	0.0030**	Supported	
Amen	166.64	<0.0001**	Supported	
Alhamdulillah	283.85	<0.0001**	Supported	Thank God
Inshaallaah	33.51	<0.0001**	Supported	In God's will
Ahhah kheir	15.16	<0.0001**	Supported	God is good
Sexually defiled	5.25	0.0220*	Supported	
Male-preferred unigrams and bigrams				
Salafi	377.17	<0.0001**	Supported	Extremist sect of Islam
Allah	290.30	<0.0001**	Supported	Allah God of Muslims
Army	66.12	<0.0001**	Supported	
Deviant	35.79	<0.0001**	Supported	
Ijtihaad	57.80	<0.0001**	Supported	Inferring or interpreting Islamic laws
Email	23.81	<0.0001**	Supported	
Great scholar	13.89	0.0002**	Supported	
Muslim intellectual	11.27	0.0008**	Supported	
Imam Nawawi	26.56	<0.0001**	Supported	Priest Nawawi
Original Arabic	3.52	0.0606	Not supported	

Note Significance levels * $\alpha = 0.05$ and ** $\alpha = 0.01$

(i.e., Allah God of Muslims), army, deviant, ijtihaad (i.e., inferring or interpreting Islamic laws), email, great scholar, Muslim intellectual, and imam Nawawi (i.e., Priest Nawawi). For the bigram "original Arabic," although men prefer to use it more frequently than women, the difference is not statistically significant ($p = 0.0606 > 0.05$). This may be because the total number of its appearances in the whole forum is small and therefore cannot show statistical significance.

The results of our experimental study show the importance of content-specific features in gender classification for web forums, which is consistent with previous gender classification studies for web blogs. For example, Nowson and Oberlander (2006) conduct a linguistic analysis of gender and personality differences in web blogs. By comparing different types of features, they find that a relatively small number of n-gram features have the best discrimination power in automatic gender detection. Schler et al. (2006) analyze a corpus of web blogs from blogger.com and find that female and male bloggers have significant differences in topics of interest. The topics significantly preferred by female bloggers include shopping, mom, cry, kiss, husband, etc. Male bloggers are more interested in topics such as Linux, Microsoft, nations, democracy, economics, etc.

As an important type of social media, political web forums have become a major communication channel for people to discuss and debate political, cultural, and social issues. More and more women are using this medium to share their political pressure and knowledge. Along with this trend, researchers have developed an increased interest in studying online gender differences. By analyzing writing styles and topics of interest, our experimental results indicate that female and male participants in political web forums do have significantly different topics of interest.

6 Conclusions and Future Directions

With the rapid development and the increasing importance of the Internet in people's daily life and work, understanding online gender differences and why they occur is becoming more and more important for both Internet service providers and users. Nowadays, more and more women are participating in cyberspace. However, this does not diminish online gender differences. In contrast, discrepancies of motivation and interest in Internet use between females and males are becoming the focus of online gender difference research. In this study, we use feature-based online social media text classification techniques to investigate the online gender differences between female and male participants in web forums by examining writing styles and topics of interests. The feature-based gender classification framework we developed can be applied to other different web forums.

In the framework, we examine different types of features that have been widely used in previous online text classification studies, including lexical features, syntactic features, structural features, and content-specific features. For content-specific features, we use unigrams and bigrams automatically extracted from the whole forum instead of those manually selected. We build five different feature sets by adding content-specific features to the basic content-free features and conducting feature selection. According to our experimental study on a large Islamic women's political forum, the feature sets combining both content-specific and content-free features perform significantly better than the ones consisting of only content-free features. In addition, feature selection on large feature sets improves the classification performance significantly. The results also indicate the existence of online gender differences in web forums. The best gender classification performance is achieved using the selected feature set F1 + F2 + F3 + F4(unigram) + F4(bigram). Through further investigation of this selected feature set, we identify different topics of interest between females and males. For example, females prefer talking about family members, God, peace, and marriage; males like to talk more about extremism, holy man, and belief.

Acknowledgments This material is based upon work supported by the National Science Foundation under Grant No. CNS-0709338, "(CRI: CRD) Developing a Dark Web Collection and Infrastructure for Computational and Social Sciences." We would also like to thank Dr. Katharina von Knop for her helpful suggestions and comments about our research test bed.

References

Abbasi, A. and H. Chen, "Applying authorship analysis to extremist-group Web forum messages," *IEEE Intelligent Systems*, vol. 20, no. 5 (Special issue on artificial intelligence for national and homeland security), 2005, pp. 67–75.

Abbasi, A. and H. Chen, "Writeprints: A stylometric approach to identity-level identification and similarity detection in cyberspace," *ACM Transactions on Information Systems*, vol. 26, no. 2, 2008, pp. 1–29.

Abbasi, H. Chen, and J.F. Nunamaker, "Stylometric identification in electronic markets: scalability and robustness," *Journal of Management Information Systems*, vol. 25, no. 1, 2008b, pp. 49–78.

Abbasi, A., H. Chen, and A. Salem, "Sentiment analysis in multiple languages: feature selection for opinion classification in Web forums," *ACM Transactions on Information Systems*, vol. 26, no. 3, 2008a, pp. 1–34.

Argamon, S., M. Koppel, and G. Avneri, "Routing documents according to style.," in *Proceedings of Proceedings of the 1st International Workshop on Innovative Information*, Pisa, Italy, 1988.

Argamon, S., M. Koppel, J. Fine, and A. Shimoni, "Gender, genre, and writing style in formal written texts," *Text*, vol. 23, no. 3, 2003a, pp. 321–346.

Argamon, S., M. Saric, and S.S. Stein, "Style mining of electronic messages for multiple authorship discrimination," in *Proceedings of Proceedings of the 9th ACM SIGKDD International Conference on Knowledge Discovery and Data Mining*, 2003b, pp. 475–480.

Baayen, R.H., H.V. Halteren, A. Neijt, and F.J. Tweedie, "An experiment in authorship attribution," in *Proceedings of Proceedings of the 6th International Conference on Statistical Analysis of Textual Data*, 2002, pp. 69–75.

Baayen, R.H., H.V. Halteren, and F.J. Tweedie, "Outside the cave of shadows: using syntactic annotation to enhance authorship attribution," *Literary and Linguistic Computing*, vol. 11, no. 3, 1996, pp. 121–132.

Bimber, B., "Measuring the gender gap on the Internet," *Social Science Quarterly*, vol. 81, no. 3, 2000, pp. 868–876.

Burrows, J.F., "'An ocean where each kind....' Statistical analysis and some major determinants of literary style," *Computers and the Humanities*, vol. 23, no. 4–5, 1989, pp. 309–321.

CommerceNet, "The CommerceNet/Nielsen Internet demographic survey (1999)," http://www.commerce.net/, 1999.

Consaluo, M. and S. Paasonen, *Women and Everyday Uses of the Internet: Agency and Identity*, New York: Peter Lang Publishing, 2002.

Corney, M., O. de Vel, A. Anderson, and G. Mohay, "Gender-preferential text mining of e-mail discourse," in *Proceedings of Proceedings of the 18th Annual Computer Security Applications Conference (ACSAC 2002)*, Las Vegas, 2002, pp. 282–292.

Dave, K., S. Lawrence, and D. Pennock, "Mining the peanut gallery: opinion extraction and semantic classification of product reviews," in *Proceedings of Proceedings of the 12th International World Wide Web Conference (WWW'03)*, 2003, pp. 519–528.

de Vel, O., "Mining E-mail Authorship," in *Proceedings of Paper presented at the Workshop on Text Mining, ACM International Conference on Knowledge Discovery and Data Mining (KDD 2000)*, Boston, MA, 2000.

de Vel, O., A. Anderson, M. Corney, and G. Mohay, "Mining e-mail content for author identification forensics," *SIGMOD Record*, vol. 30, no. 4, 2001, pp. 55–64.

Diederich, J., J. Kindermann, E. Leopold, and G. Paass, "Authorship attribution with support vector machines," *Applied Intelligence*, vol. 19, no. 1–2, 2003, pp. 109–123.

Forsyth, R.S., and D.I. Holmes, "Feature finding for text classification," *Literary and Linguistic Computing*, vol. 11, no. 4, 1996, pp. 163–174.

Fountain, J.E., "Constructing the information society: women, information technology, and design," *Technology and Society* vol. 22, no. 1, 2000, pp. 45–62.

Fuller, J.E., "Equality in cyberdemocracy? Gauging gender gaps in on-line civic participation," *Social Science Quarterly* vol. 85, no. 4, 2004, pp. 938–957.
Gamon, M., "Sentiment classification on customer feedback data: noisy data, large feature vectors, and the role of linguistic analysis," in *Proceedings of Proceedings of the 20th International Conference on Computational Linguistics*, 2004, pp. 841–847.
Grefenstette, G., Y. Qu, J.G. Shanahan, and D.A. Evans, "Coupling niche browsers and affect analysis for an opinion mining application," in *Proceedings of Proceedings of the 12th International Conference Recherche d'Information Assistee par Ordinateur*, 2004, pp. 186–194.
Guiller, J. and A. Durndell, "Students' linguistic behaviour in online discussion groups: Does gender matter?" *Computers in Human Behavior*, vol. 23, no. 5, 2007, pp. 2240–55.
Guo, B. and M.S. Nixon, "Gait feature subset selection by mutual information," *IEEE Transactions on Systems, Man, and Cybernetics – Part A: Systems and Humans*, vol. 39, no. 1, 2009, pp. 36–46.
Halbert, D. "Shulamith firestone: radical feminism and visions of the information society," *Information Communication and Society*, vol. 7, no. 1, 2004, pp. 115–136.
Harcourt, W., "The personal and the political: women using the Internet," *Cyberpsychology and Behavior* vol. 3, no. 5, 2000, pp. 693–697.
Harp, D. and M. Tremayne, "The gendered blogosphere: examining inequality using network and feminist theory," *Journalism and Mass Communication Quarterly* vol. 83, no. 2, 2006, pp. 247–264.
Holmes, D.I. and R.S. Forsyth, "The federalist revisited: new directions in authorship attribution," *Literary and Linguistic Computing*, vol. 10, no. 2, 1995, pp. 111–127.
Hota, S., S. Argamon, M. Koppel, and I. Zigdon, "Performing gender: automatic stylistic analysis of Shakespeare's characters," in *Proceedings of Proceedings of the Digital Humanities Conference (Association for Computers in Humanities and the Association for Literary and Linguistic Computing)*, 2006, pp. 100–106.
Hu, M. and B. Liu, "Mining and summarizing customer reviews," in *Proceedings of Proceedings of the ACM SIGKDD International Conference*, 2004, pp. 168–177.
Jackson, L.A., K.S. Ervin, P.D. Gardner, and N. Schmitt, "Gender and the Internet: women communicating and men searching," *Sex Roles: A Journal of Research*, vol. 44, no. 5–6, 2001, pp. 363–378.
Koppel, M., N. Akiva, and I. Dagan, "Feature instability as a criterion for selecting potential style markers," *J. Amer. Soc. Inf. Sci. Technol*, vol. 57, no. 11, 2006, pp. 1519–1525.
Koppel, M., S. Argamon, and A. Shimoni, "Automatically categorizing written texts by author gender," *Literary and Linguistic Computing*, vol. 14, no. 7, 2002, pp. 401–412.
Koppel, M. and J. Schler, "Exploiting stylistic idiosyncrasies for authorship attribution," in *Proceedings of Proceedings of the IJCAIWorkshop on Computational Approaches to Style Analysis and Synthesis*, Acapulco, Mexico, 2003.
Ledger G.R. and T.V.N. Merriam, "Shakespeare, Fletcher, and the two noble kinsmen.," *Literary and Linguistic Computing*, vol. 9, no. 4, 1994, pp. 235–248.
Li, J., Z. Zhang, X. Li, and H. Chen, "Kernel-based learning for biomedical relation extraction," *Journal of the American Society for Information Science and Technology (JASIST)*, vol. 59, no. 5, 2008, pp. 756–769.
Li, J., R. Zheng, and H. Chen, "From fingerprint to Writeprint," *Communications of the ACM*, vol. 49, no. 4, 2006, pp. 76–82.
Martindale, C. and D. McKenzie, "On the utility of content analysis in author attribution: the federalist," *Comput. Humanit.*, vol. 29, no. 4, 1995, pp. 259–270.
Mendenhall, T.C. "The characteristic curves of composition," *Science*, vol. 11, no. 11, 1887, pp. 237–249.
Mishne, G., "Experiments with mood classification," in *Proceedings of Proceedings of the 1st Workshop on Stylistic Analysis of Text for Information Access*, Salvador, Brazil, 2005.
Mitra, A., "Voices of the marginalized on the Internet: examples from a Website for women of South Asia," *Journal of Communication* vol. 54, no. 3, 2004, pp. 492–510.

Mosteller, F., *Applied Bayesian and Classical Inference: The Case of the Federalist Papers*, 2nd ed., Springer, 1964.

National Election Study, "American National Election Study. 1998 Pre- and post- election survey," *Conducted by the Center for Political Studies of the Institute for Social Research, The University of Michigan, Ann Arbor, Inter-University Consortium for Political and Social Research*, 1998.

Nowson, S. and J. Oberlander, "The identity of bloggers: openness and gender in personal Weblogs," in *Proceedings of Proceedings of the AAAI Spring Symposia on Computational Approaches to Analyzing Weblogs*, Stanford, California, 2006.

O'Reilly, T. "What Is Web 2.0? Design patterns and business models for the next generation of software," http://www.oreillynet.com/pub/a/oreilly/tim/news/2005/09/30/what-is-Web-20.html, 2005.

Ogan, C., F. Cicek, and M. Ozakca, "Letters to Sarah: analysis of email responses to an online editorial," *New Media and Society* vol. 7, no. 4, 2005, pp. 533–557.

Pang, B., L. Lee, and S. Vaithyanathain, "Thumbs up? Sentiment classification using machine learning techniques," in *Proceedings of Proceedings of the Conference on Empirical Methods in Natural Language Processing*, 2002, pp. 79–86.

Peng, F., D. Schuurmans, V. Keselj, and S. Wang, "Automated authorship attribution with character level language models," in *Proceedings of Proceedings of the 10th Conference of the European Chapter of the Association for Computational Linguistics*, Budapest, Hungary, 2003.

Pew Internet and American Life Project, http://www.pewinternet.org/trends/ User_Demo_7.22.08.htm, 2008.

Platt, J. *Fast Training on SVMs Using Sequential Minimal Optimization*, In Scholkopf, B., Burges, C., and Smola, A. (Ed.) ed., Advances in Kernel Methods: Support Vector Learning, Cambridge, MA: MIT Press, 1999.

Quinlan, J.R., "Induction of decision trees," *Machine Learning*, vol. 1, no. 1, 1986, pp. 81–106.

Schler, J., M. Koppel, S. Argamon, and J. Pennebaker, "Effects of age and gender on blogging," in *Proceedings of Proceedings of AAAI Spring Symposium on Computational Approaches for Analyzing Weblogs*, Menlo Park, California, 2006, pp. 199–205.

Seale, C., S. Ziebland, and J. Charteris-Black, "Gender, cancer experience and Internet use: a comparative keyword analysis of interviews and online cancer support groups," *Social Science and Medicine*, vol. 62, no. 10, 2006, pp. 2577–2590.

Shade, L.R., *Gender and Community in the Social Construction of the InternetGender and Community in the Social Construction of the Internet*, New York: Peter Lang Publishing, 2002.

Shannon, C.E., "A mathematical theory of communication," *Bell System Technical Journal*, vol. 27, no. 4, 1948, pp. 379–423.

Sherman, A.P., *Cybergrrl @ Work: Tips and Inspiration for the Professional You*, Berkley Trade, 2001.

Subasic, P. and A. Huettner, "Affect analysis of text using fuzzy semantic typing," *IEEE Transactions on Fuzzy Systems*, vol. 9, no. 4, 2001, pp. 483–496.

Turney, P.D., "Thumbs up or thumbs down? Semantic orientation applied to unsupervised classification of reviews," in *Proceedings of Proceedings of the 40th Annual Meetings of the Association for Computational Linguistics*, Philadelphia, Pennsylvania, 2002, pp. 417–424.

Tweedie, F.J. and R.H. Baayen, "How variable may a constant be? Measures of lexical richness in perspective.," *Computers and the Humanities*, vol. 32, no. 5, 1998, pp. 323–352.

Wiebe, J., T. Wilson, and M. Bell, "Identifying collocations for recognizing opinions," in *Proceedings of Proceedings of the ACL/EACL Workshop on Collocation*, Toulouse, France, 2001.

Wiebe, J., T. Wilson, R. Bruce, M. Bell, and M. Martin, "Learning subjective language," *Computational Linguistics*, vol. 30, no. 3, 2004, pp. 277–308.

Witten, I.H. and E. Frank, *Data Mining: Practical Machine Learning Tools and Techniques (2nd Edition)*, 2nd Edition ed., San Francisco: Morgan Kaufmann, 2005.

Yang, Y. and J.O. Pedersen, "A comparative study on feature selection in text categorization," in *Proceedings of Proceedings of the ICML97*, 1997, pp. 412–420.

Youngs, G., "Cyberspace: the new feminist frontier," in Karen Ross and Carolyn M. Byerly, ed., *Women and Media: International Perspectives* Wiley-Blackwell, 2004, pp. 185–208.

Yule, G.U., "On sentence length as a statistical characteristic of style in prose with application to two cases of disputed authorship," *Biometrika*, vol. 30, 1938, pp. 363–390.

Yule, G.U., *The Statistical Study of Literary Vocabulary*, Cambridge University Press, 1944.

Zheng, R., J. Li, H. Chen, and Z. Huang, "A framework for authorship identification of online messages: writing-style features and classification techniques," *Journal of the American Society for Information Science and Technology (JASIST)*, vol. 57, no. 3, 2006, pp. 378–393.

Chapter 20
US Domestic Extremist Groups

1 Introduction

Increasingly, extremist and hate groups are using the Internet as a powerful tool for facilitating recruitment, reaching global audiences, linking with other extremist groups, and spreading hate materials that help to persuade others to violence and terrorism. Although US extremist and hate groups may not be as well known as some of the international extremist organizations, they pose a significant threat to US homeland security as these domestic extremist organizations are based and operate entirely from within the continental United States and Puerto Rico (Blitzer 2001).

According to the Southern Poverty Law Center (SPLC), the number of active extremist and hate groups operating in the USA was 708 in 2002 (SPLC Report 2004). Their web sites increased from 443 in 2002 to 497 in 2003, a 12% increase. Researchers and watchdog organizations that monitor and analyze these web sites, such as SPLC, the Simon Wiesenthal Center, and SurfControl, are finding that keeping track of existing and new web sites and exploring their usage and content have become time consuming and challenging (CNN 1999; Gerstenfeld et al. 2003). Since such content on the Internet is expanding, it is important to develop tools that allow researchers to monitor, analyze, and predict changes and developments in extremist and hate groups' use of the web and their influences (Gerstenfeld et al. 2003).

The objectives of this chapter are twofold. First, we propose the development of automated or semiautomated procedures and systematic methodologies for capturing extremist groups' web site data and using that data for subsequent analyses. By analyzing the web sites' content and visualizing the hyperlinks at the collection level, our methodology formalizes the process of knowledge discovery. Second, we seek to broaden our understanding of how domestic extremist groups utilize the web infrastructure so that we can develop a comprehensive understanding of the extremists themselves. Because the groups are volatile and often associated with illegal activities and violence, they pose great difficulties for researchers seeking to understand the structure and dynamics of their movements (Burris et al. 2000).

H. Chen, *Dark Web: Exploring and Data Mining the Dark Side of the Web*,
Integrated Series in Information Systems 30, DOI 10.1007/978-1-4614-1557-2_20,

Since these groups are active in using the Internet, web-based research on domestic extremist groups should prove valuable for supplementing and modifying earlier findings. In this chapter, we present the related literature, proposed methodology, results, and implications of this investigation.

2 Previous Research

2.1 Social Movement Research on Extremists and the Internet

Research on social movement organizations, such as extremist and hate groups' use of the Internet, is in its early stages (Burris et al. 2000; Gustavson and Sherkat 2004). Researchers have identified a wide variety of different extremist groups, such as White Supremacist, Black Separatist, and Militia, and how they are using the Internet to support their resource mobilization strategies. Resource mobilization is a process of securing control over resources needed for collective action such as communication, money, information, human assets, and specialized skills (Gustavson and Sherkat 2004). Clandestine groups are constantly seeking ways to improve the effectiveness of their communication (Whine 1997), information operations (Burris et al. 2000), and to facilitate collective identity, solidarity, and leaderless resistance (Gerstenfeld et al. 2003; Whine 1997).

American extremist and hate groups have continuously exploited technology to enhance their operations and were among the early adopters of computer bulletin boards that eventually evolved into the Internet (Gerstenfeld et al. 2003). Stormfront.org, a neo-Nazi web site set up in 1995, is considered the first major domestic "hate site" on the World Wide Web because of its depth of content and its presentation style which represented a new period for online right-wing extremism (Whine 1997). The neo-Nazi groups share a hatred for Jews and other minorities, and a love for Adolf Hitler and Nazi Germany. A social network analysis of extremist web sites revealed that Stormfront.org served as a central node that occupied a prominent position within the White Supremacist network (Burris et al. 2000).

In addition to web sites, extremists use the Internet to access private message boards, e-mail, research, and listservs, and sell merchandise. For example, the web site of Resistance Records, the e-commerce music site of the National Alliance, is estimated to have had about $1 million in sales revenue in 2001 (Gerstenfeld et al. 2003). White Supremacist groups have a significant presence on the Internet, with several hundred sites ranging in complexity from single one-page sites to those that contain extensive documentation, discussion groups, and music collections (Gustavson and Sherkat 2004). In the literature, the White Supremacist movement is depicted as a fragmented, decentralized, and often sectarian network of organizations that can be grouped into three categories: Ku Klux Klan, neo-Nazi, and Racist Skinheads (Burris et al. 2000). An important unifying aspect of the movement is the Christian Identity theology that teaches that Whites are the only true children of God (Burris et al. 2000).

Table 20.1 Summary of research on extremist and hate groups' use of the Internet

Methodology	Finding
Observation	Tracing the early usage of the Internet by extremists, identified patterns of usage of racial computer games, Usenet, bulletin boards, and web sites (Whine 1997)
Content analysis (157 web sites)	Majority of sites contained external links to other extremist sites, half included multimedia content, and half contained racist symbols. Used web sites to expand their reach to international audiences, link to diverse extremists groups, and allow the groups to have maximum image control (Gerstenfeld et al. 2003)
Network and content analysis (80 web sites)	Internet hyperlinks appeared to provide a reasonable, accurate representation of interorganizational structure of the movement. Use of the Internet assisted in the creation of an international virtual extremist community (Burris et al. 2000)
Egocentric network and content analysis (226 web sites)	Selection of Aryan Nations web site as ego was effective and different from previous network studies. Factions within White Supremacy movement engaged in coalition building (Gustavson and Sherkat 2004)

Besides the White Supremacists, the leftist environmental and animal liberation groups also use the web as a tool for propaganda and violent leaderless resistance (Gerstenfeld et al. 2003). Table 20.1 identifies several studies that use systematic methodologies such as web content and link analysis to explore a range of research questions about domestic extremist groups' exploitation of Internet technology.

Most of the studies identified in Table 20.1 involved manual processes for gathering the web sites, classifying them, coding the web sites, and visualizing the patterns. From a post-retrieval analysis perspective, existing research tools to gather and explore the interpretations of web sites' content and usage patterns are limited as yet (Chen et al. 2004). Existing tools provided by web search engines and watchdog organizations' web sites offer limited capabilities for integrating the resources and supporting information fusion.

2.2 Web Harvesting Approaches

The first step toward studying the terrorism web infrastructure is to harvest extremist web sites back to a local repository for further analysis. web harvesting is the process of gathering and organizing unstructured information from pages and data on the web (Kay 2004). Previous studies have suggested three types of approaches to harvesting web contents in specific domains: manual, automatic, and semiautomatic.

In order to gather samples of extremist and hate groups' web sites to analyze, all previous extremist web content studies used a manual approach (Burris et al. 2000;

Gerstenfeld et al. 2003; Gustavson and Sherkat 2004; Xu and Chen 2005). For example, Burris et al. (2000) used a manual approach to collect and download seed URLs for a 2-week period in 1997. The seeds were identified using seven watchdog organizations that monitor hate and extremist groups such as Net Hate and HateWatch; HateWatch is a part of the SPLC. The limitation of such a manual approach is that it is time consuming and inefficient.

In some other relevant domains, such as e-government, automatic collection building methods were used. Albertsen used an automatic approach in the "Paradigma" project (Albertsen 2003). The goal of Paradigma is to archive Norwegian legal deposit documents on the web. It employed a focused web crawler, an automatic program that discovers and downloads web sites in particular domains by following web links found in the HTML pages of a starting set of web pages. Metadata was then extracted and used to rank the web sites in terms of relevance. The automatic approach is more efficient than the manual approach; however, due to the limitations of current focused crawling techniques, automatic approaches often introduce noise (off-topic web pages) into the harvest results.

The "Political Communications Web Archiving" group employed a semiautomatic approach to harvesting domain-specific web sites (Reilly et al. 2003). Domain experts provided seed URLs as well as typologies for constructing metadata that can be used in the crawling process. Their project's goal is to develop a methodology for constructing an archive of broad-spectrum political communications over the web. Based on our review, we believe that the semiautomatic approach is the most suitable approach for harvesting terrorism web sites because it combines the high accuracy and high efficiency of manual and automatic approaches.

2.3 Web Link and Content Analysis

Once the extremist web sites are harvested, two types of analysis methods can be applied to study the extremists' use of the web: web link analysis and web content analysis.

Web link analysis is based on hyperlink structure and has been previously used to discover hidden relationships among communities (Gibson et al. 1998; Reid 2003). Borgman and Furner (2002) define two classes of web link analysis studies: relational and evaluative. Relational analysis gives insight into the strength of relations between web entities, in particular web sites, while evaluative analysis reveals the popularity or quality level of a web entity. Terrorism research utilizes relational analysis because it provides us with insights into the nature of relations between extremist web sites and extremist organizations. The relational link analysis approach has been used in various domains outside terrorism research. For example, Gibson et al. (1998) describe an automated methodology for discerning web communities on the WWW. Their work is based on hyperlink-induced topic search (HITS), a tool that searches for authoritative hypermedia on a given broad topic. Reid (2003) made use of hyperlink-based topologies to uncover companies' noncustomer online communities. However, her approach was based on manual categorization. With the vast amount of information on the web, this qualitative methodology is difficult to apply to large-scale studies.

In order to reach an understanding of the various facets of the extremist and hate groups' web usage and communications, a systematic analysis of the web site content is required. Demchak et al. (2000) work provides a well-defined methodology for analyzing communicative content in government web sites. Their work focuses on measuring "openness" of government web sites. To achieve this goal, they developed a web site attribute system tool that is basically composed of a set of high-level attributes such as transparency and interactivity. Each high-level attribute is associated with a second layer of attributes at a more refined level of granularity. For example, the right "operational information" and "responses" on a given web page can induce an increase in the interactivity level of a government web site. Demchak and Friis' work, an example of a well-structured and systematic content analysis exercise, provides guidance for this chapter.

3 Proposed Approach

This chapter is part of a Dark Web Portal project (Chen et al. 2004) that builds on our system development experience. The goals are to understand how US domestic extremists are using the Internet and identify appropriate techniques for collecting high-quality web pages of extremist and hate groups and automating systematic procedures for analyzing and visualizing the content of individual web sites. As illustrated in Fig. 20.1, our proposed approach consists of three components: (1) collection building, (2) content analysis, and (3) link analysis.

To accomplish this, we first employ a semiautomatic procedure for harvesting and constructing a high-quality domestic extremist web site collection. We then perform link analysis and run a node clustering algorithm on the collection for the study of hyperlinked terrorism web communities. In the last step, we conduct an attribute-based systematic content analysis of our collection to study various facets of the domestic extremists' web usage.

3.1 Collection Building

Our first goal is to construct a high-quality collection of terrorism web sites. "High-quality" refers to the comprehensiveness and relevance of the collected web sites. It is desired to have a collection representing the majority of known US domestic extremist groups with a presence on the web while keeping the collection free of unrelated web sites. Because terrorism web sites are often hidden and dynamic,

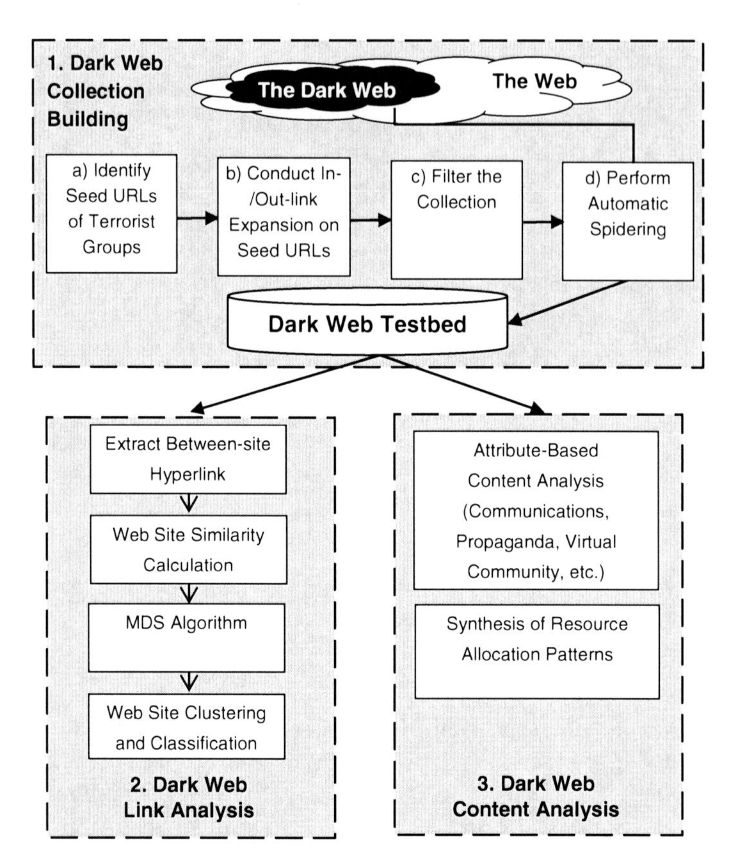

Fig. 20.1 Architecture of proposed approach

we propose to use a recursive collection building procedure which combines both manual selection and automatic web harvesting methods. We employ the following four steps:

(a) Identify seed URLs of organizations: The first task is to find an initial set of domestic terrorism web sites. We mainly search for URLs listed on the web sites of major watchdog organization, such as SPLC and the Anti-Defamation League (ADL), which continuously update their lists of domestic terrorism web sites. We obtained the lists of URLs in December 2003 which served as seeds for step 2.
(b) Conduct out-link and in-link expansion: After identifying the seed URLs, the out-links and in-links of the seed URLs were automatically extracted using link analysis programs. The out-links are extracted from the HTML contents of "favorite link" pages under the seed web sites. The in-links are extracted from Google's in-link search service through Google API. Automatic out-link and in-link expansion is an effective way to expand the scope of our collection.
(c) Filter the collection: Because bogus or unrelated sites can make their way into our collection, we have developed a robust filtering process based on evidence

and clues from the web sites. Aside from sites which explicitly identify themselves as the official sites of an extremist organization, a web site that contains even minor praise of or adopts ideologies espoused by an extremist group is included in our collection. All other web sites are excluded, for example, web sites with pure religious content with no elements of violence or hate.

(d) Perform automatic collection and processing of extremist web sites: Once the extremist web sites are identified, a spider program is used to automatically download all the contents of identified web sites. Unlike the tools used in most previous studies, in order to enable deep and comprehensive studies on the extremist web contents, our program was designed to download not only the textual files (e.g., HTML, TXT, PDF, etc.) but also multimedia files (e.g., images, video, audio, etc.) and dynamically generated web files (e.g., PHP, ASP, JSP, etc.). Moreover, because extremist organizations set up forums within their web sites whose contents are of special value to research communities, our program also can automatically log into the forums and download the dynamic forum contents.

3.2 Link Analysis

Our goal here is to shed light on the infrastructure of extremist and hate web sites and to perform a sophisticated content analysis. We believe the exploration of hidden communities over the web can give insight into the nature of relationships between web sites from the same group as well as relationships between web sites of different extremist groups. In addition, hyperlinks between web sites constitute an important cue for estimating the content similarity of any pair of web sites in our collection. Hence, we employ this cue to confirm our initial manual classification of the web sites under the categories shown in the Appendix.

Uncovering hidden web communities involves calculating a similarity measure between all pairs of web sites in our collection. We define similarity to be a real-valued multivariable function of the number of hyperlinks in web site "*A*" pointing to web site "*B*," and the number of hyperlinks in web site "*B*" pointing to web site "*A*." In addition, a hyperlink is weighted proportionally to how deep it appears in the web site hierarchy. For instance, a hyperlink appearing at the homepage of a web site is given a higher weight than hyperlinks appearing at a deeper level. Thus, the similarity between web sites "*A*" and "*B*" is calculated as follows:

$$\text{Similarity}(A,B) = \sum_{\substack{\text{All links L} \\ \text{b/w } A \text{ and } B}} \frac{1}{1+lv(L)}$$

where $lv(L)$ is the level of link L in the web site hierarchy, with the homepage as level 0 and the level increased by 1 with each level down in the hierarchy.

The similarity matrix is then fed to a multidimensional scaling (MDS) algorithm which generates a two-dimensional graph of the web sites. Multidimensional scaling is a data analysis technique which provides visual representation of proximities (dissimilarities) among objects so that objects that are more similar to each other are closer on the display and objects that are less similar to each other are farther apart (Xu and Chen 2005). This technique is often used in social network analysis (SNA). When applied to web site link analysis, the proximity of nodes (web sites) in the graph reflects the level of similarity between web sites. Gustavson and Sherkat (Gustavson and Sherkat 2004) highlight that unreciprocal ties such as friendship, resource sharing, and coordination (direction of an edge in a directed graph) can clarify the exact nature of relationships for pairs of web sites. These considerations will, however, be tackled in future extensions of this work.

3.3 *Content Analysis*

To better understand the goals and ways domestic extremists use the web, we developed an attribute-based coding scheme for methodically capturing the content. The coding scheme consists of eight high-level attributes: communications, fundraising, sharing ideology, propaganda (inside), propaganda (outside), virtual community, command and control, and recruitment and training. These attributes are of interest to terrorism researchers and were identified by a terrorism research expert who has 13 years of experience serving as a terrorism intelligence analyst in the CIA. Each high-level attribute is composed of multiple fine-grained low-level attributes. For example, level of communication is measured by the existence of e-mail contact, telephone contact, multimedia files, online feedback forms, and documentation. These low-level attributes were described in detail in the coding scheme, and they do not require any specific terrorism domain knowledge to be identified from the web sites. This attribute-based approach is similar to that employed in Demchak et al. (2000) study on government web site interactivity analysis. The Appendix shows the high-level and associated low-level attributes that were used in our study. This coding scheme tool enables the detection of the particular resource allocation patterns (e.g., fundraising, propaganda) that the domestic extremists use on the web. Moreover, it allows for measuring the levels of usage for particular purposes by assigning a weight to the low-level attributes. Gerstenfeld et al. (2003) pointed out that further research should be conducted to clarify the precise nature of the messages promoted on the web sites.

To ensure that the coding scheme is reliable, we asked four individual student coders to perform content analysis on four randomly selected US domestic extremist web sites using the coding scheme. For each of the four web sites, the corresponding sets of content analysis results were compared and a reliability score (Cronbach's alpha) was calculated. The results of the experiment are shown in Table 20.2. The high average Cronbach's alpha of 0.807 shows that the coding scheme has high reliability.

Table 20.2 Reliability test results

Web site	United Nuwaubian Nation of Moors web site	Kingdom Identity Ministries web site	Texas League of the South web site	Knights of Ku Klux Klan web site	Average
Cronbach's alpha	0.825	0.794	0.863	0.746	0.807

Table 20.3 Summary of the collection with categories

Category	Initial count before selection	Final count of URLs	Example group
Black Separatist	2	2	Nation of Islam
Christian Identity	17	13	Kinsman Redeemer Ministries
Militia	15	8	Michigan Militia
Neo-Confederate	17	4	Texas Leagues of the South
White Supremacy	29	7	Ku Klux Klan
Neo-Nazi	15	9	American Nazi Party
Ecoterrorism/Animal Rights	2	1	Earth Liberation Front
Total	97	44	

4 Test Bed: Collection of Domestic Extremist Web Sites

Following the proposed approach described in Sect. 3.1, we created a domestic extremist web site collection. We manually extracted a set of URLs from relevant literature. A total of 266 seed URLs were identified from SPLC and ADL web sites as well as in the Google directory. This procedure is similar to Gerstenfeld, Grant, and Chiang's study (2003) where they used several nonprofit watchdog organizations and Yahoo!'s category of white pride and racialism.

A link expansion of this initial set was performed and the count increased to 386 URLs. The resulting set of URLs is validated through filtering the irrelevant URLs introduced by the Google search and out-/in-link expansion. A total of 97 URLs were deemed relevant. We then used an automatic web crawling toolkit called SpidersRUs (ai.bpa.arizona.edu/research/spider/index.htm) to download all the web documents within the identified web sites. As a result, our final collection contains around 400,000 documents.

Our link analysis is based on all 97 web sites crawled. However, because of the time constraint, we could not perform content analysis on all 97 web sites. We selected the largest web sites from each category to form a subset of 44 web sites for content analysis. The 44 web sites are representative of the domestic extremist groups maintaining a presence on the web. Table 20.3 provides the summary and a categorization (based on SPLC) of the web sites. We manually coded the attributes in each web site.

5 Analysis Results

5.1 *Link Analysis Results*

Our link analysis aims to visualize and analyze hidden domestic terrorism hyperlinked communities and intercommunity relationships. Following the analysis approach proposed in Sect. 3.2, five communities were identified by a terrorism domain expert in the network shown in Fig. 20.2.

On the top left side of the network resides the Southern Separatists' cluster. This cluster mainly consists of the web sites of the New Confederate organization in the southern USA. They espouse a separatist ideology, promoting the establishment of an independent state in the south. In addition, they share White Supremacy ideas with other non-neo-Confederate racist organizations such as the Ku Klux Klan (KKK), the most prominent hate group in history. A cluster of neo-Nazi and White Supremacy web sites inhabits the top right corner of the network such as Stormfront and White Aryan Resistance (www.resist.com). In the bottom right corner, we

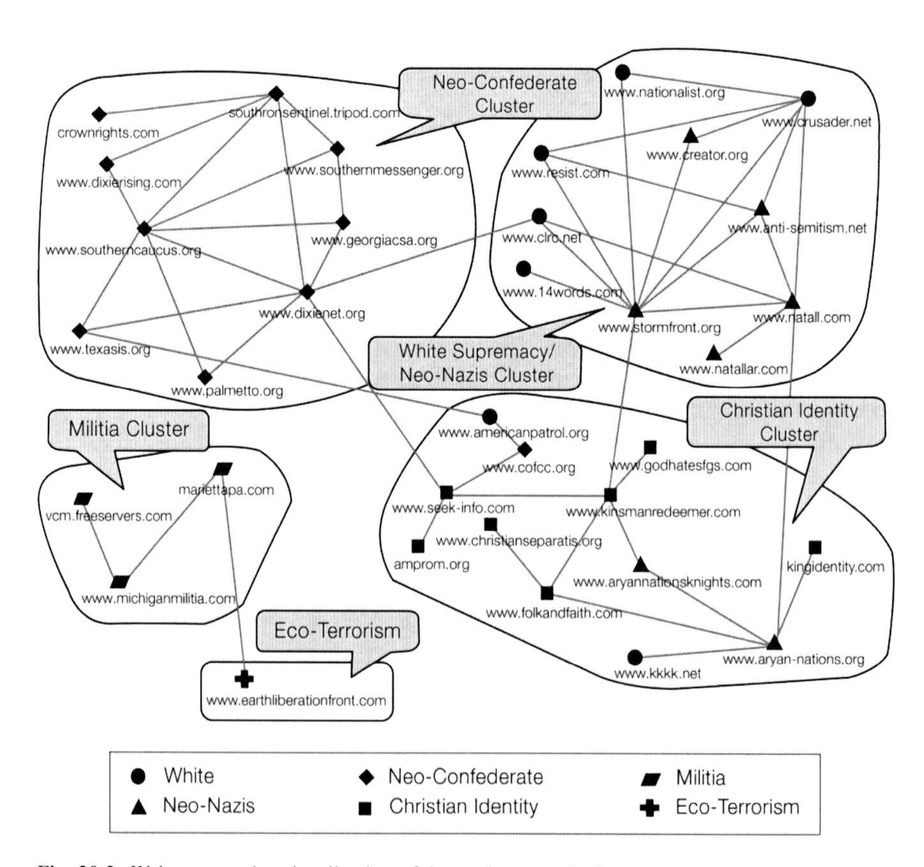

Fig. 20.2 Web community visualization of domestic extremist hate groups

identified a cluster that is primarily Christian Identity web sites. A clear separation between Christian Identity, neo-Nazi, and White Supremacy groups is, in general, hard to make. This observation agrees with previous social movement studies (Burris et al. 2000; Gustavson and Sherkat 2004). Thus, neo-Nazi web sites (www.aryannationsknight.com, www.aryan-nations.org) and a White Supremacy web site of the Knights of the Ku Klux Klan (www.kkkk.net) appear in the Christian Identity cluster.

Links between communities do not necessarily represent cooperation between them. An example is the few links between the neo-Confederates and Christian Identity/neo-Nazi/White Supremacy clusters. When investigating such links, we found that web site owners share common interests in some issues. For instance, the link between www.texasls.org (neo-Confederate) and www.americanpatrol.org (White Supremacist) reflects the common interest in "protecting" the southern border and bitterness felt toward Hispanic illegal immigrants. The numerous links between the neo-Nazi/White Supremacists and Christian Identity are, on the other hand, more likely to represent good relations between the communities. Both communities have a similar ideology, and researchers sometimes group them together. Two isolated communities can be seen on the bottom left corner of the network: the Militia and Ecoterrorism clusters. These communities have different interests and ideologies. This agrees with the results of Burris, Smith, and Strahm's research (2000) which concluded that bridges between the White Supremacy movement and other extremists such as the Militia are virtually nonexistent.

A frequently recurring question in social network analysis is that of the existence of central or prominent nodes. We identified two such nodes in our network of US domestic extremist web sites. The first node and by far the most famous among terrorism researchers is www.stormfront.org. This web site has many in-links indicating its popularity among White Supremacists, which is in agreement with results from earlier research (Burris et al. 2000). The second web site is that of National Alliance (www.natall.com), a neo-Nazi web site which also has a very high number of in-links testifying to its prominence. Owners of White Supremacy web sites tend to cite and acknowledge other Supremacists' literature which may be residing on other web sites. Their primary goal is to gain more credibility by referring to other Supremacists with whom they share the same ideology.

Another observation is the occurrence of relatively isolated web sites within a single cluster. Linking to other web sites can be of benefit to the web site owners, such that they gain credibility or they enforce the sense of solidarity within a usually geographically dispersed extremist community. However, Burris, Smith, and Strahm's study (2000) points out that this does not always hold true. In particular, some web sites may be competing over a potential population of future members and/or consumers of goods that are being sold on the web sites. For instance, we found that www.14words.com, which publishes and sells White Supremacist literature, does not have a single out-link to other White Supremacist web sites. This finding is similar to that from Burris, Smith, and Strahm's study. They posit that the site, being an e-commerce one, may not want to recommend competitors or encourage users to go to other sites.

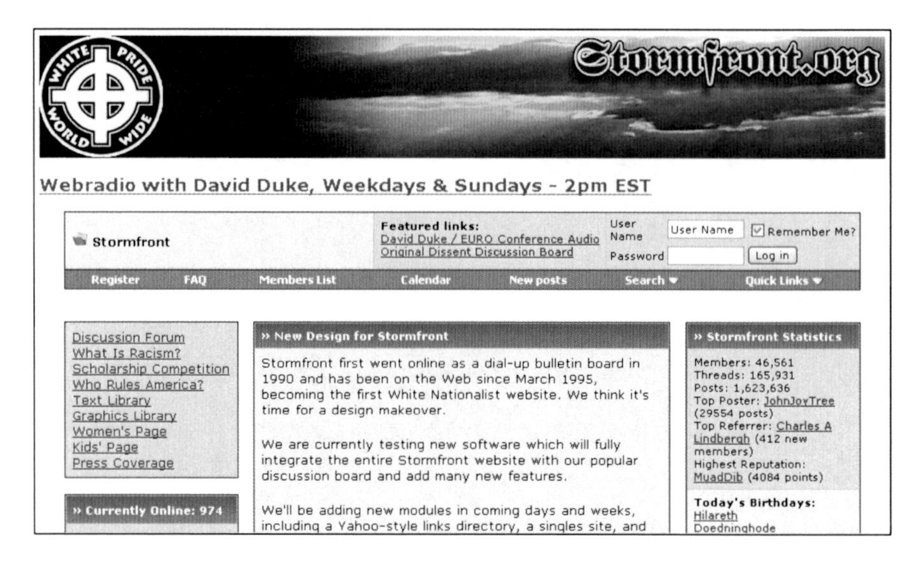

Fig. 20.3 Web site of www.stormfront.org (first White Supremacist web site)

5.2 *Content Analysis Results*

Two graduate students recruited from the business school at the University of Arizona coded each web site in our collection and recorded the presence of low-level attributes based on our coding scheme. For instance, the neo-Nazi web site www.stormfront.org, shown in Fig. 20.3, contains a forum and a bulletin board. The presence of these attributes contributes to the richness of the virtual community attribute.

After completing the coding scheme for the 44 web sites in the collection, we compared the content of each of the extremist communities described in Fig. 20.2. We aggregated data from all web sites belonging to a cluster and calculated the normalized content levels in the six dimensions. Each of these six dimensions represents a normalized activity scale between 0 and 1, showing the degree of activity on the dimensions. The activity scale of cluster *c* on dimension *d* was calculated by the following formula:

$$Activity\ scale(c,d) = \frac{\sum_{i}^{n} \sum_{j}^{m} w_{i,j}}{m \times n}$$

$$\text{where } w_{i,j} = \begin{cases} 1, & \text{Attribute } i \text{ occurs in site } j \\ 0, & \text{Otherwise} \end{cases}$$

n is the total number of attributes in dimension *d*; *m* is the total number of web sites belonging to cluster *c*.

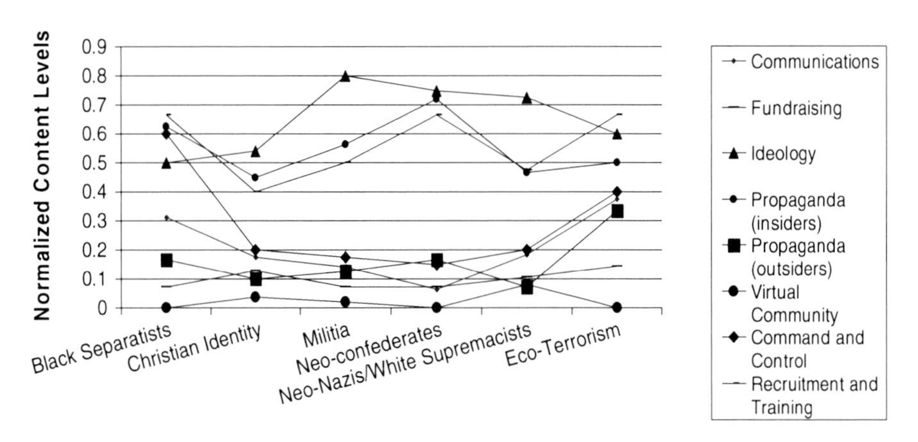

Fig. 20.4 Content analysis of web communities

Figure 20.4 shows the content levels for the six categories of extremist groups: Black Separatists, Christian Identity, Militia, neo-Confederates, neo-Nazi/White Supremacists, and Ecoterrorism.

As Fig. 20.4 shows, sharing ideology is the attribute with the highest frequency of occurrence in domestic extremist group web sites. Basically, the sharing ideology attribute encapsulates all communication media devoted to portraying the goals of the extremist group, defining its general policies, and presenting the foundational ideology. A major goal of extremist and hate groups is to expose their own definitions of the movements.

With the exception of Ecoterrorism, an interesting phenomenon in domestic terrorism web sites is the low level of content concerning the propaganda directed toward outsiders. This may be because Ecoterrorism groups have a much wider audience as compared to racist groups such as White Supremacists who only address very specific communities. For all groups, there was limited content in terms of the virtual community attribute. With their freedom of movement and speech within the USA, domestic terrorism groups are not heavily dependent on virtual communities for resources, unlike international extremist groups like al-Qaeda that depend a lot on virtual communities.

Another interesting observation is the much higher levels of communications and command and control attributes in the case of Ecoterrorism/Animal Rights and Black Separatist groups. The communications attribute tells the extent to which the owners and users of the web site rely on communication resources such as e-mail and chat. In general, most web masters provide an e-mail for feedback purposes. Moreover, some web sites, like those maintained by the Nation of Islam, reach a higher level of sophistication through posting recordings and videos of group leaders. These multimedia resources also contribute to the communication attribute, as they constitute an effective method of transmitting ideas and policies from the organization's hierarchy to lieutenants and members of the extremist/hate group.

6 Conclusions and Future Work

In this chapter, we validated a systematic methodology for the study of domestic extremist web site content. We employed focused collection building techniques, web link analysis, and attribute-based quantitative content analysis. Since there were several areas in which our findings support earlier social movement research, we concluded that the topological infrastructure of the US domestic extremist and hate group web sites seems to match domain experts' knowledge very well because the communities are formed by groups that are known to share similar ideologies or have close relationships with each other. Visualizing hyperlinked communities leads to an easier and more complete understanding of the underlying web infrastructure of domestic extremist groups. In addition, it showed the existence and strength of the relationships between various hyperlinked communities and helped to identify likely relationships between extremist groups in the world.

The results of this research have also been useful for our work on the Dark Web Portal test bed, in that it provided systematic methodologies for capturing, classifying, and analyzing domestic extremist web site data. Because this study involved a sample of 97 web sites, future studies of this kind should endeavor to enlarge the sample and verify if similar outcomes can be achieved.

We have several future research directions. First, we plan to automate the content analysis process by applying data mining techniques. We also plan to apply more sophisticated link analysis algorithms and web community mining algorithms on the extremist web site analysis and experiment with other network visualization techniques. Last, we would like to conduct a similar study on other international extremist groups and compare their use of the web with that of US domestic extremist groups.

Acknowledgments This research has been supported in part by the following grants: (1) DHS/CNRI, "BorderSafe Initiative," October 2003–March 2005, and (2) NSF/ITR, "COPLINK Center for Intelligence and Security Informatics – A Crime Data Mining Approach to Developing Border Safe Research," EIA-0326348, September 2003–August 2005. We would like to thank all members of the Artificial Intelligence Lab at the University of Arizona who have contributed to the project, in particular Wei Xi, Feng Huang, Homa Atabakhsh, Cathy Larson, Chun-Ju Tseng, and Shing Ka Wu.

References

Albertsen, K. (2003). "The Paradigma Web Harvesting Environment," *3 rd ECDL Workshop on Web Archives*, Trondheim, Norway.

Blitzer, R. (2001). "Domestic Preemption," Terrorism Threat and U.S. Government Response: Operational and Organizational Factors. J.M. Smith and W.C. Thomas, eds. Colorado: U.S. Air Force Academy, INSS, http://www.usafa.af.mil/nss/terrorism.htm. Accessed January 15, 2005.

Borgman, C. L., Furner, J. (2002). "Scholarly Communication and Bibliometrics," Annual Review of Information Science and Technology. B. Cronin, ed. Information Today, Inc.

Burris, V. Smith, E., Strahm, A. (2000). "White Supremacist Networks on the Internet," Sociological Focus, vol. 33, 2:215–235.
Chen, H., Qin, J., Reid, E. et al. (2004). "Dark Web Portal: Collecting and Analyzing the Presence of Domestic and International Terrorist Groups on the Web," IEEE Intelligent Transportation System Conference, Washington, D.C.
CNN (July 1999). U.S. Hate Groups Hard to Track.
Demchak, C. C., Friis, C., La Porte, T. M. (2000). "Webbing Governance: National Differences in Constructing the Face of Public Organizations," Handbook of Public Information Systems, G. David Garson, ed., New York: Marcel Dekker Publishers.
Gerstenfeld, P. B., Grant, D. R., Chiang, C. (2003). "Hate Online: a Content Analysis of Extremist Internet Sites," Analysis of Social Issues and Public Policy, 3(1):29–44.
Gibson, D., Kleinberg, J., Raghavan, P. (1998). "Inferring Web Communities from Link Topology," Proceedings of the 9th ACM Conference on Hypertext and Hypermedia.
Gustavson, A. T., Sherkat, D. E. (2004). "Elucidating the Web of Hate: the Ideological Structuring of Network Ties Among Right Wing Hate Groups on the Internet," Annual Meetings of the American Sociological Association.
Kay, R. (June 21, 2004). "Web Harvesting," Computer World, http://www.computerworld.com. Accessed January, 15, 2005.
Reid, E. O. F. (2003). "Identifying a Company's Non-Customer Online Communities: a Proto-typology," Proceedings of the 36th Hawaii International Conference on System Sciences, Springer, http://e-business.fhbb.ch/eb/publications.nsf/id/214. Accessed June 18, 2004.
Reilly, B., Tuchel, G., Simon, J., Palaima, C., Norsworthy, K., and Myrick, Leslie (2003). "Political Communications Web Archiving: Addressing Typology and Timing for Selection, Preservation and Access," 3rd ECDL Workshop on Web Archives, Trondheim, Norway.
SPLC Report (2004). "Hate Groups, Militias on Rise as Extremists Stage Comeback," http://www.splcenter.org/center/splcreport/article.jsp?aid=71. Accessed June 02, 2005.
Whine, M. (1997). "Far Right on the Internet," Governance of Cyberspace. B. Loader, ed., London: Routledge.
Xu, J. and Chen, H. (April 2005). "CrimeNet Explorer: A Framework for Criminal Network Knowledge Discovery," *ACM Transactions on Information Systems*, Vol. 23, No. 2, 201–226.

Chapter 21
International Falun Gong Movement on the Web

1 Introduction

The Internet nowadays is not merely a digital platform for exchanging data or information but is also a communication channel for individuals to share and promulgate their beliefs and ideas. People discuss topics of interest in forums or create their own blogs for posting experiences and opinions. Through hyperlinks, individuals can link their own posts to other web pages to cite the comments that they agree with or oppose. Gradually, "ideological networks" are formed on the Internet in which web sites with similar ideas are connected together via hyperlinks. One may find a number of web sites relevant to her/his interests by just following the hyperlinks of a few seed web sites.

The Internet has also changed how social movement organizations (SMOs) advocate their ideology and mobilize their resources. They are no longer restricted by time or space. Their web sites hold permanent campaigns appealing to a global audience (Bennett 2003, 2005; Keck and Sikkink 1998). The low cost of communication makes them more easily align together for large public actions (Bennett 2003, 2005; Myers 1994). For example, a demonstration against the war in Iraq in Washington, DC, in 2003 gathered around 100,000 people with different protest positions (Bennett 2005). Furthermore, activist groups can increase individual participation by hyperlinking an inclusive ideological network which provides multiple entry points for potential supporters to join in (Bennett 2003). The anti-Microsoft network, for instance, involves a great diversity of interest groups, including corporations, consumer protection organizations, and labor alliances (Bennett 2003).

In the traditional social movement theory, framing a coherent identity is a crucial process for an SMO to establish itself in social movements (Langman 2005). The collective identity helps participants to develop a trust relationship with their members and creates an informal interaction network for circulating important information and material resources (Porta and Diani 1999). The failure to achieve common framing can create tension and fragmentation within coordinated SMOs (Bennett 2005). Global or cyber activism, conversely, is flexible in identity framing (Bennett 2005).

H. Chen, *Dark Web: Exploring and Data Mining the Dark Side of the Web*, Integrated Series in Information Systems 30, DOI 10.1007/978-1-4614-1557-2_21,

Its advocate network may contain diverse social justice agendas, rich in personal appeal but thin in ideology (Bennett 2003, 2005). The ease of individuals joining and leaving a given network damages solidarity in SMOs and makes campaigns difficult to control (Bennett 2003). To assess the sustainability of an SMO in the Internet era, it is important to understand how the SMO constructs identity and social codes within its associated cyber societies (Bennett 2005).

In this chapter, we took a cyber-archaeology approach to investigate the framing of collective identity on the Internet and used the International Falun Gong (FLG) movement as a case study. The cyber-archaeology approach adapts methods from archaeology and anthropology to analyze a community's cultural cyber-artifacts (Jones and Rafaeli 2000; Paccagnella 1997). In applying the cyber-archaeology approach to the study of SMOs, the approach involves (1) the identification of their associated cyber societies, (2) collection of cyber-artifacts with automated procedures, and (3) analysis of cyber-artifacts from the perspective of the social movement theory. The FLG was chosen as our case study because it involves various identities, including Qi-Gong exercises, new religion, and activism (Penny 2003; Rahn 2002; Tong 2002; Lu 2005), and heavily uses the Internet as a vehicle for information dissemination (Bell and Boas 2003). Our goal was to investigate how FLG comprehensively maintains these three identities simultaneously within its cyber societies.

The remainder of this chapter is organized as follows. We first review social movement theory and cyber-archaeology and introduce two analytical tools used for examining cyber-artifacts: Social Network Analysis (SNA) and Writeprint. Then, we describe our research design which was based on the cyber-archaeology research framework, covering link, web content, and forum content analyses. Finally, we present our results and conclusions.

2 Literature Review

2.1 Social Movement Theory

Social movements are often recognized as irrational and disorganized activities working against social injustices. Unlike official political participation or lobbying, activists organize a broad range of individuals and seek to build a radically new social order through public protest (Cohen and Rai 2000). However, the study of social movements reveals that they have much deeper organizational and psychological foundations than would appear. Activists do not just blindly deploy protest actions: They calculate the costs and benefits of the actions with present resources before initiation (Porta and Diani 1999). People are not irrationally supporting a social movement: they seek an identity and share a sense of belonging through their participation (Porta and Diani 1999).

Collective action and resource mobilization are two intellectual currents which dominated the early development of social movement theory (Langman 2005; Porta and Diani 1999). Collective action is based on the idea that social movements are triggered by an overly rapid social transformation. In this school of thought, a society consisted of several balanced subsystems and a movement reflective of the failure of social institutions or control mechanisms to rebalance them after a dramatic change (Porta and Diani 1999). In such a moment, the existing norms no longer provide a sufficient rationale for social action, and a group of individuals sees the situation as injustice and reacts to it by developing shared beliefs and new norms for behavior. The American Civil War and Civil Rights Movement are significant examples illustrating this point of view (Porta and Diani 1999). The framing of collective identity is an essential process for collective action (Langman 2005; Porta and Diani 1999; Larana et al. 1994). For participants, the identity helps establish a trust relationship with others in the same group and excludes those whom they oppose (Langman 2005; Porta and Diani 1999). For the movement, it defines the orientation of public action and constrains where the action will take place (Langman 2005; Larana et al. 1994).

Resource mobilization examines the strategic components in social movements. It is based on two main assertions: (1) movement activities are not spontaneous and disorganized and (2) their participants are not irrational (Morris and Mueller 1992). In other words, social movements are meaningful actions with specific purposes (Porta and Diani 1999). In this school of thought, the movements involve so-called social movement entrepreneurs who call public attention to problems and who recruit and mobilize supporters for action (Langman 2005). Public protests are derived from a careful calculation of costs and benefits as measured against present resources (Porta and Diani 1999). To achieve their goals, activists need to control the costs of their actions, organize discontent to reach social consensus, and create solidarity networks for circulating information and resources. From this point of view, social movements are an extension of formal political actions to pursue social interests (Porta and Diani 1999).

2.2 *Social Movement Organizations and the Internet*

SMOs are nongovernmental organizations which seek to realize the ideology of a social movement with clear goals (Porta and Diani 1999). Their coordination in a movement is suggested following a SPIN model: (1) segmented, (2) polycephalous, (3) integrated, and (4) networked (Bennett 2003; Porta and Diani 1999; Gerlach 2001). According to the model, a social movement is composed of many SMOs which continuously die or rise (Porta and Diani 1999). It has multiple and sometimes competing leaders or influential centers (Bennett 2003; Gerlach 2001). Those SMOs form a loose but integrated network with overlapping membership and shared ideas (Bennett 2003; Porta and Diani 1999; Gerlach 2001).

The Internet has changed the ways SMOs operate (Bennett 2003, 2005; Myers 1994). One advantage that the Internet brought to SMOs is the reduction of communication costs (Bennett 2003; Myers 1994). In the pre-Internet era, SMOs relied on informal interaction networks of their supporters to circulate information and mobilize participants for action (Porta and Diani 1999). They now use computer-mediated communication (CMC), including e-mails and forums, to promote their ideology, recruit new members, and coordinate campaigns. Another significant change is that activists can hold permanent campaigns via their web sites, thereby appealing to a global audience and transforming domestic social movements into global or cyber activism (Bennett 2003, 2005; McCaughey and Ayers 2003). Compared to traditional activism, global activism depends on the availability of technology networks in order to expand, involves diverse social justice agendas and multiple issues, and promotes personal involvement in direct action (Bennett 2005). To reflect this transformation, Gerlach revised his SPIN model from "polycephalous" to "polycentric," meaning that global activist networks have many centers or hubs for supporters to join or leave and are less likely to be managed by permanent leaders (Bennett 2003). Therefore, they are thin in ideology but rich in personal identity and lifestyle narratives (Bennett 2003). To assess the sustainability and quality of an SMO in the global movement, it is crucial to identify the social codes and values embedded in its CMC cyber-artifacts (Bennett 2005).

2.3 *Social Network Analysis*

The central theme in social movement studies surrounds how activists "organize" themselves in campaigns to achieve impact on governments and societies, which makes SNA perfectly suitable for this kind of investigation. SNA is a graph-based methodology used to analyze social relations and their influence on individual behavior and organizational structure. It was developed by sociologists and has been applied in several academic fields, such as epidemiology and CMC. In SNA, individuals or actors are represented as nodes in a simple network graph, called a social network, and tied with edges indicating relationships. The visualization of a social network can provide a basic sense of how actors affiliate with each other and what their roles are in the group.

Centrality measures are quantitative indicators for finding those "central" individuals from a network originally developed in communication scenarios. From a topological perspective, people who are able to receive or control the mainstream of message flow typically stand in a position similar to the central point of a star (Freeman 1978/1979), such as the location of person A in the sample network. Various centrality measures, such as degree and betweenness, can be employed to determine the importance of a node within a network. Degree is the number of edges that a node has. Since the central point of a star has the largest number of edges connecting it to the other nodes, a node with a higher degree is topologically considered to be more central to its network (Freeman 1978/1979; Wasserman and Faust 1994). Betweenness measures "the extent to which a particular node lies between the various other nodes" (Scott 2000) because the central point also sits between pairs. The higher betweenness a node has, the more potential it has to be a gatekeeper controlling the connections (such as communications) between the others (Scott 2000).

Through reconstructing the network of nineteenth-century women reform leaders in New York State, Rosenthal et al. (1985) reported that weak ties played an important role in the women's movement: while strong ties linked major women's organizations in a movement, weak ties bridged several clusters and channeled the communication to diverse audiences. After the emergence of global activism, many researchers shifted their focus from SMO physical connections to SMO web site hyperlinks. Ackland et al. (2006) used their VOSON system to demonstrate the usefulness of network visualization in the analysis of linkage between environmental groups on the Internet. Garrido and Halavais (2003) used hyperlink analysis to map the Zapatista online network and examine its affiliation with other SMOs. They found that the secondary tier of Zapatista-related web sites played a bridging role linking the Zapatista network to the global SMO network.

2.4 *Writeprints*

Because of its anonymous nature, the Internet has become a major medium for cybercrime ranging from illegal sales and phishing to terrorist communication. In order to increase the awareness and accountability of users, many studies have been devoted to developing techniques to identify authors in the online environment (Abbasi and Chen 2005). Authorship identification is a process of matching unidentified writings to an author based on the similarity of writing styles between the known works of the author and unidentified pieces (Abbasi and Chen 2005, 2006). Four major categories of style features have been extensively used to identify writing styles: lexical, syntactic, structural, and content specific (Abbasi and Chen 2005). Lexical features include total number of words, words per sentence, and word-length distribution. Syntax refers to the patterns used for the formation of sentences, such as punctuation and function/stopwords. Structural features deal with the organization and layout of the text, such as the use of greetings and signatures, the number of paragraphs, and average paragraph length. Content-specific features are key words that are important within a specific topic domain. Among these four categories, lexical and syntactic features are frequently used because of their high discriminatory ability and portability across domains (Abbasi and Chen 2005, 2006).

In 2006, Abbasi and Chen (2006) proposed a visualization technique for authorship called Writeprints, which is analogous to the fingerprint biometric system. Unlike other studies of authorship visualization merely using n-gram features for discrimination, Writeprints were designed to apply to a large number of authors in an online setting. It uses all four major types of style features: lexical, syntactic, structural, and content-specific features (Abbasi and Chen 2006). The generation of a "Writeprint" consists of two main steps: (1) reduce dimensionality and (2) create visualizations. After extracting features from a set of documents, the Writeprint adopts principal component analysis (PCA) to reduce dimensionality of feature usage vectors by using the two principal components or eigenvectors with the largest eigenvalues. Once the eigenvectors have been computed with PCA, a sliding window algorithm is used to extract the feature usage vector for the text region inside a window, which slides over the text. For each window instance, the sum of the product of the principal component (primary eigenvector) and the feature vector represents the x-coordinate of the pattern point, while the sum of the product of the second component (secondary eigenvector) and the feature vector represents the y-coordinate of the data point. Each data point generated is then plotted onto a two-dimensional space to create the Writeprint. Abbasi and Chen (2006) reported that Writeprints outperformed the support vector machine (SVM) in classification of online messages in their evaluation.

3 Research Design: A Case Study of the International FLG Movement

This study employed the proposed cyber-archaeology framework and used the Falun Gong (FLG, 法輪功) movement as a case study to investigate SMOs' collective identity on the Internet and their ideological ties with others.

3.1 The Falun Gong Movement

FLG was founded by Hongzhi Li (李洪志) and introduced to the public in Mainland China in 1992 (Penny 2003). It focuses on the concept of cultivation and has two main components: practice and principle. The practice component includes five sets of gentle exercise movements which are similar to Qi-Gong (氣功) exercises. The principle component emphasizes the importance of truthfulness, compassion, and forbearance (真善忍). FLG practitioners believe they can enhance their physical health and elevate their mind at the same time.

On July 20, 1999, the Chinese government designated FLG as an evil cult and banned all its public activities (Tong 2002). This suppression is widely believed to be related to the mass petition of Falun Gong practitioners in Zhongnanhai, Beijing, on April 25, 1999 (Penny 2003). After the official ban, Hongzhi Li stayed in the United States and used FLG web sites, such as Clearwisdom.net (法輪大法明慧網), to release his articles and continue his teaching. Currently, FLG has local web sites of practitioners in over 38 countries and 5 continents. It holds several conferences annually in North America and Europe.

Before the suppression, there was no evidence that FLG or Hongzhi Li had any political agenda against the Chinese government. On January 23, 2001, a self-immolation incident by five FLG practitioners in Tiananmen Square in Beijing was reported by the international news media. In 2002, FLG web sites began releasing accounts of persecution against practitioners in Mainland China, including pictures, stories, and persecution methods. In late 2004, the Epoch Times (大紀元), which is related to FLG, published "Nine Commentaries on the Communist Party" (九評共產黨) and held a "Quitting the CCP (Chinese Communist Party)" (退黨) campaign.

Several studies, each using a different approach, have investigated how FLG transformed from a simple Qi-Gong exercise group to a religious and social movement organization (Penny 2003; Rahn 2002; Tong 2002; Lu 2005). Lu (2005) applied the religious economy model to the formation of new religions to interpret the shift of FLG from a healing system to a religion. According to Lu's analysis (Lu 2005), Hongzhi Li purposely introduced his own theory of salvation, Falun Dafa, to differentiate FLG from other competing Qi-Gong groups. In addition, he used various organizational and doctrinal mechanisms to keep his practitioners and avoid schisms. For example, he claimed that he was the incarnation of the highest supernatural force and the only master in Dafa (Lu 2005). Rahn (2002) used the conflict between FLG and the Chinese government to explain FLG's role in social movements. He suggested that the persecution of FLG in China was the key to establishing FLG's identity as a human rights movement. By examining Hongzhi Li's messages on Clearwisdom.net, he brought out a concern that the Fa-rectification teaching may induce violent behavior in FLG practitioners because the frustration at achieving this ultimate goal may intensify "the battle between good and evil" (Rahn 2002). But Rahn (2002) asserted that FLG's positive public image as a human rights group can decrease the chances of a violent act.

The widespread use of the Internet for organizing practitioners is another research focus of those studying FLG. Bell and Boas (2003) summarized three important functions of the Internet in the FLG movement: (1) disseminating Hongzhi Li's teachings, (2) strengthening the integrity of a globally dispersed community, and (3) bringing pressure on the Chinese government for lifting the ban. They also concluded that the use of the Internet might bring splinter sects challenging Li's authority (Bell and Boas 2003). This is consistent with Bennett's point of view that "the ease of joining and leaving polycentric issue networks means that it becomes difficult to control campaigns or to achieve coherent collective identify frames" (Bennett 2003).

3.2 Research Design

Figure 21.1 shows our research design.

Our interest was in two types of cyber-artifacts in the FLG movement: web sites and forums. SMO web sites typically are official "entry points" of SMO networks advocating their ideologies and campaigns. Collecting those web sites' hyperlinks and contents allowed us to investigate how FLG officially deploys itself on the Internet and connects to other SMOs. SMO forums, on the other hand, provide a relatively intimate view of how members interact with each other and discuss their

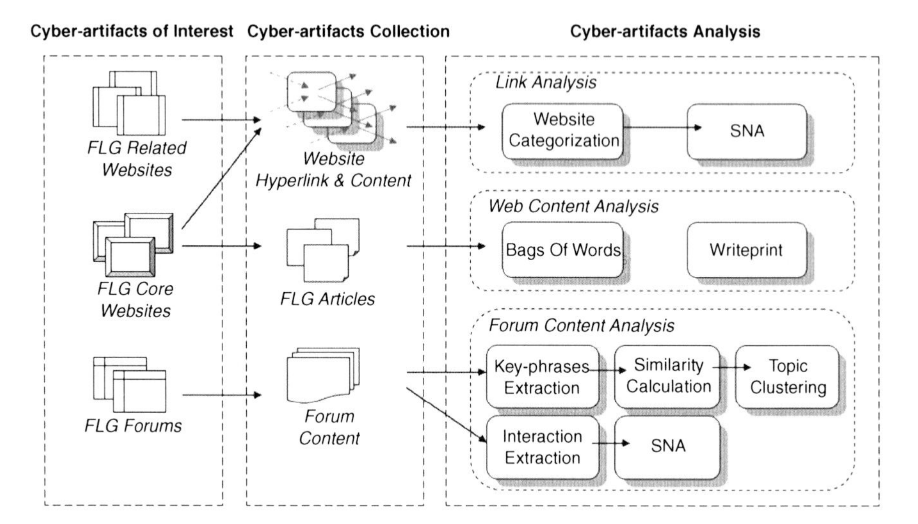

Fig. 21.1 Research design

SMO ideology. We used automatic programs to collect those cyber-artifacts and performed three analyses for our inquiries: link, web content, and forum content analyses.

3.2.1 Cyber-Artifacts Collection

Web site hyperlinks and content: FLG has four core web sites, as listed in Table 21.1, distributing FLG news, Hongzhi Li's articles, and accounts of persecution of FLG practitioners in Mainland China. These four core web sites are Clearwisdom.net (法輪大法明慧網), FalunInfo.net (法輪大法新聞社), FalunDafa.org (法輪大法), and EpochTimes.com (大紀元). Each web site offers more than ten language versions and has multiple domain names, a reflection of the fact that the FLG movement is organized on a global scale.

We automatically collected (spidered) FLG relevant web sites, including those of other activist groups having hyperlinks to FLG web sites, with two levels of in-links and out-links via 31 seed web sites, which included the four core web sites and another 27 FLG web sites identified by Google search. A total of 425 relevant domain names were found during spidering, and 172 were deemed relevant. Most of the relevant web sites were found to be directly linked to the core web sites, as shown in Table 21.2.

FLG articles: For the web content, we were particularly interested in studying the role of Hongzhi Li's articles in the FLG movement. From Clearwisdom.net, we collected 135 articles from Hongzhi Li and 74 articles from the editors for later comparison. Those articles concentrated on the discussion of three topics: teaching/principles of FLG, the position of FLG on political issue s especially related to

Table 21.1 Four core FLG web sites

Web site	Content	Language	Domain name
Clearwisdom.net (法輪大法明慧網)	1. Falun Gong information 2. Hongzi Li's articles 3. Persecution accounts 4. Practitioners' sharing	10	27
FalunDafa.org (法輪大法)	1. Falun Gong information 2. Local contact information and web sites	36	4
FalunInfo.net (法輪大法新聞社)	1. Persecution accounts 2. Falun Gong news	12	14
EpochTimes.com (大紀元)	1. World news 2. Persecution accounts 3. CCP criticism	15	36

Table 21.2 Collecting FLG-related web sites via seed web sites

Seed web site	Out-link Level 1	Out-link Level 2	In-link Level 1	In-link Level 2	Total
FalunDafa.org (法輪大法)	85	3	3	0	91
EpochTimes.com (大紀元)	26	0	1	1	28
Clearwisdom.net (法輪大法明慧網)	15	4	1	1	21
GuangMing.org (澳洲光明網)	12	0	1	0	13
FalunInfo.net (法輪大法新聞社)	4	2	1	0	7
SoundOfHope.org (希望之聲電台)	2	2	0	0	4
GrandTrial.org (全球公審江澤民)	3	0	0	0	3
GlobalRescue.net (全球營救FLG學員)	2	0	0	0	2
ZhuiChaGouJi.org (追查迫害FLG組織)	0	1	0	0	1
NtdTV.com (新唐人電視台)	0	1	0	0	1
Minghui-School.org (明慧學校)	1	0	0	0	1
Total	150	13	7	2	172

Mainland China, and summaries of various FLG conferences. The summary of these two sets of articles is in Table 21.3. Compared to the editors' articles, Hongzhi Li's are much longer in length.

Forum content: We used Google search and web site linkage to find FLG forums. Four forums were found, but only one forum, Falun Dafa Universal (世界法輪大法研究會), was, and is, still active and has more than 50 authors contributing to it.

Table 21.3 Summary of articles in Clearwisdom.net

Source	Number of articles	Average words per article	Duration
Hongzhi Li	135	1,430	5/1999~2/2007
Editors	74	670	3/2000~12/2006
Total	209	1,161	5/1999~2/2007

Therefore, we concentrated on the analysis of threads and messages in this forum. Falun Dafa Universal, located at city.udn.com (網路城邦), was established in 2005. It has 120 members and 28 discussion boards covering Hongzhi Li's articles, persecution accounts, and the FLG universal and science database. This forum circulates many articles from the four core web sites. Thus, the average length of messages is long, 1,288 characters per messages, but the average reply rate is low, 0.89 reply messages per thread. A total of 740 threads and 1,399 messages were collected for this forum.

3.2.2 Cyber-Artifact Analysis

Link analysis: In order to understand the main ideas of these web sites and how they linked together, we first classified their ideological types and performed SNA to analyze their network structure. Two measures of centrality in SNA are used to investigate which web sites are prominent in this network: degree and betweenness. The degree of a node is the number of links it has, reflecting its activity level. Betweenness is a measure of the frequency with which a node lies on the shortest geodesic paths of pairs of other nodes. It can be used to detect the extent to which a node plays the role of a gatekeeper in controlling the communication of others (Scott 2000).

Web content analysis: In order to highlight the characteristics of Hongzhi Li's writing, we used bag-of-words and Writeprints, developed by Abbasi and Chen (2006), and compared his articles with other articles written by the editors of Clearwisdom.net.

Forum content analysis: At the forum level, we performed two types of analysis: thread topic and author interaction. In the thread topic analysis, we investigated how many topics are covered in this forum and how those topics relate to each other. Since Falun Dafa Universal is a Chinese forum, we first used MI, a Chinese phrase extraction tool developed by Ong and Chen (1999), to extract key Chinese phrases from the threads and convert those threads into vectors of those key phrases. The top 20 key phrases based on frequency of appearance are shown in Table 21.4. We then used the cosine coefficient to calculate the similarity between threads and displayed those threads in a two-dimensional map. For author interaction analysis, we extracted the authors' responses to other's threads and perform SNA based on their interaction history.

Table 21.4 Key Chinese phrases of forum messages

Rank	Phrase	Rank	Phrase
1	法輪 (Falun)	11	政府 (Government)
2	法輪功 (Falun Gong)	12	醫院 (Hospital)
3	中國 (China)	13	修煉 (Cultivation)
4	學員 (Practitioner)	14	問題 (Problem)
5	器官 (Organ)	15	國際 (International)
6	迫害 (Persecution)	16	人類 (Human)
7	美國 (the United States)	17	個人 (Individual)
8	蘇家屯 (Sujiatun Camp)	18	國家 (Country)
9	大法 (Dafa)	19	集中營 (Labor Camp)
10	社會 (Society)	20	人民 (People)

4 Research Results

In this section, we present our research results.

4.1 *Link Analysis*

The 203 FLG relevant web sites, including seed web sites and those identified via the seeds, are classified into five main categories based on their web content: FLG cultivation, human rights, democracy, anti-FLG, and mixture (topics of more than one category). Two coders were hired for the web site classification. The network of these web sites, displayed with a spring-embedded algorithm, is shown in Fig. 21.2. The network has three main components: human rights and democracy on the left-hand side, FLG cultivation on the right-hand side, and mixture (and anti-FLG) in the middle. The mixture web sites, including Clearwisdom.net and EpochTimes.com, act as bridges connecting the other two main components. The human rights and democracy web sites are somewhat mixed together.

We used two centrality measures, degree and betweenness, to identify the most prominent web sites within this network. Here, the degree, or in-degree, is calculated by the number of in-links and reflects the popularity of a web site. The betweenness measures the potential that a web site may be a gatekeeper controlling the interaction with other web sites. The top 10 prominent web sites in this network are listed in Table 21.5. The four core FLG web sites are at the top of the list.

We used the in-links and out-links of the connected web sites to check the role of the FLG core web sites in this network. The results are shown in Fig. 21.3. EpochTimes.com is mainly responsible for the linkage of human rights and democracy web sites. Clearwisdom.net is located in the middle of the network and connects other major mixture web sites. FalunDafa.org focuses on FLG cultivation and links local FLG practitioners' web sites.

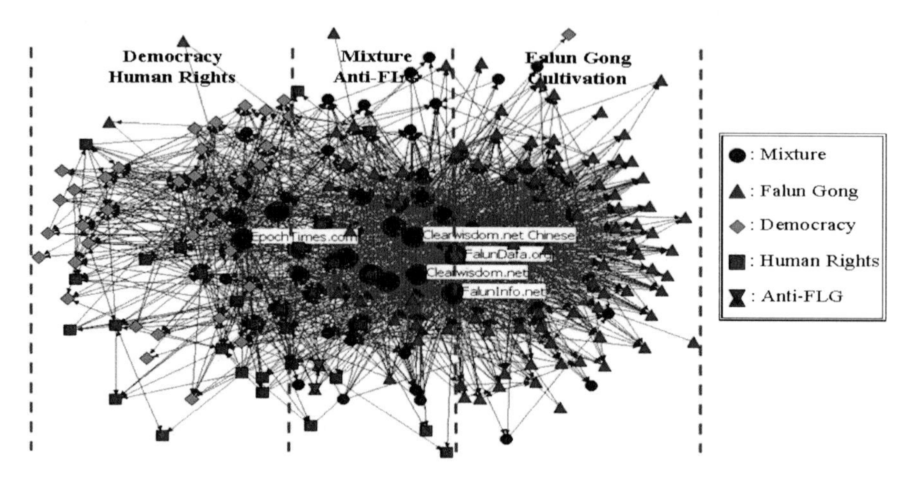

Fig. 21.2 Network of FLG relevant web sites

Table 21.5 Top ten web sites in FLG network based on centrality

Rank	Web site	In-degree	Web site	Betweenness
1	FalunDafa.org (法輪大法)	113	FalunDafa.org (法輪大法)	14657.33
2	FalunInfo.net (法輪大法新聞社)	99	EpochTimes.com (大紀元)	6166.15
3	Clearwisdom.net (法輪大法明慧網英文)	90	Clearwisdom.net Chinese version (法輪大法明慧網中文)	4318.54
4	Clearwisdom.net Chinese version (法輪大法明慧網中文)	88	GuangMing.org (澳洲光明網)	2533.59
5	EpochTimes.com (大紀元)	78	Clearwisdom.net (法輪大法明慧網英文)	2298.31
6	ZhengJian.org (正見)	65	FalunInfo.net (法輪大法新聞社)	2014.37
7	Fofg.org (法輪功之友)	54	ZhengJian.org (正見)	1335.54
8	ClearHarmony.net (歐洲圓明網)	50	SoundOfHope.org (希望之聲電台)	1276.49
9	SoundOfHope.org (希望之聲電台)	50	ClearHarmony.net (歐洲圓明網)	1077.36
10	NtdTV.com (新唐人電視台)	48	HriChina.org (中國人權)	792.51

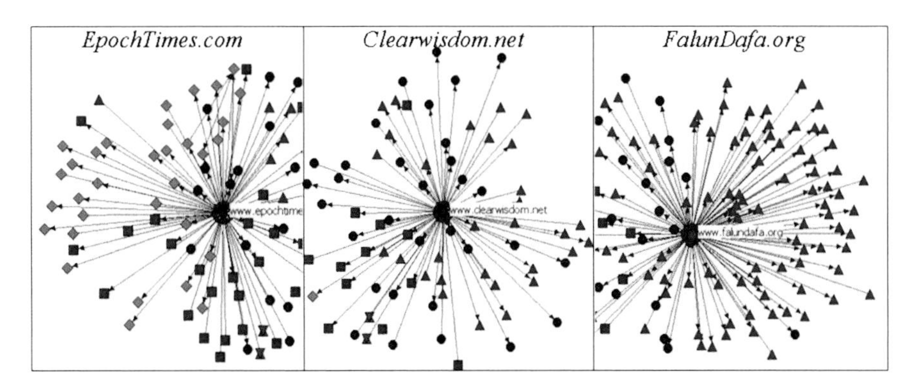

Fig. 21.3 Roles of FLG core web sites in FLG network

4.2 Web Content Analysis

Writeprints illustrate the characteristics of words that authors frequently use to express their opinions or ideas. Figure 21.4 shows the Writeprint of Hongzhi Li. His discussion revolves around the teachings of Falun Dafa (法輪大法) and has neither significant temporal variation nor concentration on subtopics, as shown in Fig. 21.4a. However, in his bag-of-words, as shown in Fig. 21.4c, the word "evil" is used frequently.

The Writeprints of the editors of Clearwisdom.net, as shown in Fig. 21.5a and b, had three significantly deviated areas which represented the topics of Dafa rectification (正法), righteous thoughts (發正念), and persecution of FLG practitioners (學員迫害真相). Their writings consistently revolved around these three major topics between 2000 and 2006. Comparing the Writeprints of Hongzhi Li and the editors allowed us to see their roles in the FLG movement. Hongzhi Li's articles focused on the central concepts of FLG cultivation but provided some hints of his political attitude (e.g., against evil). The editors' articles, on the other hand, provided their interpretations of Hongzhi Li's teaching.

4.3 Forum Content Analysis

4.3.1 Thread Topics

In the 740 threads collected from the forum "Falun Dafa Universal," ten main topics were identified based on the content of the threads: persecution accounts (學員迫害真相), FLG success story sharing (修煉心得分享), FLG ideology (法輪功哲學), FLG articles (法輪功書籍文獻), anti-Chinese Communist Party (anti-CCP, 反對中國共產黨), life philosophy (生活哲學), mysterious phenomena (宇宙科學與神秘

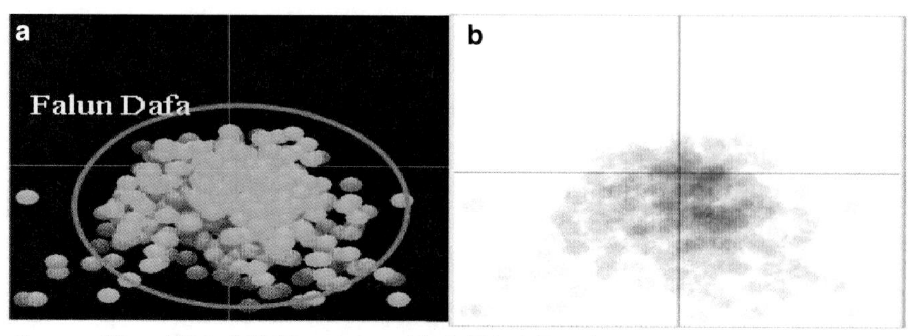

c Key Word N-Gram and Ink Blot Features

No.	Description	Usage	Mean	E
0	DAFA	0.868	0.037	0
1	FA	0.853	0.036	-
2	PEOPLE	0.732	0.025	0
3	BEINGS	0.663	0.017	-
4	THINGS	0.568	0.015	0
5	DISCIPLES	0.555	0.021	0
6	HUMAN	0.47	0.013	-
7	EVIL	0.397	0.018	0
8	RECTIFICAT...	0.35	0.013	-
9	COSMOS	0.263	0.0070	-
10	CULTIVATION	0.26	0.01	0
11	TIME	0.252	0.011	-

No.	Description	Weight	Blot
0	RIGHTEOUS	63.61	
0	HISTORY	25.81	
0	BEINGS	13.87	
0	DON	8.4	
0	GODS	5.52	
0	STUDENTS	3.94	
0	MASTER	3.09	
0	THINGS	2.7	
0	SITUATION	2.62	
0	MATTER	2.76	
0	ORDINARY	3.06	
0	HUMAN	3.47	
0	SOCIETY	3.97	
0	EVIL	2.47	

No.	Description	Usage	Mean
0	THE FA	0.348	0.014
1	OF THE	0.328	0.013
2	IN THE	0.323	0.012
3	DAFA DISCIPLES	0.299	0.011
4	FA RECTIFICA	0.223	0.008
5	THE EVIL	0.158	0.007
6	AND THE	0.133	0.005
7	THE COSMOS	0.118	0.003
8	TO THE	0.107	0.005
9	THE OLD	0.105	0.003
10	TO DO	0.1	0.003
11	SENTIENT BEI...	0.093	0.003

No.	Description	Usage
11	OF THE COSMOS	0.029
12	SENTIENT BEINGS	0.0
13	A DAFA DISCIPLE	0.027
14	VALIDATING THE FA	0.027
15	THE EVIL	0.0
16	THE HUMAN WORLD	0.024
17	IN OTHER WORDS	0.024
18	OTHER WORDS	0.0
19	BEINGS IN THE	0.022
20	NO MATTER HOW	0.022
21	DAFA DISCIPLES ARE	0.022
22	OLD FORCES	0.0
23	THE FUTURE	0.0

Fig. 21.4 Writeprint of Hongzhi Li

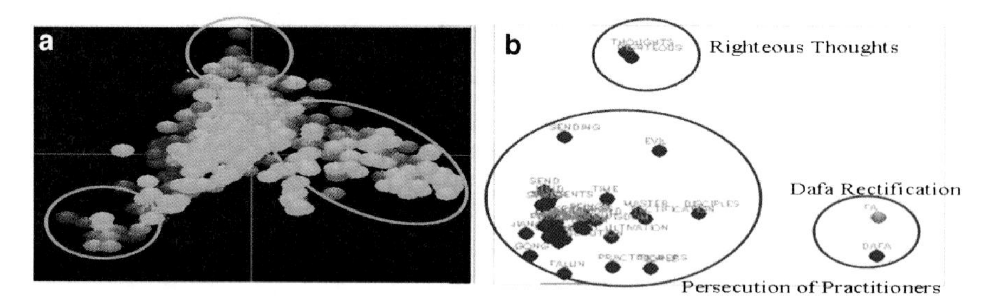

Fig. 21.5 Writeprints of the editors of Clearwisdom.net

現象), social issues (社會議題), health issues (健康議題), and general messages (網站管理訊息). The ten topics and their descriptions are listed in Table 21.6.

The distribution of threads over these ten topics is summarized in Table 21.7. Although life philosophy has the highest number of threads, anti-CCP has the highest average reply rate. Major discussions in this forum are often about anti-CCP topics.

Table 21.6 Ten main topics of forum threads

Main topic	Description
Persecution accounts (學員迫害真相)	Detailed description of torture process and methods
FLG success story sharing (修煉心得分享)	Describe how a practitioner benefits from FLG
FLG ideology (法輪功哲學)	Share personal beliefs about Dafa rectification, cultivation, and righteous thoughts
FLG articles (法輪功書籍文獻)	Include articles and books from Hongzhi Li and FLG
Anti-CCP (反對中國共產黨)	Criticize CCP for organ harvest, human rights, and religious freedom
Life philosophy (生活哲學)	Share inspired life stories and words of wisdom
Mysterious phenomena (宇宙科學與神秘現象)	Distribute articles about the origin of the cosmos and unexplainable phenomena
Social issues (社會議題)	Discuss social issues, such as the role of news press and impact of violent video games
Health issues (健康議題)	Distribute health-related news and healthy recipes
General message (網站管理訊息)	Messages about the forum management and arguments

Table 21.7 Distribution of threads over ten topics

Main topic	Threads	Messages	Reply rate
Persecution accounts	100	214	1.14
FLG success story sharing	10	16	0.6
FLG ideologies	112	256	1.28
FLG articles	29	29	0
Anti-CCP	112	336	2
Life philosophy	166	255	0.54
Mysterious phenomena	87	107	0.23
Social issues	16	39	1.44
Health issues	86	105	0.22
General message	22	42	0.9

Figure 21.6 displays the threads based on their similarity. We can see that the persecution accounts and anti-CCP are aligned and, to some degree, mixed together on the upper parts of the circle. Such a mixture is due to high usage of the same key phrases, such as organ harvest (器官活摘) and labor camp (集中營). In the lower part, Falun Gong ideology is closely aligned with life philosophy and mysterious phenomena. From the relative positions of these three topics, we may infer that Falun Gong is similar to a religion, which not only teaches a certain life philosophy but also explains the origin of life.

From the display of threads, we can further see to which topics an author primarily contributes. Figure 21.7 shows the distribution of threads of the top two active authors in the forum. The author "Sujcs888" focused on Falun Gong ideology, life philosophy, and mysterious phenomena, while the author "LoveTender" targeted the topics of persecution and anti-CCP.

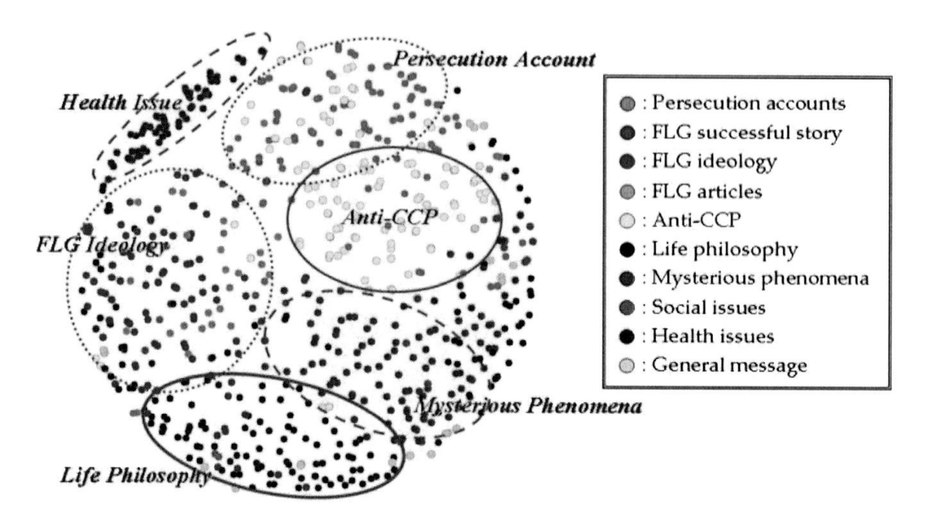

Fig. 21.6 Display of threads according to their similarity

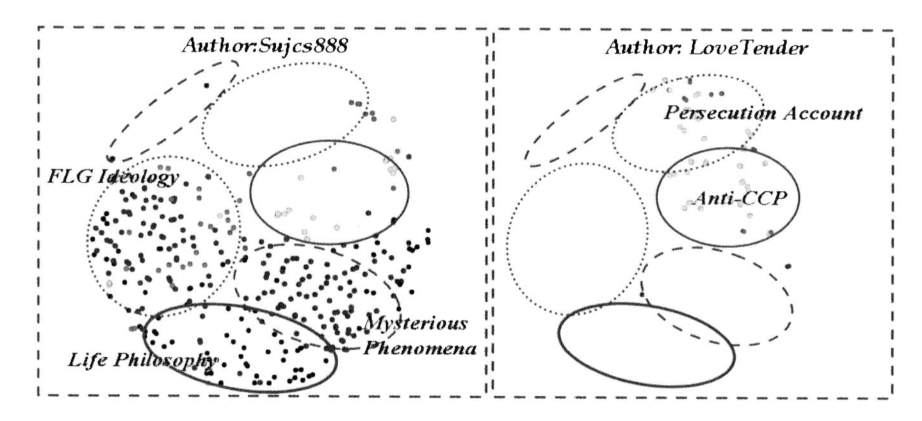

Fig 21.7 Distribution of topics of the top two active authors in the forum

4.3.2 Author Interaction

In order to see which topics provoked more intense discussion among authors, we measured the average degree and clustering coefficient of the interaction networks of the top five topics. The average degree shows the overall activity or interaction density of authors in a network. The clustering coefficient reflects clusters, which can indicate cliques or groups (Wasserman and Faust 1994). The results of these two measures are summarized in Table 21.8. The most intense interaction occurred in the discussion of FLG ideology and anti-CCP. However, compared to FLG ideology, anti-CCP had lower average degree but a much higher clustering coefficient.

Table 21.8 Degree and clustering coefficient of five main topics

Main topic	Degree	Clustering coefficient
FLG ideology	2.400	0.086
Life philosophy	2.080	0.093
Mysterious phenomena	1.00	0.000
Persecution	1.733	0.000
Anti-CCP	2.261	0.128

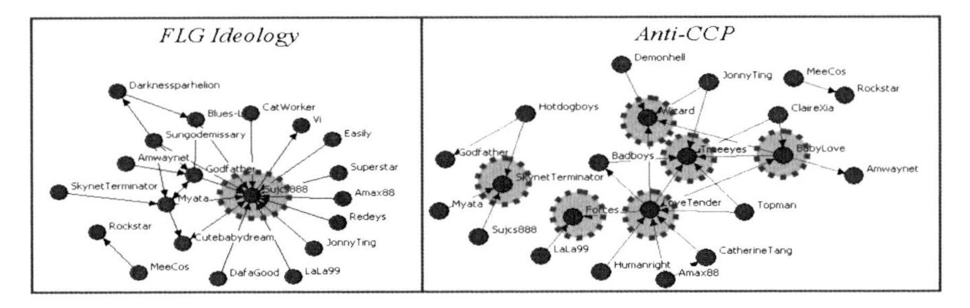

Fig. 21.8 Interaction network of authors in FLG ideology and anti-CCP

This implies that those authors were more likely to create discussion groups around this topic.

Figure 21.8 shows the interaction networks relating to FLG ideology and the anti-CCP topic. Authors discussing FLG ideology seemed to follow the ideas and preaching of master practitioners (the node in the center of the network). In anti-CCP discussions, authors were more likely to share their opinions and interact with each other freely (as shown in small clusters of interactions).

5 Conclusions

In this chapter, we took a cyber-archaeology approach and performed three separate analyses on the cyber-artifacts, including link, web content, and forum content analyses, to investigate the framing of FLG's identity in social movements. First, it is not surprising to see that FLG's web sites closely link to two types of SMOs: Chinese democracy and human rights. This affiliation can be explained with the conflicts between FLG and the Chinese government: the official ban in 1999 and subsequent persecution of its practitioners. As Rahn (2002) pointed out, those persecution accounts gave FLG a justified identity as a human rights movement. Second, by taking the cyber-archaeology perspective, we can see how FLG "strategically" deploys its cyber-artifacts to form/define its inclusive identity on the Internet. Not only are its four core web sites located in the center of the FLG network, but they also bridge two seemingly unrelated groups of web sites for different roles: activist and Qi-Gong groups. Each of the core web sites plays a specific role in joining web sites of different attributes. For example, most of the human rights– and democracy-related web sites in the FLG network are connected via EpochTimes.com. In addition, we find a trace of its religious role in its forum: it tries to explain the origin and meaning of life. From topic clustering, we see how FLG includes and organizes several different concepts on a continuum: from FLG ideology to life philosophy and mysterious phenomena, and from mysterious phenomena to anti-CCP and persecution, conceptualizing the CCP as "evil." By deploying its cyber-artifacts, FLG smoothly connects different ideologies and establishes its inclusive role as a Qi-Gong, religious, and activist group.

As a religious group, Hongzhi Li without question is the spiritual leader of FLG. The Writeprint revealed a structural pattern resembling a religious hierarchy in the writings of Hongzhi Li and of the editors: Hongzhi Li's centered on the ideological teaching of Falun Dafa while the editors' posted specific programs outlining Li's teaching for the practitioners to follow. In the FLG forum, we also found that the authors exhibited different interaction patterns in the discussions of FLG ideology and of anti-CCP. They followed the monologue pattern of master preaching in the FLG ideology but had more interaction in the anti-CCP discussion. Does the structural difference between a religious and an activist organization cause tension in FLG's cooperation with other SMOs? While we do not find any evidence to support this from analyzing its cyber-artifacts, the human rights torch relay, mainly hosted by CIPFG.org (Coalition to Investigate the Persecution of Falun Gong in China) between 2007 and 2008, revealed potential coordination problems between FLG and its allies. The relay was widely conceived as an FLG campaign against the Beijing 2008 Olympic Games and used as its slogan, "The Olympics and crimes against humanity cannot coexist in China." However, Hongzhi Li posted a message on April 4, 2008, explaining that the human rights torch relay is for everyday people and "so it is not Dafa disciples that this event is for." This message can be interpreted as a call for FLG practitioners to focus on FLG persecution and truth-clarification work rather than involvement in human rights activities. How the FLG is going to reframe its identity in social movements will require careful monitoring and future studies.

Acknowledgments This material is based upon work supported by the National Science Foundation under Grant No. CNS-0709338, "CRI: CRD-Developing a Dark Web Collection and Infrastructure for Computational and Social Sciences."

References

Abbasi, A., Chen, H. (2005). "Applying Authorship Analysis to Extremist-Group Web Forum Messages," *IEEE Intelligent Systems*, 20: 1541–1672.

Abbasi, A., Chen, H. (2006). "Visualizing Authorship for Identification," In: *Intelligence and Security Informatics*, pp. 60–71.

Ackland, R., O'Neil, M., Bimber, B., Gibson, R.K., Ward, S. (2006). "New Methods for Studying Online Environmental-Activist Networks," In: *26th International Sunbelt Social Network Conference*, Vancouver.

Bell, M.R., Boas, T.C. (2003). "Falun Gong and the Internet: Evangelism, Community, and Struggle for Survival," *Nova Religio*, 6: 277–293.

Bennett, W. (2003). "Communicating Global Activism," *Information, Communication and Society*, 6: 143–168.

Bennett, W.L. (2005), "Social Movements beyond Borders: Understanding Two Eras of Transnational Activism," In: *Porta Dd and Tarrow S (eds) Transnational Protest and Global Activism*, Rowman and Littlefield Publishers, INC, New York, pp. 203–226.

Cohen, R., Rai, S.M., eds. (2000), Global Social Movements. *Athlone Press,* New Brunswick.

Freeman, L.C. (1978/79), "Centrality in Social Networks: Conceptual Clarification." *Social Networks* 1: 215–239.

Garrido, M., Halavais, A. (2003), "Mapping Networks of Support for the Zapatista Movement: Applying Social-Network Analysis to Study Contemporary Social Movements," In: *McCaughey M, Ayers MD (eds) Cyberactivism: Online Activism in Theory and Practice*. Routledge, New York.

Gerlach, L.P. (2001), "The Structure of Social Movements: Environmental Activism and its Opponents," In: *Arquilla J, Ronfeldt DF (eds) Networks and Netwars: The Future of Terror, Crime, and Militancy.* Rand, Santa Monica, pp. 289–309.

Jones, Q., Rafaeli, S. (2000), "What Do Virtual "Tells" Tell? Placing Cybersociety Research into a Hierarchy of Social Explanation," In: *Hawaii International Conference on System Sciences*, Hawaii.

Keck, M.E., Sikkink, K. (1998), "Activists Beyond Borders: Advocacy Networks in International Politics," Cornell University Press, New York.

Langman, L. (2005), "From Virtual Public Spheres to Global Justice: A Critical Theory of Internetworked Social Movements," *Sociological Theory* 23: 42–74.

Larana, E., Johnston, H., Gusfield, J.R. (1994), "New Social Movements: From Ideology to Identity," Temple University Press, Philadelphia.

Lu Y (2005), "Entrepreneurial Logics and the Evolution of Falun Gong," *Journal for the Scientific Study of Religion* 44: 173–185.

McCaughey, M., Ayers, M.D. (2003), "Cyberactivism: Online Activism in Theory and Practice," *Routledge*, New York.

Morris, A.D., Mueller CM (1992), "Frontiers in Social Movement Theory," *Yale University Press*, New Haven.

Myers, D.J. (1994), "Communication Technology and Social Movements: Contributions of Computer Networks to Activism," *Social Science Computer Review* 12: 250–260.

Ong, T.H., Chen, H. (1999), "Updateable PAT-Tree Approach to Chinese Key Phrase Extraction using Mutual Information: A Linguistic Foundation for Knowledge Management," In: *Proceedings of the Second Asian Digital Library Conference*, Taipei.

Paccagnella, L. (1997), "Getting the Seats of Your Pants Dirty: Strategies for Ethnographic Research on Virtual Communities," *Journal of Computer-Mediated Communication* 3.

Penny, B. (2003), "The Life and Times of Li Hongzhi: Falun Gong and Religious Biography," *The China Quarterly* 175: 643–661.

Porta, D., Diani, M. (1999), "Social Movements: An Introduction," *Blackwell Publishers Ltd*, Malden.

Rahn, P. (2002), "The Chemistry of a Conflict: The Chinese Government and the Falun Gong," *Terrorism and Political Violence* 14: 41–65.

Rosenthal, N., Fingrutd, M., Ethier, M., Karant, R., McDonald, D. (1985), "Social Movements and Network Analysis: A Case Study of Nineteenth-Century Women's Reform in New York State," *The American Journal of Sociology* 90: 1022–1054.

Scott, J. (2000), "Social Network Analysis: A Handbook, 2nd edn," Sage, London.

Tong, J. (2002), "An Organizational Analysis of the Falun Gong: Structure, Communications, Financing," *The China Quarterly* 171: 636–660.

Wasserman, S., Faust, K. (1994), "Social Network Analysis: Methods and Applications," *Cambridge University Press*, New York.

Chapter 22
Botnets and Cyber Criminals

1 Introduction

Botnets are the latest stage in the evolution of cyber criminal operations online. The earliest malware was typically used for pranks, either for damaging systems or printing taunting messages to the user. Traditional computer viruses were engineered to self-copy themselves to other legitimate host programs on the system. These infected host files could then be carried to other computers by unknowing computer users or by the underlying infected programs. Trojan horses arose to infiltrate computer systems via social engineering: by disguising themselves as harmless files until executed by an unknowing user. These early forms of malware were relatively easy to combat and had low spreading rates.

The arms race continued with the birth of worms. Network-bound worms do not rely on infected software or human factors to propagate as they are engineered to exploit network services. Scanning and infecting new victims is a task easy to automate, and consequently, the propagation times are significantly faster. If utilizing an unpatched "zero-day" exploit, a network worm can be released in the wild and infect millions of vulnerable systems before vendors or authorities can respond.

Viruses, Trojans, and worms are usually preprogrammed with their mischievous deeds: performing their malicious duties immediately, or relying on a cue from the system in the form of time delays or logic bombs. While attackers could, in principle, gain control of their targets via backdoors, such practice was not common as it could not be scaled to many victims.

Before the bot era, malicious software was predominantly intended to attack, destroy, and cause general mischief in an automated fashion. The paradigm of modern cybercrime is recruitment and utilization of infected computers as a strategic asset. Infected victims are no longer disposable casualties in an attack, but a resource to be collected and built upon.

A botnet is a collection of victim computers, known as bots, drones, or zombies, that have been infected and assimilated into a greater collective through a centralized command and control (C&C) infrastructure (Nazario 2007). The simple addition of this

H. Chen, *Dark Web: Exploring and Data Mining the Dark Side of the Web*,
Integrated Series in Information Systems 30, DOI 10.1007/978-1-4614-1557-2_22,

communication back channel enables the attacker (also botmaster or bot herder) to issue commands to the infected drone collective and perform a variety of attacks that rely on significant computational or network resources.

Botnets can grow to millions of drones. Infected populations span the commercial and home sectors and, in some cases, may even find their way into military and government networks. The wide base of infected personal computers and the large coverage among the world Internet infrastructure give botnet owners unprecedented power and resources to acquire identities or wage digital war.

Botnets can be utilized for many nefarious purposes:

- *DDoS attacks*: Distributed Denial of Service (DDoS) attacks enable attackers to incapacitate a variety of network services by using thousands of compromised machines to flood servers with countless useless requests for information.

The excessive bandwidth or computational requirements to respond to such a deluge of illegitimate clients cripples a server's capacity to deliver information to legitimate users. DDoS attacks have been shown to be immensely effective against entire corporations and nations. In late April 2007, the nation of Estonia sustained a massive distributed cyber attack that disabled critical infrastructure and media.

- *Infection*: Like traditional worms, bots can scan for and infect computers over a network to assimilate them into the drone collective.

Increasingly, web-based infection mechanisms have been observed, whereby drone populations actively spider web sites for known web application vulnerabilities. Legitimate web sites are infected with a drive-by exploit that attacks and infects vulnerable web browsers. Such attacks can leverage a given web site's popularity to attack most of its frequently visiting customers.

- *Spamming*: Botnets have risen to become a significant source of the worlds e-mail spam. Increasingly, bot herders are relying on spam-related services as a primary income source (Smith 2008). Such spam can also bring phish attacks and additional infections through Trojan horse attachments.

Many botnets are actually rented to other criminal organizations, usually for spamming-related activities. While ad-laden spam can be annoying, phish attacks and stock market pump-and-dump scams can bring untold losses to individuals and profits to criminal syndicates (Smith 2008).

- *Espionage*: Any victim of a bot infestation becomes the target of spying. Bot software is capable of searching the local file system for documents to send back to the attackers, as well as extracting all saved Internet passwords from Internet Explorer's protected storage.

Many bots are additionally equipped with keylogging modules, capable of intercepting all keystrokes from within the Windows API. Some bots have even been observed with features to access webcams and take full screenshots of the desktop. Features such as these can be exploited for identity theft or corporate espionage.

- *Proxies*: Many bots have proxy capabilities. Utilizing forward proxies, a criminal can access any resource online through his infected drone population to hide his communications and evade capture from authorities.

Reverse proxy systems can also be built to protect secret web servers by hiding them behind an ever-changing network of decoy drones. Fast-flux DNS techniques are used to point a web site's domain name to a large set of expendable infected computers, making the real web server difficult to discover and take down (Smith 2008).

- *Click-through fraud*: Many pop-ups have been attributed to botnet-related fraud, and several herders have made small fortunes by instructing their army of drones to visit web sites with affiliate advertisements. Bots can also be commanded to download and install adware to earn additional profits.

Bots continue to evolve. Bot software can be downloaded for free from many online hacking web sites. New exploits and features are added continuously as a form of collaboration between criminals. Consequently, engaging in modern cyber criminal enterprise is easier than ever.

1.1 The Underground Economy

Beneath the infected botnet computer networks lies a hidden social network of individuals engaged in cybercrime.

Entire chat networks have been located whereby spammers, bot herders, malware authors, web site crackers, and other criminals gather to cooperate and sell their services in a thriving black market economy. Among the exchanges that occur in these covert channels is a vibrant market of stolen data, new zero-day exploits, bot payloads, and botnet rental services for spammers.

Stolen personal information such as financial account passwords and credit card numbers can be acquired in bulk. Such information is easily lifted from bot infestations with keyloggers and password stealers. Criminals who compromise these accounts electronically are usually unable to extract the funds, so they sell them for pennies on the dollar to hardened criminals who can (Thomas and Martin 2006).

The underground is protected with reputation-based trust networks. Operators encourage the reporting of suspicious individuals or those known to conduct fraudulent deals. These chat communities are now diving further and further underground and utilizing encryption techniques to hide from authorities who are becoming more aware of their existence (Thomas and Martin 2006). Adequate investigation of these underground channels requires both technological sophistication and careful human intelligence.

While many participants in the underground economy are acting alone, several organized crime syndicates have formed to take advantage of the profits to be made. The rise of these organizations has spawned botnet turf wars in which criminals fight over compromised hosts, leaving innocent victims in the crossfire. The famous SpamThru Trojan, a spambot, was found to install the Kaspersky antivirus on compromised machines in order to secure it from other criminal groups. Many other bots have been found in the wild that monitor the compromised host's Internet connection for the presence of other bot infections (Holtz 2005).

Perhaps the most well-known example of organized cybercrime is the infamous Russian Business Network. It functions as a service provider for criminals who engage in illegal activities, such as the distribution of child pornography and launch of phishing attacks, as well as the distribution of malware and spam.

1.2 Terrorism Online

The World Wide Web provides an efficient means to learn and communicate. The ability to share information quickly and overcome geographic isolation enables individuals to cluster and thrive in collaborative efforts. It should be no surprise that it is also an effective platform to spread violent and extremist agendas. In recent years, many radical Islamic groups have utilized the web to disseminate jihadist ideologies and recruit new members.

Jihadist propaganda can be found in thousands of individual web sites, personal blogs of sympathizers, social network groups, and large forum communities (Weimann 2006). In general, the content can be difficult to monitor; however, several university projects and private intelligence groups have risen to study the phenomenon. The UA Dark Web project, for one, uses focused web spiders to download relevant content and sophisticated analysis techniques to characterize it.

Using link analysis, the Dark Web team has studied the structure of groups of web sites within the jihadist propaganda sphere (Xu et al. 2006). Affect and sentiment analysis techniques have been developed to statistically classify web-gathered text for emotional strength, as well as racist and violent speech patterns (Abbasi and Chen 2008). Additionally, analysis of web forum posts reveals social interactions between individual sympathizers.

Aside from web-based jihadist propaganda, terrorist groups are beginning to take notice in the payoff of cyber criminal operations. In October 2005, Waseem Mughal, Younis Tsouli, and Tariq al-Daour were found by investigators in Britain to be engaging in cyber criminal operations to support terror-related activities. The group had established an online network of jihadist propaganda web sites and forums hosted on the servers of compromised web sites (Krebs 2007).

The group had possession of over 37,000 (Krebs 2007) stolen credit card numbers, along with personal information from victims. These credit cards had been obtained through online phishing attacks and the distribution of malware. The trio had made more than $3.5 million in charges on these compromised accounts and had purchased prepaid credit cards, cell phones, and more than 250 airline tickets (Krebs 2007).

The arrests of these three individuals confirmed that conventional cybercrime tactics can be applied to support terrorist activities.

2 Research Test Bed: The Shadowserver Foundation

The Shadowserver Foundation is a nonprofit group of volunteer watchdog security professionals whose mission is to gather intelligence from the darker side of the Internet. The team follows a rigid methodology of threat detection, analysis, investigation, and mitigation, whereby new threats are carefully considered and, if necessary, brought to the appropriate authorities to assist in incident response.

Among its many research projects, Shadowserver actively detects and monitors thousands of botnets, and it is this monitored botnet data that this chapter is based upon.

2.1 Honeypots

In computing, a honeypot is a network resource that is either vulnerable to or emulates security vulnerabilities with the express purpose of attracting attacks by cyber criminals (McCarthy 2003). Honeypots are utilized to detect security threats, understand the behaviors and motivations of attackers, and to collect malicious software (hereafter known as malware). Much honeypot research has been conducted by the Honeynet Project (Spitzner 2003).

Shadowserver makes use of malware honeypots to passively collect malware from attacking hosts online, including the freely available malware-collecting honeypot known as Nepenthes, which is developed and maintained by the mwcollect alliance (mwcollect alliance). The honeypot is low interaction, in that it passively emulates known network service vulnerabilities to successfully extract malware samples from attacks.

Furthermore, we have made use of e-mail spamtraps and active honeyclient spiders, which are live Windows systems which crawl the net looking for web sites that engage in drive-by-downloads. New technologies are being investigated, including p2p-based malware harvesters to look for malicious files passed on as Trojan horses on popular file-trading platforms. Furthermore, preliminary analysis of malicious files distributed on warez sites is being performed.

Having placed passive honeypots in key networks around the world, and periodically harvesting malware from known malicious sites, we centrally gather all malware for passive processing and active sandbox analysis.

2.2 Malware Analysis

- *Passive:* Passive malware analysis involves scrutinizing each malware sample without actually executing it. We scan all incoming malware with many of the most popular antivirus engines, including Panda, AntiVir, F-Secure, Kaspersky,

Norman, Vexira, VirusBuster, F-Prot6, ClamAV, NOD32, AVG, BitDefender, Avast, DrWeb, VBA32, and McAfee. Each engine's detection rate is reported on the Shadowserver web site (www.shadowserver.org).

Increasingly, malware authors are relying on advanced polymorphic techniques, in which the malware rewrites itself upon execution. Such practices make malware virtually undetectable by most antivirus engines. Consequently, most malware is not easily characterized without live behavioral analysis.

- *Active*: Sandbox analysis is a technique which enables a malware researcher to actively execute untrustworthy malicious code in a carefully controlled and monitored environment. We make use of Sunbelt Software's CWSandbox (Sunbelt Software), a commercial sandboxing solution. CWSandbox executes malware samples within a virtualized Windows environment. System calls made to the underlying Windows API are recorded and included in a detailed report.

All sandbox output is parsed for any relevant malware statistics. Our primary interest is whether or not the malware sample generated any network traffic, including the IRC protocols used by the majority of botnet-related malware.

2.3 *Snooping*

Any IRC servers accessed during sandbox analysis are assumed to be malicious, and thus closely monitored. We connect to and monitor newly discovered IRC networks and record all IRC traffic. The IRC logs are analyzed by a pattern matching signature system, whereby events and commands are classified by type. The extracted information is eventually disseminated in reports to various parties around the world, including CERT groups and vendors.

Outside of IRC communications, a small subset of botnets today is beginning to utilize HTTP and p2p communications protocols for C&C mechanisms (Holtz 2005). A few bots have been discovered using completely encrypted communications which preclude simple monitoring and require live behavioral analysis. These new communications technologies are making botnets more difficult to track.

3 Investigating the Botnet World

An analysis was performed on Shadowserver's collected botnet logfiles to prototype new threat classification techniques. The investigation examined the recorded events of 3,611 unique botnet C&C channels with upward of 4,000 distinct monitored criminals, spanning April 2006 to July 2007.

A prime motivation for the study was to increase our intelligence on the botnet criminal underground and the general state of the botnet threat at hand. How many criminals were we dealing with? Of those we do successfully monitor, who are the most prolific among the criminal subculture? Do these prominent criminals form gangs, and if so, what are those groups primarily engaged in? Who should law enforcement focus on first, given the severity of crimes? Who is the most dangerous?

3.1 Dataset Processing

The dataset consists of raw ASCII text, split among separate logfiles for each monitored C&C channel. The contents of each log consist of IRC protocol traffic structured with a simple "event-type" format. As IRC clients enter or exit the chat channel, JOIN and QUIT events are printed, detailing the nicknames and IP addresses of the connecting clients. Among these nicknames are the many infected drones, as well as the criminal herders that control them.

In addition to JOIN/QUIT events, PRIVMSG events are displayed in the logs whenever a message was sent to the chat channel. In the case of IRC-based botnets, it is within these "PRIVMSGs" that most controlling commands are issued to the infected drones. Additionally, drones communicate back to the criminals via PRIVMSGs in the form of bot status messages.

Several other event types are supported by the IRC protocol. The TOPIC keyword, for example, enables an operator of the channel to set the channels current topic of conversation. While in normal IRC chat communication this directive has a well-defined purpose, in IRC botnet communications, it is often used to assign work to infected drones as they connect to the C&C server. The NICK event signifies when a client, using a particular nickname, is switching to a different nickname.

To process the contents of each logfile, a signature system was used to parse metadata of events as they occur. A large set of regular expression signatures have been written by Shadowserver volunteers over the years to scrutinize the botnet logs. Not only do these signatures recognize the IRC protocol events, but they also detect and classify commands and status messages typed by criminals or printed by infected bot software.

While the IRC protocol is well defined, the range of botnet messages encountered in the wild is incredibly diverse. Not only can different bot software packages speak different command sets, but the individual criminals will additionally modify their infectious payloads to avoid detection and classification.

The signature set grows each day and recognizes many bot commands and status messages encountered in the wild. Events are parsed and classified into several categories, such as "DDoS command," or "infection event," or a "password-theft event." The entire contents of the botnet backlog were analyzed and classified. This produced a compendium of what events transpired on each C&C server.

1. *Entity resolution:* In the IRC chat protocol, all connecting clients communicate with a nickname. In botnet logs, nicknames of channel participants can be parsed from PRIVMSG, JOIN, QUIT, and NICK events. Most nicknames encountered

```
<E8159> :root!hackMe@333.ip JOIN #dvdripfr
<E7534> :[CHN]17!gu@60.2.12.29 JOIN :#pwn
<E8159> :s1ns!u@cz.org PRIVMSG #dvd :.login
<E7534> :[USA]56!ve@rs.rr.com JOIN :#pwn
```

Fig. 22.1 A truncated botnet log snippet. In the JOIN messages, nicknames are connecting to channels. Lines 2 and 4 are likely connecting drones, as the nicknames look like random IDs. Line 3 is a herder issuing a log in command to his drones. Due to this activity, our signature system will select "s1ns" as a herder nickname for the C&C ID "E8159" and reject the other nicknames as drones

in C&C channels are indeed drones; however, a small subset of nicknames may issue commands to operate the botnet. These are the criminals. Human criminals must be isolated from the infected drones to perform analysis of the criminal element behind any botnet.

Many bots, when connecting to their host C&C channels, will assume a random numeric ID code. This convention is not valued by all bots however, and in many cases, drone nicknames will originate from a dictionary built into the malware. Thus, identifying drones through their nicknames alone can be less than trivial and certainly not easy to automate.

The solution to the problem was to utilize the aforementioned signature system to pick out nicknames found issuing bot command strings. Drones could also be positively confirmed as "nonhuman" by a large set of characteristic drone status messages. A set of heuristics was used to analyze these patterns and produce a sanitized list of the human leaders of each and every monitored C&C.

2. *Population modeling:* While the IRC protocol natively reports the occupancy of a given chat channel, we have found that most botnet C&C servers have explicitly disabled this feature. Separate indicators are needed to establish a botnet's infection size. A system was devised in which nicknames of channel participants are tracked via JOIN, QUIT, PRIVMSG, and NICK events.

A simple approach can be taken whereby each nickname encountered in the channel has its connection state tracked (Fig. 22.1). JOIN and QUIT events will increment and decrement a population counter depending on whether or not the nickname has been detected before. While JOIN events will always indicate a channel population increase, a QUIT event will only signify a population decrease if we saw the nickname JOIN previously in time.

A more refined approach was taken to arrive at more accurate population estimates. When an unobserved nickname is witnessed through any non-JOIN nickname-related IRC event (PRIVMSG, NICK, QUIT), the estimator increments the botnet population for all time prior to the occurrence of that event. This compensates for having not witnessed the drone's connection to the channel before we began monitoring the botnet C&C server.

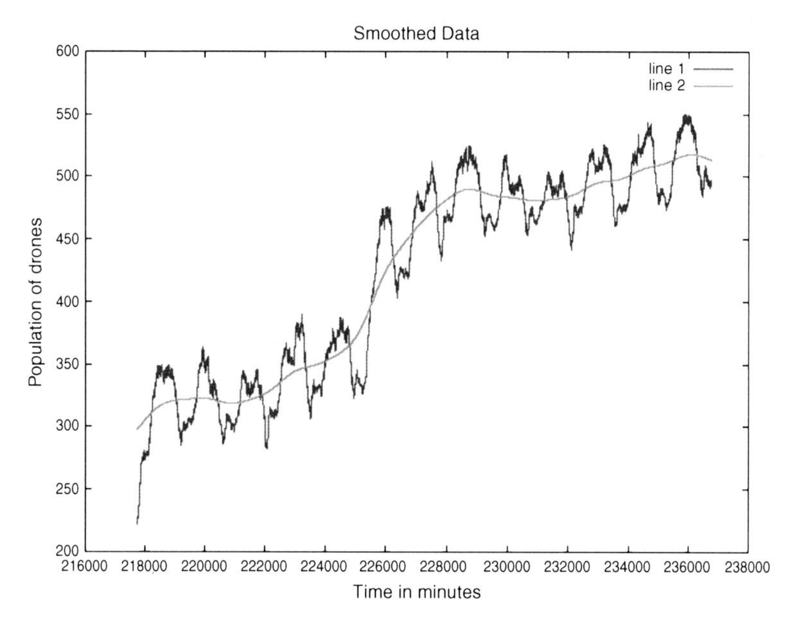

Fig. 22.2 Results of botnet population estimation. The botnet illustrated is experiencing diurnal activity patterns and linear infectious growth after the initiation of an infectious scan. The fit line is the result of Gaussian kernel-based smoothing with a full width half max of 24 h

The successfully computed drone population estimates gave us a rough measure for the infection base of each monitored channel, as well as an evolution curve for the population of each botnet (Fig. 22.2). The results also confirmed the diurnal population hypothesis discussed in previous literature (Dagon et al. 2006). The population size of a botnet oscillates within a 24-h period, as some fractional subset of victims will power down their computers at night.

3.2 Criminal Social Networks

The simplest metric by which we can gauge a criminal's influence in the underground community would be to count the number of C&C channels he was observed participating in. The results provide a staggering look into the assets controlled by some of these criminals.

While this simple analysis gives insight into who is potentially the most successful of the monitored botnet herders, it fails to explore any inherent structure found among entire criminal communities. The aforementioned signature system enables tagging of specific crimes to specific channels and, in some cases, to individual nicknames. While the enumeration of these threats is useful in dealing with specific incidents as they occur, they fail to illuminate the long-term motivations of entire

Fig. 22.3 Top 30 most prolific botnet herders

nickname	# C&C's
always	65
joeblow	60
emr3	60
KoRn	59
lol	58
asd	57
Marvin_	53
bill	52
root	51
process	48
hidden	47
xx	46
mr'bet—	42
edzy	40
D—_PaLo	40

nickname	# C&C's
aaa	40
evil	39
MArian0z	36
gu3sT	33
CeeK	33
Busy	33
Source	31
JuMp	31
aa	31
Lindi_Cracker	30
mjd	29
KaHiN	29
as	29
tonii	28
ILGuardiano	28

groups of criminals who may be collaborating with each other among many C&C servers.

Analysis was performed on the community structure of the underlying social network. All prefiltered "human" nicknames were taken as nodes in a large social network. Links were defined between any two nodes found collaborating in a single C&C channel (Fig. 22.3). Weights were assigned to each link with a simple Jaccard metric measuring the percentage of channels the nicknames shared in common divided by the total number of channels occupied by either nickname. That is, the weight can be represented:

$$w(ij) = \frac{|T_i \cap T_j|}{|T_i \cup T_j|}$$

where Tn is the set of C&C channels that nickname n was found occupying.

A hierarchical agglomerative clustering algorithm was used to cluster vertices together based under an average group linkage heuristic requiring a minimum similarity of 50%. The resulting dendrogram (Fig. 22.4) has 104 clusters containing 957 nicknames.

For each resulting cluster of criminals, a set of associated botnets was assembled. From this combination of criminal groups and their associated crime scenes, the combined criminal events for each cluster can be taken in aggregate to reveal a groups' overall criminal focus.

Of the 104 clusters, a small subset of significant groups is listed in the table shown in Fig. 22.5, along with a summary of the criminal activity detected by the aforementioned signature and bot-counting systems.

The maximum estimated drone count is listed, as is both the number of unique DDoS targets victimized and the number of passwords stolen from unsuspecting

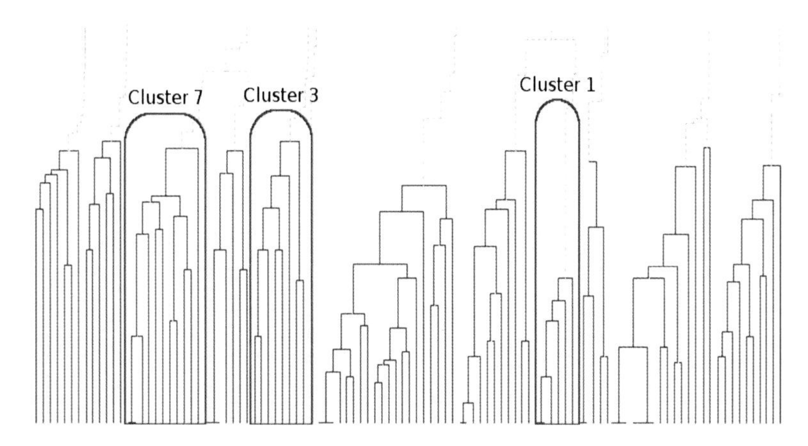

Fig. 22.4 A small component of the hierarchical clustering results showing the inherent structure in three of the discussed criminal gangs

Herders	C	D	B	P
[0]USA—2KSP3[Om]824584, creature, edzy, fri, frioz, wejbwfe, wloo, BlaCkD̂3v—L	51	1263	235713	
bill gu3sT Besi D—_PaLo hidden load process tonii	88	3310	30140	
Albania DaddyCooL[a] jelo jeloo [KleviS] Opium Silv3r-ArRoW waleed	44	730	252969	
ILGuardiano liga MArian0z PepP0z JuMp	56	6698	256193	
xRaZoRx xBreaKx xxDCxx vDCv xGoDx xCKx xBeNx xBrandoNx xAmplifyx xSKYx xToaDx xTiMx	15	2350	6094	1988
Max hans matrix toxic abc Peter bob home Andy dan Jack blbla billy mark xxx sss mr	15	2615	303286	
StRuGaNi_0̂07 bostss Heropos niggaz yeste Pacino NhG Ld fada pilz AsC [a] bAcaRdI dRiVeR alejandro mut hook Dritton ArditS Corrupted	32	730	220703	
Attacker hh	12	479	239708	4484

Fig. 22.5 Several interesting groups found based on concentrations of criminal events and membership size. The columns, from left to right, represent the criminal nicknames, the number of C&C channels controlled by the gang (C), the number of unique DDoS targets (D), estimated number of bots (B), and number of PSTORE compromises of victim passwords (P)

users. Overall, the results show some of the largest gangs personally observed by the author to be prominent in the botnet scene. While the first 5 groups were well known by the author, the last 3 shown, as well as some 20 other groups not listed, were brought to light as potential future targets requiring investigative scrutiny. The overall hierarchy suggests that most criminals do cooperate within private trusted syndicates; however, almost all individuals monitored are loosely connected through covert black market hub channels found sprawled among the underground economy.

4 Future Work

While the analysis techniques outlined above are an excellent start for obtaining a big-picture look at the botnet underground, several additional steps must be taken to better filter and condition the data. Furthermore, additional investigation is needed to better scrutinize the motivations behind the logged crimes.

To better profile the DDoS attack motivations of these gangs, the actual DDoS targets must be individually scrutinized. Current work is in progress to quantify the distribution of targets among geographic and national network ranges. Additionally, IP addresses could be correlated with latitude and longitude to determine any geo-political or regional focus.

Ideally, attacked web sites should be automatically text mined to determine the subject matter. If correlations between many attacked sites are found over a given time period, coordinated hacktivism could potentially be automatically detected and tied to the correct criminal gangs.

Work is being done to utilize graph-mining approaches to assist in the investigation of cybercrime data. By considering observed entities such as nicknames, IP addresses, DNS and whois records, and other classifiable features within botnet logs, sandnet analysis, and web page spidering, it is hoped that link analysis will discover hidden connections/causations between seemingly unrelated events. Additionally, we are interested in additional scrutiny of any links between extremist forums and botnet-related technologies.

5 Conclusions

The Cyber Underground has dramatically evolved in recent years to encompass world-influencing crimes with large sums of money at stake. Some prolific players in the botnet scene have built, upon the resources of countless infected and unwilling drone machines, computerized armies with unbelievable power. Behind these spreading networks is a community of cyber criminals engaging in serious crimes that influence people, markets, and entire nations.

As these communities dive deeper underground and as encryption technologies become more widespread, tracking these miscreants and their botnets will become more and more challenging. Although the importance of computer security has finally entered the public consciousness, the problem of ubiquitous cybercrime will likely not be significantly combated in the coming years without substantial policy change. The only defense we have against this new wave of cybercrime is for individuals to secure themselves at the personal level and for researchers and policy makers to keep an open mind concerning the complexity of the problems at hand.

Acknowledgments The author would like to thank the UA Dark Web team for their support of ongoing terrorism and cybercrime research, and the volunteer effort of The Shadowserver Foundation for helping to make the Internet a safer place.

References

Abbasi, A. and Chen, H. (2008), "Analysis of Affect Intensities in Extremist Group Forums," In: *Terrorism Informatics*, E. Reid and H. Chen, Eds., Springer, pp. 285–307.

Dagon, D., Zou, C. and Lee, W. (2006), "Modeling Botnet Propagation Using Time Zones." In *Proceedings of the 13th Network and Distributed System Security Symposium* (NDSS).

Holtz, Thorsten (2005), "A Short Visit to the Bot Zoo," *IEEE Security and Privacy*, Vol. 3, No. 3 pp. 76–79.

Krebs, Brian (2007). "Terrorism's Hook Into Your Inbox." *Washington Post*, July 5, 2007. http://www.washingtonpost.com/wp-dyn/content/article/2007/07/05/AR2007070501153.html

McCarthy, Bill (2003). "Botnets: Big and Bigger," *IEEE Security and Privacy*, Vol. 1, No. 4, pp. 15–23.

Nazario, Jose (2007). "Botnet Tracking: Tools, Techniques, and Lessons Learned," *Black Hat DC 2007 Presentations*, https://www.blackhat.com/presentations/bh-dc-07/ Nazario/Paper/bh-dc-07-Nazario-WP.pdf.mwcollect Alliance, Nepenthes honeypot. http://mwcollect.org/; http://nepenthes.carnivore.it/.

Smith, Brad (2008), "A Storm (Worm) Is Brewing," *IEEE Technology News*, Vol. 41, No. 2, pp. 20–22.

Spitzner, L. (2003), "The Honeynet Project: Trapping the Hackers," *IEEE Security and Privacy*, Vol. 1, No. 2, pp. 15–23.

Sunbelt Software. CWSandbox, http://www.cwsandbox.org.

The Shadowserver Foundation, http://www.shadowserver.org.

Thomas, Rob and Martin, Jerry (2006). "The Cyber Underground Economy: Priceless,"*;login: The USENIX Magazine,* Vol. 31, No. 6. http://www.usenix.org/ publications/login/2006–12/openpdfs/cymru.pdf.

Weimann, Gabriel. (2006), *Terror on the Internet: The New Arena, the New Challenges*. Washington, D.C.: United States Institute of Peace Press.

Xu, Jennifer, Chen, Hsinchun, Zhou, Yilu and Qin, Jialun (2006). "On the Topology of the Dark Web of Terrorist Groups," *Intelligence and Security Informatics*, ISI 2006, LNCS 3975, pp. 367–376.

Index

H. Chen, *Dark Web: Exploring and Data Mining the Dark Side of the Web*, Integrated Series in Information Systems 30, DOI 10.1007/978-1-4614-1557-2,

G

H

I